Heat Transfer

Macmillan Publishing Company
866 Third Avenue, New York, New York 10022

Collier Macmillan Canada, Inc.

Library of Congress Cataloging in Publication Data

Chapman, Alan J. (date)
  Heat transfer.

  Includes bibliographical references and index.
  1. Heat—Transmission. I. Title.
QC320.C5 1984   536'.2   83-144
ISBN 0-02-321470-8

Printing: 2 3 4 5 6 7 8   Year: 4 5 6 7 8 9 0 1

ISBN   0-02-321470-8

ALAN J. CHAPMAN

Rice University

HEAT TRANSFER

4th EDITION

Macmillan Publishing Company

New York

Collier Macmillan Publishers

London

Dedicated to the memory of
Isabel and Wallace Chapman

PREFACE

This fourth edition follows the basic philosophy set forth in the earlier editions of this book. Heat transfer, a field comprising a synthesis of a number of engineering-science disciplines, is significantly influenced by the rapidly occurring changes in modern technology. To provide sufficient flexibility to permit the modern engineer to work effectively in a great variety of new and emerging fields, as well as in the mature, established areas of engineering practice, it is important to teach the fundamental concepts underlying a subject as well as the details of application. This text presents these fundamental concepts in a fairly rigorous manner, while showing how the analytically obtained facts can be applied with meaningful results to a real physical problem. Sound physical insight and the way in which it supplements analysis are stressed as being of equal importance.

This book, then, is meant to be a teaching text and not a treatise on heat transfer. As such, many worked-out examples are included throughout the work, and numerous problems for student exercises are supplied with most chapters. As a reflection of the continual move of American engineering practice to the use of SI units, this system of units is stressed and forms the basic system for the text. However, modern practice will require engineers to be able to work with the English system of units as well for the foreseeable future. Thus, answers to all examples are given in both systems of units and occasional examples and problems are given in the English system.

In keeping with the basic idea that heat transfer is a meaningful subject only when real problems are solved, the appendices provide extensive collections of the thermal properties of a wide variety of gases, liquids, and solids frequently encountered in engineering practice. Most physical properties are given in SI units; however, the data for the important engineering fluids air, water, and steam are given in both SI and English units to avoid extensive conversion calculations.

The approach of the presentation presumes that the subject is being studied at the introductory level by advanced undergraduates. Mathematical preparation through advanced calculus and differential equations is presumed, as well as basic thermodynamics and fluid dynamics. However, some appendix material is provided on certain mathematical techniques of heat conduction, and Chapter 6 is devoted to a brief treatment of those topics of viscous fluid dynamics pertinent to convective heat transfer.

The work on conduction includes one-dimensional cases having engineering significance in addition to multidimensional and transient problems. Extended surfaces are treated thoroughly, and the treatment of arrays of fins is presented in a manner to lead into heat exchanger applications in later chapters. The numerical analysis of conduction is the subject of an entire chapter. While specific computer programs are not presented, the numerical techniques described are those most commonly applied in practice through the application of digital computers.

Convective heat transfer offers an excellent opportunity to show how powerful analytic techniques can be combined with sound engineering judgment and intelligent empiricism to solve complex physical problems. The chapters on convection have been extensively revised in this edition. The importance of boundary layer theory is emphasized, and the treatment of turbulent flow has been unified and made more complete than in earlier editions. More modern correlations, for both forced and free convection, are presented than in the past. Particular attention has been paid to the careful delineation of the applicable limits of each convective correlation. This edition also stresses more fully the relation between fluid friction and heat transfer than has been the case in earlier editions, and extensive use of the Reynolds analogy and its variations is made.

Radiative heat transfer is explained in terms of the behavior of real surfaces and includes the idealized analysis of black or gray surfaces. Included are treatments of enclosure theory, solar radiation, and application to industrial and aerospace problems of significance. The often-tedious problem of shape factor algebra is presented in an appendix. The exchange factor analysis for black and gray enclosures emphasized in the earlier editions has been dropped in favor of the more direct solution of systems of equations, relying on the use of hand calculators or digital computers.

The final two chapters are devoted to the ultimate aim of a heat transfer course: the solution of problems involving the combined modes of conduction, convection, and radiation. Included here are the iterative calculation techniques so often required in such problems. Again, advantage is taken of the current availability of modern computational devices. A significant amount of space is devoted to the analysis of heat exchangers, and the material in this area has been extensively reorganized. The equivalence of the mean-temperature-difference and the effectiveness formulations is stressed, and the motivation for the use of each is emphasized. A more careful distinction of the difference between the performance and design calculations of heat exchangers is presented. The analysis of thermometric errors incurred by conductive, convective, and radiative interactions is presented. Modern applications such as space radiators, heat pipes, and solar collectors are explained. The latter subject, one of the most actively pursued developments in recent years, is new in this edition.

Throughout the text, acknowledgment has been made of the many authors of books and journal articles that formed the sources from which many data were taken. The author would also like to acknowledge the many helpful suggestions of students and other users of earlier editions. Their comments were invaluable in the preparation of this volume.

Houston, Texas   A. J. C.

CONTENTS

3 STEADY STATE HEAT CONDUCTION IN ONE DIMENSION

4 HEAT CONDUCTION IN TWO OR MORE INDEPENDENT VARIABLES

12   HEAT TRANSFER BY COMBINED CONDUCTION AND CONVECTION—HEAT EXCHANGERS   439

13   ADDITIONAL CASES OF COMBINED HEAT TRANSFER   509

CHAPTER 1

Introduction

INTRODUCTORY REMARKS

In the science of thermodynamics, which deals with energy in its various forms and with its transformation from one form to another, two particularly important *transient* forms are defined, work and heat. These energies are termed transient since, by definition, they exist only when there is an exchange of energy between two systems or between a system and its surroundings, i.e., when an energy form of one system (such as kinetic energy, potential energy, internal energy, flow energy, chemical energy, etc.) is transformed into an energy form of another system or of the surroundings. When such an exchange takes place without the transfer of mass from the system and not by means of a temperature difference, the energy is said to have been transferred through the performance of *work*. If, on the other hand the exchange is due to a temperature difference, the energy is said to have been transferred by a flow of *heat*. It is this form, heat, and the basic physical laws governing its exchange with which this book is concerned. It should be noted that the existence of a temperature difference is a distinguishing feature of the energy form known as heat.

In most instances the problems of engineering importance involving an exchange of energy by the flow of heat are those in which there is a transfer of internal energy (or enthalpy in the case of flow processes) between two systems.

The two systems may be different parts of the same body. In general this internal energy transfer is called *heat transfer,* although, thermodynamically speaking, this is incorrect. The flow of heat is the mechanism of the transfer of the internal energy, not the quantity transferred. However, it is convenient to use this expression and to speak of heat as "flowing," as has been done here in spite of the implied contradiction with thermodynamics. Indeed, the old and discredited caloric theory of heat, in which heat is defined as a weightless, colorless, odorless fluid flowing from one body to another, would form an adequate basis for the study of heat transfer.

When such exchanges of internal energy or heat take place, the first law of thermodynamics requires that the heat given up by one body must equal that taken up by the other. The second law of thermodynamics demands that the transfer of heat take place from the hotter system to the colder system.

1.2

THE IMPORTANCE OF HEAT TRANSFER

The importance of a thorough knowledge of the science of heat transfer and the necessity of being able to analyze, quantitatively, problems involving a transfer of heat have become increasingly important as modern technology has become more and more complex. In almost every phase of scientific and engineering work, processes involving the exchange of energy through a flow of heat are encountered.

Mechanical and chemical engineers are particularly concerned with problems of heat transfer. Modern power generation involves the production of work from either a combustible fuel or a nuclear reaction. This energy is converted into useful work by means of boilers, turbines, condensers, air heaters, water pre-heaters, pumps, etc. All these pieces of apparatus involve a transfer of heat by one means or another, as does almost every piece of apparatus found in a chemical process industry or a petroleum refinery. Certainly, designing the familiar internal combustion engine, gas turbine, and jet engine requires a complete understanding of heat transfer for a thorough analysis of the combustion and cooling processes.

The so-called "thermal barrier" in aerodynamics involves finding means of transferring away from the aircraft the enormous amounts of heat produced by the dissipative effect of the viscosity of the air. Indeed, since all processes in nature have been observed to be irreversible, it follows that all natural processes involve a dissipation of the various forms of mechanical energy into thermal energy with consequent heat transfer processes taking place.

It is this dissipative aspect of the second law of thermodynamics which leads to the much-discussed problem of "thermal pollution," created by the inevitable discharge of waste heat into the environment (air and water). The development of effective solutions to the problems of thermal pollution is perhaps one of the most challenging applications of the knowledge of heat transfer at the present time.

The importance of heat transfer in the production of comfort cooling or comfort heating is readily apparent. This influences the design of building structures of all kinds.

THE FUNDAMENTAL CONCEPTS AND THE BASIC MODES OF HEAT TRANSFER

It is customary to categorize the various heat transfer processes into three basic types or modes, although, as will become apparent as one studies the subject, it is certainly a rare instance when one encounters a problem of practical importance which does not involve at least two, and sometimes all three, of these modes occurring simultaneously. The three modes are conduction, convection, and radiation.

Heat *conduction* is the term applied to the mechanism of internal energy exchange from one body to another, or from one part of a body to another part, by the exchange of the kinetic energy of motion of the molecules by direct communication or by the drift of free electrons in the case of heat conduction in metals. This flow of energy or heat passes from the higher energy molecules to the lower energy ones (i.e., from a high temperature region to a low temperature region). The distinguishing feature of conduction is that it takes place within the boundaries of a body, or across the boundary of a body into another body placed in contact with the first, without an appreciable displacement of the matter comprising the body.

A metal bar heated on one end will, in time, become hot at its other end. This is an example of conduction. The laws governing conduction can be expressed in concise mathematical terms, and the analysis of the heat flow can be treated analytically in many instances.

Convection is the term applied to the heat transfer mechanism which occurs in a fluid by the mixing of one portion of the fluid with another portion due to gross movements of the mass of fluid. The actual process of energy transfer from one fluid particle or molecule to another is still one of conduction, but the energy may be transported from one point in space to another by the displacement of the fluid itself.

The fluid motion may be caused by external mechanical means (e.g., by a fan, pump, etc.), in which case the process is called *forced convection*. If the fluid motion is caused by density differences which are created by the temperature differences existing in the fluid mass, the process is termed *free convection* or *natural convection*. The circulation of the water in a pan heated on a stove is an example of free convection. The important heat transfer problems of condensing and boiling are also examples of convection—involving the additional complication of a latent heat exchange.

It is virtually impossible to observe pure heat conduction in a fluid because as soon as a temperature difference is imposed on a fluid, natural convection currents will occur as a result of density differences.

The basic laws of heat conduction must be coupled with those of fluid motion in order to describe, mathematically, the process of heat convection. The mathematical analysis of the resulting system of differential equations is perhaps one of the most complex fields of applied mathematics. Thus, for engineering applications, convection analysis will be seen to be a subtle combination of powerful mathematical techniques and the intelligent use of empiricism and experience.

Thermal *radiation* is the term used to describe the electromagnetic radiation which has been observed to be emitted at the surface of a body which has been thermally excited. This electromagnetic radiation is emitted in all directions; and when it strikes another body, part may be reflected, part may be transmitted, and part may be absorbed. If the incident radiation is thermal radiation (i.e., if it is of the proper wavelength), the absorbed radiation will appear as heat within the absorbing body.

Thus, in a manner completely different from the two modes discussed above, heat may pass from one body to another without the need of a medium of transport between them. In some instances there may be a separating medium, such as air, which is unaffected by this passage of energy. The heat of the sun is the most obvious example of thermal radiation.

There will be a continuous interchange of energy between two radiating bodies, with a net exchange of energy from the hotter to the colder. Even in the case of thermal equilibrium, an energy exchange occurs, although the net exchange will be zero.

1.4

THE FUNDAMENTAL LAWS OF CONDUCTION

Thermal Conductivity and Thermal Conductance

The basic law governing heat conduction may best be illustrated by considering the simple, idealized situation shown in Fig. 1.1. Consider a plate of material having a surface area A and a thickness Δx. Let one side be maintained at a

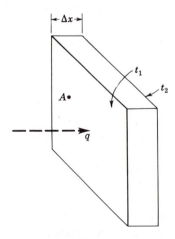

FIGURE 1.1

tempurature t_1, uniformly over the surface, and the other side at temperature t_2. Let q denote the rate of heat flow (i.e., energy per unit time) through the plate, neglecting any edge effects. Experiment has shown that the rate of heat flow is directly proportional to the area A and the temperature difference $(t_1 - t_2)$ but inversely proportional to the thickness Δx. This proportionality is made an equality by the definition of a constant of proportionality k. Thus,

$$q = kA \frac{t_1 - t_2}{\Delta x} \tag{1.1}$$

The constant of proportionality, k, is called the *thermal conductivity* of the material of which the plate is composed. It is a property dependent only on the composition of the material, not its geometrical configuration. While more details concerning thermal conductivity are given in Chapter 2, it is worth noting here the typical order of magnitude of this property. Values of k may range from 0.008 W/m-°C (0.005 Btu/h-ft-°F) for gases such as Freon-12 to as great as 417 W/m-°C (241 Btu/h-ft-°F) for metals such as silver.

Sometimes a gross quantity, *unit thermal conductance,* is used to express the heat conducting capacity of a given physical system, so that if $C = k/\Delta x$ denotes the unit thermal conductance,

$$q = CA(t_1 - t_2).$$

Thus, it is seen that thermal conductance is the conductivity of a substance divided by its thickness. It is no longer a physical property but depends, as well, on the geometrical configuration at hand and, thus, is a less general quantity than is thermal conductivity.

Equation (1.1) forms the basis for the fundamental relation of heat conduction. Consider now a homogeneous, isotropic solid as depicted in Fig. 1.2. If the solid is subjected to certain known boundary temperatures, what is the rate at which heat is conducted across a surface, S, within the solid? Selecting a point P on the surface S, one can select a wafer of material having an area δA, which is part of the surface S containing P, and having a thickness δn in the direction

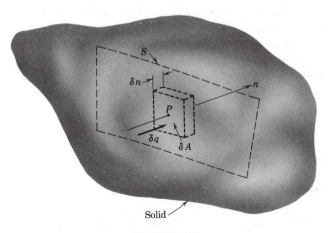

FIGURE 1.2

of the normal drawn to the surface at P. If the difference between the temperature of the back face of the wafer and its front face is δt, and if δA is chosen small enough so that δt is essentially uniform over it, the rate of heat flow across the wafer, δq, is, by Eq. (1.1),

$$\delta q = -k\delta A \frac{\delta t}{\delta n}.$$

The minus sign is due to the convention that the heat flow is taken to be positive if δt is negative in the direction of increasing n, the normal displacement. Forming the ratio $\delta q/\delta A$ and allowing the area $\delta A \to 0$, one obtains what is termed the *flux* of heat conducted through the thickness δn at the point P,

$$f_n = \frac{dq}{dA} = -k \frac{\delta t}{\delta n}.$$

Further, allowing $\delta n \to 0$, one arrives at the flux of heat across S at the point P in terms of the *temperature gradient* at P in the n direction, $\partial t/\partial n$:

$$f_n = -k \frac{\partial t}{\partial n}, \tag{1.2}$$

wherein the notation f_n is used to denote the flux in the n direction. The statement in Eq. (1.2) is called Fourier's conduction law after the French mathematician who first made an extensive analysis of heat conduction. It states that the flux of heat conducted (energy per unit time per unit area) across a surface is proportional to the temperature gradient taken in a direction normal to the surface at the point in question. It should be emphasized that Fourier's law is based on *empirical* observation and is not derived from other physical principles.

Upon returning to the situation pictured in Fig. 1.2, the total rate of heat transferred across the finite surface S would be

$$q = -\int_s k \frac{\partial t}{\partial n} \, dA.$$

Generally speaking, the normal gradient $\partial t/\partial n$ may vary over the surface, but in many instances it is possible to select the surface as one on which the gradient is everywhere the same. This is the situation in the case depicted in Fig. 1.1, in which every plane normal to Δx is such a surface. In the case of a hollow cylinder with uniform outside and inside surface temperatures, every concentric interior cylindrical surface is isothermal with a uniform temperature gradient normal to it. In such cases, then,

$$q = -kA \frac{\partial t}{\partial n}, \tag{1.3}$$

where A is the total area of the finite surface.

The General Heat Conduction Equation

The above relations may be used to develop an equation describing the distribution of the temperature throughout a heat-conducting solid. In general, a heat conduction problem consists of finding the temperature at any time and at any point within a specified solid which has been heated to a known initial temperature distribution and whose surface has been subjected to a known set of boundary conditions.

The development that follows will consider the case in which the heat-conducting solid may also have internal sources of heat generation. Such sources may be the result of chemical or nuclear reaction, electrical dissipation, etc., and may be either concentrated at certain locations or distributed throughout the solid.

To develop the differential equation governing this problem, consider a solid, as shown in Fig. 1.3, and select arbitrarily three mutually perpendicular coordinate directions, x, y, and z. Select in the solid a parallelepiped of dimensions Δx, Δy, and Δz. By making an energy balance on this element between the heat conducted in and out of its six faces, the heat stored within, and the heat generated within the element, an expression interrelating the temperatures throughout the solid is obtained. First, consider heat conduction in the x direction only. Let f_x denote the flux of heat conducted into the element through the left yz face and f_x' the flux of heat conducted out the right yz face. If f_x' is then written in terms of f_x by use of a Taylor's expansion, then

$$f_x' = f_x + \left(\frac{\partial f_x}{\partial x}\right)\Delta x + \left(\frac{\partial^2 f_x}{\partial x^2}\right)\frac{(\Delta x)^2}{2!} + \cdots.$$

The excess rate at which heat is conducted into the element over that conducted

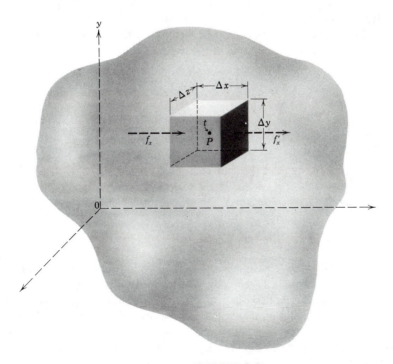

FIGURE 1.3

out is $(f_x - f_x') \, \Delta y \, \Delta z$. Thus, the net energy flow into the element due to conduction in the x direction is

$$-\left[\frac{\partial f_x}{\partial x} + \left(\frac{\partial^2 f_x}{\partial x^2}\right)\frac{\Delta x}{2!} + \cdots\right]\Delta x \, \Delta y \, \Delta z.$$

Similar expressions may be written for conduction in the other two coordinate directions, so that the total *net* rate of conduction *into* the element is

$$-\left[\frac{\partial f_x}{\partial x} + \frac{\partial f_y}{\partial y} + \frac{\partial f_z}{\partial z} + \frac{\partial^2 f_x}{\partial x^2}\frac{\Delta x}{2!} + \frac{\partial^2 f_y}{\partial y^2}\frac{\Delta y}{2!}\right.$$
$$\left. + \frac{\partial^2 f_z}{\partial z^2}\frac{\Delta z}{2!} + \cdots\right]\Delta x \, \Delta y \, \Delta z. \quad (1.4)$$

If q^* is used to denote the rate at which heat is being internally generated in the element, *per unit volume,* then

$$q^*(\Delta x \, \Delta y \, \Delta z) \quad (1.5)$$

is the total heat released in the element.

The energies introduced in the element as given in Eqs. (1.4) and (1.5) must equal the rate of heat storage which is reflected in the time rate of change of the average temperature of the element, t_{av}, as given by

$$\rho c_p \frac{\partial t_{av}}{\partial \tau} \Delta x \, \Delta y \, \Delta z. \quad (1.6)$$

Energy conservation requires that Eq. (1.6) equal the sum of Eqs. (1.4) and (1.5). Upon making this equality; dividing by the volume of the element, $\Delta x \, \Delta y \, \Delta z$; allowing Δx, Δy, and $\Delta z \rightarrow 0$; and noting that under these conditions $t_{av} \rightarrow t$ (the temperature at point P), one finally obtains

$$\rho c_p \frac{\partial t}{\partial \tau} = -\left(\frac{\partial f_x}{\partial x} + \frac{\partial f_y}{\partial y} + \frac{\partial f_z}{\partial z}\right) + q^*.$$

According to Eq. (1.2), each of the fluxes in the above equation is proportional to the temperature gradient in each particular direction. Thus,

$$\rho c_p \frac{\partial t}{\partial \tau} = \frac{\partial}{\partial x}\left(k\frac{\partial t}{\partial x}\right) + \frac{\partial}{\partial y}\left(k\frac{\partial t}{\partial y}\right) + \frac{\partial}{\partial z}\left(k\frac{\partial t}{\partial z}\right) + q^*, \quad (1.7)$$

or, using the vector operator ∇,

$$\rho c_p \frac{\partial t}{\partial \tau} = \nabla \cdot (k \, \nabla t) + q^*. \quad (1.8)$$

Equation (1.7) or (1.8) represents a *volumetric* heat balance which must be satisfied at each point in the body. This expression, known as the *general heat conduction equation,* describes, in differential form, the dependence of the temperature in the solid on the spatial coordinates and on time. It should be noted that the heat generation term, q^*, may be a function of position or time, or both.

In the form given in Eqs. (1.7) and (1.8) allowance is made for the possible dependence of the thermal conductivity k on position within the solid (or on temperature and, hence, on position). If the thermal conductivity is treated as constant, an assumption the significance of which will be discussed in more detail later, the following special forms are obtained:

$$\rho c_p \frac{\partial t}{\partial \tau} = k\left(\frac{\partial^2 t}{\partial x^2} + \frac{\partial^2 t}{\partial y^2} + \frac{\partial^2 t}{\partial z^2}\right) + q^*$$

$$= k\nabla^2 t + q^*.$$

(1.9)

Frequently, Eq. (1.9) is written in the following form, which, although useful, destroys the volumetric representation of each term:

$$\frac{\partial t}{\partial \tau} = \alpha\left(\frac{\partial^2 t}{\partial x^2} + \frac{\partial^2 t}{\partial y^2} + \frac{\partial^2 t}{\partial z^2}\right) + \frac{q^*}{\rho c_p}$$

$$= \alpha\nabla^2 t + \frac{q^*}{\rho c_p}.$$

(1.10)

The quantity

$$\alpha = \frac{k}{\rho c_p}$$

(1.11)

is called the *thermal diffusivity* and is seen to be a physical property of the material of which the solid is composed.

Other Coordinate Systems

The above discussion was carried out in terms of rectangular coordinates. It is often useful to write the equation in cylindrical or spherical coordinates. The results are, for the case of constant k:

Cylindrical coordinates, r (radius), z (axis), θ (longitude):

$$\frac{\partial t}{\partial \tau} = \alpha\left(\frac{\partial^2 t}{\partial r^2} + \frac{1}{r}\frac{\partial t}{\partial r} + \frac{1}{r^2}\frac{\partial^2 t}{\partial \theta^2} + \frac{\partial^2 t}{\partial z^2}\right) + \frac{q^*}{\rho c_p}.$$

(1.12)

Spherical coordinates, r (radius), φ (longitude), θ (colatitude):

$$\frac{\partial t}{\partial \tau} = \alpha\left[\frac{1}{r^2}\frac{\partial}{\partial r}\left(r^2\frac{\partial t}{\partial r}\right) + \frac{1}{r^2 \sin\theta}\frac{\partial}{\partial \theta}\left(\sin\theta\frac{\partial t}{\partial \theta}\right)\right.$$

$$\left. + \frac{1}{r^2 \sin^2\theta}\frac{\partial^2 t}{\partial \varphi^2}\right] + \frac{q^*}{\rho c_p}.$$

(1.13)

The proof of these equations will be left as exercises for the reader.

The Steady State

A particularly useful special case of Eq. (1.10) is one which has a very wide range of application in engineering. This is the *steady state,* in which there is no dependence on time. The heat conduction equation then reduces to Poisson's equation: In rectangular coordinates this is

$$\alpha\left(\frac{\partial^2 t}{\partial x^2} + \frac{\partial^2 t}{\partial y^2} + \frac{\partial^2 t}{\partial z^2}\right) + \frac{q^*}{\rho c_p} = 0.$$

$$\frac{\partial^2 t}{\partial x^2} + \frac{\partial^2 t}{\partial y^2} + \frac{\partial^2 t}{\partial z^2} + \frac{q^*}{k} = 0.$$

(1.14)

In the absence of internal heat generation, Laplace's equation is obtained:

$$\frac{\partial^2 t}{\partial x^2} + \frac{\partial^2 t}{\partial y^2} + \frac{\partial^2 t}{\partial z^2} = 0.$$

(1.15)

1.5

THE FUNDAMENTAL LAWS OF CONVECTION

The Boundary Layer Concept

The discussion of Sec. 1.3 defined "convection" as the term applied to the heat transfer mechanism which takes place in a fluid because of a combination of conduction within the fluid and energy transport which is due to the fluid motion itself—the fluid motion being produced either by artificial means or by density currents.

Since fluid motion is the distinguishing feature of heat convection, it is necessary to understand some of the principles of fluid dynamics in order to describe adequately the processes of convection. When any real fluid moves past a solid surface, it is observed that the fluid velocity varies from a zero value immediately adjacent to the wall to a finite value at a point some distance away.

Considering the simplest case of flow past a plane surface as shown in Fig. 1.4(a), the fluid velocity will vary from a uniform, "free stream" value at points far away from the wall to zero at the wall—somewhat in the way shown in the figure. For fluids of low viscosity, such as air or water, the region near the surface, in which most of the velocity variation occurs, may be quite thin— depending on the free stream velocity of the fluid. In many applications—such as low speed aerodynamics, hydraulics, etc.—it is possible to obtain satisfactory results by assuming that the fluid is inviscid (i.e., without viscosity). Hence, the flow may be treated as though it slips past the surface with no viscous retardation. However, since the process of convection of heat away from the wall (if the wall is at a temperature different from the free stream of the fluid) is intimately concerned with thermal conduction and energy transport due to motion in the fluid layers in the immediate vicinity of the wall, the simplification of assuming the fluid to be inviscid may not be made when analyses of heat convection are undertaken.

Since the region in which the retarding effect of the fluid viscosity plays a dominant role will often be a very thin layer near the wall, it is possible to simplify the description of the convection process by introducing the concept of the *velocity boundary layer*. The velocity boundary layer is defined as the thin

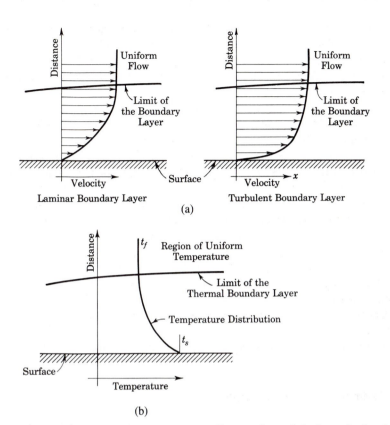

FIGURE 1.4. Boundary layer flow past a flat surface: (a) the velocity boundary layer; (b) the thermal boundary layer.

layer near the wall in which one assumes that viscous effects are important. Within this region the effect of the wall on the motion of the fluid is significant. Outside the boundary layer it is assumed that the effect of the wall may be neglected. The exact limit of the boundary layer cannot be precisely defined because of the asymptotic nature of the velocity variation. The limit of the boundary layer is usually taken to be at the distance from the wall at which the fluid velocity is equal to a predetermined percentage of the free stream value. This percentage depends on the accuracy desired, 99 or 95% being customary.

Outside the boundary layer region the flow is assumed to be inviscid. Inside the boundary layer the viscous flow may be either *laminar* or *turbulent*. In the case of laminar boundary layer flow, adjacent fluid layers slide relative to one another but do not mix in the direction normal to the fluid streamlines. Thus, any heat that flows from the surface to the free stream fluid does so mostly by conduction—although a transport of energy is also accomplished by virtue of the fact that the fluid has a velocity component normal to the surface. This normal velocity component is caused by the fact that the boundary layer must become progressively thicker as it moves along the surface.

In the event that the fluid motion in the boundary layer is turbulent, the mean flow is essentially parallel to the surface, but it has superimposed upon it a fluctuating motion—in directions both parallel and normal to the surface. The transverse fluctuations cause additional fluid mixing, which increases the rate at which heat is transported in the direction perpendicular to the surface. Typical velocity variations through laminar and turbulent boundary layers are illustrated in Fig. 1.4(a).

If the solid surface is maintained at a temperature, say t_s, which is different from the fluid temperature, t_f, measured at a point far removed from the surface, a variation of the temperature of the fluid is observed which is similar to the velocity variation described above. That is, the fluid temperature varies from t_s at the wall to t_f far away from the wall—with most of the variation occurring close to the surface. This is illustrated in Fig. 1.4(b), where it has been assumed that the surface is hotter than the fluid. The fluid temperature approaches t_f asymptotically. However, a *thermal boundary layer* may be defined (in the same sense that the velocity boundary layer was defined above) as the region between the surface and the point at which the fluid temperature has reached a certain percentage of t_f. Outside the thermal boundary layer the fluid is assumed to be a heat sink at a uniform temperature of t_f. The thermal boundary layer is generally not coincident with the velocity boundary layer, although it is certainly dependent on it—i.e., the velocity boundary layer thickness, the variation of the velocity, whether the flow is laminar or turbulent, etc., are all factors which determine the temperature variation in the thermal boundary layer.

Newton's Law of Cooling

Chapters 6 through 10 consider, in some detail, the present state of knowledge of convective heat transfer—from both the theoretical and empirical points of view. As should be apparent from the discussion, the prediction of the rates at which heat is convected away from a solid surface by an ambient fluid involves a thorough understanding of the principles of heat conduction, fluid dynamics, and boundary layer theory. All the complexities involved in such an analytical approach may be lumped together in terms of a single parameter by introduction of Newton's law of cooling:

$$\frac{q}{A} = h(t_s - t_f). \tag{1.16}$$

The quantity h in this equation is variously known as the *heat transfer coefficient, film coefficient,* or *unit thermal conductance.* It is seen to be (refer to Sec. 1.4) a unit conductance, not a material property as is thermal conductivity. It is a complex function of the composition of the fluid, the geometry of the solid surface, and the hydrodynamics of the fluid motion past the surface. The typical range of values one might expect to encounter for h quoted in Table 1.1 indicates the complexity of the convective process and the difficulty in determining h. Chapters 6 through 10 present detailed analyses of the determination of the heat transfer coefficient.

Before proceeding to the study of the determination of the heat transfer coefficient h, Chapters 3 through 5 will concentrate on the solution of the heat

TABLE 1.1 Typical Values of the Convective Heat Transfer Coefficient

Situation	h W/m²-°C	h Btu/h-ft²-°F
Free convection in air	5–25	1–5
Free convection in water	500–1000	100–200
Forced convection in air	10–500	2–100
Forced convection in water	100–15,000	20–3000
Boiling water	2500–25,000	500–5000
Condensing steam	5000–100,000	1000–20,000

conduction equation subjected to various boundary conditions, including the condition of convection to a fluid adjacent to the boundary. In these cases, the boundary condition will be imposed, using Newton's law of cooling, presuming h to be known and given. Following the study of convection in Chapters 6 through 10, some general problems involving the simultaneous solution of the conduction and convection processes will be studied.

The boundary condition to be applied to the general conduction equation, Eq. (1.10), when one has a solid bounded by a convecting fluid of known temperature is simply obtained by equating, at the surface, the heat fluxes given by Newton's law in Eq. (1.16) and Fourier's law in Eq. (1.3). Thus, at the convecting surface, the condition to be satisfied by the temperature field in the solid is

$$-\left(\frac{\partial t}{\partial n}\right)_s = \frac{h}{k}(t_s - t_f). \tag{1.17}$$

In Eq. (1.17), n denotes the normal direction to the surface (positive away from the surface), and the subscript s indicates quantities evaluated at the surface. It is important to note that the k in Eq. (1.17) is the thermal conductivity of the *solid,* not the surrounding fluid.

1.6

THE FUNDAMENTAL LAWS OF RADIATION

As mentioned in Sec. 1.3, it has been observed that a body may lose or gain thermal energy without the need of a physical medium of transport. That is, a heated body placed in the presence of cooler surroundings but having no physical contact with them (imagine it suspended in a vacuum) is observed to lose energy. This loss is due to the electromagnetic emission known as thermal radiation. It has been pointed out that this absence of a medium of transport is a distinguishing feature of the thermal radiation mode of heat transfer.

A second distinguishing feature of thermal radiation is the effect of the level of the temperature of the emitting bodies. As the discussions of Secs. 1.4 and 1.5 have shown, the rate of heat transfer by the modes of conduction and convection is proportional to the difference of the temperature between the heat source and the heat sink. Thus, regarding physical properties to be constant for the moment, the amount of heat transferred is independent of the absolute magnitude of the temperature as long as the difference is the same. This is not the case for thermal radiation. The quantity of heat exchanged by radiation is proportional to the difference of the fourth power of the absolute temperatures of the radiating bodies. Thus, for a given temperature difference, the heat transferred is much greater at high temperatures than at low temperatures.

The exact character of radiant heat emission is not completely agreed upon, but it is known that the rate at which energy is emitted per unit area of emitting surface is given by the Stefan–Boltzmann law:

$$E = \epsilon\sigma T^4.$$

In this equation E is the rate of energy emission per unit area, T is the absolute temperature of the body, σ is a universal physical constant, and ϵ is a property of the particular emitting surface known as the *emissivity*. This law was originally proposed by Stefan based on experimental evidence. Boltzmann showed later that the law is derivable from the laws of thermodynamics and hence is not based just on experimental data as are Fourier's law [Eq. (1.3)] and Newton's law of cooling [Eq. (1.16)].

Because of the extremely difficult character of this mode of heat transfer, further discussion will be reserved for Chapter 11.

1.7

DIMENSIONS AND UNITS

Careful attention to dimensions and units is necessary if one is to obtain correct results in any engineering problem. The English engineering system of units is still widely used in the practice of engineering in the United States. However, the use of SI units (Système International d'Unités) is increasing rapidly in view of the almost universal adoption of these units in the rest of the world. Consequently, the current engineering practitioner must be versatile in both systems. This text will concentrate on the use of SI units, but occasional examples and problems will be given in English units.

The main source of difficulty in dealing with systems of units results from the interrelation of *force* and *mass* through Newton's second law of motion. It is convenient to express Newton's law as

$$F = \frac{1}{g_c} ma,$$

where F denotes force, m denotes mass, a is the acceleration of the mass, and g_c is a dimensional constant. The four physical dimensions: force, mass, length, and time are interrelated in the above law. Presuming, as is always the case, the length and time are fundamental quantities, then one can either make $g_c = 1$ (without dimensions) and define either mass or force in terms of the other, or one can define mass and force as primitive quantities and derive dimensions for

g_c. The foregoing possibilities lead to several feasible systems of units. While all of these possibilities have been used in various systems, only those of interest to this text will be discussed: the English engineering system and the SI system.

In addition to the mechanical and kinematic dimensions just discussed, every system of dimensions and units must also include a temperature scale. While not involved in the confusion that often results between the dimensions of mass and force, completeness in the definition of a system of units requires the inclusion of a temperature scale.

The English Engineering System.   In the English engineering system of units all four of the dimensions in Newton's law are taken as fundamental, making it necessary to give dimensions to g_c. This, then, is not a consistent system of units. In this system one takes the following fundamental dimensions with the units indicated:

Dimension	Unit	Unit Abbreviation
Force	1 pound force	lb_f
Mass	1 pound mass	lb_m
Length	1 foot	ft
Time	1 second	s
Temperature	degree Rankine	°R

The derived dimensions of g_c are then

$$g_c = \frac{\text{mass} \times \text{length}}{\text{force} \times (\text{time})^2}.$$

If one, in addition, makes the standard arbitrary definition

$$1 \text{ lb}_f \text{ will accelerate } 1 \text{ lb}_m \text{ at } 32.1739 \text{ ft/s}^2,$$

the units of g_c become

$$g_c = 32.1739 \frac{lb_m\text{-ft}}{lb_f\text{-s}^2}.$$

Usually, a value of $g_c = 32.2(lb_m\text{-ft})/(lb_f\text{-s}^2)$ is accurate enough. One of the main difficulties encountered in the English system is the fact that when objects are weighed, the weight is a force and the acceleration is that of gravity. At sea level the acceleration of gravity is $g = 32.2 \text{ ft/s}^2$. This is the same magnitude as g_c, but certainly does not have the same dimensions. This means that, at sea level, 1 lb_m will weigh 1 lb_f—numerically the same, but with different units! This fact produces confusion in the dimensions of certain "specific" quantities such as density, ρ, and specific weight, γ. Density has the dimensions of lb_m/ft^3, whereas specific weight has the dimensions of lb_f/ft^3. Again one has the situation of two quantities being numerically equal but having different dimensions. The relation between the two is

$$\gamma = \frac{g}{g_c} \rho.$$

The same is true for other quantities based on mass or weight, such as specific heat, etc.

In terms of the dimensions and units defined above, the English system adopts a thermal measure of energy. That is, in terms of the above, one can define energy as force $\times$ length so that the fundamental measure of energy is the ft-lb$_f$. However, the English system defines an equivalent thermal unit, the British thermal unit, Btu:

$$1 \text{ Btu} = 778.16 \text{ ft-lb}_f,$$

introducing, then, the additional dimensional constant:

$$J = 778.16 \frac{\text{ft-lb}_f}{\text{Btu}}.$$

Finally, the English system uses the Fahrenheit temperature scale, based on an ice point of 32°F and a steam point of 212°F.

Thus, the *derived* quantities in the English system are:

Quantity	Unit	Unit Abbreviation
Energy	1 foot pound force	ft-lb$_f$
Energy	1 British thermal unit	Btu
g_c	32.2 lb$_m$-ft/lb$_f$-s^2	
J	778 ft-lb$_f$/Btu	
Temperature	degree Fahrenheit = degree Rankine − 459.69	°F

The SI System. In the SI system of units, the dimensions of length, time, and mass are taken as fundamental, the quantity g_c is taken as 1, without dimensions, and force becomes a derived quantity. Further, the Kelvin scale is taken as the measure of temperature. Thus, the SI system takes as fundamental the dimensions and units:

Dimension	Unit	Unit Abbreviation
Mass	1 kilogram	kg
Length	1 meter	m
Time	1 second	s
Temperature	degree Kelvin	°K

Mechanical energy is, then, measured in terms of the derived force times length; however, unlike the English system no thermal energy unit is introduced and all energy terms are expressed in mechanical units (i.e., the calorie used in the metric system is not used in the SI system). Thus, both force (from Newton's law) and energy (from force × length) are *derived* quantities; although they are given special names for their units. Thus, the *derived* units in the SI system are:

Quantity	Unit	Unit Abbreviation
Force	$1 \text{ newton} = 1 \dfrac{\text{kg-m}}{\text{s}^2}$	N
Energy	$1 \text{ joule} = 1 \text{ N-m}$	J
Power	$1 \text{ watt} = 1 \text{ J/s}$	W
Temperature	degree Celsius = degree Kelvin − 273.15	°C

As may be noted the time rate of energy is given a special unit name, the watt.

The advantages of using the SI system are apparent. No dimensional conversion constants, such as g_c and the mechanical equivalent of heat, are necessary. All units are expressed in terms of the four fundamental units: kilogram, meter, second, and °K.

As noted earlier, the principal emphasis of this text is in the SI system. Because of the still extensive use of English units, some problems and examples will use them. Most of the tables of physical properties, discussed in the next chapter will be given in SI units; however, some of the tabular presentations in the appendices are given in both SI and English units to avoid the necessity of extensive conversions between the systems. Appendix F at the end of the book presents a tabulation of the most commonly needed conversions between these systems of units for quantities commonly encountered in heat transfer calculations.

1.8

THE DIMENSIONS AND UNITS OF HEAT FLOW, CONDUCTIVITY, CONDUCTANCE, AND DIFFUSIVITY

Following from the foregoing discussion, the units to be used in this text for heat flow rate, q, and heat flux, q/A, are

$$q = \text{W, Btu/s, Btu/h}$$

$$\frac{q}{A} = \text{W/m}^2, \text{Btu/s-ft}^2, \text{Btu/h-ft}^2.$$

Referring to the definition of thermal conductivity, k, in Eq. (1.1) or (1.2) this quantity is expressed in the following units:

$$k = \text{W/m-°C, Btu/h-ft-°F},$$

while conductance, most notably the heat transfer coefficient h, is expressed as

$$h = \text{W/m}^2\text{-°C, Btu/h-ft}^2\text{-°F}.$$

Equation (1.11) defines the thermal diffusivity as $\alpha = k/\rho c_p$. Since ρ is the mass density, c_p must be the *mass* specific heat (i.e., energy per unit *mass* per unit temperature difference). Thus, the dimensions and units of diffusivity are

$$\alpha = \frac{(\text{length})^2}{\text{time}}$$

$$= \text{m}^2/\text{s}, \ \text{ft}^2/\text{h}.$$

REFERENCES

1. CARSLAW, H. S., and J. C. JAEGER, *Conduction of Heat in Solids,* 2nd ed., New York, Oxford U.P., 1959.
2. ARPACI, V., *Conduction Heat Transfer,* Reading, Mass., Addison-Wesley, 1966.
3. ÖZISIK, M. N., *Heat Conduction,* New York, Wiley, 1980.

PROBLEMS

1.1 Obtain Eq. (1.12) by making a coordinate transformation of Eq. (1.10).

1.2 Obtain Eq. (1.13) by making a coordinate transformation of Eq. (1.10).

1.3 For steady state heat flow in a plane (i.e., the temperature depends only on two coordinates), show that Eq. (1.15) is satisfied by

$$t = (c_1 e^{-\lambda y} + c_2 e^{\lambda y})(c_3 \sin \lambda x + c_4 \cos \lambda x).$$

1.4 If steady heat conduction without internal heat generation in a hollow cylinder of very long length can be considered as radial only [i.e., if t in Eq. (1.12) is independent of τ, θ, or z], show that the solution of Eq. (1.12) is

$$t = c_1 \ln r + c_2.$$

1.5 Show that for nonsteady heat flow in one dimension only, Eq. (1.10) is satisfied by

$$t = e^{-\lambda^2 \alpha \tau}(c_1 \sin \lambda x + c_2 \cos \lambda x)$$

if there is no internal heat generation.

CHAPTER 2

Material Properties of Importance in Heat Transfer

INTRODUCTORY REMARKS

Any application to real problems of the results of the physical laws and mathematical analyses related to heat transfer will require that one have available numerical values of the necessary physical properties of the substance under consideration. This chapter is devoted to a discussion of these important properties. The effects of various factors, such as pressure, temperature, density, porosity, etc., are also discussed.

As shown in Chapter 1, in the derivation of the general heat conduction equation, Eq. (1.9), the properties of importance in conduction are thermal conductivity, density, and specific heat. For problems involving convection these properties are still important, but, due to the fluid motion involved, the additional property of the fluid viscosity becomes extremely important.

In the mechanism of free convection, the fluid motion, being caused by density differences, is very much influenced by the coefficient of the thermal expansion of the fluid.

THE THERMAL CONDUCTIVITY
OF HOMOGENEOUS MATERIALS

As discussed in Sec. 1.4, heat conduction is, basically, the transmission of energy by molecular motion. Thermal conductivity is, then, the physical property denoting the ease with which a particular substance can accomplish this transmission. The thermal conductivity of a material is found to depend on the chemical composition of the substance, or substances, of which it is composed, the phase (i.e., gas, liquid, or solid) in which it exists, its crystalline structure if a solid, the temperature and pressure to which it is subjected, and whether or not it is a homogeneous material.

These factors will be discussed in the following sections, with attention first directed to homogeneous materials. The nonhomogeneous materials of engineering importance are, principally, those used for insulating purposes and will be described after the general principles have been established for homogeneous materials.

The discussion of the relative order of magnitude of the thermal conductivities of various homogeneous substances can best be illustrated by examination of Table 2.1, in which some typical values are tabulated.

One immediately observes a vast range of possible values of thermal conductivity, with silver having a conductivity almost 50,000 times as great as that of Freon.

One also sees that, generally speaking, a liquid is a better conductor than a gas and that a solid is a better conductor than a liquid. These facts are best illustrated by considering the three phases of a single substance, such as mercury. As a solid at $-193°C$, mercury has a thermal conductivity of 48 W/m-°C, as a liquid at 0°C, the conductivity has dropped to 8.0 W/m-°C, whereas as a gas at 200°C its thermal conductivity is reported to be 0.0341 W/m-°C. These differences can be explained partially by the fact that while in a gaseous state, the molecules of a substance are spaced relatively far apart and their motion is random. This means that energy transfer by molecular impact is much more slow than in the case of a liquid, in which the motion is still random but in which the molecules are more closely packed. The same is true concerning the difference between the thermal conductivity of the liquid and solid phases; however, other factors become important when the solid state is formed.

Referring again to Table 2.1, one sees that a solid having a crystalline structure, such as quartz, has a higher thermal conductivity than a substance in an amorphous solid state, such as glass. Also, metals, crystalline in structure, are seen to have greater thermal conductivities than do nonmetals. In the case of the amorphous solids, the irregular arrangement of the molecules inhibits the effectiveness of the transfer of the energy by molecular impact, and, hence, the thermal conductivity is of the same order of magnitude as that observed for liquids. On the other hand, in a solid having a crystalline structure, there is an additional transfer of heat energy, as a result of a vibratory motion of the crystal lattice as a whole, in the direction of decreasing temperature. Imperfections in

TABLE 2.1 The Thermal Conductivity of Various Substances*

Substance	Thermal Conductivity W/m-°C
Gases	
Freon-12 (0°C, 1 atm)	0.0083
Air (0°C, 1 atm)	0.0241
Liquids	
Carbon dioxide (sat. liq., 0°C)	0.105
Glycerine, pure (0°C)	0.282
Water (sat liq., 0°C)	0.562
Solids	
Glass, plate (20°C)	0.76
Ice (0°C)	2.22
Magnesite brick (204°C)	3.81
Quartz (20°C)	7.6
Stainless steel (18% Cr, 8% Ni)(0°C)	16.3
Iron, pure (0°C)	73
Zinc, pure (0°C)	112
Aluminum, pure (0°C)	202
Copper, pure (0°C)	386
Silver, pure (0°C)	417

*Abstracted mainly from tables in Appendix A.

the lattice structure tend to distort and scatter these "thermoelastic" vibrations and, hence, tend to decrease their intensity.

In the case of metallic conduction, still a third mechanism of energy transfer, in addition to the molecular communication and lattice vibration mentioned above, comes into play. When the crystal of a nonmetallic substance is formed, the valence electrons (i.e., the outermost electrons) are shared among atoms to form the chemical bond which holds the atoms together as a molecule. In a metal crystal, however, these valence electrons become detached and are free to move within the lattice formed by the remaining positive ions of the metal atoms. When a difference exists between the temperatures of different parts of the metal, a general drift of these free electrons occurs in the direction of the decreasing temperature. It is this drift of free electrons which makes the metals so much better as conductors than other solids. These free electrons account for the observed proportionality between the thermal and electrical conductivities of pure metals.

With these brief remarks concerning the mechanism of conduction in gases, liquids, and solids, the following three sections consider each phase separately. The effects of pressure and temperature are also discussed, as are the tabulations in Appendix A of the thermal conductivity of various substances of importance in engineering.

The Thermal Conductivity of Homogeneous Solids

Table A.1 of Appendix A tabulates, among other properties, the thermal conductivity of various pure metals and metal alloys. Many factors are known to influence the thermal conductivity of metals, such as chemical composition, atomic structure, phase changes, grain size, temperature, pressure, and deformation. The factors with the greatest influence are the chemical composition, phase changes, and temperature. Usually the first two of these do not enter a case in which one is interested in a particular material, and, hence, only the temperature effect has to be accounted for.

It is known (Ref. 1) that the thermal conductivity of metals is directly proportional to the absolute temperature and the mean free path of the molecules. The mean free path tends to decrease with increasing temperature so that the net variation is the result of opposing influences. Pure metals generally have thermal conductivities which decrease with temperature, but the presence of impurities or alloying elements, even in minute amounts, may reverse this trend. These effects may be observed by examination of Table A.1. The data of Table A.1. for chrome steels are plotted in Fig. 2.1, which illustrates typical effects of temperature and composition on the thermal conductivity of a metal.

It is usually possible to represent the temperature dependence of the thermal conductivity of a metal by a linear relation of the form $k = k_0(1 + bt)$, in which k_0 is the thermal conductivity at $t = 0°C$ and b is a constant.

The thermal conductivities of other homogeneous solids are presented in Table A.2, along with many nonhomogeneous solids. Generally speaking, the thermal conductivities of these materials do not vary with pressure, only with temperature. The temperature variation is usually approximately linear, with the conductivity increasing with temperature.

The Thermal Conductivity of Liquids

Tables A.3 and A.4 present the thermal conductivities of various saturated liquids as functions of temperature. Other properties to be discussed later are given also. In general, the thermal conductivities of liquids are relatively insensitive to pressure, particularly at pressures not too close to the critical pressure. For this reason the temperature variation is usually the only influence which is taken into account, and the saturated state is the condition at which the thermal conductivity is reported because of the uniqueness of this state.

Most liquids exhibit a decreasing thermal conductivity with temperature, although water is, as usual, a notable exception.

Because of its importance as an engineering fluid, the properties of water are presented separately in the more extensive tabulation of Table A.3 in both SI and English units. Table A.4 presents data for several other liquids of importance.

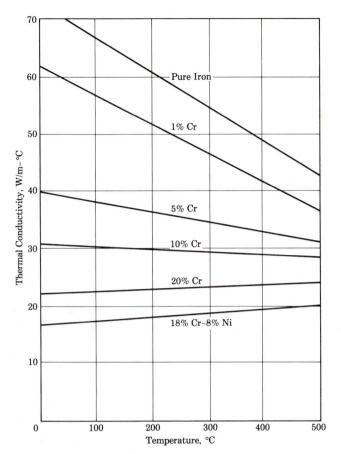

FIGURE 2.1 The effect of temperature and composition on the thermal conductivity of steel.

Of the nonmetallic liquids, water has the highest thermal conductivity, with a maximum value occurring at about 450°C. Some of the data of Tables A.3 and A.4 are plotted in Fig. 2.2 for purposes of comparison. Table A.9 gives the thermal properties of some liquid metals. These data are of current interest because of their use as heat transfer media in nuclear power plants. One can note that the thermal conductivities of liquid metals are significantly greater than for other liquid substances.

The Thermal Conductivity of Gases

The thermal conductivities of several gases are given in Table A.7 for a pressure of 1 atm. Some of the values have been plotted in Fig. 2.3 for purposes of discussion. It can be noted that in general the conductivity increases with increasing temperature and decreases with increasing molecular weight.

In general, the thermal conductivity of a gas is relatively independent of pressure if the pressure is near atmospheric pressure. Vapors near the saturation point show a strong pressure dependence. In the absence of other information,

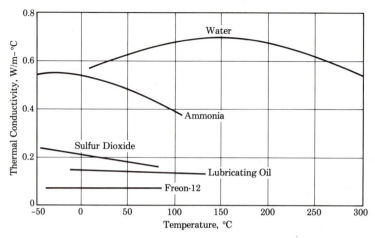

FIGURE 2.2. The effect of temperature on the thermal conductivity of several saturated liquids.

the pressure effect on the thermal conductivity may be approximated at high pressures by use of the generalized chart shown in Fig. A.1 of Appendix A. This chart plots the ratio of the thermal conductivity at high pressure to that at 1 atmosphere pressure as a function of the *reduced pressure* and *reduced temperature*. The latter quantities are the ratios of the pressure and temperature, respectively, to their values at the critical point. Table A.8 tabulates the critical constants for various substances.

Two particular gases of great engineering importance are steam and air. For this reason, more extensive presentations are made for their thermal conductivities in Tables A.5 and A.6 for both SI and English units. Other important properties, to be discussed later, are given also.

As usual, steam (Table A.5) proves to be an irregularly behaving gas, showing a rather strong pressure dependence for the thermal conductivity as well as a temperature dependence.

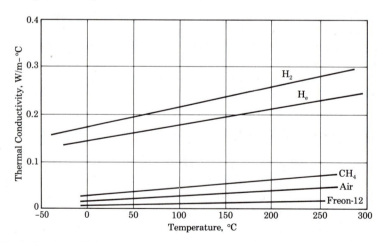

FIGURE 2.3. The effect of temperature on the thermal conductivity of several gases at atmospheric pressure.

PROPERTIES OF IMPORTANCE IN HEAT TRANSFER

THE APPARENT THERMAL CONDUCTIVITY OF NONHOMOGENEOUS MATERIALS

The above discussions concerning thermal conductivity were restricted to materials composed of homogeneous or pure substances. Many of the engineering materials encountered in practice are not of this nature. This is particularly true of building materials and insulating materials.

Some materials may exhibit nonisotropic conductivities that result from a directional preference caused by a fibrous structure (as in the case of wood, asbestos, etc.). Other materials can be discussed only from the point of view of an *apparent thermal conductivity* due to inhomogeneities present because of a porous structure (glass wool, cork, etc.) or because of a structure that is composed of different substances (concrete, stone, brick, etc.). In any of these instances the thermal conductivity may vary from sample to sample due to variations in structure, composition, density, or porosity. The values reported in Table A.2 are only typical values encountered for such substances, and a considerable deviation from these values in practice should be expected.

Insulating Materials

Insulating materials are used in cases in which one wishes to obstruct the flow of heat between an enclosure and its surroundings. Low temperature insulations are used in instances in which the enclosure in question is at a temperature lower than the ambient temperatures and when one wishes to prevent the enclosure from gaining heat. High temperature insulations are used in the reverse case when one wishes to prevent an enclosure at a temperature higher than the ambient from losing heat to its surroundings.

Cork, rock wool, glass wool, etc., are examples of low temperature insulations. Asbestos, diatomaceous earth, magnesia, etc., are examples of high temperature insulations. The thermal conductivities of these materials may be found in Table A.2. One observes that for a given material there is a strong dependence of the thermal conductivity with density as well as temperature. The term *apparent bulk density* should actually be used. This is the mass of the substance divided by its total volume (including the volume of pores in the case of a porous material). The low conductivity of these insulating materials is due primarily to the air (a poorly conducting gas) that is contained in the pores rather than to a low conductivity of the solid substance itself. Generally speaking, the apparent thermal conductivity varies directly with the apparent bulk density, although a limit exists at which the temperature of the material becomes great enough that convection and radiation within the pore may increase the transfer of heat through the material, reversing the effect of the apparent bulk density.

Refractory Materials

Refractory materials are used in instances in which it is desired to have a material capable of withstanding high temperatures without physical deterioration. The various fire-clay bricks (kaolin, magnesite, chrome, etc.) are usually encountered in this application. Some of these are included in Table A.2.

The refractories, as may be seen by consulting Table A.2, are not necessarily insulating materials as well. Further examination of this table will show the dependence on apparent density, as in the case of the insulations, as well as a dependence on the temperature at which the brick is fired. The variation of the thermal conductivity with temperature depends on whether or not a predominantly crystalline structure is formed. Usually the thermal conductivity increases with temperature for such substances, although an exception may be noted in the case of magnesite brick, which, like a pure metal, shows a decrease in conductivity with temperature.

2.4

SPECIFIC HEAT

The thermal conductivity discussed the foregoing relates to the facility with which heat is propagated through a material due to a temperature gradient. The variation in temperature of a material with the amount of heat stored in it is expressed in terms of the *specific heat* of the substance. Since heat is, thermodynamically, a path-dependent quantity, so is specific heat. Most heat transfer processes occur at virtually constant pressure; hence, the constant pressure specific heat, c_p, is invariably used.

The dimensions of c_p are energy per unit mass per unit temperature change, and in the SI system the units usually employed are kJ/kg-°C. In general the specific heat of a substance is a function of its thermodynamic state as given by pressure and temperature. Some simplifications may be made of this fact, particularly in the case of solids and liquids.

Specific Heats of Solids and Liquids

The theories of the specific heat of solids and liquids are highly developed and summaries may be found in Refs. 1 and 2. These theories show, and experiment verifies, the fact that there is little pressure dependence of specific heat of solids and liquids until extremely high pressures are encountered. Hence for ordinary engineering applications any pressure effect may be ignored, and the tables of Appendix A report only the temperature dependence. Even the temperature dependence of c_p is slight for solids and liquids, for moderate changes in temperature.

Table A.1 gives c_p for the various metals and alloys listed but only at the single temperature of 20°C. In the absence of any other information these values may be used at other temperatures without grave error. For those solids listed in Table A.2 which are not homogeneous, the values of c_p listed are apparent values. Tables A.3, A.4, and A.9 give the dependence of c_p on temperature for saturated water, several other saturated liquids, and for liquid metals. In most instances this temperature dependence is seen to be modest.

Specific Heats of Gases

In the case of gases, the temperature dependence of the specific heat is much more pronounced than in the case of solids and liquids. A dependence on pressure is also sometimes present, but for gases far removed from saturation or the critical point, this dependence may often be neglected. In general, the higher the temperature, the less is the effect of pressure. Reference 2 presents a very comprehensive review of the molecular theory of the specific heats of gases.

Again, steam is an ill-behaved gas, and the effect of both temperature and pressure may be significant. Table A.5 presents data for steam, both in SI and English units. Table A.6 presents data for air at atmospheric pressure, and these data may be used at other commonly encountered pressures without much error. Table A.7 gives c_p data for several other gases, again at atmospheric pressure. In the absence of other data, these may be used for pressures up to, say, 1400 kN/m^2 (200 psia).

2.5

THERMAL DIFFUSIVITY

In Chapter 1, during the derivation of the general heat conduction equation, the property of *thermal diffusivity* was defined:

$$\alpha = \frac{k}{\rho c_p}.$$

As noted in Sec. 1.8 the units to be used for α in this work are m^2/s (or ft^2/h). Since the thermal diffusivity involves the properties of thermal conductivity and specific heat discussed above, no further discussion here is necessary other than to note that since α involves the density, ρ, it may be quite pressure dependent for gases even though k and c_p are not. Thus, in the tables of Appendix A, values are tabulated only for the liquids and solids for which ρ may be taken as virtually constant. For the gases shown in Appendix A, one should determine α from its definition, using k and c_p as tabulated (and nearly pressure independent) and taking ρ as pressure dependent from available equations of state or thermodynamic tabulations.

2.6

THE COEFFICIENT OF THERMAL EXPANSION

As noted in Chapter 1, the driving force in the process of free convection is that of gravity acting on fluid regions of differing density. These density differences exist because of differential thermal expansion resulting from temperature differences in the fluid field. The physical property of the fluid of significance in this process is the *coefficient of thermal expansion*. Since free convection processes of significance occur only in liquids and gases, the following discussion does not include the consideration of solids.

The coefficient of thermal expansion is defined, thermodynamically, as

$$\beta = \frac{1}{v}\left(\frac{\partial v}{\partial t}\right)_P$$

$$= -\frac{1}{\rho}\left(\frac{\partial \rho}{\partial t}\right)_P,$$

(2.1)

in which $v = 1/\rho$ is the specific volume. Thus, β is determined by the thermal behavior of the fluid as given by the governing P–ρ–t relation.

The Coefficient of Expansion of Gases

For gases, the coefficient of expansion may be calculated from the definition in Eq. (2.1) if the thermal equation of state

$$\rho = f(P, t)$$

is known. For many gases, such equations of state are available in thermodynamic texts. In the event such an equation of state is not available, tabulated values of ρ (or v), as functions of P and t, may be used by applying Eq. (2.1) in a finite difference form.

Many gases (e.g., air) may be adequately represented by the ideal gas equation

$$\rho = \frac{P}{RT}$$

in which T is the absolute temperature. Application of Eq. (2.1) gives, for the ideal gas,

$$\beta = \frac{1}{T}.$$

(2.2)

For work in this text, Eq. (2.2) may be used for most of the gases tabulated (hence values of β are not given in Appendix A) except for steam. In this latter instance, approximate values of β will have to be calculated from the tabulated values of ρ, although the need for β of steam is rare.

The Coefficient of Expansion of Liquids

The case of liquids is somewhat more complex than that of gases because of the lack of equations of state of sufficient generality to permit evaluation of β in general forms. In general, as long as extreme pressures are avoided, one may consider the coefficient of expansion of liquids to be dependent on temperature only. Table A.4 tabulates β for various liquids at a single temperature. Table A.3 presents an extensive tabulation of β for saturated water.

For a great many liquids, the approximate rule (Ref. 3)

$$\beta \cong 0.07760(T_c - T)^{-0.641}$$

(2.3)

may be used when other data are lacking. In Eq. (2.3), T and T_c are the absolute liquid temperature and critical temperature (°K) and β is in $1/°K$.

FLUID VISCOSITY

All real substances are known to offer a certain amount of resistance to deformation. Generally speaking, solids in the so-called "elastic" range offer a resistance which is proportional to the amount of deformation. Fluids, both gases and liquids, are observed to resist deformation in a manner which is proportional to the time rate of deformation.

When, as in the processes of heat convection (free or forced), a fluid is caused to move past solid bodies, velocity gradients are set up within the fluid because of the relative motion of various parts of the fluid. This relative motion may be a sliding motion of adjacent fluid elements producing shearing resistance, or it may be a relative expanding or contracting motion producing tensile or compressive resistance.

In general, a fluid motion will be a complex combination of such movements with a correspondingly complex set of resisting forces acting within the fluid. The resistive forces, the inertia force due to the motion, and the applied external forces (pressure, gravity, etc.) determine the movements of the various portions of the fluid.

In order to analyze this motion and to apply the results to cases of convective heat transfer, it is necessary to be able to express the forces of resistance in terms of the velocity field of the fluid. The resistance to the shearing motion is the dominant resistive force and is expressed in terms of the fluid property known as *viscosity*.

Various concepts and definitions exist for the viscosity of a fluid, but the most useful is that of Newton, as discussed in the next section.

Dynamic Viscosity

The fundamental definition of viscosity is best illustrated by considering "parallel," laminar fluid motion such as that past a plane wall as depicted in Fig. 2.4. Some distribution in the velocity (parallel to the wall) is presumed to exist, varying from zero at the wall to some value far removed from the wall.

By selecting a plane, S, parallel to the wall, experiment has shown that the fluid layers on either side of S experience a shearing force or viscous drag which is due to their relative motion. Newton postulated that the shearing stress, τ, produced by this relative motion is directly proportional to the velocity gradient taken in a direction normal to the plane S. The constant of proportionality, μ, is termed the *coefficient of dynamic viscosity*. Thus,

$$\tau = \mu \frac{dv}{dy}, \qquad (2.4)$$

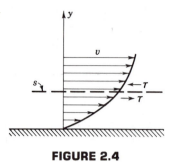

FIGURE 2.4

in which v is the fluid velocity, parallel to the wall, and y is the coordinate measured normal to the wall.

The physical concept behind this definition of viscosity is that of a transfer of momentum across S due to an exchange of molecules between the fluid layers on either side. As a molecule (moving because of thermal excitation) crosses the plane S, it will transfer an amount of momentum which is proportional to the difference in the flow velocity between the point on one side of S from which the molecule started and the point on the other side of S where it was stopped by the next molecular collision. The velocity difference, and hence the momentum exchanged, is directed parallel to S. According to Newton's law of motion, this change in momentum is accompanied by a resultant force in the same direction. It is this latter shearing force which is the viscous resistance to deformation offered by the fluid. This physical interpretation requires the exist-ence of a velocity gradient in the fluid, and Newton's definition of the dynamic viscosity coefficient in Eq. (2.4) relates the viscous shear to the velocity gra-dient. This concept of viscosity is based on a molecular exchange. The existence of any transport of mass in a direction normal to the velocity on a scale larger than a molecular one is excluded in this definition. Thus, the above definition of viscosity applies only to laminar flow—not to turbulent flow, in which gross transverse fluctuations and mixing occur.

Extensive experiments have confirmed the law stated in Eq. (2.4), and most of the commonly encountered fluids are known to be very nearly Newtonian. The coefficient of dynamic viscosity is then a characteristic molecular property of the fluid and is not dependent upon the fluid motion. Reference 3 presents an extensive summary of the theoretical molecular theory of the dynamic vis-cosity.

The Dimensions and Units of Dynamic Viscosity

The defining relation for the dynamic viscosity, Eq. (2.4) shows its dimensions to be:

$$\frac{(\text{force}) \times (\text{time})}{(\text{length})^2}.$$

Generally, the units used are

$$\text{SI:} \qquad \text{N-s/m}^2,$$

$$\text{English:} \quad \text{lb}_f\text{-s/in.}^2, \ \text{lb}_f\text{-h/ft}^2.$$

These units are convenient in many cases; however, it is often desirable to express the dimensions of viscosity in terms of the fundamental dimensions of mass rather than force. Since Newton's second law expresses (force) = (mass) × (length)/(time)^2, the dimensions of viscosity may also be expressed as

$$\frac{\text{mass}}{(\text{length}) \times (\text{time})},$$

and the units most commonly used are

$$\text{SI:} \qquad \text{kg/m-s,}$$

$$\text{English:} \quad \text{lb}_m\text{/ft-s, } \text{lb}_m\text{/ft-h.}$$

The defining relation in the SI system that 1 newton = 1 kg-m/s^2 shows that in this system, the dynamic viscosity is identical in either force or mass units:

$$1 \text{ N-s/m}^2 = 1 \text{ kg/m-s} = 10.0 \text{ poise.}$$

As noted above, the unit *poise* is often used to express viscosity in metric units and it is equal to 0.1 N-s/m^2 = 0.1 kg/m-s.

As is rather apparent, there are a great number of viscosity units in use. Various conversion factors are given in Appendix F.

Kinematic Viscosity

Since the forces of viscosity act directly on a fluid, and since the fluid inertia resists these forces, the ratio of the viscous force to the inertia force would be expected to be an important parameter in the analysis of fluid motion. Thus, the *kinematic viscosity,* defined as the ratio of the dynamic viscosity to the fluid density (or specific weight), becomes an important fluid property. The kinematic viscosity is denoted by ν:

$$\nu = \frac{\mu}{\rho}. \qquad (2.5)$$

Thus, the dimensions of kinematic viscosity are

$$\frac{(\text{length})^2}{\text{time,}}$$

with the usual units being

$$\text{SI:} \qquad \text{m}^2\text{/s}$$

$$\text{English:} \quad \text{ft}^2\text{/s, ft}^2\text{/h.}$$

Dynamic and Kinematic Viscosity of Liquids

The molecular theory of the viscosity of gases and liquids (Ref. 2) predicts that the dynamic viscosity depends primarily on temperature and, to a lesser degree, on pressure. These facts are borne out by experimental observation.

In liquids, as in the case of thermal conductivity and specific heat, the pressure dependence of the dynamic viscosity is found to be quite slight unless extreme pressures are reached. Usually, then, this effect may be ignored and only the temperature dependence need be considered. Likewise, since liquid densities are not very pressure sensitive, the kinematic viscosity may usually be taken as only temperature dependent. Tables A.3, A.4, and A.9 give data for the dynamic and kinematic viscosity of several liquids.

Dynamic and Kinematic Viscosity of Gases

For gases, the picture is somewhat different than that noted above for liquids. Generally speaking, the dynamic viscosity of gases is quite a bit smaller than that for liquids; although the kinematic viscosity may be greater because of lower densities. In most gases the temperature is, again, the most significant factor influencing the dynamic viscosity. The pressure effect may be quite small, as long as the saturation state or critical state is not approached.

Steam, as usual, is an exceptional gas and the dependence of μ on both pressure and temperature may be significant as shown in Table A.5, in both SI and English units.

Table A.6 gives values of μ for air at atmospheric pressure (in both SI and English units) and Table A.7 gives μ at atmospheric pressure for several other gases. As in the case of thermal conductivity, the behavior of viscosity at high pressures may be estimated rather well by the generalized correlation in Fig. A.2 of Appendix A. This figure relates the ratio of the dynamic viscosity at high pressures to that at 1 atm in terms of the reduced pressure, P_r, and reduced temperature, T_r, defined in Sec. 2.2.

Kinematic viscosity is most certainly pressure dependent as well as temperature dependent, owing to the pressure dependence of the density. Thus, one normally needs to calculate ν after finding μ and ρ separately. The value of ν given for air in Table A.6 apply only to atmospheric pressure and would need to be recalculated at other pressures after appropriately determining ρ. Likewise, values for the kinematic viscosity of the gases in Table A.7 can be found from the tabulated values of μ after finding ρ from an appropriate equation of state or table of thermodynamic properties.

2.8

THE PRANDTL NUMBER

In problems of convection in which one must consider the simultaneous exchange of momentum through viscosity and heat through conduction, it is found that the ratio of the kinematic viscosity to the thermal diffusivity plays a most significant role. The physical meaning of this parameter will be discussed at some length in Chapter 7, but it seems appropriate to mention it here since it is purely a fluid property.

This ratio is called the *Prandtl number* in honor of the German engineer who contributed extensively to the knowledge of fluid motion and heat transfer. The Prandtl number is, then, defined

$$\text{Pr} = \frac{\nu}{\alpha} \tag{2.6}$$

$$= \frac{\mu c_p}{k}.$$

Since both ν and α have the dimensions of $(\text{length})^2/\text{time}$, Pr has no dimensions although it is a fluid property.

Quite apparently, values of Pr can be calculated from the properties already discussed. Its main dependence is on temperature, with little, if any, dependence on pressure. Tables A.3 through A.7 and A.9 tabulate values of Pr for the fluids involved.

REFERENCES

1. MOELWYN-HUGHES, E. A., *Physical Chemistry,* 2nd ed., New York, Macmillan, 1964.
2. HIRSCHFELDER, J. O., C. F. CURTISS, and R. B. BIRD, *Molecular Theory of Gases and Liquids,* New York, Wiley, 1954.
3. SMITH W. T., S. GREENBAUM, and G. P. RUTLEDGE, "Correlation of Critical Constants with the Thermal Expansion of Organic Liquids," *J. Phys. Chem.,* Vol. 58, 1954, p. 443.

PROBLEMS

2.1   Using Table A.7 and Fig. A.1, find the thermal conductivity of the following gases at the conditions indicated:
(a) Nitrogen at 14,000 kN/m², 100°C.
(b) Helium at 350 kN/m², 0°C.
(c) Methane at 10,000 kN/m², 0°C.

2.2   Using Table A.7 and Fig. A.2, find the dynamic viscosity of the gases given in Prob. 2.1.

2.3   Using Table A.7, Fig. A.1, and the ideal gas law, find the thermal conductivity and the thermal diffusivity of the following gases at the conditions indicated:
(a) Methane at 1 atm, 240°C.
(b) Carbon dioxide at 1 atm, 50°C.
(c) Oxygen at 8200 kN/m², 100°C.

2.4 Using Table A.7 and Fig. A.2, find the dynamic viscosity of the gases given in Prob. 2.3.

2.5 Using the results of Probs. 2.3 and 2.4, calculate the Prandtl number of the gases given and compare the results with values given in Table A.7.

2.6 Using Table A.7 and Fig. A.1, find the thermal conductivity of the following gases at the conditions indicated:
(a) Ammonia at 12,000 kN/m^2, 160°C.
(b) Nitrogen at 5000 kN/m^2, 0°C.
(c) Freon-12 at 5000 kN/m^2, 200°C.

2.7 Using Table A.7 and Fig. A.2, find the dynamic viscosity of the gases given in Prob. 2.6.

2.8 Using Table A.7, Fig. A.1, and the ideal gas law, find the thermal conductivity and the thermal diffusivity of the following gases at the conditions indicated:
(a) Carbon dioxide at 10,000 kN/m^2, 500°C.
(b) Ethane at 1 atm, 100°C.
(c) Propane at 6000 kN/m^2, 160°C.

2.9 Using Table A.7 and Fig. A.2, find the dynamic viscosity of the gases given in Prob. 2.8.

2.10 Using the results of Probs. 2.8 and 2.9, calculate the Prandtl number of the gases given and compare the results with values given in Table A.7.

2.11 Using Table A.7, Fig. A.1, and the ideal gas law, find the thermal conductivity and the thermal diffusivity of the following gases at the conditions indicated:
(a) Methane at 14.7 psia, 500°F.
(b) Carbon dioxide at 14.7 psia, 250°F.
(c) Oxygen at 1500 psia, 150°F.

2.12 Using Table A.7 and Fig. A.2, find the dynamic viscosity and Prandtl number for the gases given in Prob. 2.11.

CHAPTER 3

Steady State Heat Conduction in One Dimension

3.1

THE MEANING OF "ONE-DIMENSIONAL" CONDUCTION

The term "steady state" conduction was defined in Chapter 1 as the condition which prevails in a heat conducting body when temperatures at fixed points do not change with time. The term "one-dimensional" is applied to a heat conduction problem when only one space coordinate is required to describe the distribution of temperature within the body. Such a situation rarely exists precisely in real problems, but a great number of problems of practical engineering interest may be approximated rather well by making the assumption of one-dimensional conduction.

The flow of heat through a plane wall, at regions far removed from the edges, depends only on the coordinate measured normal to the plane of the wall. Hence, neglecting edge effects in a wall leads to a one-dimensional heat conduction problem. Similarly, conduction through a very long, hollow cylinder (such as a pipe) which is maintained at uniform temperature on its inner and outer surfaces may be considered to depend only on a single coordinate, the radial distance. A very thin rod, or wire, maintained at different temperatures on its ends may be considered to conduct heat one-dimensionally if it is sufficiently thin so that its temperature may be taken as uniform over any cross section.

For the case of steady state one-dimensional conduction, the general conduction equation, Eq. (1.8), reduces to

$$\frac{d}{dx}\left(k\,\frac{dt}{dx}\right) + q^* = 0, \tag{3.1}$$

or its equivalent in other coordinate systems. The solution of one-dimensional steady state problems involves the integration of the above equation and the evaluation of the resultant integration constants by use of the imposed boundary conditions. The simplest boundary condition is that of specified boundary temperatures, and certain examples are presented in the next few sections. Cases with and without internal heat generation are treated. Then problems in which the one-dimensional solid is bounded by convecting fluids of known temperature are discussed.

3.2

THE PLANE WALL WITH SPECIFIED BOUNDARY TEMPERATURES

An elementary conduction problem is that of a plane wall of finite thickness but infinite in extent in all other directions (so that the conduction may be taken as one-dimensional), with each face maintained at a uniform temperature, and with a constant conductivity.

The expressions for the temperature distribution (linear) and the rate of heat flow through the wall may be readily deduced from Fourier's law, Eq. (1.3), without overly elaborate mathematical analysis. However, in order to illustrate principles to be used in later, more complex, problems, the general approach will be used here.

Figure 3.1 illustrates a plane wall of thickness Δx and composed of a material of constant thermal conductivity k. The single coordinate is the displacement x, and x_1, denotes the wall surface maintained at a uniform temperature t_1, while x_2 is the surface maintained at t_2. Since there is assumed to be no internal heat generation, Eq. (3.1) reduces to

$$\frac{d^2t}{dx^2} = 0, \tag{3.2}$$

with

$$t = t_1 \text{ at } x = x_1,$$

$$t = t_2 \text{ at } x = x_2.$$

The rather elementary integration of Eq. (3.2) with the resulting constants evaluated by use of the stated boundary conditions yields the expected linear variation of temperature through the wall:

$$t = t_1 + (t_2 - t_1)\frac{x - x_1}{x_2 - x_1}. \tag{3.3}$$

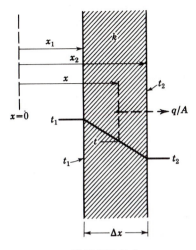

FIGURE 3.1

Fourier's law in the form of Eq. (1.3) is applicable since the temperature gradient is uniform over every plane drawn normal to x. Thus,

$$\frac{q}{A} = -k\frac{dt}{dx} = k\frac{t_1 - t_2}{x_2 - x_1} = k\frac{t_1 - t_2}{\Delta x}. \tag{3.4}$$

The above result depends on the assumption that the thermal conductivity, k, is constant. The discussion of Chapter 2 makes it apparent that this is rarely the case in real materials. The method of applying the result of Eq. (3.4) for cases of linearly varying k is discussed in a later section.

3.3

THE MULTILAYER WALL WITH SPECIFIED BOUNDARY TEMPERATURES

Consider now a plane wall composed of layers of materials having different thicknesses and thermal conductivities. Figure 3.2 illustrates a wall of three layers for the purposes of discussion, although any number of layers may be considered.

Denote each junction of two different materials by a number (1, 2, 3, 4 in Fig. 3.2) and adopt the following notational convention:

Δx_{mn} = thickness of material between plane m and plane n,

k_{mn} = thermal conductivity of material between plane m and plane n.

If the outside temperatures are specified (i.e., t_1 and t_4) and if one knows the thickness and conductivities of all the materials between the planes where these temperatures are specified, what is the distribution of the temperature through the wall and what is the rate of heat flow through the wall?

Since the steady state is assumed to exist, Eq. (3.4) may be applied to obtain

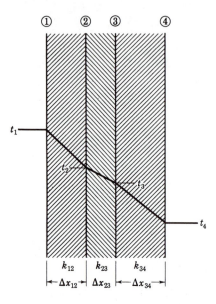

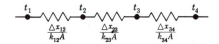

FIGURE 3.2

the following set of equations since the rate of heat flow per unit area, q/A, is the same for each layer:

$$\frac{q}{A} = \frac{t_1 - t_2}{\Delta x_{12}/k_{12}} = \frac{t_2 - t_3}{\Delta x_{23}/k_{23}} = \frac{t_3 - t_4}{\Delta x_{34}/k_{34}}.$$

Solving for each of the temperature differences, one obtains

$$t_1 - t_2 = \frac{q}{A}\frac{\Delta x_{12}}{k_{12}},$$

$$t_2 - t_3 = \frac{q}{A}\frac{\Delta x_{23}}{k_{23}}, \tag{3.5}$$

$$t_3 - t_4 = \frac{q}{A}\frac{\Delta x_{34}}{k_{34}}.$$

Upon addition of these equations the unknown temperatures are eliminated, and the following expression for the heat flow through the wall in terms of the overall temperature difference may be obtained:

$$\frac{q}{A} = \frac{t_1 - t_4}{\dfrac{\Delta x_{12}}{k_{12}} + \dfrac{\Delta x_{23}}{k_{23}} + \dfrac{\Delta x_{34}}{k_{34}}}. \tag{3.6}$$

Once the heat flow rate is obtained from Eq. (3.6), one may determine the temperatures t_2 and t_3 from Eq. (3.5). The distribution of temperature is then known to be linear between the known values.

Thermal Resistance and Thermal Conductance

In Sec. 1.4 a unit thermal conductance was defined:

$$C = \frac{k}{\Delta x}.$$

A unit thermal resistance R_k could be defined as the reciprocal of the unit conductance:

$$R_k = \frac{1}{C} = \frac{\Delta x}{k},$$

which would lead to an analogy between electrical and thermal conduction. The subscript k is used to emphasize the fact that the resistance defined is a *conduction* resistance. However, a clearer analogy may be obtained if one defines a *total* thermal resistance

$$\mathcal{R}_k = \frac{1}{CA} = \frac{\Delta x}{kA}.$$

On the basis of this definition, the total heat flow through a multilayer wall is

$$q = \frac{\Delta t_{\text{overall}}}{\mathcal{R}_{k_{12}} + \mathcal{R}_{k_{23}} + \cdots},$$

$$\frac{q}{A} = \frac{\Delta t_{\text{overall}}}{A(\mathcal{R}_{k_{12}} + \mathcal{R}_{k_{23}} + \cdots)}, \tag{3.7}$$

$$\frac{q}{A} = \frac{\Delta t_{\text{overall}}}{R_{k_{12}} + R_{k_{23}} + \cdots}.$$

Introduction of the definition of $\mathcal{R}$ or R' will yield Eq. (3.6).

Figure 3.2 depicts the equivalent electrical network implied by the above analysis. One sees that thermal resistance in a series should be added—as in the electrical equivalent. For the plane wall, the area, A, is the same for each layer, and the unit resistance could have been employed. However, for instances in which the areas might not be the same (in the case of parallel circuit, for instance), the formulation given above is necessary.

EXAMPLE 3.1

A house wall consists of an outer layer of common brick 10 cm thick ($k = 0.69$ W/m-°C), followed by a 1.25-cm layer of Celotex sheathing ($k = 0.048$ W/m-°C). A 1.25-cm layer of sheetrock ($k = 0.744$ W/m-°C) forms the inner surface and is separated from the sheathing by 10 cm of air space—as provided by the wall studs. The air space has a *unit conductance* of

6.25 W/m²-°C. The outside brick temperature is 5°C; the inner wall surface is maintained at 20°C. What is the rate of heat loss, per unit area of wall? What is the temperature at a point midway through the Celotex layer?

Solution.   From Eq. (3.6) the heat loss is

$$\frac{q}{A} = \frac{20 - 5}{\dfrac{0.0125}{0.744} + \dfrac{1}{6.25} + \dfrac{0.0125}{0.048} + \dfrac{0.10}{0.69}} = 25.77 \text{ W/m}^2 \ (8.17 \text{ Btu/h-ft}^2).$$

A point midway through the Celotex is 0.00625 m from the outside surface. The heat flow is the same between these two points since the steady state exists. Thus, the desired temperature is

$$\frac{q}{A} = 25.77 = \frac{t - 5}{\dfrac{0.00625}{0.048} + \dfrac{0.10}{0.69}},$$

$$t = 12.1°C \ (53.8°F). \qquad \blacksquare$$

Thermal Contact Resistance

The discussion of the multilayer wall in the foregoing presumed perfect contact between adjacent layers so that it could be assumed that the temperatures of the layers were the same at the plane of contact. In real systems such is rarely the case, and the contacting surfaces touch only at discrete locations, due to surface roughness, interspersed with void spaces. The void spaces are usually filled with air. Figure 3.3 depicts such a situation in an exaggerated scale. Thus, there is not a single plane of contact, and heat may flow across the interface by parallel

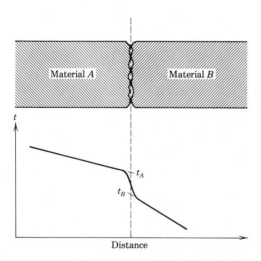

FIGURE 3.3.   Typical temperature distribution due to thermal contact resistance.

STEADY CONDUCTION IN ONE DIMENSION

TABLE 3.1 Typical Values of the Unit Contact Resistance for 75S-T6 Aluminum Surfaces at 93°C

RMS Surface Roughness μm	$R_{tc} \times 10^4$ m²·°C/W	
	Contact Pressure 100 kN/m²	Contact Pressure 1000 kN/m²
3.05	3.5	1.6
1.65	2.0	0.98
0.25	0.98	0.59

paths: that through the contact spots by conduction and that across the voids by convection and, perhaps, radiation. Thus, an apparent temperature drop may be presumed to occur between the two materials at the interface, as suggested in Fig. 3.3. If t_A and t_B represent the temperatures at the theoretical plane interface obtained by extrapolation of the temperature gradients in the materials on either side, then a *unit thermal contact resistance* may be defined as

$$R_{tc} = \frac{t_A - t_B}{q/A}.$$

Correspondingly, for a given wall area A, a *total* thermal contact resistance

$$\mathcal{R}_{tc} = \frac{R_{tc}}{A}$$

can be defined.

The effect of interfacial contact can be introduced into Eqs. (3.6) and (3.7) by including values of R_{tc} or $\mathcal{R}_{tc}$ in the resistance summations. The utility of the contact resistance concept depends on the availability of reliable numerical values. In most instances such values have to be obtained experimentally; however, in the case of metal-to-metal contact many theories have also been developed to predict values of R_{tc}. As would be expected, metallic contact resistance depends on the metals involved, the surface roughness, the contact pressure, the material occupying the void spaces, and temperature. Reference 1 may be consulted as a source of both theoretical and experimental values of R_{tc}. As a guide to the typical orders of magnitude and the effect of various parameters, sample experimental data are given in Table 3.1.

3.4

THE SINGLE-LAYER CYLINDER WITH SPECIFIED BOUNDARY TEMPERATURES

A geometrical configuration which is mathematically simple and also of great engineering importance is that of a hollow cylinder, as pictured in Fig. 3.4. The cylindrical system of coordinates shown is the natural one to use.

If the inner surface of radius r_1 and the outer surface of radius r_2 are main-

FIGURE 3.4

tained at uniform temperatures t_1 and t_2, respectively, then for a sufficiently long cylinder the end effects may be ignored. One may thus eliminate any dependence of the temperature on the axial coordinate, z, or the circumferential coordinate, θ. Thus, the problem is reduced, in the steady state, to a one-dimensional situation, with the radial distance r as the coordinate. Under these conditions the general heat conduction equation in cylindrical coordinates [Eq. (1.12)] reduces the following expression if the generation term is omitted and k is taken as constant:

$$\frac{d^2t}{dr^2} + \frac{1}{r}\frac{dt}{dr} = 0 \quad \text{or} \quad \frac{d}{dr}\left(r\frac{dt}{dr}\right) = 0. \tag{3.8}$$

The boundary conditions are

$$\text{At } r = r_1: \quad t = t_1.$$
$$\text{At } r = r_2: \quad t = t_2. \tag{3.9}$$

The solution of Eq. (3.8) under the conditions of Eq. (3.9) gives a logarithmic temperature distribution

$$t = t_1 + (t_2 - t_1)\frac{\ln (r/r_1)}{\ln (r_2/r_1)}. \tag{3.10}$$

Again, since the steady state exists, the het flow across any cylindrical surface of arbitrary radius r is constant and given by Fourier's law to be

$$q = -kA_r \frac{dt}{dr},$$

where $A_r = 2\pi r L$, L being the length of the cylinder. Since the cylinder is taken as infinite in length to achieve one-dimensional conduction, it is better to work in heat flow per unit length. Thus,

$$\frac{q}{L} = -k2\pi r \frac{dt}{dr},$$

from which Eq. (3.10) yields

$$\frac{q}{L} = \frac{2\pi k(t_1 - t_2)}{\ln (r_2/r_1)}.$$ (3.11)

THE MULTILAYER CYLINDER WITH SPECIFIED BOUNDARY TEMPERATURES

A pipe covered with insulation is a perfect example of the next configuration to be discussed—that of a long cylinder composed of two or more layers of materials having different thermal conductivities. For purposes of discussion consider the three-layer cylinder shown in Fig. 3.5. The same notational scheme employed in the case of the multilayer wall is used here.

The same procedure used for the plane wall may be applied in this case. Writing three expressions for the heat flow through each of the layers, in terms of unspecified temperatures t_2 and t_3:

$$\frac{q}{L} = \frac{2\pi k_{12}(t_1 - t_2)}{\ln (r_2/r_1)} = \frac{2\pi k_{23}(t_2 - t_3)}{\ln (r_3/r_2)} = \frac{2\pi k_{34}(t_3 - t_4)}{\ln (r_4/r_3)}.$$

Elimination of the unknown t_2 and t_3 gives

$$\frac{q}{L} = \frac{2\pi(t_1 - t_4)}{\dfrac{\ln (r_2/r_1)}{k_{12}} + \dfrac{\ln (r_3/r_2)}{k_{23}} + \dfrac{\ln (r_4/r_3)}{k_{34}}}$$ (3.12)

as the heat flow per unit length in terms of the overall temperature difference. Once q/L is known, the temperatures t_2 and t_3 may be found from the expressions for the heat flow through each layer.

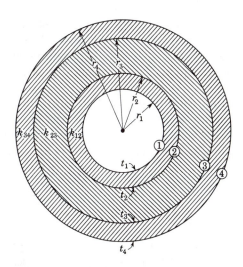

FIGURE 3.5

EXAMPLE 3.2

A 4-in. schedule 40 wrought iron pipe ($k = 55$ W/m-°C) is covered with 2.5 cm of magnesia insulation ($k = 0.071$ W/m-°C). If the inside pipe wall and outer insulation surface temperature are 150°C and 30°C, respectively, find the heat loss per meter of pipe length.

Solution. Appendix B gives the inside and outside radii of a 4-in. schedule 40 pipe to be $10.226/2 = 5.113$ cm, and $11.43/2 = 5.715$ cm, respectively. Hence, Eq. (3.12) gives the heat loss to be

$$\frac{q}{L} = \frac{2\pi(150 - 30)}{\dfrac{\ln (5.715/5.113)}{55} + \dfrac{\ln [(5.715 + 2.5)/5.715]}{0.071}}$$

$$= 147.5 \text{ W/m} (153.4 \text{ Btu/h-ft}). \qquad \blacksquare$$

Equation (3.11) suggests that for a cylindrical wall, a *total* thermal resistance may be defined

$$\mathcal{R}_k = \frac{\ln (r_2/r_1)}{2\pi L k}. \qquad (3.13)$$

Based on this definition, the electrical network analogy

$$q = \frac{\Delta t_{overall}}{\mathcal{R}_{k_{12}} + \mathcal{R}_{k_{23}} + \cdots}$$

yields the same result as that given by Eq. (3.12).

3.6

THE EFFECT OF VARIABLE THERMAL CONDUCTIVITY

All the cases discussed above involve the assumption of constant thermal conductivity. Chapter 2 noted the fact that almost all substances show some dependence of thermal conductivity on temperature. It is the purpose of this section to illustrate the effect of this dependence on the temperature distribution and the heat flow rate.

Examination of the tables of Appendix A or Figs. 2.1 through 2.3 will show that for most substances, particularly solids, a *linear* dependence of the thermal conductivity on temperature exists, at least for limited ranges of temperatures. Thus, let the thermal conductivity be expressed as a function of temperature in the following way:

$$k = k_0(1 + bt). \qquad (3.14)$$

The symbol k_0 is the value of the conductivity at $t = 0$, and b is a constant property of the material.

For simplicity of discussion, consider conduction through a plane wall (depicted in Fig. 3.6) of thickness Δx and with uniform surface temperatures t_1 and t_2.

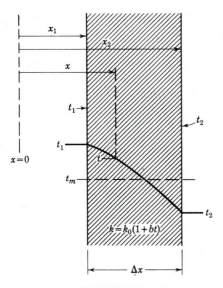

FIGURE 3.6

Again using Fourier's law, the heat flow per unit area of wall is

$$q = -kA \frac{dt}{dx} = -k_0(1 + bt) \frac{dt}{dx} A.$$

Since the steady state exists, q/A is a constant, and the equation above may be separated and integrated directly:

$$\frac{q}{A} \int_{x_1}^{x_2} dx = -k_0 \int_{t_1}^{t_2} (1 + bt)dt,$$

$$\frac{q}{A}(x_2 - x_1) = -k_0 \left[\left(t_2 + \frac{b}{2} t_2^2 \right) - \left(t_1 + \frac{b}{2} t_1^2 \right) \right].$$

(3.15)

Equation (3.15) may be factored and written

$$\frac{q}{A} = -k_0 \left[1 + \frac{b}{2} (t_1 + t_2) \right] \frac{t_2 - t_1}{\Delta x}.$$

(3.16)

If one now evaluates the thermal conductivity at the arithmetic mean of the surface temperature,

$$t_m = \frac{t_1 + t_2}{2},$$

Eq. (3.14) gives the mean conductivity to be

$$k_m = k_0 \left[1 + \frac{b}{2} (t_1 + t_2) \right].$$

Equation (3.16) for the heat flux then becomes

$$\frac{q}{A} = k_m \frac{t_1 - t_2}{\Delta x}.$$

(3.17)

This is identical in form to Eq. (3.4) for the wall of constant thermal conductivity except for the fact that the mean thermal conductivity replaces the constant value.

Thus, as far as the rate of heat transfer is concerned, it is adequate to employ the relations developed for constant thermal conductivity if the conductivity is evaluated at the arithmetic mean temperature. This is true, of course, only if the linear variation assumed adequately represents the actual temperature dependence of the conductivity.

In some instances it may not be possible to know in advance the mean temperature at which the thermal conductivity should be evaluated. This would be the case when plane or cylindrical walls of multiple layers are encountered. Only the overall temperature range for the entire structure is usually known—not the surface temperature of each layer of material. In such cases one would have to assume reasonable values of the thermal conductivities of the materials involved and use these values to determine the temperature of the boundaries of each layer. Then the mean temperature of each layer may be computed, and the correctness of the assumed conductivity may be determined. This would then be repeated until a satisfactory check is obtained. However, unless there is a strong dependence of the thermal conductivity on temperature, and unless the temperature differences are large, such a procedure is usually not worthwhile since thermal conductivities are not generally known accurately enough to warrant such precision. Furthermore, in many problems of practical engineering importance, the accuracy of the given data (e.g., the temperature) may not justify such procedures.

3.7

CASES INVOLVING INTERNAL HEAT GENERATION

The cases considered thus far have been those in which the conducting solid is free of internal heat generation. In this section two instances in which internal generation is present will be considered, and both will treat cases in which the heat generation is distributed throughout the solid.

Numerous practical instances exist in which such generation must be accounted for. The dissipative processes in current-carrying electrical conductors result in heat generation. Induction heating produces distributed heat additions as do certain exothermic chemical reactions such as the curing of concrete.

Uniformly Distributed Generation in a Plane Wall

Consider the instance in which a plane wall, for which the surface temperatures are specified, has a uniformly distributed heat generation rate of q^* (per unit volume) throughout its interior. The steady state temperature distribution and heat flow rates will be found and compared with the results of Sec. 3.2, in which generation is absent. For convenience, then, use the geometry and boundary conditions of that section as depicted in Fig. 3.1. In this instance the governing differential equation, Eq. (3.1), and the associated boundary conditions become, for constant k,

$$\frac{d^2t}{dx^2} + \frac{q^*}{k} = 0.$$

$$\text{At } x = x_1: \quad t = t_1.$$ (3.18)

$$\text{At } x = x_2: \quad t = t_2.$$

In this instance the generation rate q^* is taken as constant. Twice integrating Eq. (3.18), one obtains

$$t = \left[t_1 + \frac{t_2 - t_1}{x_2 - x_1}(x - x_1) \right]$$

$$+ \left[\frac{q^*(x_2 - x_1)^2}{2k} \right]\left[\frac{x - x_1}{x_2 - x_1} - \left(\frac{x - x_1}{x_2 - x_1}\right)^2 \right]$$ (3.19)

One notes when comparing this result with that for the wall without generation [Eq. (3.3)] that the linear temperature variation is altered by the presence of the second term involving the internal heat generation rate q^*. If one were to consider as an example, the instance in which $t_1 > t_2$, it may be noted that if the parameter

$$\frac{q^*(x_2 - x_1)^2}{2k(t_1 - t_2)} = \frac{q^*(\Delta x)^2}{2k(t_1 - t_2)} > 1,$$ (3.20)

temperatures in excess of either t_1 or t_2 will occur within the wall. This is more easily seen, perhaps, by considering the heat flux. Since

$$\frac{q}{A} = -k\frac{dt}{dx},$$

Eq. (3.19) gives

$$\frac{q}{A} = k\frac{t_1 - t_2}{x_2 - x_1} - q^*(x_2 - x_1)\left(\frac{1}{2} - \frac{x - x_1}{x_2 - x_1} \right).$$ (3.21)

Once again one may note how the heat generation term modifies the uniform heat flux of the nongeneration case. It is instructive to observe the heat flux evaluated at each surface ($x = x_1$ and $x = x_2$):

$$\left(\frac{q}{A} \right)_{x_1} = k\frac{t_1 - t_2}{\Delta x} - \frac{q^*\Delta x}{2},$$

$$\left(\frac{q}{A} \right)_{x_2} = k\frac{t_1 - t_2}{\Delta x} + \frac{q^*\Delta x}{2}.$$ (3.22)

One sees that one-half of the total generated heat flows out each face of the wall—appropriately adding to or subtracting from the nongeneration heat flow, depending on the surface in question. It is also apparent that, for example, in the case when $t_1 > t_2$, the condition stated in Eq. (3.20) reverses the direction of heat flow at the face, $x = x_1$.

Uniformly Distributed Generation in a Cylinder

The case of uniformly distributed generation may be treated for the cylindrical geometry much in the same fashion as was just done for the plane wall. Using the same geometry and boundary conditions as for the case without generation shown in Fig. 3.4, Eq. (1.12) becomes

$$\frac{d^2t}{dr^2} + \frac{1}{r}\frac{dt}{dr} + \frac{q^*}{k} = 0,$$

(3.23)

$$\frac{d}{dr}\left(r\frac{dt}{dr}\right) + r\frac{q^*}{k} = 0.$$

The boundary conditions are

$$\text{At } r = r_1: \quad t = t_1.$$
$$\text{At } r = r_2: \quad t = t_2.$$

Twice integrating Eq. (3.23), one obtains

$$t = -\frac{r^2}{4}\frac{q^*}{k} + B \ln r + C.$$

(3.24)

Application of the boundary conditions yields the temperature distribution:

$$t = t_1 + (t_2 - t_1)\frac{\ln (r/r_1)}{\ln (r_2/r_1)}$$

(3.25)

$$+ \frac{q^*}{4k}\left[(r_2^2 - r_1^2)\frac{\ln (r/r_1)}{\ln (r_2/r_1)} - (r^2 - r_1^2)\right].$$

Although the above expression might be written more concisely, it is left in this form in order that it may be compared to the result in which generation is absent [see Eq. (3.10)].

The heat flow, per unit length, may be found at the two surfaces by using

$$\frac{q}{L} = -k2\pi r\frac{dt}{dr}.$$

The results may be found to be

$$\left(\frac{q}{L}\right)_{r_1} = \frac{2\pi k(t_1 - t_2)}{\ln (r_2/r_1)} - \pi r_1^2 q^*\left[\frac{\frac{1}{2}\left(\frac{r_2^2}{r_1^2} - 1\right)}{\ln (r_2/r_1)} - 1\right],$$

(3.26)

$$\left(\frac{q}{L}\right)_{r_2} = \frac{2\pi k(t_1 - t_2)}{\ln (r_2/r_1)} + \pi r_1^2 q^*\left[1 - \frac{\frac{1}{2}\left(1 - \frac{r_1^2}{r_2^2}\right)}{\ln (r_2/r_1)}\right].$$

(3.27)

As in the plane case, one may ascertain the magnitude of the generation rate which reverses the heat flow at r_1 when $t_1 > t_2$.

The special case that results when the cylinder is a solid bar is most easily found by returning to Eq. (3.24) and noting that t must remain finite as $r \to 0$. Hence, $B = 0$. The remaining boundary condition of $t = t_2$ at $r = r_2$ gives

$$t = t_2 + \frac{q^* r_2^2}{4k}\left(1 - \frac{r^2}{r_2^2}\right). \tag{3.28}$$

The centerline temperature, call it t_0, is then

$$t_0 = t_2 + \frac{q^* r_2^2}{4k}. \tag{3.29}$$

The heat flow at the surface must necessarily be the total heat generated in the bar:

$$\left(\frac{q}{L}\right)_{\text{surf}} = \pi r_2^2 q^*. \tag{3.30}$$

3.8

BOUNDARIES SURROUNDED BY FLUIDS OF SPECIFIED TEMPERATURES

The examples presented thus far have been restricted to cases in which certain boundary temperatures of the bodies in question were assumed to be known. In many problems of practical importance this is not the case. Usually, the configurations mentioned above are encountered in situations in which they separate fluids of different temperatures—these fluid temperatures being specified or known.

If fluid motion exists, as it invariably will because of either free or forced convection, the resulting thermal and velocity boundary layers cause a temperature difference to exist between the main bulk of the fluid (essentially at a uniform temperature) and the surface. This was discussed in Sec. 1.5. In that discussion the heat transfer coefficient, h, was defined in Eq. (1.16) in order to relate the rate of heat flow per unit area to the difference between the surface and fluid temperatures:

$$\frac{q}{A} = h(t_s - t_f).$$

Application of this definition will be made in the following sections for various geometrical configurations.

Since this chapter is meant to deal primarily with conduction, the mechanism of convection at the boundary of a solid is introduced mainly to illustrate the application of this important boundary condition to a conduction problem. The actual mechanism of heat convection is not of interest at this point. Hence, it will be assumed that the heat transfer coefficient h is a known quantity. Chapters 6 through 10 are devoted to a detailed analysis of convection and the determination of the coefficient h. In Chapters 12 and 13 the problems of combined conduction and convection will be reconsidered.

The Plane Wall Bounded by Fluids of Different Temperature

Figure 3.7 shows a plane wall (composed of two solid layers for purposes of discussion) bounded on each side by convecting fluids. By denoting the fluid regions of uniform temperature and the various junctures between different materials by numbers, the subscript notation introduced in Sec. 3.3 may be used to denote the thermal conductances and conductivities separating the numbered regions: h_{12}, k_{23}, k_{34}, and h_{45}.

The heat flow per unit wall area may be written for each "layer" by application of the defining relation for h given above the equations obtained in Sec. 3.3. Thus,

$$\frac{q}{A} = h_{12}(t_1 - t_2),$$

$$\frac{q}{A} = \frac{t_2 - t_4}{\dfrac{\Delta x_{23}}{k_{23}} + \dfrac{\Delta x_{34}}{k_{34}}},$$

$$\frac{q}{A} = h_{45}(t_4 - t_5).$$

Again, by combining these relations to eliminate t_2 and t_4, the following expression for the heat flux is obtained in terms of the overall temperature difference and the thermal properties of the matter in between:

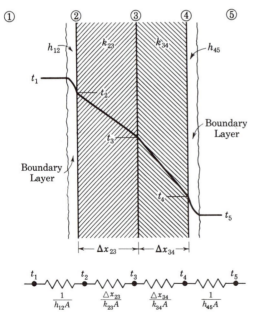

FIGURE 3.7

$$\frac{q}{A} = \frac{t_1 - t_5}{\dfrac{1}{h_{12}} + \dfrac{\Delta x_{23}}{k_{23}} + \dfrac{\Delta x_{34}}{k_{34}} + \dfrac{1}{h_{45}}}.$$ (3.31)

In keeping with the thermal resistance defined above,

$$\mathcal{R}_k = \frac{1}{CA},$$

a *convective* thermal resistance may be defined:

$$\mathcal{R}_c = \frac{1}{hA}.$$

The application of this definition, along with that already used for conduction resistances, to the network shown in Fig. 3.7 leads to the result given by Eq. (3.31).

EXAMPLE 3.3
The masonry wall of a building consists of an outer layer of facing brick ($k = 1.32$ W/m-°C) 10 cm thick, followed by a 15-cm thick layer of common brick ($k = 0.69$ W/m-°C), followed by a 1.25-cm layer of gypsum plaster ($k = 0.48$ W/m-°C). An outside coefficient of 30 W/m²-°C may be expected, and a coefficient of 8 W/m²-°C is a reasonable value to use for the inner surface of a ventilated room. What will be the rate of heat gain, per unit area, when the outside air is at 35°C and the inside air is conditioned to 22°C? What will be the temperature of the surface of the plaster?

Solution. From Eq. (3.31) the heat flow is

$$\frac{q}{A} = \frac{35 - 22}{\dfrac{1}{30} + \dfrac{0.1}{1.32} + \dfrac{0.15}{0.69} + \dfrac{0.0125}{0.48} + \dfrac{1}{8}} = 27.2 \text{ W/m}^2 \ (8.63 \text{ Btu/h-ft}^2).$$

From the definition of h, the inside surface temperature may be found.

$$\frac{q}{A} = 8(t - 22) = 27.2,$$

$$t = 25.4°C \ (77.7°F). \quad \blacksquare$$

Cylindrical Surfaces Bounded by Fluids of Fixed Temperatures

A cylindrical surface separating fluids of different temperatures is one of immense practical importance since fluids are transported, heated, cooled, evaporated, and condensed in cylindrical pipes, tubes, and vessels. Such processes are encountered in almost every phase of engineering work involving fluids or heat transfer.

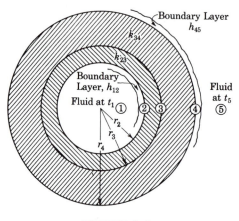

FIGURE 3.8

The notation and numbering scheme introduced above is used in Fig. 3.8, which shows a double-layer pipe separating fluids of different fixed temperatures.

The rate of heat flow through the various layers of the configuration shown in Fig. 3.8 may be written by use of the equations of Sec. 3.5 and Eq. (1.16), remembering that the area through which the heat is convected is different at the two surfaces. Thus,

$$q = 2\pi L r_2 h_{12}(t_1 - t_2) = \frac{2\pi L(t_2 - t_4)}{\dfrac{\ln (r_3/r_2)}{k_{23}} + \dfrac{\ln (r_4/r_3)}{k_{34}}} = 2\pi L r_4(t_4 - t_5)h_{45}.$$

The above set of equations yields the following expression for the rate of heat flow through the cylinder per unit length:

$$\frac{q}{L} = \frac{2\pi(t_1 - t_5)}{\dfrac{1}{r_2 h_{12}} + \dfrac{\ln (r_3/r_2)}{k_{23}} + \dfrac{\ln (r_4/r_3)}{k_{34}} + \dfrac{1}{r_4 h_{45}}} \tag{3.32}$$

EXAMPLE 3.4

A $\frac{3}{4}$-in. 18-gage brass condenser tube ($k = 115$ W/m-°C) is used to condense steam on its outer surface at 10 kN/m² pressure. Cooling water at 18°C flows through the tube. If the inside and outside heat transfer coefficients are 1700 W/m²-°C and 8500 W/m²-°C, respectively, find the mass of steam condensed per hour per unit of tube length.

Solution. Appendix B gives the outside and inside diameters of the tube to be 1.905 cm and 1.656 cm, respectively. At a pressure of 10 kN/m², the Steam

Tables give the saturation temperature to be 45.8°C and the latent heat of vaporization h_{fg} to be 2392.8 kJ/kg. Thus, Eq. (3.32) gives the heat flow to be

$$\frac{q}{L} = \frac{2\pi(45.8 - 18)}{\dfrac{1}{(0.01656/2)1700} + \dfrac{\ln(0.01905/0.01656)}{115} + \dfrac{1}{(0.01905/2)8500}}$$

$$= 2064 \text{ W/m}.$$

Then, the hourly rate of steam condensed is $q = \dot{m}h_{fg}$, or

$$\dot{m} = \frac{2064}{1000 \times 2392.8} \times 3600 = 3.11 \text{ kg/h-m (2.09 lb}_m\text{/h-ft).} \quad \blacksquare$$

3.9

THE CRITICAL THICKNESS OF PIPE INSULATION

An interesting application of the above relations having some practical significance is found in the case of insulation of small pipes or electrical wires. Given a pipe of fixed size, let it be desired to examine the variation in heat loss from the pipe as the thickness of insulation is changed. As insulation is added to the pipe, the outer exposed surface temperature will decrease, but at the same time the surface area available for convective heat dissipation will increase. These two opposing effects produce some interesting optimization effects.

For ease of analysis, some simplifying assumptions will be made. As noted in Fig. 3.9, let the pipe radius be R and the insulation radius r, so that $(r - R)$ represents the insulation thickness. For fixed values of the temperature of the fluid carried by the pipe and the ambient air at t_a, the addition of insulation will alter the pipe surface temperature, T. However, the variation of T is generally so slight that one may take it to be constant. If one also assumes that the heat

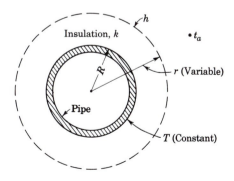

FIGURE 3.9

transfer coefficient at the exposed insulation surface, h, is also constant, then the heat lost from the pipe is

$$\frac{q}{L} = \frac{2\pi(T - t_a)}{\dfrac{1}{hr} + \dfrac{\ln(r/R)}{k}} \tag{3.33}$$

in which r is the only variable.

Differentiation of Eq. (3.33) with respect to r will show that the heat loss, q/L, reaches a *maximum* when the insulation radius is equal to

$$r = r_c = \frac{k}{h}. \tag{3.34}$$

The symbol r_c denotes the "critical radius" of the insulation. The fact that q/L attains a maximum at $r = r_c$ is the result of the above-mentioned opposing effects: increasing r increases the thermal resistance of the insulation layer but decreases the thermal resistance of the surface coefficient because of the increasing surface area. At $r = r_c$, the total resistance reaches a minimum.

Thus, it appears possible to *increase* the heat loss from a pipe by the addition of insulation. This is illustrated in Fig. 3.10. As shown, the critical radius r_c is a quantity fixed by the thermal properties involved. If the pipe size is such that $R < r_c$, then initial addition of insulation will increase the heat loss until $r = r_c$ after which it will begin to decrease. The bare pipe heat loss is again reached at some radius, shown as r^* in Fig. 3.10(a). For a larger pipe, as suggested in Fig. 3.10(b) for which $R > r_c$, any insulation added will decrease the heat loss.

The main purpose of introducing this example is to illustrate the opposing effects a geometry change may introduce in the thermal resistance and available convective area of a given situation. Similar effects will be seen to occur in more subtle ways in other more complex cases to be studied later. This example was not shown as a definitive study of insulation optimization. Indeed, such optimization studies must also include other considerations such as the effect of radiation as well as economic factors such as the cost of the insulation.

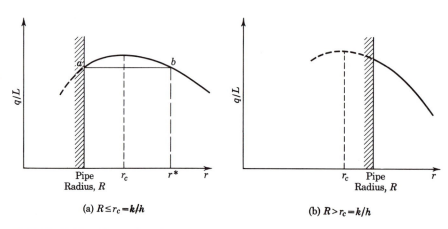

(a) $R \leq r_c = k/h$ (b) $R > r_c = k/h$

FIGURE 3.10. The critical thickness of pipe insulation.

THE OVERALL HEAT TRANSFER COEFFICIENT

In many instances it is customary to express the heat flow rate in the cases of the plane wall and cylinder (single or multilayered) with convection at the boundaries in terms of an "overall conductance" or "overall heat transfer coefficient." This overall heat transfer coefficient, symbolized by U, is simply defined as a quantity such that the rate of heat flow through a configuration is given by taking a product of U, the surface area, and the overall temperature difference:

$$q = UA(\Delta t)_{\text{overall}}. \tag{3.35}$$

The dimensions of U are seen to be those of a conductance.

If one utilizes the equivalent resistance concept of the electical networks introduced earlier, the overall conductance U is simply related to the total resistance between the points at which the overall potential is applied:

$$U = \frac{1}{\mathcal{R}_{\text{total}}A}.$$

The Plane Wall

The case of the plane wall is easily deduced from the discussion of Sec. 3.8. For a wall of n layers bounded on either side by fluids of temperatures t_1 and t_{n+3},

$$q = UA(t_1 - t_{n+3}), \tag{3.36}$$

$$U = \frac{1}{\dfrac{1}{h_{12}} + \dfrac{\Delta x_{23}}{k_{23}} + \dfrac{\Delta x_{34}}{k_{34}} + \cdots + \dfrac{1}{h_{(n+2)(n+3)}}}.$$

This representation is not essentially different from that already discussed, but it is found to be useful in the calculation of heat flow rates in the determination of heating or cooling needs of buildings. In such cases the overall heat transfer coefficient of standard structural walls, floors, roofs, etc., may be tabulated for ready reference. This has been done, and the results are available in various handbooks.

The Cylinder

Much use is made of the overall heat transfer coefficient for the cylindrical case in expressing the heat transfer capacity of heat exchangers. Frequently heat exchangers use a bundle of cylindrical tubes to provide the surface area for the transfer of heat between two fluids.

In the case of the cylindrical configuration, the overall heat transfer coefficient depends on what surface area (inside or outside) is used in the defining relation given by Eq. (3.35), although the product UA is always the same. It is customary to use the outside surface area, and the associated overall coefficient is sometimes called the "outside overall heat transfer coefficient." This is best illustrated by use of the double-layered cylinder discussed in Sec. 3.8 and pictured in Fig. 3.8. If U_4 is used to denote the overall heat transfer coefficient based on the outside area A_4 of radius r_4, then (see Fig. 3.8)

$$q = U_4 A_4 (t_1 - t_5).$$ (3.37)

Since $A_4 = 2\pi r_4 L$, Eqs. (3.32) and (3.37) combine to give the following expression for U_4:

$$U_4 = \cfrac{1}{\cfrac{r_4}{r_2 h_{12}} + \cfrac{r_4 \ln (r_3/r_2)}{k_{23}} + \cfrac{r_4 \ln (r_4/r_3)}{k_{34}} + \cfrac{1}{h_{45}}}.$$ (3.38)

EXAMPLE 3.5

The overall heat transfer coefficient for the configuration of Example 3.3 is

$$U = \cfrac{1}{\cfrac{1}{30} + \cfrac{0.1}{1.32} + \cfrac{0.15}{0.69} + \cfrac{0.0125}{0.48} + \cfrac{1}{8}}$$

$$= 2.09 \ \text{W/m}^2\text{-}°\text{C} \ (0.369 \ \text{Btu/h-ft}^2\text{-}°\text{F}).$$ ■

EXAMPLE 3.6

The overall heat transfer coefficient, based on the outside surface area, of the condenser tube in Example 3.4 is

$$U = \cfrac{1}{\cfrac{0.01905}{0.01656 \times 1700} + \cfrac{(0.01905/2) \ln (0.01905/0.01656)}{115} + \cfrac{1}{8500}}$$

$$= 1241 \ \text{W/m}^2\text{-}°\text{C} \ (218.5 \ \text{Btu/h-ft}^2\text{-}°\text{F}).$$ ■

3.11

EXTENDED SURFACES

When it is desired to increase the heat removal between a structure and a surrounding ambient fluid, it is common practice to utilize "extended surfaces" attached to the primary surface. In such instances the extended surfaces are provided to increase artificially the surface area of heat transmission, although the average surface temperature may be decreased by so doing. If the surface is proportioned properly, the net result will be an increase in the heat transmission rate between the structure and the ambient fluid.

The uses of extended surfaces in applications of practical importance are numerous. Examples may be found in the cooling fins of air-cooled engines, the fin extensions to the tubes of radiators and other heat exchangers, the "pins" or "studs" attached to boiler tubes, etc. The extended surface applications noted above are all cases in which one purposely wishes to increase the rate of heat

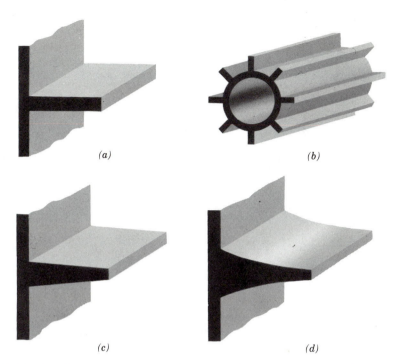

(a)

(b)

(c)

(d)

FIGURE 3.11. Examples of extended surfaces: (a) and (b), straight fins of uniform thickness; (c) and (d), straight fins of nonuniform thickness.

exchange between a source and an ambient fluid. Similar extended surface configurations may occur in other instances where the exchange of heat with the ambient fluid may be a disadvantage. Such instances are encountered in the measurement of temperature—the conduction of heat along thermocouple wires attached to a heated surface being an example.

An extended surface configuration is generally classed as a straight fin, an annular fin, or a spine. The term *straight fin* is applied to the extended surface attached to a wall which is otherwise plane, whereas an *annular fin* is one attached, circumferentially, to a cylindrical surface. A *spine* or *pin fin* is an extended surface of cylindrical or conical shape. These definitions are illustrated in Figs. 3.11 and 3.12.

The basic problem with which the designer is faced is: Given a fin of a certain configuration and size attached to a surface of a fixed temperature (i.e., the fin base temperature) and surrounded by a fluid of fixed temperature, what is the rate of heat dissipated by the fin, and what is the variation in the temperature of the fin as one proceeds from the base to the tip?

In most instances the fin proportions used in practice are such that the length of the fin (its dimension measured normal to the primary surface to which it is attached) is large compared to its maximum thickness. When this is the case, one may assume that the temperature of the fin depends on only the single coordinate measured in the direction of the fin's length. For example, if a cylindrical rod protruding from a heated wall is very long compared to its diameter, one may assume that the temperature is uniform over any cross section

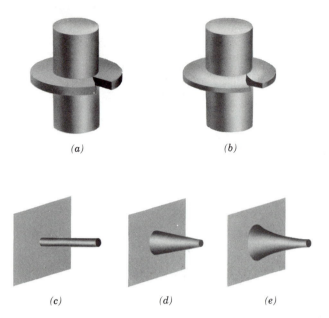

FIGURE 3.12. Examples of extended surfaces: (a) and (b), annular fins; (c), (d), and (e), spines.

taken normal to the axis. Hence, the temperature in the rod depends only on the axial coordinate measured along the rod. Similarly, if an annular fin is sufficiently slender, the distribution of the temperature may be taken to depend only on the radial coordinate.

If the above simplification can be made, the problem for the solution of the temperature distribution and heat flux rate becomes a one-dimensional conduction problem. A few shapes of practical importance will be considered in detail in this chapter. Only the steady state case will be considered.

3.12

THE STRAIGHT FIN OF UNIFORM THICKNESS AND THE SPINE OF UNIFORM CROSS SECTION

The straight fin of uniform thickness and the spine of uniform cross section may be treated identically since in either case the cross-sectional area for heat flow (normal to the length coordinate) is constant. Also, in either case, the exposed surface area for heat convection is a linear function of the distance measured along the length; i.e., the perimeter of the cross section is constant.

Figure 3.13 illustrates these two configurations and shows the notation to be used in the following analysis. The symbol L will denote the length of the fin, k the thermal conductivity, h the heat transfer coefficient at the exposed surface, t_0 the fixed temperature at the fin base, and t_f the temperature of the bulk of the ambient fluid. The coordinate distance along the fin length will be symbolized by x. The symbols A and C will be used to denote the area of a cross section normal to x and the perimeter of this section, respectively.

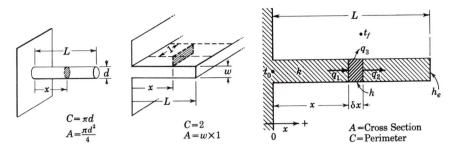

$$C = \pi d$$
$$A = \frac{\pi d^2}{4}$$

$$C = 2$$
$$A = w \times 1$$

A = Cross Section
C = Perimeter

FIGURE 3.13. The straight fin of uniform thickness or spine of constant cross section.

Usually when dealing with straight fins it is customary to neglect heat losses from the side edges of the fin and to express the heat flow, etc., per unit width. In this case, as illustrated in Fig. 3.13, the perimeter C will be 2 and the cross-sectional area will equal the thickness, w.

If one now considers an element of the fin δx in length (see Fig. 3.13) and defines q_1, to be the heat conducted into the element at x, q_2 the heat conducted out of the element at $x + \delta x$, and q_3 the heat convected out of the surface of the element, then energy conservation requires, for the steady state, that

$$q_1 = q_2 + q_3.$$

Fourier's law and the definition of the heat transfer coefficient renders the above equation into the form

$$-kA\left(\frac{dt}{dx}\right)_x = -kA\left(\frac{dt}{dx}\right)_{x+\delta x} + h(C\,\delta x)(t_{av} - t_f),$$

where t_{av} is the average temperature of the element. If the temperature gradient at $x + \delta x$ is written in terms of that at x by a Taylor's expansion and the equation is divided by the element volume, $A\,\delta x$, one has

$$-\frac{k}{\delta x}\left(\frac{dt}{dx}\right)_x = -\frac{k}{\delta x}\left(\frac{dt}{dx}\right)_x - k\left(\frac{d^2t}{dx^2}\right)_x - k\left(\frac{d^3t}{dx^3}\right)_x \frac{\delta x}{2} + \cdots + \frac{hC}{A}(t_{av} - t_f).$$

This energy conservation must be satisfied for all elements, so upon letting $\delta x \to 0$, the governing differential equation for the temperature distribution in the fin is obtained:

$$\frac{d^2t}{dx^2} - \frac{hC}{kA}(t - t_f) = 0.$$

If the thermal conductivity and the heat transfer coefficient are taken as constant, the above equation can be easily solved.

The differential equation may be written more concisely by defining the temperature difference variable

$$\theta = t - t_f$$

and a parameter

$$m = \sqrt{hC/kA}. \tag{3.39}$$

Then the governing equation is

$$\frac{d^2\theta}{dx^2} - m^2\theta = 0$$

with the solution

$$\theta = Be^{-mx} + De^{mx}. \tag{3.40}$$

The B and D in Eq. (3.40) are arbitrary constants determined by the boundary conditions imposed at the ends of the fin. At the base of the fin, the temperature is fixed at t_0 if the fin material is integral with the material of the primary surface. If, as is sometimes the case, the fin is a separate piece attached to the primary surface, some contact resistance may exist at the fin base and its temperature may be different from that of the primary surface. In such a case t_0 should be interpreted as the fin base temperature, not the temperature of the primary surface. At the outer end of the fin, convection is taking place. In other instances some other condition may be imposed—such as another fixed temperature if the fin extends between two heat sources.

For the present analysis let the conditions be those indicated by Fig. 3.13—an imposed temperature at the base and convection at the free end. Then the conditions determining B and D are,

$$At\ x = 0: \quad t = t_0.$$

$$At\ x = L: \quad -k\frac{dt}{dx} = h_e(t - t_f).$$

It should be noted that the boundary condition of convection at the end of the fin as stated above assumes that the heat transfer coefficient at the end, h_e, is different from that at the other surface, h. This is done for two reasons. First, as will be shown later, the convective heat transfer coefficient depends on the orientation of the surface, and it is quite likely that the coefficient at the fin end will differ from that on the other surfaces. Second, this distinction between the heat transfer coefficient will permit easy simplification of futher results to the special case in which the heat convected out the end is considered negligible by merely setting $h_e = 0$.

Written in terms of the temperature difference variable, θ, defined above, these boundary conditions are

$$At\ x = 0: \quad \theta = \theta_0 = t_0 - t_f.$$

$$At\ x = L: \quad \frac{d\theta}{dx} = -\frac{h_e}{k}\theta. \tag{3.41}$$

Application of these boundary conditions to Eq. (3.40) to determine the constants B and D yields

$$\frac{\theta}{\theta_0} = \frac{t - t_f}{t_0 - t_f} = \frac{[e^{m(L-x)} + e^{-m(L-x)}] + \dfrac{h_e}{km}[e^{m(L-x)} - e^{-m(L-x)}]}{(e^{mL} + e^{-mL}) + \dfrac{h_e}{km}(e^{mL} - e^{-mL})}. \quad (3.42)$$

This is, perhaps, more conveniently expressed in terms of the hyperbolic functions:

$$\frac{\theta}{\theta_0} = \frac{t - t_f}{t_0 - t_f} = \frac{\cosh m(L - x) + H \sinh m(L - x)}{\cosh mL + H \sinh mL}, \quad (3.43)$$

where

$$H = \frac{h_e}{km} \quad \text{and} \quad m = \sqrt{\frac{hC}{kA}}.$$

The rate of heat flow from the fin could be evaluated by integrating the convected heat over the fin surface. However, the result is obtained more directly by noting the fact that all the heat dissipated by the fin must be *conducted* past the point where $x = 0$. Thus, if q symbolizes the heat flow rate,

$$q = -kA\left(\frac{dt}{dx}\right)_{x=0}.$$

Introduction of the temperature distribution from Eq. (3.43) yields

$$q = kmA\theta_0 \frac{\sinh mL + H \cosh mL}{\cosh mL + H \sinh mL}. \quad (3.44)$$

In the development of these relations for the temperature distribution and heat dissipation, it was assumed that the fin was slender enough for a one-dimensional condition to prevail. If this condition is met, it is very likely that the amount of heat convected out the end of the fin is a small fraction of the heat convected out the other surfaces. If this is the case, the above relations may be simplified by neglecting this heat loss at the end. This implies that $h_e = 0$, or $H = 0$. Then Eqs. (3.43) and (3.44) reduce to

$$\frac{\theta}{\theta_0} = \frac{t - t_f}{t_0 - t_f} = \frac{\cosh m(L - x)}{\cosh mL}, \quad (3.45)$$

$$q = kmA\theta_0 \tanh mL. \quad (3.46)$$

Other Considerations for Straight Fins of Uniform Cross Section. The relations just developed for straight fins of uniform cross section are limited to the case of a fin heated on one end, with the other end free. This is just one set of boundary conditions that can be applied to this configuration. Other cases of practical significance are those in which the fin may be treated as infinitely long or in which the fin is maintained at fixed temperatures on each end. The solutions to these two problems are reserved as problems at the end of the chapter, and the reader is referred to Probs. 3.41 and 3.45 for the solutions.

An additional problem of some interest is that posed in Prob. 3.58. For a straight fin of uniform thickness in which $m^2 = 2h/kw$, Eq. (3.44) may be interpreted as the dependence of the heat loss on the fin length L for fixed values of h, k, w, and θ_0. If one evaluates $\partial q/\partial L$ from this equation it is not difficult to show that q increases with L only if $hw/2k < 1$. That is, unless $hw/2k < 1$, the addition of fins will *inhibit* heat flow to the ambient fluid, not increase it! This observation is the inverse of the critical insulation thickness problem discussed in Sec. 3.9. Unless $hw/2k < 1$, addition of fins increases the surface area available for heat flow, but decreases the average surface temperature more—resulting in a lowered heat transfer rate. Thus, the requirement that $hw/2k < 1$ establishes the conditions that must be met to make a straight fin useful. It implies, depending on h, that fins generally must be thin, of high conductivity, and used only when h is not large.

EXAMPLE 3.7

Three rods, one made of glass ($k = 1.09$ W/m-°C), one of pure aluminum ($k = 228$ W/m-°C), and one of wrought iron ($k = 57$ W/m-°C), all have diameters of 1.25 cm, lengths of 30 cm, and are heated to 120°C at one end. The rods extend into air at 20°C, and the heat transfer coefficient on the surfaces is known to be 9.0 W/m²-°C. Find (a) the distribution of temperature in the rods if the heat loss from the ends is neglected; (b) the total heat flow from the rods, neglecting the end heat loss; (c) the heat flow from the rods if the end heat loss is not neglected, and the heat transfer coefficient at the ends is also 9.0 W/m²-°C.

Solution

(a) For a cylindrical rod, the parameter $m = \sqrt{hC/kA}$ is $m = \sqrt{4h/kd}$. Thus,

$$m = \sqrt{\frac{4 \times 9.0}{1.09 \times 1.25 \times 100}} = 0.5140 \, \frac{1}{\text{cm}}, \text{ for glass.}$$

$$m = 0.07108 \, \frac{1}{\text{cm}}, \text{ for iron.}$$

$$m = 0.03554 \, \frac{1}{\text{cm}}, \text{ for aluminum.}$$

Then, by Eq. (3.45), the temperature distribution in the rods is given by

$$\frac{t - 20}{120 - 20} = \frac{\cosh m(L - x)}{\cosh mL},$$

in which $L = 30$ cm, and x is allowed to vary from 0 to 30 cm. The temperature distributions given by the above expression are plotted in Fig. 3.14 for comparison.

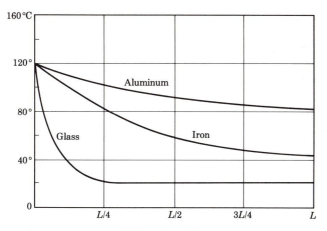

FIGURE 3.14

(b) Equation (3.46) gives the heat flow to be

$$q = (1.09)(0.5140) \frac{\pi(1.25)^2}{4 \times 100} (120 - 20) \tanh (0.514 \times 30)$$

$$= 0.688 \text{ W } (2.35 \text{ Btu/h), for glass.}$$

$$q = 4.83 \text{ W } (16.49 \text{ Btu/h), for iron.}$$

$$q = 7.84 \text{ W } (26.74 \text{ Btu/h), for aluminum.}$$

(c) If the heat loss out of the end is to be accounted for, the parameter $H = h_e/km$ is needed. Thus,

$$H = \frac{9.0}{1.09 \times 0.514 \times 100} = 0.1606, \text{ for glass.}$$

$$H = 0.0222, \text{ for iron.}$$

$$H = 0.0111, \text{ for aluminum.}$$

Thus, Eq. (3.44) gives the heat flow to be

$$q = 0.688 \text{ W } (2.35 \text{ Btu/h), for glass.}$$

$$q = 4.84 \text{ W } (16.5 \text{ Btu/h), for iron.}$$

$$q = 7.88 \text{ W } (26.9 \text{ Btu/h), for aluminum.} \blacksquare$$

3.13

EXTENDED SURFACES OF NONUNIFORM CROSS SECTION—GENERAL CONSIDERATIONS

Some of the other extended surfaces shown in Figs. 3.11 and 3.12 differ from the straight fin of uniform thickness or the spine of uniform cross section in that the cross-sectional area for heat conduction will not be constant, nor will the surface area for convection to an ambient fluid vary linearly with the distance from the fin base. For example the annular fin of uniform thickness (for radial conduction) has a cross-sectional area which varies linearly with the radius and a surface area which varies with the square of the radius.

The differential equation of the temperature distribution in these extended surfaces may be deduced in the same manner as was done above (i.e., by making an energy balance on an element of the fin), but certain general facts may be deduced before treating specific configurations. As before, a one-dimensional problem will result if the analysis is limited to those fins in which the thickness is quite small compared to the fin length. Imagine, then, an extended surface (geometry unspecified, for the present) so proportioned that the heat conduction within may be treated as one-dimensional. That is, let the temperature be expressible as a function of a single coordinate, say x. Such an arrangement is depicted in Fig. 3.15. In general, the cross-sectional area (normal to x) for heat conduction will be a function of x, as will be the surface area for heat convection to the surroundings. Let $A(x)$ represent the cross-sectional area at any value of x and let $S(x)$ represent the surface area exposed for convection between $x = 0$ and $x = x$.

Select now an element of the fin contained between the cross sections at x, and $x + \delta x$. Thus, δx is the thickness of this finite element. Denote, as before, the heat conducted into the element across the area at x by q_1, the heat conducted out of the element across the area at $x + \delta x$ by q_2 and the heat convected out of the surface of the element and into the surrounding fluid by q_3. Then the requirement of the steady state gives

$$q_1 = q_2 + q_3.$$

The conducted heat, q_1 or q_2, depends on the temperature gradient in the x direction and on the cross-sectional area. This area is itself a function of x, and hence the conducted heat is purely a function of x. Thus, one can express q_2 in terms of q_1 by use of Taylor's expansion:

$$q_2 = q_1 + \frac{d}{dx}(q_1)\,\delta x + \frac{d^2}{dx^2}(q_1)\frac{(\delta x)^2}{2!} + \cdots.$$

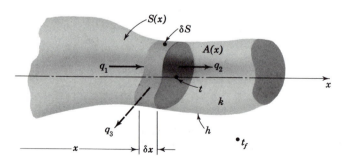

FIGURE 3.15. A generalized extended surface.

Thus, the heat balance above becomes

$$\frac{d}{dx}(q_1)\,\delta x + \frac{d^2}{dx^2}(q_1)\frac{(\delta x)^2}{2!} + \cdots + q_3 = 0. \tag{3.47}$$

By again using θ to denote the difference between the fin temperature and the surrounding fluid temperature, Eq. (3.47) is expressible as

$$\frac{d}{dx}\left(-kA\frac{d\theta}{dx}\right)\delta x + \frac{d^2}{dx^2}\left(-kA\frac{d\theta}{dx}\right)\frac{(\delta x)^2}{2!} + \cdots + h\theta(\delta S) = 0,$$

where δS denotes the exposed surface area of the chosen element.

In the above expression h and k, as usual, denote the convective heat transfer coefficient and the thermal conductivity. It should be noted that A and S are to be treated as functions of the coordinate x. Rewriting the above heat balance on a per-unit-volume basis, one finds that

$$\frac{1}{A}\frac{d}{dx}\left(-kA\frac{d\theta}{dx}\right) + \frac{1}{A}\frac{d^2}{dx^2}\left(-kA\frac{d\theta}{dx}\right)\frac{\delta x}{2} + \cdots + \frac{h\theta}{A}\frac{\delta S}{\delta x} = 0.$$

Now, this expression must be satisfied for *all* elements, no matter how large or small. Hence, it must be satisfied as $\delta x \to 0$. Allowing, then, $\delta x \to 0$ in this equation, one obtains

$$\frac{1}{A}\frac{d}{dx}\left(-kA\frac{d\theta}{dx}\right) + \frac{h\theta}{A}\frac{dS}{dx} = 0,$$

or

$$\frac{d^2\theta}{dx^2} + \frac{1}{A}\frac{dA}{dx}\frac{d\theta}{dx} - \frac{h}{k}\frac{1}{A}\frac{dS}{dx}\theta = 0. \tag{3.48}$$

This equation may be applied generally to all extended surface configurations for which the one-dimensional assumption is valid. For example, if the special case of the straight fin or uniform spine treated in Sec. 3.12 is considered, $A = $ constant and $S = C \cdot x$. In this instance Eq. (3.48) becomes

$$\frac{d^2\theta}{dx^2} - \frac{hC}{kA}\theta = 0.$$

This equation is identical to the equation obtained earlier.

This expression for the general extended surface [Eq. (3.48)] will now be applied to certain other configurations of practical importance.

3.14

THE ANNULAR FIN OF UNIFORM THICKNESS

As mentioned earlier, a second extended surface configuration of considerable engineering importance is that of an annular fin of constant thickness attached circumferentially to a circular cylinder. Such a configuration is found to be utilized on certain liquid-to-gas heat exchanger tubes and on the cylinders of air-cooled engines.

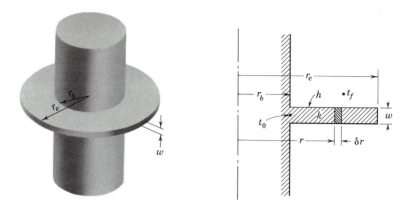

FIGURE 3.16. The annular fin of uniform thickness.

Figure 3.16 shows the notation to be used in the analysis of an annular fin. The symbols r_b and r_e denote the radii of the base and end of the fin, respectively, whereas w denotes the constant fin thickness. By assuming that there is symmetry of the temperature distribution with respect to a circumferential coordinate and that $w \ll (r_e - r_b)$, the heat conduction within the fin may be taken to depend on the radial coordinate r only.

The cross-sectional area and surface area are functions of this radial coordinate:

$$A = 2\pi r w,$$

$$S = 2[\pi(r^2 - r_b^2)].$$

Thus, Eq. (3.48) leads to the following expression for the temperature distribution as a function of r:

$$\frac{d^2\theta}{dr^2} + \left(\frac{1}{2\pi r w} \times 2\pi w\right)\frac{d\theta}{dr} - \frac{h}{k}\left(\frac{1}{2\pi r w} \times 4\pi r\right)\theta = 0, \tag{3.49}$$

$$\frac{d^2\theta}{dr^2} + \frac{1}{r}\frac{d\theta}{dr} - \frac{2h}{kw}\theta = 0.$$

This is recognized as Bessel's equation of zero order. A summary of the essential features of this equation and of the properties of the functions which are its solutions is given in Appendix C. Appendix C shows that the solution to Eq. (3.49) is

$$\theta = BI_0(nr) + CK_0(nr), \tag{3.50}$$

in which

$$n = \sqrt{\frac{2h}{kw}},$$

and I_0 and K_0 represent the modified Bessel functions of the first and second kind, respectively.

The constants B and C in Eq. (3.50) are determined by the boundary conditions to be imposed at r_b and r_e. For a fin maintained at a fixed temperature at the base and for which the end heat loss is neglected, one has

$$\text{At } r = r_b: \quad t = t_0 \text{ or } \theta = \theta_0 = t_0 - t_f.$$

$$\text{At } r = r_e: \quad \frac{dt}{dr} = \frac{d\theta}{dr} = 0.$$

Using the properties of the Bessel functions given in Appendix C, the constants B and C may be evaluated and substituted into Eq. (3.50) to yield the temperature distribution as a function of the radial position r:

$$\frac{\theta}{\theta_0} = \frac{I_0(nr)K_1(nr_e) + K_0(nr)I_1(nr_e)}{I_0(nr_b)K_1(nr_e) + K_0(nr_b)I_1(nr_e)}. \tag{3.51}$$

The rate of heat dissipation from an annular fin may be found by evaluating the heat conducted past the base radius:

$$q = -k\left(A \frac{d\theta}{dr}\right)_{r=r_b}$$

Introduction of Eq. (3.51) gives

$$q = 2\pi k n w \theta_0 r_b \frac{K_1(nr_b)I_1(nr_e) - I_1(nr_b)K_1(nr_e)}{K_0(nr_b)I_1(nr_e) + I_0(nr_b)K_1(nr_e)}. \tag{3.52}$$

Equations (3.51) and (3.52) may be used, with the tabulated values of the Bessel functions in Appendix C to determine the temperature in an annular fin and the heat loss from it.

3.15

THE STRAIGHT FIN OF TRIANGULAR PROFILE

As a second example of a fin with a nonuniform cross section, consider a straight fin having a tapered or trapezoidal profile. This is illustrated in Fig. 3.17, wherein the dimensions and coordinate system are shown. It is more convenient to place the origin of the distance coordinate at the point of intersection of the extension of the sides of the fin. L and x_e represent the distances of the base and fin end, respectively, from this origin, and w is the fin thickness at the base.

Again one assumes that the fin is sufficiently thin [i.e., $w \ll (L - x_e)$] so that a one-dimensional situation exists. Now, for a *unit width* of the fin (neglecting side losses) the cross-sectional area A and surface area S vary with x in the following ways:

$$A = \frac{wx}{L} \times 1 \tag{3.53}$$

$$S = 2(x - x_e)\sqrt{1 + \left(\frac{w}{2L}\right)^2} \tag{3.54}$$

$$= 2(x - x_e)f,$$

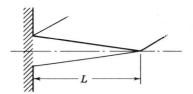

Full Triangular Fin, $x_e = 0$

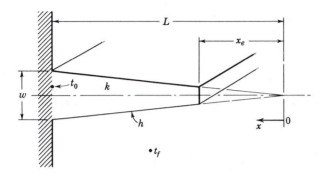

FIGURE 3.17. The straight fin of trapezoidal profile.

where $f = \sqrt{1 + (w/2L)^2}$ is simply a geometrical property of the fin. For most cases of engineering application, the fin is sufficiently thin that $f \approx 1$.

Equation (3.48) for the general extended surface may now be applied, and Eqs. (3.53) and (3.54) give

$$\frac{d^2\theta}{dx^2} + \frac{1}{x}\frac{d\theta}{dx} - p^2\frac{\theta}{x} = 0 \qquad (3.55)$$

in which

$$p = \sqrt{\frac{2fhL}{kw}}. \qquad (3.56)$$

Comparison with the generalized Bessel equation in Appendix C gives the solution for the temperature:

$$\theta = BI_0(2px^{1/2}) + CK_0(2px^{1/2}) \qquad (3.57)$$

Again, the constants B and C are determined by applying boundary conditions at the two ends of the fin. For the special case in which the fin is a full triangle, coming to a point (i.e., $x_e = 0$), one notes that the constant C must be zero since the Bessel function K_0 is infinite for zero argument. Thus, for the complete triangular fin,

$$\theta = BI_0(2px^{1/2}),$$

and for a fixed base temperature of t_0 (or $\theta = \theta_0 = t_0 - t_f$) at $x = L$,

$$\frac{\theta}{\theta_0} = \frac{I_0(2px^{1/2})}{I_0(2pL^{1/2})}. \qquad (3.58)$$

Again the heat flow from the fin is obtained by finding the heat conducted at the base:

$$q = -k\left(A\,\frac{d\theta}{dx}\right)_{x=L}$$

$$= -\frac{kw\theta_0 p}{L^{1/2}}\,\frac{I_1(2pL^{1/2})}{I_0(2pL^{1/2})}.$$

(3.59)

The negative sign in Eq. (3.59) arises from the choice of the direction of the coordinate x. Equations (3.58) and (3.59) constitute, then, the solution for the temperature in a triangular fin and its heat loss.

3.16

OTHER SHAPES OF NONUNIFORM CROSS SECTION

The cases discussed above (i.e., the uniform thickness straight fin, the triangular straight fin, the uniform annular fin) are only a few of the possible extended surface configurations. Extended surfaces, in which the thickness (or diameter in the case of spine) varies linearly, parabolically, or hyperbolically with the fin length, have also been analyzed and used in practice. This is equally true for straight fins, annular fins, or spines.

All these possible configurations will not be discussed here. The reader is referred to Refs. 2 through 4. References 3 and 4 also discuss the important problem of the optimum shape, for a given amount of material, that should be employed in straight or annular fins for the maximum rate of heat dissipation.

The problems at the end of the chapter, exercises for the reader, include some of these other shapes.

3.17

FIN EFFECTIVENESS

The purpose of adding fins to a surface is to increase the surface area available for convective heat transfer to the surrounding fluid. However, the addition of the extended surface lowers the mean surface temperature to a value below that which it had before the fins were added. If the increase of the surface area is greater than the decrease of the mean surface temperature, the fins will enhance the exchange of heat.

In order to express the heat-exchanging capacity of an extended surface relative to the heat-exchanging capacity of the primary surface with no fins, it is useful to define the *fin effectiveness* as the ratio of the heat transfer rate from a fin to the heat transfer rate that would be obtained if the entire fin surface area were to be maintained at the same temperature as the primary surface. It is assumed that no contact resistance exists at the fin base in order that the fin base temperature and the primary surface temperature may be taken to be the same. The relations obtained earlier for the heat flow rate from various fin shapes may now be used to deduce equations for the fin effectiveness.

The relations to be developed in this section will simply present the fin effectiveness as functions of the pertinent thermal and geometric parameters. The use of these relations will be made in the following section on total surface temperature effectiveness.

The Straight Fin of Uniform Thickness

Neglecting the heat loss from the end of a straight fin of uniform thickness, one may use Eq. (3.46) to obtain the rate of heat flow from the fin:

$$q = kmA\theta_0 \tanh mL.$$

The surface area of the fin, per unit width, is $2L$, so the fin effectiveness, κ_u, is given as

$$\kappa_u = \frac{kmA\theta_0 \tanh mL}{2Lh\theta_0}.$$

Since Eq. (3.39) gives $m^2 = 2h/kw$,

$$\kappa_u = \frac{1}{mL} \tanh mL. \tag{3.60}$$

The subscript u is used to denote a straight fin of uniform thickness.

The Straight Fin of Triangular Profile

If the same procedure is followed, Eq. (3.59) leads to the following expression for the fin effectiveness, κ_t, for a triangular fin in which f [see Eq. (3.54)] is taken as unity.

$$\kappa_t = \frac{1}{pL^{1/2}} \frac{I_1(2pL^{1/2})}{I_0(2pL^{1/2})}. \tag{3.61}$$

Figure 3.18 is a plot of Eq. (3.61).

The Annular Fin of Uniform Thickness

The effectiveness of an annular fin of uniform thickness is readily obtained from Eq. (3.52) by dividing q by $2\pi(r_e^2 - r_b^2)h\theta_0$. Thus, the effectiveness, κ_{an}, is

$$\kappa_{an} = \frac{kwnr_b}{(r_e^2 - r_b^2)h} \frac{K_1(nr_b)I_1(nr_e) - I_1(nr_b)K_1(nr_e)}{K_0(nr_b)I_1(nr_e) + I_0(nr_b)K_1(nr_e)}.$$

The above expression is more conveniently expressed in terms of two dimensionless parameters:

$$\alpha = \frac{r_b}{r_e}, \qquad \beta = r_e n = r_e \sqrt{\frac{2h}{kw}}. \tag{3.62}$$

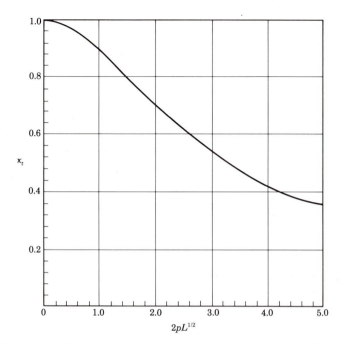

FIGURE 3.18. Fin effectiveness for a straight fin of triangular profile.

Then

$$\kappa_{an} = \frac{2\alpha}{\beta(1 - \alpha^2)} \frac{K_1(\alpha\beta)I_1(\beta) - I_1(\alpha\beta)K_1(\beta)}{K_0(\alpha\beta)I_1(\beta) + I_0(\alpha\beta)K_1(\beta)}. \tag{3.63}$$

The above expression for the effectiveness of an annular fin of uniform thickness is plotted in Fig. 3.19 as a function of the geometric parameter $\alpha = r_b/r_e$ and the thermal parameter $\beta = r_e\sqrt{2h/kw}$.

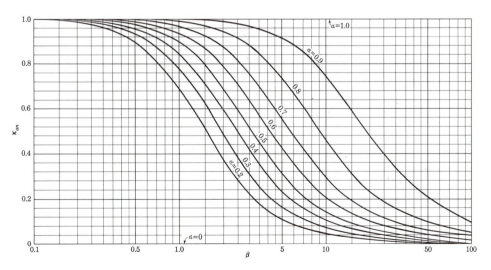

FIGURE 3.19. The effectiveness of the annular fin of uniform thickness.

TOTAL SURFACE TEMPERATURE EFFECTIVENESS

The fin effectiveness discussed in the foregoing section is concerned with describing the performance of a single fin. However, many applications employing extended surfaces involve the use of an array of fins attached to the primary surface. Figure 3.20 depicts such arrays for straight and annular fins. In such applications it is useful to define a total surface temperature effectiveness which gives a measure of the performance of the total exposed surface of the array— that is, both the finned and unfinned surface. Let

A_f = exposed surface area of fins only,

A_t = total exposed surface area, including the fins and unfinned surface,

κ = effectiveness for the particular fin shape involved.

Then if the total surface temperature effectiveness, η, is defined as the ratio of the actual heat transferred by the array to that it would transfer if its entire surface were maintained at the base temperature,

$$\eta = \frac{(A_t - A_f)h\theta_0 + A_f h\theta_0}{A_t h\theta_0},$$

$$(3.64)$$

$$\eta = 1 - \frac{A_f}{A_t}(1 - \kappa).$$

Since $A_f/A_t < 1$ and $\kappa \le 1$, it is apparent that $\eta \le 1$. The ratio A_f/A_t is readily evaluated from the geometry of the array. For example, for an array of uniform fins of length L, thickness w, spacing on centers δ, as shown in Fig. 3.20(a), one may deduce

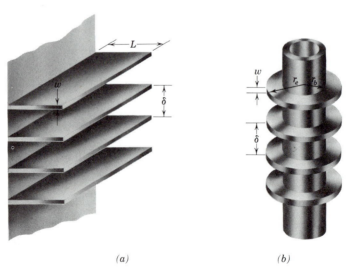

(a)   (b)

FIGURE 3.20

$$\left(\frac{A_f}{A_t}\right)_u = \frac{2L + w}{2L + w + (\delta - w)} = \frac{2L + w}{2L + \delta}. \tag{3.65}$$

Similarly, one may show for an array of triangular straight fins:

$$\left(\frac{A_f}{A_t}\right)_t = \frac{2L}{2L + \delta - w}. \tag{3.66}$$

The purpose of adding fins to a given primary surface is to increase the surface area available for heat transfer to the ambient fluid; however, the average temperature of the exposed surface is less than that of the original primary surface because of the gradient in temperature along the fin. The total surface effectiveness is a measure of how well the total exposed surface is being used for heat transfer with respect to the original primary surface. If A_p represents the surface area of the primary surface *before* the fin array is added, then the total thermal resistance was

$$A_p h,$$

while *after* the fins are added, the resistance becomes

$$A_t h \, \eta. \tag{3.67}$$

Hopefully, then, the product $A_t\eta$ is greater than A_p. These points are best illustrated by the calculations given in the following example.

EXAMPLE 3.8

As depicted in Fig. 3.21, heat is to be transferred through a plane wall, 1.25 cm thick, composed of a material with $k_{23} = 200$ W/m-°C. On the left side of the wall is an ambient fluid at $t_1 = 120$°C and the heat transfer coefficient there for all exposed surface is $h_{12} = 450$ W/m²-°C. On the right side of the wall there is another ambient fluid at $t_4 = 20$°C and the heat transfer coefficient there for all exposed surfaces is $h_{34} = 25$ W/m²-°C. It is desired to enhance the heat transfer between the two fluids by using straight, rectangular fins having a length

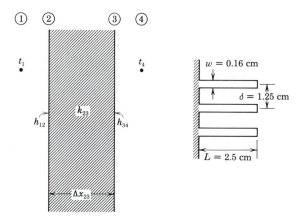

FIGURE 3.21

$L = 2.5$ cm, thickness $w = 0.16$ cm, and spaced $\delta = 1.25$ cm, on centers, and a k the same as the wall. Assuming that a one-dimensional representation may be used, find the heat transfer rate between the two fluids, per unit of primary wall area, if (a) there are no fins; (b) the fins are added to the right side only; (c) the fins are added to the left side only.

Solution

(a) For the bare wall, with primary area A_p, the heat flow is

$$\frac{q}{A_p} = \frac{t_1 - t_2}{\dfrac{1}{h_{12}} + \dfrac{\Delta x_{23}}{k_{23}} + \dfrac{1}{h_{34}}} = \frac{120 - 20}{\dfrac{1}{450} + \dfrac{0.0125}{200} + \dfrac{1}{25}}$$

$$= 2364.9 \text{ W/m}^2 \ (749.6 \text{ Btu/h-ft}^2).$$

(b) If the fins are added to the right side, the heat flow is, using Eq. (3.67),

$$\frac{q}{A_p} = \frac{t_1 - t_2}{\dfrac{1}{h_{12}} + \dfrac{\Delta x_{23}}{k_{23}} + \dfrac{1}{(A_t/A_p)_3 \eta_3 h_{34}}}.$$

The effectiveness of the surface 3 is, by Eq. (3.64),

$$\eta_3 = 1 - \left(\frac{A_f}{A_t}\right)_3 (1 - \kappa_3),$$

and κ_3 is found from Eq. (3.60):

$$\kappa_3 = \frac{1}{mL} \tanh mL.$$

For the conditions stated on the right side,

$$mL = \sqrt{\frac{2h}{kw}} \, L = \sqrt{\frac{2 \times 25}{200 \times 0.0016}} \times 0.025 = 0.3125,$$

$$\kappa_3 = 0.9687,$$

$$\frac{A_f}{A_t} = \frac{2L + w}{2L + \delta} = 0.8256,$$

$$\frac{A_t}{A_p} = \frac{2L + \delta}{\delta} = 5.0,$$

$$\eta_3 = 1 - 0.8256(1 - 0.9687) = 0.9742,$$

$$\frac{q}{A} = \frac{120 - 20}{\dfrac{1}{450} + \dfrac{0.0125}{200} + \dfrac{1}{5.0 \times 0.9742 \times 25}}$$

$$= 9526.6 \text{ W/m}^2 \ (3019 \text{ Btu/h-ft}^2).$$

Thus, the heat flow is increased by 300%.

(c) Identical calculations for fins attached to the left side, where $h = h_{12} = 450$ W/m²-°C, yields

$$mL = 1.3258,$$

$$\kappa_2 = 0.6549,$$

$$\eta_2 = 0.7151,$$

$$\frac{q}{A} = \frac{120 - 20}{\dfrac{1}{5.0 \times 0.7151 \times 450} + \dfrac{0.0125}{200} + \dfrac{1}{25}}$$

$$= 2458.0 \text{ W/m}^2 \text{ (779.2 Btu/h-ft}^2).$$

Fins applied to the left side increase the heat flow by only 4%. The poor performance of the fins results from the large value of h on the left side, resulting in a relatively poor surface effectiveness. ∎

REFERENCES

1. ROHSENOW, W. M., and J. P. HARTNET, eds., *Handbook of Heat Transfer*, New York, McGraw-Hill, 1973, pp. 3–14.

2. GARDNER, K. A., "Efficiency of Extended Surface," *Trans. ASME*, Vol. 67, No. 8, 1945, pp. 621–631.

3. JAKOB, M., *Heat Transfer*, Vol. I, New York, Wiley, 1949.

4. SCHNEIDER, P. J., *Conduction Heat Transfer*, Reading, Mass., Addison-Wesley, 1955.

PROBLEMS

3.1 It is desired to limit the heat loss through a boiler furnace wall to 2000 W/m². The wall is composed of a material having a thermal conductivity of 1.15 W/m-°C. If the inner surface temperature is 1100°C and the outer surface 370°C, what thickness should be used?

3.2 If the wall of Prob. 3.1 has 5 cm of insulation ($k = 0.09$ W/m-°C) added to its outer surface and reduces the exposed surface temperature to 180°C, what will be the rate of heat loss through the wall per unit of area?

3.3 A composite wall is made of 20 cm of fire-clay brick (burnt at 1330°C), 15 cm of fired diatomaceous earth brick, and an outer layer of common brick 10 cm thick. If the inside surface is at 1050°C and the outside surface is held at 150°C, find (a) the heat loss, per unit area; (b) the temperatures at the junction between the different layers of brick; and (c) the temperature at a point 15 cm from the outer surface into the wall.

3.4   A composite wall is made of 7 in. of fire-clay brick (burnt at 2426°F), 5 in. of fired diatomaceous earth brick, and an outer 4-in. layer of common brick. If the inside surface temperature is 1800°F and the outside surface is held at 250°F, find (a) the heat loss rate per square foot of wall area; (b) the temperature at the junction between the different layers of brick; and (c) the temperature at a point 6 in. from the outer surface, into the wall.

3.5   A wall is composed of a 12-cm-thick layer of material A and a 20-cm-thick layer of material B. The temperature of the outer surface of layer A is known to be 260°C when the outer temperature at layer B is at 30°C. A layer of 2.5-cm-thick insulation ($k = 0.09$ W/m-°C) is added to the outer surface of layer B. Under these conditions it is observed that the surface of layer A rises to 310°C, the junction between layer B and the insulation (formerly the outer surface of layer B) becomes 220°C. If the insulation surface is measured to be 25°C, what is the rate of heat flow, per square meter of wall area, *before* and after the insulation is added?

3.6   A wall of a house is composed of 10 cm of common brick, 1.25 cm of Celotex, a 9-cm air space created by the studs, and 1.25 cm of asbestos cement board. If the outside brick surface is at 30°C, and the inner wall board surface is at 20°C, what is the heat flow rate, per unit area of wall surface, if (a) the air space *conductance* is 7.0 W/m²-°C, and (b) the air space is filled with glass wool at a density of 64 kg/m³?

3.7   A nominal 5-in. wrought iron schedule 40 pipe is covered with a 5-cm layer of 85% magnesia insulation. If the pipe carries steam so that the inner surface is at 425°C and the outer insulation surface is 40°C, what is the heat loss per hour from 30 m of this pipe?

3.8   If an 8-in.-nominal schedule 80 steel pipe is covered first with a 7.5-cm layer of 85% magnesia insulation and then a 2.5-cm layer of air-cell insulation ($k = 0.064$ W/m-°C), find the heat loss, per unit length of pipe, if the inside pipe surface is at 530°C and the outside insulation surface is at 90°C. What error is incurred if the thermal resistance of the pipe wall is neglected?

3.9   Find the temperature at the junction between the two insulations in Prob. 3.8.

3.10   A pipe with an outside diameter of 6 cm and a surface temperature of 0°C is to be covered with a 2.5-cm thickness of expanded cork insulation and a 2.5-cm thickness of rock wool (64 kg/m³ density). Which insulation should be placed next to the pipe surface to achieve the maximum insulating effect, if the outer surface temperature is 35°C in either instance?

3.11   A nominal-4-in. wrought iron schedule 40 pipe is covered with a 2-in. layer of 85% magnesia insulation. If the pipe carries steam so that the inner surface is at 700°F and the outer insulation surface is 150°F, what is the heat loss per hour from 100 ft of this pipe?

3.12 Derive the following expression for the heat flow per unit surface area from a sphere of inside radius r_1 and outside radius r_2. The sphere is heated to uniform inside and outside surface temperatures of t_1 and t_2, respectively.

$$\frac{q}{A} = \frac{k(t_1 - t_2)}{\dfrac{r_2}{r_1}(r_2 - r_1)}.$$

3.13 A hollow cylinder of inside and outside radii r_1 and r_2, respectively, is heated such that its inner and outer surfaces are at uniform temperatures t_1 and t_2. If the material of which the cylinder is composed has a thermal conductivity which varies with temperature in the following way:

$$k = k_0(1 + bt),$$

find the rate of heat flow through the cylinder. At what mean temperature should one evaluate k in order to use the constant-thermal-conductivity formula given in Eq. (3.11)?

3.14 Find the relation for the rate of heat flow through a single-layered plane wall, the thermal conductivity of which varies quadratically:

$$k = k_0(1 + bt + ct^2).$$

3.15 Verify the algebra leading to Eqs. (3.19) and (3.21).

3.16 Verify the algebra leading to Eq. (3.25).

3.17 A large slab of concrete, 1 m thick, has both surfaces maintained at 20°C. During the curing process a uniform internal heat generation of 60 W/m³ occurs throughout the slab. If the concrete thermal conductivity is 1.13 W/m-°C, find the steady temperature which results at the center of the slab.

3.18 A copper rod ($k = 380$ W/m-°C) 0.5 cm in diameter and 30 cm long has its two ends maintained at 20°C. The lateral surface of the rod is perfectly insulated, so conduction may be taken as one dimensional along the length of the rod. Find the maximum electrical current that the rod may carry if the temperature is not to exceed 120°C at any point and the electrical resistivity (resistance $\times$ cross section/length) is 1.73×10^{-6} Ω-cm.

3.19 A bare copper wire 0.3 cm in diameter has its outer surface maintained at 25°C while carrying an electrical current. The electrically generated heat is conducted one-dimensionally in a radial direction. If the centerline temperature of the wire is not to exceed 120°C, find the maximum current the wire can carry. Use the thermal and electrical properties for copper given in Prob. 3.18.

3.20 Imagine that the wall described in Prob. 3.3, rather than having its surface temperature specified is surrounded on its fire-brick side by hot gases at 1150°C with a heat transfer coefficient of 32.0 W/m²-°C and is surrounded on its other side by air at 38°C with a heat transfer coefficient of 8.5 W/m²-°C. Find (a) the rate of heat flow per unit area, and (b) the temperature of the two surfaces.

3.21 Imagine that the wall described in Prob. 3.4, rather than having its surface temperatures specified, is surrounded on its fire-brick side by hot gases at 2000°F with a heat transfer coefficient of 5.4 Btu/h-ft²-°F and is surrounded on its other side by air at 100°F with a heat transfer coefficient of 1.5 Btu/h-ft²-°F. Find (a) the rate of heat flow through each square foot of wall, and (b) the temperature of the two surfaces.

3.22 Let the wall of the house in Prob. 3.6 be subjected to an outdoor wind velocity of 16 km/h so that a heat transfer coefficient of 30 W/m²-°C exists on its outer surface. Assuming that the inside heat transfer coefficient is 10 W/m²-°C, find the rate of heat flow and the wall surface temperatures for inside and outside air temperatures of 18°C and 38°C, respectively.

3.23 A steel ($k = 43$ W/m-°C) plate 1.25 thick is exposed on one side to steam at 650°C through a heat transfer coefficient of 570 W/m²-°C. It is desired to insulate the outer surface so that the exposed surface of the outer insulation does not exceed 38°C. To minimize cost, an expensive high temperature insulation ($k = 0.26$ W/m-°C) is applied to the steel surface, and then a less-expensive insulation ($k = 0.09$ W/m-°C) is placed on the outside. The maximum allowable temperature of the less-expensive insulation is 315°C. The heat transfer coefficient at the outermost surface is 11.3 W/m²-°C, and the ambient air there is at 30°C. Find the thickness of the two layers of insulation.

3.24 A nominal-4-in. schedule 40 wrought iron pipe is covered with 3.8 cm of 85% magnesia insulation. The pipe carries superheated steam at a temperature of 400°C, and the outer insulation surface is exposed to air at 25°C. If the inside and outside heat transfer coefficients are 1400 and 10 W/m²-°C, respectively, find the rate of heat loss per unit of pipe length, and find the temperatures of the inside pipe surface, outside pipe surface, and outside insulation surface.

3.25 The pipe configuration described in Prob. 3.8 carries steam at 650°C with an inside heat transfer coefficient of 2800 W/m²-°C. The outer insulation surface is exposed to air at 50°C with a surface coefficient of 10 W/m²-°C. Find the heat loss per unit length.

3.26 Calculate the heat loss per unit of length of a 4-in. schedule 40 pipe covered with a 1.25-cm layer of insulation ($k = 0.09$ W/m-°C) if the inside pipe surface temperature is 200°C and the outside air temperature is 20°C. Assume an outside heat transfer coefficient of 170 W/m²-°C.

3.27 A bare 2.5-cm-diameter pipe has a surface temperature of 175°C and is placed in air at 30°C. The convective heat transfer coefficient between the surface and the air is 5.6 W/m²-°C. It is desired to reduce the heat loss to 50% of its present value by the addition of an insulation with $k = 0.17$ W/m-°C. Assuming that the pipe surface temperature and the exposed surface heat transfer coefficient remain unchanged as insulation is added, find the required thickness of insulation. Is this thickness an economically reasonable value?

3.28 Repeat Prob. 3.27 for a 10-cm-diameter pipe, all other data remaining unchanged.

3.29 Repeat Prob. 3.24 for a 10-in.-nominal schedule 40 steel pipe subjected to identical conditions. Take all conductivities and film coefficients to be the same.

3.30 A 6-in. schedule 40 steel pipe carries steam at 200 psia, 400°F. Minimum cost is to determine the selection of the thickness of magnesia insulation to be used in insulating the pipe. The air temperature in the room through which the pipe passes is 90°F and a film coefficient of 1.5 Btu/h-ft²-°F exists at the exposed insulation surface. The cost of generating the steam is $1.00 per million Btu. The cost of the insulation, installed, depends on the thickness:

> 1 in.: $5.00 per foot of length
>
> 2 in.: $8.00
>
> 3 in.: $14.00
>
> 4 in.: $20.00
>
> 5 in.: $30.00

Annual fixed charges for interest, repairs, etc., are 10% of the initial cost. The steam line operates 8000 h per year. Recommend the thickness of insulation to be used based on an estimated life of 10 years.

3.31 A 12-in. schedule 40 steel pipe has a surface temperature (outside) of 480°C. It is desired to insulate the pipe so that the exposed insulation surface does not exceed 40°C. Two insulating materials are to be used. First a high temperature insulation ($k = 0.26$ W/m-°C) is applied next to the pipe and then a less-expensive insulation ($k = 0.073$ W/m-°C) is placed on the outside. The maximum temperature of the less-expensive insulation is 315°C. The outermost insulation surface is exposed to ambient air at 24°C through a surface heat transfer coefficient of 9.65 W/m²-°C. Assuming that the pipe surface temperature remains constant, find the thickness of the two types of insulation.

3.32 An electrical wire has a diameter of 0.32 cm. If it is to be covered with an electrical insulation having a thermal conductivity of 0.09 W/m-°C, find the thickness that will give the maximum rate of heat dissipation for a surface heat transfer coefficient of 22.7 W/m²-°C.

3.33 Find the overall heat transfer coefficient, U, for the situations described in Probs. 3.21 and 3.22.

3.34 Find the overall heat transfer coefficient, U, based on the outside exposed surface area, for the situations described in Probs. 3.24 and 3.25.

3.35 A $\frac{3}{4}$-in 18-gage brass condenser tube has a heat transfer coefficient at its inner surface of 5400 W/m²-°C and one of 6800 W/m²-°C at its outer surface. The cooling water flowing in the tube is at 30°C. Find (a) the overall heat transfer coefficient, U, and (b) the mass of saturated steam at 10 kN/m² pressure that will be condensed for each meter of tube length.

3.36 A water-to-water heat exchanger is made of brass tubes, 1 in. O.D., 16 gage. The inside and outside heat transfer coefficients are 800 and 1200 Btu/h-ft²-°F, respectively. Find the overall heat transfer coefficient, U.

3.37 Using the result of Prob. 3.12, write an expression for the overall heat transfer coefficient of a two-layer sphere with inside and outside convective heat transfer coefficients. Base the overall coefficient on the outside surface area.

3.38 A sphere of fixed outside radius and fixed surface temperature is to be insulated with a material of known thermal conductivity. For fixed values of an ambient fluid temperature and surface film coefficient, find if a "critical thickness" of insulation exists (as in the cylindrical case considered in Sec. 3.9) and, if so, what its value is.

3.39 Show, for a single-layered cylinder, that the equation for the rate of heat flow through a plane wall may be used if one uses for the wall area the log-mean of the inner and outer cylindrical surface areas. That is,

$$A_m = \frac{A_2 - A_1}{\ln (A_2/A_1)}.$$

3.40 Repeat Prob. 3.39 for a single-layered sphere, showing that the geometric mean area should be used:

$$A_m = \sqrt{A_2 A_1}.$$

3.41 Show, for a straight fin of constant cross section, that as $L \to \infty$, the temperature distribution and heat flow are given by

$$\theta = \theta_0 e^{-mx},$$

$$q = kmA\theta_0.$$

3.42 A 1.25-cm-diameter rod of iron ($k = 45$ W/m-°C) is heated to 260°C at its base and protrudes into air at 38°C where $h = 8.5$ W/m²-°C. How long must the rod be so that if its end temperature is computed using the equations of Prob. 3.41 with $x = L$, the error will be less than 2.5°C as compared to the exact solution given in Eq. (3.45)? What error is made at this length in the total heat flow from the rod?

3.43 A cylindrical rod of 1.9 cm diameter, 24 cm long, protrudes from a heat source at 150°C into air at 32°C. The film coefficient of convective heat transfer is known to be 5.6 W/m²-°C on all exposed surfaces. Find the following information for the three cases of the rod being composed of copper, cast iron, and glass:

(a) The temperature at points located $\frac{1}{4}$, $\frac{2}{4}$, $\frac{3}{4}$, $\frac{4}{4}$ the distance from the source to the rod end. Neglect end heat loss.

(b) The rate of heat flow out of the source if the end heat loss is neglected, and if the end heat loss is not neglected.

3.44 A 0.6-cm-diameter pure copper rod is 46 cm long and is heated to 90°C at each end. If the fluid which surrounds the rod is at 25°C and a heat transfer coefficient of 25 W/m²-°C exists at the surface, find the temperature at the midpoint of the rod and at a point 10 cm from one end. How much heat is dissipated to the surroundings by the first 10 cm of the rod?

3.45 A straight fin of uniform cross section A, length L, perimeter C, conductivity k is maintained at a temperature, above the ambient fluid, of θ_0 at the end where $x = 0$ and at θ_L at the end where $x = L$. The heat transfer coefficient at the exposed surface is h. Derive the following expressions for the rate of heat flow at the two ends (positive from $x = 0$ to $x = L$):

$$q_0 = kmA \, \frac{\theta_0 \cosh mL - \theta_L}{\sinh mL},$$

$$q_2 = kmA \, \frac{\theta_0 - \theta_L \cosh mL}{\sinh mL}.$$

3.46 In Prob. 3.45, let the fin be a circular rod with a diameter of 1.25 cm, and a length of 45 cm. The left end is maintained at 175°C, the right end at 50°C, and the ambient fluid is at 20°C. For $h = 11$ W/m²-°C and $k = 43$ W/m-°C, find the heat flow rate out of each source.

3.47 A Chromel–Alumel thermocouple (wire diameters $= 0.125$ cm) is attached to a surface at 120°C and extends into air at 25°C ($h = 10$ W/m²-°C). Estimate the rate of heat loss from the surface due to the attachment of the thermocouple.

3.48 For free convection coefficients in air of the order of magnitude of 1 Btu/h-ft²-°F, make a reasonable estimate for the minimum length of a wiener-roasting wire made of an old coat hanger in order to avoid an uncomfortably hot temperature on the end held by the user.

3.49 Two rods of identical size and shape are both supported between two heat sources at 100°C and are surrounded by air at 25°C. One rod is known to have a thermal conductivity of 43 W/m-°C and its midpoint temperature is measured to be 49°C. If the midpoint temperature of the other rod is measured to be 75°C, what is its thermal conductivity?

3.50 An annular fin of uniform thickness has an inner radius of 7.5 cm and an outer radius of 12.7 cm. The fin has a uniform thickness of 0.5 cm and is composed of a material with $k = 43$ W/m-°C. The base of the fin is maintained at 200°C and the surrounding fluid is at 35°C. The heat transfer coefficient between the fin surface and the fluid is 56.8 W/m²-°C. Find (a) the rate at which heat is dissipated by the fin, (b) the temperature at the fin end and at a point midway between the base and the end, and (c) the heat dissipated by the last 2.6 cm of fin.

3.51 An annular web connects two concentric cylinders. The web is 0.65 cm thick and has a thermal conductivity of 62 W/m-°C. The outer cylinder has an I.D. of 15 cm and its inner surface is maintained at 390°C; the inner cylinder has an O.D. of 7.5 cm and a surface temperature of 100°C. The web is surrounded by a convecting fluid at 50°C with $h = 75$ W/m²-°C. Assuming that the temperature in the web is a function of the radius only, find the rate at which heat is being given up by the web to the fluid.

3.52 A spine protruding from a wall at temperature t_0 has the shape of a circular cone. The radius of the cone base is R. The spine comes to a point at its tip, and its length is L.

    (a) If t_f denotes the temperature of the ambient fluid, h the surface heat transfer coefficient, and k the thermal conductivity of the cone, show that the differential equation of the temperature distribution is as follows, where x is the distance measured from the cone tip:

$$\frac{d^2\theta}{dx^2} + \frac{2}{x}\frac{d\theta}{dx} - l^2\frac{\theta}{x} = 0,$$

$$l^2 = \frac{2hL}{kR}\sqrt{1 + \left(\frac{R}{L}\right)^2}.$$

    (b) Show that the general solution this equation is

$$\theta = \frac{BI_1(2lx^{1/2}) + DK_1(2lx^{1/2})}{x^{1/2}}.$$

3.53 For the conical spine described in Prob. 3.52, show that as $x \to 0$, D must be 0 and that the ratio $I_1(2lx^{1/2})/x^{1/2} \to l$. Then show, for $\theta = \theta_0$ at $x = L$, that the temperature distribution in the spine is given by

$$\frac{\theta}{\theta_0} = \left(\frac{L}{x}\right)^{1/2}\frac{I_1(2lx^{1/2})}{I_1(2lL^{1/2})}.$$

3.54 Show that the rate of heat flow from the spine in Probs. 3.52 and 3.53 is given by

$$q = -k\pi R^2\theta_0\left[\frac{l}{L^{1/2}}\frac{I_0(2lL^{1/2})}{I_1(2lL^{1/2})} - \frac{1}{L}\right].$$

3.55 A straight fin has a thickness which varies parabolically—i.e., its half thickness at the base is $w/2$, at the tip it is 0, and it varies as x^2 between the tip and the base. If its total length is L and if its base is maintained at a uniform temperature, deduce the equation of the temperature distribution and the equation for the heat loss per unit width if the fin is very thin.

3.56 A spine of circular cross section has a base radius of R and a tip radius of 0. The total length is L and the radius varies as $x^{1/2}$, where x is the coordinate measured from the tip to the wall. Write the expression for the temperature distribution in the spine if the spine is very thin.

3.57 An annular fin has a form which varies hyperbolically from its base to its tip. If its base thickness is w at r_b and if the thickness varies as $1/r$ (r = radius between r_b and r_e), deduce the equation for the temperature distribution in the fin if the fin is very thin.

3.58 For a straight fin of uniform thickness, the parameter m is, for unit width, $m^2 = 2h/kw$. If such a fin, of given k and given width, w, is to be utilized under fixed service conditions (i.e., θ_0, h), the heat loss from the fin may be expressed as a function of the fin length L. Write such an expression, including the loss of heat from the end with $h_e = h$, and show that if $(hw/2k) > 1$, $dq/dL < 0$ for *all* L; and if $hw/2k < 1$, $dq/dL > 0$. This shows that unless $hw/2k < 1$, the application of such fins will produce an insulating effect.

3.59 Consider a straight fin of uniform thickness. Let the "profile area" be that area taken in a plane that is parallel to the fin length and normal to the width, $A_p = w \times L$. For a fixed amount of material (i.e., A_p = constant), show that if the heat loss from the fin end is negligible, the fin dissipates the maximum amount of heat if its length, L, and thickness, w, are related by the following condition:

$$\tanh \xi = 3\xi \operatorname{sech}^2 \xi,$$

$$\xi = L \sqrt{\frac{2h}{kw}}.$$

3.60 For a set of circumstances similar to those given in Prob. 3.59, the profile area of a fully triangular fin is $A_p = \frac{1}{2} \times w \times L$. Show, for a fixed A_p, that the maximum heat is dissipated when w and L are related by

$$\frac{4}{3} \frac{I_1(\eta)}{I_0(\eta)} = 4 \left[1 - \left(\frac{I_1(\eta)}{I_0(\eta)} \right)^2 \right],$$

$$\eta = 2L \sqrt{\frac{2h}{kw}}.$$

3.61 In Prob. 3.59 the value of ξ which satisfies the optimum condition is $\xi = 1.4192$, and in Prob. 3.60 the optimum condition is given by $\eta = 2.6188$. With these data, show that for equal amounts of heat dissipated, the optimum proportioned triangular fin has a profile area 0.690 times that of the optimum proportioned rectangular fin. Assume that θ_0 is the same in both instances.

3.62 A cylindrical spine (radius R, length L) is maintained at a fixed base temperature,

t_0, and extends into a fluid at temperature t_f. A heat transfer coefficient, h, exists at the exposed surfaces. For a fixed volume of material, find the proportions of L and R for the spine which dissipates the maximum amount of heat. Neglect end heat losses.

3.63 Write an expression for the total surface temperature effectiveness of the annular array shown in Fig. 3.20(b) in terms of the geometry shown.

3.64 A plane wall with a surface temperature of 120°C is exposed to an ambient fluid at 25°C through a heat transfer coefficient of 20 W/m²-°C. If the surface is equipped with an array of straight fins of triangular profile (length 3.8 cm, thickness 0.32 cm, spaced 1.5 cm on centers) what is the total surface effectiveness? The fins are made of steel with $k = 43$ W/m-°C.

3.65 A plane surface is equipped with an array of straight fins of rectangular profile. The fins are 1.9 cm long, 0.13 cm thick, and spaced 1.9 cm on centers. The fins are made of Duralumin and the surface heat transfer coefficient is 142 W/m²-°C. What is the total surface effectiveness?

3.66 Heat is being transferred through a plane wall, 1.25 cm thick, composed of a material with $k = 17.3$ W/m-°C. On the left side of the wall is a fluid of temperature 90°C, and the heat transfer coefficient there is 284 W/m²-°C. On the right side of the wall is another fluid of temperature 38°C, and the heat transfer coefficient to all exposed surfaces there is 17.0 W/m²-°C. It is desired to enhance the heat transfer between the two fluids by adding to the right surface either (1) straight rectangular fins, 0.13 cm thick, 2.5 cm long, spaced on 1.25-cm centers; or (2) straight triangular fins, 0.13 cm thick at the base, 2.5 cm long, spaced on 1.25-cm centers. Assume that the fins are made of the same material as the wall, that the heat transfer coefficient is unchanged by adding the fins, and that a one-dimensional representation may be used. Based per unit of area of plane, unfinned, wall, find (a) the heat transfer between the fluids when the wall is bare, (b) the total surface temperature effectiveness of the right surface and the heat exchange between the two fluids when fins are used as in case (1), and (c) same as (b) when fins are used as in case (2).

3.67 Heat is to be transferred through a plane wall, 1.25 cm thick, composed of a material with $k = 43$ W/m-°C. On the left side of the wall is an ambient fluid of specified temperature, and the heat transfer coefficient there between the fluid and any exposed surface is 182 W/m²-°C. On the right side of the wall there is also an ambient fluid of known temperature, and the fluid-to-surface heat transfer coefficient is 17 W/m²-°C. It is contemplated to enhance the heat transfer between the two fluids by using straight, rectangular fins having the same conductivity as the wall. The fins are to be 0.064 cm thick, 2.29 cm long, and spaced 1.5 cm on centers. Presuming that the addition of the fins will not alter the existing heat transfer coefficients, that these coefficients apply to fin surfaces as well as the unfinned surface, and that a one-dimensional representation may be used, find the percent increase in the heat transfer between the two fluids (based per unit of area of the original wall without fins) over the case without fins if (a) fins are added to the left side only, (b) fins are added to the right side only, and (c) fins are added to both sides.

3.68 Two circular rods, both of diameter D and length L, are joined at one end and both heated to the same temperature, t_0, at the free ends. The film coefficient, h, is the same for all surfaces. If t_f is the temperature of the surrounding fluid and if the thermal conductivities of the two rods are k_a and k_b, show that the temperature of the junction, t_j, is

$$\frac{t_j - t_f}{t_0 - t_f} = \frac{\sqrt{\dfrac{k_a}{k_b}}\, \sinh m_b L + \sinh m_a L}{\sqrt{\dfrac{k_a}{k_b}}\, (\cosh m_a L)(\sinh m_b L) + (\sinh m_a L)(\cosh m_b L)}.$$

3.69 As shown in the accompanying figure, a straight fin protruding from a wall at 200°C consists of a rectangular portion, 12.5 cm long and 1.25 cm thick, capped by a triangular fin 10 cm long. The fin material has a $k = 86.5$ W/m-°C and the surface heat transfer coefficient between the fin and the ambient fluid at 38°C is 34 W/m²-°C. By appropriately combining the solution for a rectangular straight fin maintained at known temperatures at each end, as given in Prob. 3.45, with that for a triangular straight fin given in Sec. 3.15, find the temperature at the section where the fin changes from the rectangular shape to the triangular shape. What is the total heat transferred to the fluid from the fin?

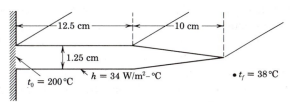

PROBLEM 3.69

3.70 A spine protruding from a wall at 375°F has the shape of a cylindrical rod ($\frac{1}{2}$ in. diameter, 7 in. long) capped by a cone 5 in. long. The entire spine is made of a material with $k = 25$ Btu/h-ft-°F and a surface heat transfer coefficient of 5.4 Btu/h-ft²-°F exists between the spine and the ambient fluid at 125°F. Using the solution of Prob. 3.45 with that of Prob. 3.54, find (a) the temperature at the section where the spine changes from the cylindrical to the conical shape, and (b) the total rate of heat transfer to the fluid.

3.71 Because of their very small size, transistors have little surface area for the dissipation of internally generated heat. As a consequence, they are often provided with a cap which not only protects the transistor but also provides additional heat transfer area to ambient air. Such a cap is shown in the accompanying figure, with all dimensions given. The cap has cylindrical symmetry with respect to the centerline shown—a hollow cylindrical wall topped with a circular disk. The walls of the cap are so thin that one may imagine that the heat is conducted up the cylindrical walls (one-dimensionally, but losing heat to the surroundings) and thence radially in the top disk (again one-dimensionally with heat loss to the surroundings). Presumably no heat is lost to the cap interior. The cap illustrated is made of steel with $k = 20$ W/m-°C. The ambient air is at 32°C and the heat transfer coefficient at all exposed surfaces is 25.5 W/m²-°C. The transistor is dissipating 375 mW. Find (a) the surface temperature of the transistor— i.e., the temperature at the junction between the cap and the transistor; and (b) the surface temperature of the transistor if no cap were provided and the 375 mW had to be transferred from the circular transistor surface with the same heat transfer coefficient and ambient air temperature.

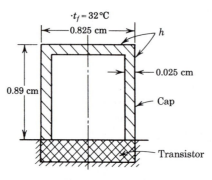

PROBLEM 3.71

3.72 The cross section of the core of a plate-fin heat exchanger is shown in the accompanying figure. The two primary heated surfaces are maintained at the same temperature, t_0, as shown. A fluid, of temperature t_f, flows in the open passages between the fin-matrix array shown. The fluid flows in a direction normal to the plane of the paper, and the fins are straight in that direction. The heat transfer coefficient, h, is presumed to be the same for all exposed surfaces.

(a) Find the total surface effectiveness for the primary surfaces maintained at t_0 in terms of the dimensions and parameters shown in the figure.

(b) If $h = 170$ W/m²-°C, $k = 52$ W/m-°C, $t_0 = 50$°C, $t_f = 30$°C, $d = 1.0$ cm, $L = 1.25$ cm and $w = 0.125$ cm, find the total heat transferred to the fluid for a section of the core 0.5 m in depth (normal to the paper) and 0.5 m in width.

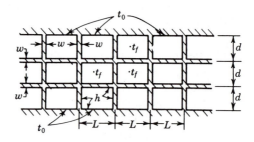

PROBLEM 3.72

CHAPTER 4

Heat Conduction in Two or More Independent Variables

4.1

INTRODUCTORY REMARKS

The cases of heat conduction discussed in Chapter 3 were all ones in which the temperature distribution throughout the bodies could be expressed in terms of a single variable. This meant that only steady state, one-dimensional systems could be discussed.

The present chapter is devoted to discussions of conduction problems that have to be described in two or more independent variables. The major portion of the chapter will be used to describe systems in two independent variables. This means either steady conduction in two space dimensions or nonsteady conduction in one dimension (the second variable being time). The two-dimensional, steady state systems will be treated first. The particular cases to be studied will be the rectangular plate and the circular bar because of the simplicity in describing the boundary conditions. One-dimensional, transient conduction in the plane wall and circular cylinder will be treated to illustrate nonsteady conduction problems. Finally, a section will be devoted to some special cases of transient conduction in two or three space dimensions.

No attempt will be made to present a comprehensive coverage of the mathematical theory of multidimensional heat conduction. Primary attention will be directed toward geometrical shapes of frequent occurrence and subjected to boundary conditions which are significant from the point of view of practical applications. In order to establish the mathematical techniques to be used later in the chapter, Secs. 4.2 and 4.3 will be devoted to two elementary systems subjected to simple boundary conditions. Although the solutions to these first few cases have little practical value, they will prove useful to the understanding of later applications.

For the rapid application of the results of the analyses to real physical problems, the solutions of problems having practical value have been reduced to relatively simple graphical presentations.

4.2

STEADY STATE CONDUCTION IN RECTANGULAR PLATES

If attention is now directed to steady state conduction in rectangular plates, such as shown in Fig. 4.1, it is most convenient to use cartesian coordinates to describe the temperature distribution in the plate. The plate will be considered to be in the x-y plane with the origin of the coordinates at one corner. No conduction will be considered in the z direction normal to the plate. This may be imagined to be the case if the plate has such a great extent in the z direction that no end effects exist or if the x-y faces of the plate are insulated so that no heat will pass in the z direction.

The heat conduction equation for the steady state [Eq. (1.15)] is, for cartesian coordinates in two dimensions,

$$\frac{\partial^2 t}{\partial x^2} + \frac{\partial^2 t}{\partial y^2} = 0. \tag{4.1}$$

Since this steady state heat conduction equation is a linear differential equation, the principle of superposition of solutions is applicable. This fact will be used later to build the solutions to more complex situations by adding the solutions of simpler problems.

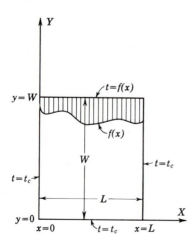

FIGURE 4.1. Rectangular plate with a specified temperature distribution on one edge, the other edges at constant temperature.

The solution of Eq. (4.1) is obtained by assuming the temperature distribution to be expressible as the product of two functions, each of which involves only one of the independent variables. That is, if $X(x)$ is a function of x only and if $Y(y)$ is a function of y only, one assumes that the temperature, t, is given by

$$t = X(x)Y(y). \tag{4.2}$$

When this is substituted into Eq. (4.1) and the resulting expression is rearranged, one obtains

$$-\frac{1}{X}\frac{d^2X}{dx^2} = \frac{1}{Y}\frac{d^2Y}{dy^2}.$$

Since each side of this equation involves only one of the independent variables, they may be equal only if they are both equal to the same constant. Calling this constant λ^2, one finds that

$$-\frac{1}{X}\frac{d^2X}{dx^2} = \frac{1}{Y}\frac{d^2Y}{dy^2} = \lambda^2.$$

This equation is equivalent to the following two ordinary differential equations:

$$\frac{d^2X}{dx^2} + \lambda^2 X = 0,$$

$$\frac{d^2Y}{dy^2} - \lambda^2 Y = 0.$$

The solutions of these equations are

$$Y = B_1 \sinh \lambda y + B_2 \cosh \lambda y,$$

$$X = B_3 \sin \lambda x + B_4 \cos \lambda x.$$

When these results are introduced into the assumed form of the solution given in Eq. (4.2), one finally obtains the following general solution of Eq. (4.1):

$$t = (B_1 \sinh \lambda y + B_2 \cosh \lambda y)(B_3 \sin \lambda x + B_4 \cos \lambda x). \tag{4.3}$$

The λ's and B's are constants to be determined by application of boundary conditions. An example of this will now be illustrated.

The Rectangular Plate with a Specified Temperature Distribution on One Edge—The Other Edges at Constant Temperature

Let the configuration under consideration be a rectangular plate of finite width L in the x-coordinate direction and of width W in the y-coordinate direction. This is illustrated in Fig. 4.1.

Let the temperature of the edges of the plate at $x = 0$, $x = L$, and $y = 0$ be maintained at a constant temperature t_c and that of the edge at $y = W$ be maintained at values that vary along that edge. Let this variation be represented as $f(x)(0 \le x \le L)$, remembering that $f(x)$, although unspecified, is to be treated as known.

The differential equation and the boundary conditions to be satisfied here are

$$\frac{\partial^2 t}{\partial x^2} + \frac{\partial^2 t}{\partial y^2} = 0.$$

At $x = 0$: $\quad t = t_c$.

At $x = L$: $\quad t = t_c$.

At $y = 0$: $\quad t = t_c$.

At $y = W$: $\quad t = f(x)$.

The problem is, then, for the system of equations noted above: What is the temperature at any specified point within the plate? That is, find the function $t = t(x, y)$ that gives the distribution of temperature in the plate. This problem may be simplified somewhat by rendering the boundary conditions homogeneous through the introduction of the temperature difference variable.

$$\theta = t - t_c.$$

Then the system to be solved is

$$\frac{\partial^2 \theta}{\partial x^2} + \frac{\partial^2 \theta}{\partial y^2} = 0.$$

At $x = 0$: $\quad \theta = 0$. $\hspace{2cm}$ (1)

At $x = L$: $\quad \theta = 0$. $\hspace{2cm}$ (2) $\hspace{1.5cm}$ (4.4)

At $y = 0$: $\quad \theta = 0$. $\hspace{2cm}$ (3)

At $y = W$: $\quad \theta = f(x) - t_c$. $\hspace{1.1cm}$ (4)

The solution of the differential equation in Eq. (4.4) is of the same form as Eq. (4.3), which is the solution to Eq. (4.1). Thus,

$$\theta = (B_1 \sinh \lambda y + B_2 \cosh \lambda y)(B_3 \sin \lambda x + B_4 \cos \lambda x).$$

Application of condition (3) of Eq. (4.4) shows that $B_2 = 0$, so

$$\theta = B_1 \sinh \lambda y(B_3 \sin \lambda x + B_4 \cos \lambda x).$$

In order to satisfy condition (1) of Eq. (4.4), $B_4 = 0$. Thus,

$$\theta = \sinh \lambda y \, B \sin \lambda x, \hspace{2cm} (4.5)$$

where B replaces the product $B_1 B_3$. Substitution of condition (2) gives

$$0 = \sinh \lambda y \, B \sin \lambda L.$$

The only way that this may be satisfied for *all* values of y is for

$$\sin \lambda L = 0.$$

This expression is satisfied for $\lambda = 0, \pi/L, 2\pi/L, \ldots$, or, in general,

$$\lambda_n = \frac{n\pi}{L}, \hspace{1cm} n = 0, 1, 2, 3, \ldots \hspace{2cm} (4.6)$$

Each of the λ's of Eq. (4.6) gives rise to a separate solution of Eq. (4.5), and since the general solution will be the sum of these individual solutions, one has

$$\theta = \sum_{n=0}^{\infty} B_n(\sinh \lambda_n y)(\sin \lambda_n x).$$

The symbol B_n represents the constant B for each of the solutions. Since $\lambda_n = 0$ for $n = 0$, no contribution is made by the first term,

$$\theta = \sum_{n=1}^{\infty} B_n(\sinh \lambda_n y)(\sin \lambda_n x). \tag{4.7}$$

Applying, finally, condition (4) of Eq. (4.4) for $y = W$, one obtains

$$[f(x) - t_c] = \sum_{n=1}^{\infty} B_n \sinh \lambda_n W \sin \lambda_n x, \tag{4.8}$$

$$\lambda_n = \frac{n\pi}{L}; \qquad n = 1, 2, 3, \ldots; \qquad 0 \le x \le L.$$

Comparing this result with the facts concerning the orthogonal functions in Eqs. (D.9), (D.10), and (D.11) of Appendix D, and recognizing that $(B_n \sinh \lambda_n W)$ is a constant, one sees that the constants B_n of Eq. (4.8) may be expressed in terms of the constants C_n of Eq. (D.10):

$$B_n \sinh \lambda_n W = C_n = \frac{2}{L} \int_0^L [f(x) - t_c] \sin \lambda_n x.$$

Since $f(x)$, unspecified here as to form, is a *known* boundary condition, the above integration could be performed [perhaps numerically, depending on the form of $f(x)$]. Still being general in not specifying $f(x)$ yet, one finds that the final solution of Eq. (4.7) is

$$\theta = \frac{2}{L} \sum_{n=1}^{\infty} \frac{\sinh \left(\dfrac{n\pi y}{L} \right)}{\sinh \left(\dfrac{n\pi W}{L} \right)} \sin \left(\frac{n\pi x}{L} \right) \int_0^L [f(x) - t_c] \sin \left(\frac{n\pi x}{L} \right) dx. \tag{4.9}$$

One Edge at a Uniform Temperature.   A special case in which the edge at $y = W$ is maintained at a constant temperature [i.e., $f(x) = t_0$, a constant] is illustrated in Fig. 4.2. For this case Eq. (4.9) becomes

$$\frac{t - t_c}{t_0 - t_c} = \frac{\theta}{t_0 - t_c} = 2 \sum_{n=1}^{\infty} \frac{1 - (-1)^n}{n\pi} \frac{\sinh \left(\dfrac{n\pi y}{L} \right)}{\sinh \left(\dfrac{n\pi W}{L} \right)} \sin \left(\frac{n\pi x}{L} \right). \tag{4.10}$$

Equation (4.10) would then enable one to compute the temperature at any desired point in the plate. As an illustration, Fig. 4.3 shows a rectangular plate, 10 by

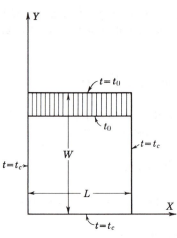

FIGURE 4.2. Rectangular plate with one edge at a uniform temperature, all other edges at constant temperature.

6, with one edge held at 100° and all others held at 0°. The isothermal lines within the plate were plotted by use of Eq. (4.10).

The Rectangular Plate with a Specified Temperature Distribution on More Than One Edge

A rectangular plate with specified temperature functions on more than one edge may be reduced to the above problem by a simple superposition. Consider, for

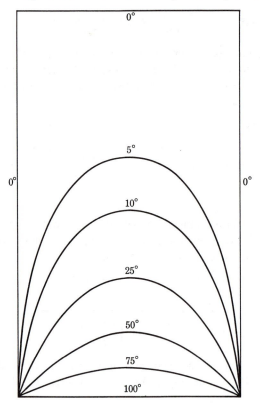

FIGURE 4.3. Isotherms in a rectangular plate heated on one edge.

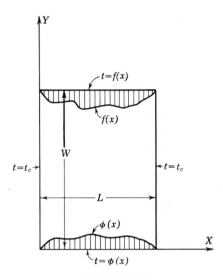

FIGURE 4.4

example, the situation illustrated in Fig. 4.4, in which the plate has a distribution $t = f(x)$ specifed at $y = W$, and $t = \varphi(x)$ is imposed at $y = 0$. Let the other edges be maintained at $t = t_c$. Then one must solve

$$\frac{\partial^2 \theta}{\partial x^2} + \frac{\partial^2 \theta}{\partial y^2} = 0,$$

$$\theta = f(x) - t_c \quad \text{at } y = W,$$

$$\theta = \varphi(x) - t_c \quad \text{at } y = 0, \qquad\qquad (4.11)$$

$$\theta = 0 \qquad\qquad \text{at } x = 0,$$

$$\theta = 0 \qquad\qquad \text{at } x = L,$$

where $\theta = t - t_c$ has again been introduced. Because the differential equation in Eq. (4.11) is linear, it may be reduced to two simpler systems by defining u and v such that

$$\theta = u + v. \qquad\qquad (4.12)$$

The symbols u and v are used to denote the solutions to the following two systems:

$$\frac{\partial^2 u}{\partial x^2} + \frac{\partial^2 u}{\partial y^2} = 0,$$

$$u = f(x) - t_c \quad \text{at } y = W,$$

$$u = 0 \qquad\qquad \text{at } y = 0, \qquad\qquad (4.13)$$

$$u = 0 \qquad\qquad \text{at } x = 0,$$

$$u = 0 \qquad\qquad \text{at } x = L.$$

$$\frac{\partial^2 v}{\partial x^2} + \frac{\partial^2 v}{\partial y^2} = 0,$$

$$v = 0 \qquad \text{at } y = W,$$

$$v = \varphi(x) - t_c \quad \text{at } y = 0,$$

$$v = 0 \qquad \text{at } x = 0,$$

$$v = 0 \qquad \text{at } x = L. \tag{4.14}$$

The solution to the system of Eq. (4.13) is the same as that expressed in Eq. (4.9) with u replacing θ. By a simple change of variable the solution given in Eq. (4.9) may be made to apply to the determination of v from Eq. (4.14):

$$v = \frac{2}{L} \sum_{n=1}^{\infty} \frac{\sinh\left(\dfrac{n\pi(W - y)}{L}\right)}{\sinh\left(\dfrac{n\pi W}{L}\right)} \sin\left(\frac{n\pi x}{L}\right) \int_0^L [\varphi(x) - t_c] \sin\left(\frac{n\pi x}{L}\right) dx. \tag{4.15}$$

Then, by Eq. (4.12), the solution to the system of Eq. (4.11) is the sum of Eqs. (4.9) and (4.15).

4.3

STEADY CONDUCTION IN A CIRCULAR CYLINDER OF FINITE LENGTH

As a final example of steady conduction in two space dimensions consider, as depicted in Fig. 4.5, a solid circular cylinder of finite length. The symbol R denotes the outer radius of the cylinder and L denotes its length.

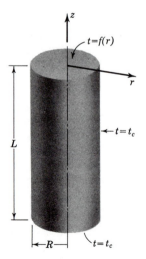

FIGURE 4.5

HEAT CONDUCTION IN TWO OR MORE VARIABLES

Cylindrical coordinates are the natural ones to use in this case because of the ease in specification of the boundary conditions. Of special interest are steady state conduction problems in such a cylindrical configuration but with the boundary conditions so chosen that *axial symmetry* exists. That is, if one considers problems (in the steady state) in which there is no dependence on the circumferential coordinate, the distribution of temperature in the cylinder depends only on the radial coordinate, r, and the axial coordinate, z. Thus, the general conduction equation in cylindrical coordinates, Eq. (1.12), reduces to the following:

$$\frac{\partial^2 \theta}{\partial r^2} + \frac{1}{r}\frac{\partial \theta}{\partial r} + \frac{\partial^2 \theta}{\partial z^2} = 0. \tag{4.16}$$

In Eq. (4.16) anticipation has been made of the eventual introduction of the temperature difference variable $\theta = t - t_c$. The imposed boundary conditions must, of course, be independent of the circumferential coordinate.

One obtains the general solution to Eq. (4.16) by use of the procedure that was employed in Sec. 4.2 to arrive at the solution of Eq. (4.1). That is, one seeks a solution in the form

$$\theta = \mathcal{R}(r)Z(z), \tag{4.17}$$

in which $\mathcal{R}(r)$ and $Z(z)$ denote the functions of r and z only, respectively. Substitution of Eq. (4.17) into Eq. (4.16) and subsequent rearrangement leads to

$$\frac{1}{\mathcal{R}}\frac{d^2\mathcal{R}}{dr^2} + \frac{1}{r}\frac{1}{\mathcal{R}}\frac{d\mathcal{R}}{dr} = -\frac{1}{Z}\frac{d^2Z}{dz^2}.$$

In the equation, the variables have been separated—the right side being a function of z alone and the left side a function of r alone. Since the two are equal and since each is a function of a single independent variable, the two sides can equal only a constant. Calling the constant λ^2, one obtains two ordinary differential equations in place of the original partial differential equation:

$$\frac{d^2\mathcal{R}}{dr^2} + \frac{1}{r}\frac{d\mathcal{R}}{dr} + \lambda^2\mathcal{R} = 0, \tag{4.18}$$

$$\frac{d^2Z}{dz^2} - \lambda^2Z = 0. \tag{4.19}$$

Equation (4.18) is recognized as Bessel's equation of zero order (Appendix C), whereas Eq. (4.19) leads to the hyperbolic functions. Thus, upon applying the definition of $\mathcal{R}$ and Z in Eq. (4.17), the solution of the differential equation in Eq. (4.16) may be expressed as

$$\theta = [B_1 J_0(\lambda r) + B_2 Y_0(\lambda r)](B_3 \sinh \lambda z + B_4 \cosh \lambda z). \tag{4.20}$$

Taking a particular problem, let the surface of the cylinder noted in Fig. 4.5 be maintained at a constant temperature t_c at every surface except for the circular end at $z = L$. On this surface let the temperature be specified as a known function of the radius r [i.e., $f(r)$]. Thus, the problem is now: Find an equation giving the distribution of the temperature throughout the cylinder for the following conditions:

$$\text{At } z = 0: \quad t = t_c \quad \text{or } \theta = 0. \tag{1}$$

$$\text{At } z = L: \quad t = f(r) \text{ or } \theta = f(r) - t_c. \tag{2} \qquad (4.21)$$

$$\text{At } r = R: \quad t = t_c \quad \text{or } \theta = 0. \tag{3}$$

One additional condition exists: that the temperature must be finite at $r = 0$. Knowing that $Y_0(\lambda r) \to \infty$ as $\lambda r \to 0$ (see Table C.1), one obtains $B_2 = 0$. Condition (1) of Eq. (4.21) makes B_4 also vanish, so one has

$$\theta = B \sinh \lambda z \, J_0(\lambda r),$$

in which B replaces $B_1 B_3$. Application of condition (3) demands that

$$B \sinh \lambda z \, J_0(\lambda R) = 0.$$

The only way that this latter condition can be satisfied for all values of z between 0 and L is for

$$J_0(\lambda R) = 0.$$

Examination of the tables of $J_0(\lambda R)$ in Appendix C shows that J_0 has a succession of zeros that differ by an interval approaching π as $\lambda R \to \infty$. Hence, there are a countably infinite number of λ's satisfying the defining relation:

$$J_0(\lambda_n R) = 0. \tag{4.22}$$

The first five are (Ref. 1) $\lambda_1 R = 2.4048$, $\lambda_2 R = 5.5201$, $\lambda_3 R = 8.6537$, $\lambda_4 R = 11.7915$, and $\lambda_5 R = 14.9309$.

Hence, the general solution is the sum of all the solutions corresponding to each of the λ_n's:

$$\theta = \sum_{n=1}^{\infty} (B_n \sinh \lambda_n z) J_0(\lambda_n r).$$

Finally, application of condition (2) of Eq. (4.21) determines the unknown B_n's since this gives

$$f(r) - t_c = \sum_{n=1}^{\infty} (B_n \sinh \lambda_n L) J_0(\lambda_n r).$$

Comparison of this last equation with Eq. (D.27) indicates that, since the λ's are defined by $J_0(\lambda_n R) = 0$, the constant coefficients $(B_n \sinh \lambda_n L)$ are given by

$$B_n \sinh \lambda_n L = \frac{\displaystyle\int_0^R r[f(r) - t_c] J_0(\lambda_n r) \, dr}{\dfrac{R^2}{2} J_1^2(\lambda_n R)}.$$

Finally, then, the solution to the problem giving the distribution of the temperature through the cylinder is

$$\theta = \frac{2}{R^2} \sum_{n=1}^{\infty} \frac{\sinh \lambda_n z}{\sinh \lambda_n L} \frac{J_0(\lambda_n r)}{J_1^2(\lambda_n R)} \int_0^R r[f(r) - t_c] J_0(\lambda_n r) \, dr. \tag{4.23}$$

FIGURE 4.6. Axial temperature distribution in solid cylinders of finite length.

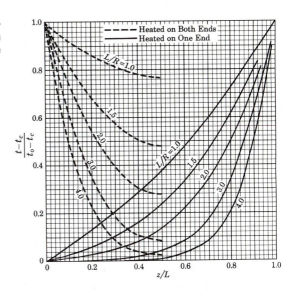

One End at Uniform Temperature. For the special case in which $f(r) = t_0$ (a constant), i.e., if the cylinder is maintained at zero temperature on all surfaces except one end where the temperature is t_0, the solution reduces to

$$\frac{t - t_c}{t_0 - t_c} = \frac{\theta}{t_0 - t_c} = 2 \sum_{n=1}^{\infty} \frac{1}{\lambda_n R} \frac{\sin \lambda_n z}{\sinh \lambda_n L} \frac{J_0(\lambda_n r)}{J_1(\lambda_n R)}. \qquad (4.24)$$

The relations of Appendix D have been used to show that

$$\int_0^R r J_0(\lambda_n r)\, dr = \frac{R}{\lambda_n} J_1(\lambda_n R).$$

The principle of superposition may be used to build up other cases. For example, the temperature in a cylinder heated to uniform temperatures at each end and zero temperature on its other surface may be found by adding two solutions of the form of Eq. (4.24).

Figure 4.6 plots the axial temperature distribution ($r = 0$) for the solid cylinder heated to constant temperature on one end. Also, the case of a cylinder heated on both ends is plotted.

4.4

NONSTEADY CONDUCTION IN ONE SPACE DIMENSION

The problem of describing the temperature distribution and its variation with time for *nonsteady* heat conduction in only *one space dimension* has much similarity to the two-dimensional steady conduction problems discussed in the foregoing in that both cases involve the determination of the temperature in terms of two independent variables. While many nonsteady cases may be posed, only those cases having some practical engineering application will be described here. The reader is referred to extensive works on conduction such as Refs. 2 through 4, for additional cases.

Some of the most often encountered nonsteady situations are those in which a given body at a uniform temperature suddenly has its boundary conditions altered—either by changing its boundary temperature or by changing the temperature of a surrounding, convecting, fluid. If such is the case, certain generalizations of the form of the expected result may be made before carrying out any particular solution.

For the purposes of discussion, imagine a nonsteady conduction problem, with constant properties, in which the single one-dimensional coordinate is the cartesian variable x. (An analogous discussion could be carried out in any other one-dimensional coordinate system, e.g., the cylindrical coordinates with r as the only variable.) The general conduction equation, for constant properties and no internal heat generation, Eq. (1.8), is

$$\frac{1}{\alpha} \frac{\partial t}{\partial \tau} = \frac{\partial^2 t}{\partial x^2}. \tag{4.25}$$

Let the body be initially at a uniform temperature, t_i, and imagine that the temperature of a surrounding fluid (with heat transfer coefficient h) is suddenly changed to some value, t_f. It is then desired to find the temperature within the body at any subsequent time. (The alternative case in which the surface temperature of the body itself is changed to some value will be treated as a special case below.) Thus, the boundary conditions on Eq. (4.25) are

$$\text{At } \tau = 0: \quad t = t_i, \text{ for all } x,$$

$$\text{For } \tau > 0: \quad -k\left(\frac{\partial t}{\partial x}\right)_s = h(t - t_f)_s, \tag{4.26}$$

where the subscript s denotes the body surface.

The solution to the system of eqs. (4.25) and (4.26) may be generalized by introducing certain nondimensional variables. If l represents some characteristic dimension of the body (e.g., slab thickness, cylinder radius, etc.) then the linear variable may be nondimensionalized by defining

$$\xi = \frac{x}{l}.$$

The temperature may be nondimensionalized by measuring all temperatures above that of the ambient fluid and referencing to the initial uniform temperature:

$$\mathbf{T} = \frac{t - t_f}{t_i - t_f}. \tag{4.27}$$

Additionally, the time and heat transfer coefficient may be rendered dimensionless by the following definitions:

$$\text{Dimensionless time} = \text{Fourier number} = \text{Fo} = \frac{\alpha^2 \tau}{l^2}. \tag{4.28}$$

$$\text{Dimensionless coefficient} = \text{Biot number} = \text{Bi} = \frac{hl}{k}. \tag{4.29}$$

With these definitions, the differential equation and its boundary conditions reduce to

$$\frac{\partial T}{\partial Fo} = \frac{\partial^2 T}{\partial \xi^2},$$

At Fo $= 0$: $T = 0$, for all ξ, (4.30)

for Fo > 0: $\left(\frac{\partial T}{\partial \xi}\right)_s = -BiT_s.$

The solution of Eq. (4.30) will then be of the form

$$T = fn\ (\xi,\ Fo,\ Bi).$$

Thus, all geometrically similar bodies with identical Bi will have the same dimensionless temperature response at geometrically similar points according to time measured as Fo.

Thus, the parameters Fo and Bi are important measures as to how a body responds to temperature changes. The Fourier number, Fo, is a measure of time, and its definition in Eq. (4.28) shows that bodies with high diffusivity respond faster than those with a low diffusivity; large bodies respond more slowly than small bodies. The significance of the Biot number can be more readily seen if it is rewritten

$$Bi = \frac{l/k}{1/h}.$$

It is seen to be a measure of the ratio of the thermal resistance of the body, l/k, and that of the surface film, $1/h$. Thus, the temperature response of bodies with a low Bi is dominated by the surface resistance while those with a large Bi are dominated by the internal resistance. The further significance of these parameters will be seen as specific solutions are obtained.

The case in which the surface temperature is suddenly changed to a fixed value, t_s, was not discussed above. This case may be viewed as that in which the heat transfer coefficient $h \to \infty$, or Bi $\to \infty$. In such a case the surface temperature, t_s, becomes equal to the fixed fluid temperature, i.e., $t_s \to t_f$. Thus, the dimensionless temperature becomes $T = (t - t_s)/(t_i - t_s)$ and the solution form becomes

$$T = \frac{t - t_s}{t_i - t_s} = fn\ (\xi,\ Fo).$$

In the solutions presented in the remainder of this chapter, cases for Bi $\to \infty$ (i.e., fixed surface temperature) will be considered first. Then cases with finite Bi will be studied, and, finally, cases for Bi $\to 0$ will be examined.

4.5

TRANSIENT CONDUCTION IN THE INFINITE SLAB

In order that the conduction of heat will depend on only one *space* variable, it will be necessary to select geometrical configurations of special types. In the

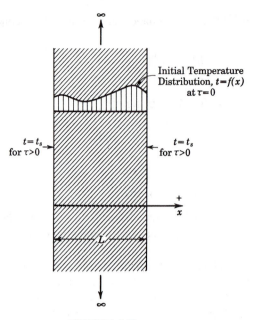

FIGURE 4.7

case in which rectangular coordinates are to be used, one considers conduction through a plane slab having a finite thickness in one direction but having an infinite extent in the other directions. That is, edge effects are to be neglected so that only the coordinate measured in the direction of the finite thickness is needed to describe positions. Take, as illustrated in Fig. 4.7, the x coordinate as the one-dimensional coordinate. Then, with time, τ, as a variable the general conduction equation, Eq. (1.10), reduces to the following with internal heat generation absent:

$$\frac{\partial t}{\partial \tau} = \alpha \frac{\partial^2 t}{\partial x^2}.$$

As in the cases considered in Sec. 4.4, it will prove desirable to express solutions in terms of a temperature difference variable, θ, to be defined specifically in each case which follows, so that the basic differential equation becomes

$$\frac{\partial \theta}{\partial \tau} = \alpha \frac{\partial^2 \theta}{\partial x^2} \tag{4.31}$$

where x is the single space variable and α is the thermal diffusivity defined in Chapter 1.

The solution to this equation can be found in the same manner as used in Sec. 4.3. Seeking solutions of the separable form, one assumes that the solution is representable as

$$\theta = X(x) \cdot T(\tau), \tag{4.32}$$

in which $X(x)$ and $T(\tau)$ represents functions of x and τ only. Introducing this into Eq. (4.31), one obtains

$$X \frac{dT}{d\tau} = \alpha T \frac{d^2X}{dx^2}.$$

This equation separates into

$$\frac{1}{\alpha T} \frac{dT}{d\tau} = \frac{d^2X}{dx^2} \frac{1}{X} = -\lambda^2.$$

The separation parameter, λ^2, has been introduced in the above equality since each side is a function of only a single variable. The sign of this constant has been taken to be negative to ensure a negative exponential solution in time. This separation of variables leads to the following two ordinary differential equations with the indicated solutions:

$$\frac{dT}{d\tau} + \alpha\lambda^2 T = 0, \qquad T(\tau) = B_1' e^{-\alpha\lambda^2\tau}$$

$$\frac{d^2X}{dx^2} + \lambda^2 X = 0, \qquad X(x) = B_2' \sin \lambda x + B_3' \cos \lambda x.$$

The B's are arbitrary constants. From the definition in Eq. (4.32) the general solution to Eq. (4.31) is

$$\theta = e^{-\lambda^2\alpha\tau}(B_1 \sin \lambda x + B_2 \cos \lambda x). \qquad (4.33)$$

Certain particular problems in which the boundary conditions are used to evaluate the constants will now be discussed.

Sudden Changes in the Boundary Temperature of an Infinite Slab

Consider first the infinite slab which is initially heated to some known, but for the present unprescribed, temperature distribution. Call this initial distribution $f(x)$. Then at time $\tau = 0$ let the bounding surfaces have their temperatures reduced to and maintained at a constant temperature t_s. This is shown in Fig. 4.7.

If the temperature-difference variable is defined as $\theta = t - t_s$, the conditions on Eq. (4.33) are:

$$\text{For } \tau = 0: \quad t = f(x) \text{ or } \theta = f(x) - t_s. \qquad (1)$$

$$\text{For } \tau \geq 0: \quad \text{at } x = 0, t = t_s \text{ or } \theta = 0 \qquad (2) \qquad (4.34)$$

$$\text{at } x = L, t = t_s \text{ or } \theta = 0. \qquad (3)$$

Application of conditions (2) and (3) to Eq. (4.33) yields

$$B_2 = 0 \qquad \text{and} \qquad \sin \lambda_n L = 0.$$

Thus, following the procedure used in the earlier discussions, one obtains the solution

$$\theta = \sum_{n=1}^{\infty} e^{-\lambda_n^2 \alpha \tau} B_n \sin \lambda_n x,$$

$$\lambda_n = \frac{n\pi}{L}; \qquad n = 1, 2, 3, \ldots .$$

(4.35)

Application of the initial condition at $\tau = 0$ gives

$$f(x) - t_s = \sum_{n=1}^{\infty} B_n \sin \lambda_n x.$$

Upon comparing these results with Eq. (D.9), Eq. (D.10) shows that

$$B_n = \frac{2}{L} \int_0^L [f(x) - t_s] \sin \lambda_n x \, dx.$$

Thus, the equation of the distribution of temperature throughout the slab is

$$\theta = \frac{2}{L} \sum_{n=1}^{\infty} e^{-(n\pi/L)^2 \alpha \tau} \sin \left(\frac{n\pi x}{L}\right) \int_0^L [f(x) - t_s] \sin \left(\frac{n\pi x}{L}\right) dx, \quad (4.36)$$

where $f(x)$ is the known distribution of temperature in the slab at the time $\tau = 0$. This last expression is a general one in that it gives the temperature distribution in the slab as a function of time and as dependent on the specified initial temperature distribution.

Uniform Initial Temperature Distribution. For the special case of a uniform initial distribution [i.e., $f(x) = t_i$ (a constant)], one has the problem of a slab initially heated to a uniform temperature and which suddenly has its surface temperatures reduced to and maintained at constant temperature. Equation (4.36) for the temperature-time history of the slab becomes

$$\frac{t - t_s}{t_i - t_s} = \frac{\theta}{t_i - t_s} = 2 \sum_{n=1}^{\infty} e^{-(n\pi/L)^2 \alpha \tau} \frac{1 - (-1)^n}{n\pi} \sin \left(\frac{n\pi x}{L}\right).$$

Examination of the above equation shows that if the slab thickness L is taken as the characteristic dimension, then the dimensionless parameters

$$\xi = \frac{x}{L}$$

$$\text{Fo} = \frac{\alpha \tau}{L^2}$$

enter naturally. The solution then devolves to the form predicted in Sec. 4.4:

$$\frac{t - t_s}{t_i - t_s} = \frac{2}{\pi} \sum_{n=1}^{\infty} e^{-(n\pi)^2 \text{Fo}} \frac{1 - (-1)^n}{n} \sin n\pi\xi. \qquad (4.37)$$

The rate of heat flow out of the slab is (per unit area normal to the x direction)

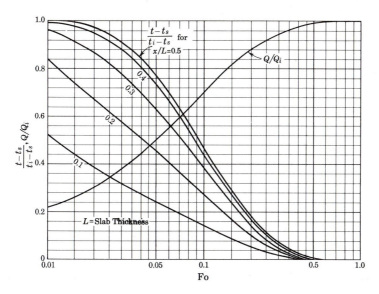

FIGURE 4.8. Time variation of the temperature distribution and heat flow in an infinite slab initially at a constant temperature which has its surface temperature suddenly changed.

$$\frac{q}{A} = \left[-k\left(\frac{\partial t}{\partial x}\right)_{x=0} \right] 2.$$

The factor 2 appears due to symmetry and the fact that heat flows out *two* faces of the slab, at $x = 0$ and $x = L$. The total heat flow up to time τ is then

$$\frac{Q}{A} = \int_0^\tau \frac{q}{A}\, d\tau.$$

Introduction of Eq. (4.37) gives, since $\alpha = k/\rho c_p$,

$$\frac{Q}{AL\rho c_p(t_i - t_s)} = 4 \sum_{n=1}^\infty \frac{1 - (-1)^n}{(n\pi)^2} [1 - e^{-(n\pi)^2 \text{Fo}}].$$

The denominator of the left side of the above equation is recognizable as the total heat stored in the slab initially—measured relative to the fixed boundary temperature. Call this $Q_i = AL\rho c_p(t_i - t_s)$, so the ratio of the total heat flow out of the slab up to time τ to that initially in the slab is

$$\frac{Q}{Q_i} = 4 \sum_{n=1}^\infty \frac{1 - (-1)^n}{(n\pi)^2} [1 - e^{-(n\pi)^2 \text{Fo}}]. \tag{4.38}$$

Figure 4.8 shows plots of Eqs. (4.37) and (4.38) so that the temperature–time history and the total heat flow from the slab· may be readily determined as functions of time.

EXAMPLE 4.1

A wall 25 cm thick, made of common brick, is initially at 80°C. Its surface temperatures are suddenly reduced to 15°C. Find the temperature at a plane

10 cm from the surface after 2 h have passed. How much heat has been conducted out of the wall during that time?

Solution. Table A.2 gives

$$c_p = 0.84 \text{ kJ/kg-°C}, \qquad \rho = 1602 \text{ kg/m}^3,$$

$$k = 0.69 \text{ W/m-°C}, \qquad \alpha = 5.2 \times 10^{-7} \text{m}^2/\text{s}.$$

Thus, the Fourier number is

$$\text{Fo} = \frac{5.2 \times 10^{-7} \times 2 \times 3600}{(0.25)^2} = 0.0599,$$

so that at $\xi = x/L = 10/25 = 0.4$, Fig. 4.8 gives

$$\frac{t - t_s}{t_i - t_s} = 0.669, \qquad \frac{Q}{Q_i} = 0.551.$$

Thus,

$$t = 0.669(80 - 15) + 15 = 58.5°C \ (137.3°F),$$

$$\frac{Q}{A} = \frac{Q_i}{A} \frac{Q}{Q_i} = L\rho c_p(t_i - t_s) \frac{Q}{Q_i}$$

$$= 0.25 \times 1602 \times 0.84 \times (80 - 15) \times 0.551$$

$$= 12.05 \times 10^3 \text{ kJ/m}^2 \ (1061 \text{ Btu/ft}^2).$$

■

Sudden Changes in the Temperature of the Fluid Surrounding an Infinite Slab

A situation having more practical significance than that discussed above is one in which the slab is surrounded by a convecting fluid. For example, the quenching of a heated solid in a liquid bath is a nonsteady conduction case in which the temperature of the surrounding fluid, not the surface of the solid, is suddenly changed to and maintained at some temperature different from the initial solid temperature.

Consider, then, an infinite slab, as shown in Fig. 4.9(a), of thickness 2L in which at time $\tau = 0$ a known temperature distribution, $\varphi(x)$, exists. The slab is surrounded on both sides by a convecting fluid so that a heat transfer coefficient, h, exists at the surface. Then let the temperature of this surrounding fluid be suddenly reduced to t_f and maintained at that temperature for all subsequent times. This is equivalent to plunging a heated bar into a quenching bath.

The condition to impose on the differential equation at the boundary is obtained from Newton's law of cooling as expressed in Eq. (1.17):

$$\frac{\partial t}{\partial x} = -\frac{h}{k}(t - t_f).$$

Without much loss in generality as far as practical problems are concerned, let it be assumed that the initial distribution of the slab temperature, $\varphi(x)$, is

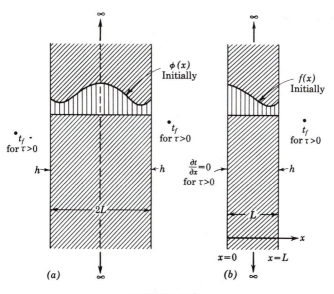

FIGURE 4.9

symmetrical with respect to the midline of the slab. If this is so, there will be *no* heat flow across the midline and the temperature history in the slab will be the same on each side of the midline since the fluid on each side is reduced to the same temperature. In such an instance the problem may be reduced to that shown in Fig. 4.9(b), in which one considers an infinite slab of thickness L which is initially heated to a known distribution $f(x)$ and which has its face at $x = 0$ insulated so that no heat can cross that boundary. Then for all subsequent times, the fluid surrounding the slab at its $x = L$ face is reduced to and maintained at temperature t_f while convection takes place at $x = L$ through a heat transfer coefficient of h. The heat transfer coefficient will be treated as a constant.

The problem then is to evaluate the solution to Eq. (4.31), i.e.,

$$\theta = e^{-\lambda^2 \alpha \tau}(B_1 \sin \lambda x + B_2 \cos \lambda x).$$

In this case the temperature difference variable is taken to be

$$\theta = t - t_f.$$

The boundary conditions are

$$\text{For } \tau = 0: \quad t = f(x), \text{ or } \theta = f(x) - t_f. \tag{1}$$

$$\text{For } \tau \geq 0: \quad \text{at } x = 0, \frac{\partial t}{\partial x} = 0 \text{ or } \frac{\partial \theta}{\partial x} = 0. \tag{2}$$

$$\tag{4.39}$$

$$\text{at } x = L, \frac{\partial t}{\partial x} = -\frac{h}{k}(t - t_f).$$

$$\text{or } \frac{\partial \theta}{\partial x} = -\frac{h}{k}\theta. \tag{3}$$

Application of condition (2) gives

$$\left.\frac{\partial \theta}{\partial x}\right|_{x=0} = 0 = e^{-\lambda^2 \alpha \tau} \lambda (B_1 \cos \lambda x - B_2 \sin \lambda x)\Big|_{x=0}$$

or

$$B_1 = 0.$$

So the solution reduces to

$$\theta = e^{-\lambda^2 \alpha \tau} B \cos \lambda x.$$

Then condition (3) gives

$$[e^{-\lambda^2 \alpha \tau} B \lambda (-\sin \lambda x)]_{x=L} = -\frac{h}{k} [e^{-\lambda^2 \alpha \tau} B \cos \lambda x]_{x=L}$$

or

$$\lambda \sin \lambda L = \frac{h}{k} \cos \lambda L.$$

This is equivalent to

$$\tan \lambda L = \frac{hL}{k} \frac{1}{\lambda L}. \tag{4.40}$$

For purposes of visualization, rewrite Eq. (4.40) as

$$\cot \lambda L = \frac{k}{hL} \lambda L. \tag{4.41}$$

Plotting each side of Eq. (4.41) as functions of λL (see Fig. 4.10), one observes

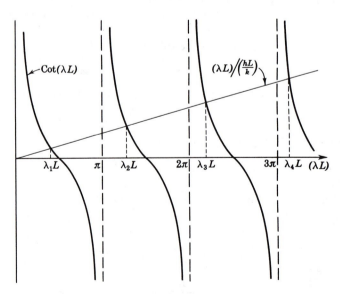

FIGURE 4.10

that the equation is satisfied for an infinite succession of values of the parameter λL so that for a given L, Eq. (4.40) defines the value of λ. This succession of values of λ will be denoted by λ_n, which are seen to depend on the magnitude of the parameter hL/k. This latter parameter is seen to be the Biot number based on the half-thickness, L, as the characteristic dimension:

$$\text{Bi} = \frac{hL}{k}.$$

The values of λ_n have been tabulated for various values of Bi (Ref. 2). For example, for Bi $= 10$, the first five roots of Eq. (4.41) are $\lambda_1 L = 1.4289$, $\lambda_2 L = 4.3058$, $\lambda_3 L = 7.2281$, $\lambda_4 L = 10.2003$, $\lambda_5 L = 13.2143, \ldots$.

Thus, one now has, as the solution,

$$\theta = \sum_{n=1}^{\infty} B_n e^{-\lambda_n^2 \alpha \tau} \cos \lambda_n x,$$

with λ_n being the nth root of

$$\lambda_n L \tan \lambda_n L - \text{Bi} = 0. \tag{4.42}$$

Application of the initial condition (1) of Eq. (4.39) gives

$$f(x) - t_f = \sum_{n=1}^{\infty} B_n \cos \lambda_n x. \tag{4.43}$$

This implies that one wishes to represent $f(x)$ as a series of cosines when λ_n is defined as a root of Eq. (4.42). This is discussed in Appendix D, where comparison of the above Eqs. (4.42) and (4.43) with Eqs. (D.19) and (D.20) shows that the coefficient, B_n, must be

$$B_n = \frac{\displaystyle\int_0^L [f(x) - t_f] \cos \lambda_n x \, dx}{\dfrac{L}{2} + \dfrac{\sin \lambda_n L \cos \lambda_n L}{2\lambda_n}}.$$

The final expression for the temperature in the slab, as a function of position and time, is then

$$\theta = 2 \sum_{n=1}^{\infty} \lambda_n e^{-\lambda_n^2 \alpha \tau} \frac{\cos \lambda_n x}{\lambda_n L + \sin \lambda_n L \cos \lambda_n L} \int_0^L [f(x) - t_f] \cos \lambda_n x \, dx. \tag{4.44}$$

Uniform Initial Temperature Distribution. Of particular practical interest is the case in which the initial distribution, $f(x)$, is a constant, say, t_i. Then Eq. (4.44) becomes

$$\frac{t - t_f}{t_i - t_f} = \frac{\theta}{t_i - t_f} = 2 \sum_{n=1}^{\infty} e^{-\lambda_n^2 \alpha \tau} \frac{\sin \lambda_n L}{\lambda_n L + \sin \lambda_n L \cos \lambda_n L} \cos \lambda_n x. \tag{4.45}$$

The total heat removed from the slab up to time τ is, as in the preceding section,

$$Q = \int_0^\tau q \, d\tau$$

$$= -kA \int_0^\tau \left(\frac{\partial t}{\partial x} \right)_{x=L} d\tau.$$

Performing the indicated differentiation and integration on Eq. (4.45) and again introducing $Q_i = kAL(t_i - t_f)/\alpha = AL\rho c_p(t_i - t_f)$ as the heat initially stored in the *half-slab* (since, due to symmetry, only half of the full slab of width $2L$ is being treated), one obtains

$$\frac{Q}{Q_i} = 2 \sum_{n=1}^{\infty} \frac{1}{\lambda_n L} \frac{\sin^2 \lambda_n L}{\lambda_n L + \sin \lambda_n L \cos \lambda_n L} (1 - e^{-\lambda_n^2 \alpha \tau}). \qquad (4.46)$$

As before, it is seen that the solution obtained is in the form of that predicted in Sec. 4.4, when one bases the dimensionless variables on the slab half-width L:

$$\xi = \frac{x}{L},$$

$$\mathrm{Fo} = \frac{\alpha \tau}{L^2},$$

$$\mathrm{Bi} = \frac{hL}{k}.$$

Then one has

$$\frac{t - t_f}{t_i - t_f} = 2 \sum_{n=1}^{\infty} e^{-\delta_n^2 \mathrm{Fo}} \frac{\sin \delta_n}{\delta_n + \sin \delta_n \cos \delta_n} \cos \delta_n \xi, \qquad (4.47)$$

$$\frac{Q}{Q_i} = 2 \sum_{n=1}^{\infty} \frac{1}{\delta_n} \frac{\sin^2 \delta_n}{\delta_n + \sin \delta_n \cos \delta_n} (1 - e^{-\delta_n^2 \mathrm{Fo}}), \qquad (4.48)$$

with δ_n being the nth root of

$$\delta_n \tan \delta_n - \mathrm{Bi} = 0. \qquad (4.49)$$

Equation (4.47) for the temperature distribution has been evaluated numerically by a number of investigators for various ranges of the three parameters ξ (position), Fo (time), and Bi (resistance). The results of Heisler (Ref. 5) are perhaps the most useful. Figure 4.11 plots the dimensionless temperature of the center of the slab (i.e., $\xi = 0$) versus Fo for various Bi. The form of Eq. (4.47) would suggest that a similar plot would be necessary for other values of ξ between 0 and 1; however, detailed study of this equation reveals that for the range of Fo involved, the ratio of the dimensionless temperature at a given ξ to that at $\xi = 0$ is virtually independent of Fo. Thus, Heisler was able to devise a ''position-correction'' chart (Fig. 4.12) which relates the temperatures at other ξ's to that at $\xi = 0$. Thus, Figs. 4.11 and 4.12 constitute a graphical presentation of Eq. (4.47). The expression for Q/Q_i given in Eq. (4.48) is displayed in Fig. 4.13. The quantity Q/Q_i may be used for purposes other than the calculation of heat flow. Since Q/Q_i represents the relative heat flow out of the slab,

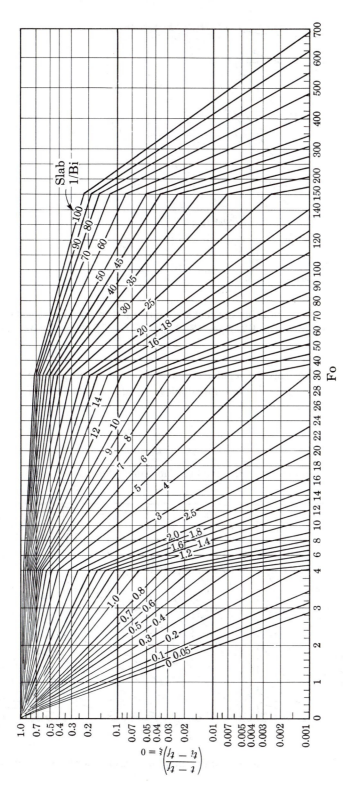

FIGURE 4.11. Heisler's chart for the temperature–time history at the center of an infinite slab initially at temperature t_i and placed in a medium at t_f. (From M. P. Heisler, *Trans. ASME*, Vol. 69, Apr. 1947, pp. 227–236. Used by permission.)

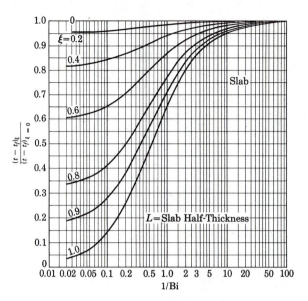

FIGURE 4.12. Heisler's position-correction chart for the infinite slab. (From M. P. Heisler, *Trans. ASME*, Vol. 69, Apr. 1947, pp. 227–236. Used by permission.)

$1 - Q/Q_i$ represents the relative amount of heat remaining in the slab. Consequently, $1 - Q/Q_i$ is the dimensionless *average* temperature of the slab at any time. That is,

$$1 - \frac{Q}{Q_i} = \frac{t_{av} - t_f}{t_i - t_f}$$

may be used to determine the average slab temperature.

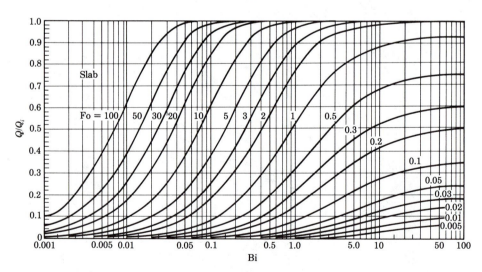

FIGURE 4.13. Heat flow from an infinite slab as a function of time and thermal resistance.

EXAMPLE 4.2

The brick wall of Example 4.1 (25 cm thick, initial temperature of 80°C) has a medium at 15°C suddenly placed in contact with both its surfaces through a heat transfer coefficient of 90 W/m²-°C. Find the temperature at a point 2.5 cm from the surface after 10 h have passed. Find the heat that has flowed out of the wall during that time and the average temperature within the wall at the end of the 10 h.

Solution.

For the properties quoted in Example 4.1, and taking $L = 25/2 = 12.5$ cm:

$$\text{Fo} = \frac{5.2 \times 10^{-7} \times 10 \times 3600}{(0.125)^2} = 1.20$$

$$\text{Bi} = \frac{90 \times 0.125}{0.69} = 16.3,$$

$$\xi = \frac{12.5 - 2.5}{12.5} = 0.80.$$

Figures 4.11 through 4.13 give

$$\left(\frac{t - t_f}{t_i - t_f}\right)_{\xi=0} = 0.092,$$

$$\frac{(t - t_f)_\xi}{(t - t_f)_{\xi=0}} = 0.38,$$

$$\frac{Q}{Q_i} = 0.94.$$

Thus,

$$t = 0.092 \times 0.38 \, (80 - 15) + 15 = 17.3°C \, (63.1°F).$$

$$\frac{Q}{A} = \frac{Q_i}{A} \frac{Q}{Q_i} = L\rho c_p (t_i - t_f) \frac{Q}{Q_i},$$

$$= 0.125 \times 1602 \times 0.84 \times (80 - 15) \times 0.94,$$

$$= 10.28 \times 10^3 \text{ kJ/m}^2 \text{ for one side,}$$

$$= 20.56 \times 10^3 \text{ kJ/m}^2 \, (1810 \text{ Btu/ft}^2)$$

for both sides.

$$t_{av} = (1 - 0.94)(80 - 15) + 15 = 18.9°C \, (66.0°F). \blacksquare$$

4.6

TRANSIENT RADIAL CONDUCTION IN A LONG SOLID CYLINDER

Equally important as the slab configuration just discussed is the solid circular cylinder. In order to ensure one-dimensional conduction so that the transient problem will involve only two independent variables, it will be assumed that the cylinder is infinitely long and that axial symmetry exists. Thus, the heat conduction can be considered to be in a radial direction only. Then the heat conduction equation in cylindrical coordinates, Eq. (1.12), reduces to the following with internal heat generation absent:

$$\frac{\partial \theta}{\partial \tau} = \alpha \left(\frac{\partial^2 \theta}{\partial r^2} + \frac{1}{r} \frac{\partial \theta}{\partial r} \right). \tag{4.50}$$

Again seeking solution of the separable form, one finds that

$$\theta = \mathcal{R}(r) \cdot T(\tau), \tag{4.51}$$

where $\mathcal{R}(r)$ and $T(\tau)$ are functions of only the indicated variable. Then Eq. (4.50) separates into

$$\frac{1}{\alpha T} \frac{dT}{d\tau} = \frac{1}{\mathcal{R}} \left(\frac{d^2 \mathcal{R}}{dr^2} + \frac{1}{r} \frac{d\mathcal{R}}{dr} \right) = -\lambda^2.$$

In the above equation λ^2 is once more used to represent the separation constant created because of the fact that each portion of this equation is a function of only one of the variables. Again, the separation constant is taken as negative to obtain a negative exponential solution in τ.

The two resulting ordinary differential equations and their solutions are

$$\frac{dT}{d\tau} + \lambda^2 \alpha T = 0,$$

$$T = B_1' e^{-\lambda^2 \alpha \tau},$$

$$\frac{d^2 \mathcal{R}}{dr^2} + \frac{1}{r} \frac{d\mathcal{R}}{dr} + \lambda^2 \mathcal{R} = 0,$$

$$\mathcal{R} = B_2' J_0(\lambda r) + B_3' Y_0(\lambda r).$$

Since the cylinder is solid and since no undefined solution can occur at $r = 0$ (the center of the cylinder), B_3' must be zero because Y_0 is undefined for zero argument. So the solution must be

$$\theta = B e^{-\lambda^2 \alpha \tau} J_0(\lambda r). \tag{4.52}$$

The constants B and λ are to be determined by the initial and boundary conditions.

Two particular cases of this configuration will now be discussed—that of transient conduction in a cylinder with specified boundary temperature and that of conduction in a cylinder of specified ambient fluid temperature.

Sudden Change in the Boundary Temperature of an Infinitely Long Cylinder

Consider now a solid cylinder of outer radius R and infinite in length. Let it be initially heated to some temperature distribution which is a function of the radial coordinate r only, $f(r)$, in keeping with the assumed axial symmetry. Then let the temperature of its outer surface be suddenly reduced to a constant, t_s, and maintained at that value for all subsequent times.

The procedure and relations used in this section are the same as those used in Secs. 4.3 and 4.5. Hence, only the briefest description will be given.

The solution is

$$\theta = Be^{-\lambda^2\alpha\tau}J_0(\lambda r),\tag{4.52}$$

and when the temperature difference variable is defined as $\theta = t - t_s$, the boundary conditions are

$$\text{For } \tau = 0: \quad t = f(r) \text{ or } \theta = f(r) - t_s \qquad (1)$$

$$\text{For } \tau \geq 0: \quad t = t_s \text{ or } \theta = 0 \text{ at } r = R. \qquad (2)$$

$$\tag{4.53}$$

Condition (2) of Eq. (4.53) determines λ_n as the roots of the equation

$$J_0(\lambda_n R) = 0, \qquad n = 1, 2, 3, \ldots,\tag{4.54}$$

as in Sec. 4.4. Thus,

$$\theta = \sum_{n=1}^{\infty} B_n e^{-\lambda_n^2\alpha\tau}J_0(\lambda_n r).$$

Then application of condition (1) leads to

$$f(r) - t_s = \sum_{n=1}^{\infty} B_n J_0(\lambda_n r).\tag{4.55}$$

Then, as in Sec. 4.3, Appendix D shows that under the conditions of Eqs. (4.54) and (4.55) the B_n's are

$$B_n = \frac{\displaystyle\int_0^R [f(r) - t_s]J_0(\lambda_n r)\, dr}{\dfrac{R^2}{2}J_1^2(\lambda_n R)}.$$

Thus, the final solution is

$$\theta = \frac{2}{R^2}\sum_{n=1}^{\infty} e^{-\lambda_n^2\alpha\tau}\frac{J_0(\lambda_n r)}{J_1^2(\lambda_n R)}\int_0^R [f(r) - t_s]J_0(\lambda_n r)\, dr.$$

Uniform Initial Temperature Distribution. For the special case of a uniform initial distribution, $f(r) = t_i$ (a constant),

$$\frac{t - t_s}{t_i - t_s} = \frac{\theta}{t_i - t_s} = 2\sum_{n=1}^{\infty} e^{-\lambda_n^2\alpha\tau}\frac{1}{\lambda_n R}\frac{J_0(\lambda_n r)}{J_1(\lambda_n R)}.\tag{4.56}$$

The total heat conducted out of a unit length of the cylinder up to a time τ is

$$\frac{Q}{L} = -k(2\pi R)\int_0^\tau \left(\frac{\partial t}{\partial r}\right)_{r=R} d\tau.\tag{4.57}$$

When Eq. (4.56) is introduced, and the total heat stored in the cylinder initially is defined as

$$\frac{Q_i}{L} = \pi R^2 \rho c_p(t_s - t_i) = \pi R^2 k \frac{t_s - t_i}{\alpha},$$

one obtains

$$\frac{Q}{Q_i} = 4 \sum_{n=1}^{\infty} \left(\frac{1}{\lambda_n R}\right)^2 (1 - e^{-\lambda_n^2 \alpha \tau}). \tag{4.58}$$

Sudden Change in the Temperature of the Fluid Surrounding an Infinitely Long Cylinder

Consider now the more practical case of the long solid cylinder (outer radius R) which is heated initially to a known, axially symmetric, distribution of temperature, $f(r)$, and which is suddenly placed in contact with a convecting fluid of constant temperature. This is illustrated in Fig. 4.14. The equation to be solved is then

$$\theta = Be^{-\lambda^2 \alpha \tau} J_0(\lambda r)$$

with the temperature difference variable defined in this case as

$$\theta = t - t_f.$$

The boundary conditions are

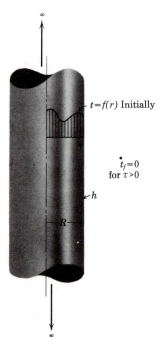

$t = f(r)$ Initially

$\dot{t}_f = 0$
for $\tau > 0$

h

R

FIGURE 4.14

For $\tau = 0$: $t = f(r)$ or $\theta = f(r) - t_f$.   (1)   (4.59)

For $\tau \geq 0$:   at $r = R$, $\dfrac{\partial t}{\partial r} = -\dfrac{h}{k}(t - t_f)$ or $\dfrac{\partial \theta}{\partial r} = -\dfrac{h}{k}\theta$.   (2)

Application of condition (2) of Eq. (4.59) leads to

$$Be^{-\lambda^2\alpha\tau}[-\lambda J_1(\lambda r)]_{r=R} = -\frac{h}{k}[Be^{-\lambda^2\alpha\tau}J_0(\lambda r)]_{r=R}.$$

This gives the following defining relation for the λ_n's:

$$\lambda_n R \frac{J_1(\lambda_n R)}{J_0(\lambda_n R)} = \frac{hR}{k}.$$

Defining the Biot number for the cylinder to be

$$\mathrm{Bi} = \frac{hR}{k},$$

one finds that the λ_n's are the roots of

$$\lambda_n R \frac{J_1(\lambda_n R)}{J_0(\lambda_n R)} - \mathrm{Bi} = 0. \tag{4.60}$$

Then the solution is

$$\theta = \sum_{n=1}^{\infty} B_n e^{-\lambda_n^2\alpha\tau} J_0(\lambda_n r).$$

Thus, upon application of initial condition (1) of Eq. (4.59),

$$f(r) - t_f = \sum_{n=1}^{\infty} B_n J_0(\lambda_n r). \tag{4.61}$$

Referring once more to Appendix D, Eqs. (D.31) and (D.32), one sees that the conditions of Eqs. (4.60) and (4.61) lead to the following relation for the B_n's:

$$B_n = \frac{\dfrac{2}{R^2} \displaystyle\int_0^R r[f(r) - t_f]J_0(\lambda_n r)\, dr}{J_0^2(\lambda_n R) + J_1^2(\lambda_n R)}.$$

Thus, one finally obtains, for the temperature distribution,

$$\theta = \frac{2}{R^2} \sum_{n=1}^{\infty} e^{-\lambda_n^2\alpha\tau} \frac{J_0(\lambda_n r)}{J_0^2(\lambda_n R) + J_1^2(\lambda_n R)} \int_0^R r[f(r) - t_f]J_0(\lambda_n r)\, dr. \tag{4.62}$$

Uniform Initial Temperature Distribution.   Just as in all the foregoing cases, the case of the circular bar initially heated to a uniform temperature is of particular practical value. So, with $f(r) = t_i$ (a constant), the above expression becomes

$$\frac{t - t_f}{t_i - t_f} = \frac{\theta}{t_i - t_f} = 2 \sum_{n=1}^{\infty} \frac{1}{\lambda_n R} e^{-\lambda_n^2\alpha\tau} \frac{J_0(\lambda_n r)J_1(\lambda_n R)}{J_0^2(\lambda_n R) + J_1^2(\lambda_n R)}. \tag{4.63}$$

Applying Eq. (4.57) for the total heat flow from the cylinder up to time τ, one obtains the result

$$\frac{Q}{Q_i} = 4 \sum_{n=1}^{\infty} \frac{1}{(\lambda_n R)^2} \frac{J_1^2(\lambda_n R)}{J_0^2(\lambda_n R) + J_1^2(\lambda_n R)} (1 - e^{-\lambda_n^2 \alpha \tau}). \qquad (4.64)$$

The foregoing expressions are subject to the definition of the λ_n's given in Eq. (4.60).

The obvious choice for the characteristic dimension is the cylinder radius, R, so that the nondimensional variables are

$$\rho = \frac{r}{R},$$

$$\text{Fo} = \frac{\alpha \tau}{R^2},$$

$$\text{Bi} = \frac{hR}{k}.$$

Then one has

$$\frac{t - t_f}{t_i - t_f} = 2 \sum_{n=1}^{\infty} \frac{1}{\delta_n} e^{-\delta_n^2 \text{Fo}} \frac{J_0(\delta_n \rho) J_1(\delta_n)}{J_0^2(\delta_n) + J_1^2(\delta_n)}. \qquad (4.65)$$

$$\frac{Q}{Q_i} = 4 \sum_{n=1}^{\infty} \frac{1}{\delta_n^2} \frac{J_1^2(\delta_n)}{J_0^2(\delta_n) + J_1^2(\delta_n)} (1 - e^{-\delta_n^2 \text{Fo}}), \qquad (4.66)$$

$$\delta_n \frac{J_1(\delta_n)}{J_0(\delta_n)} - \text{Bi} = 0. \qquad (4.67)$$

As in the case of the slab, Eqs. (4.65) and (4.66) have been evaluated numerically by many workers, and the results of Heisler will be used here. Again, while Eq. (4.65) indicates that the dimensionless temperature depends on all three variables ρ, Fo, Bi, calculations show that the ratio of the temperature at some radius ρ to that at the centerline ($\rho = 0$) is virtually independent of time (or Fo). Thus, Fig. 4.15 plots the centerline temperature at $\rho = 0$ as a function of time (Fo) and resistance (Bi). Figure 4.16, then, is the position-correction chart, giving the ratio of the dimensionless temperature at some ρ to that at $\rho = 0$, as a function of only Bi. Figure 4.17 gives Q/Q_i as a function of Bi and Fo. The use of these figures is completely analogous to that described above for the slab.

EXAMPLE 4.3

A long cylindrical bar 20 cm in diameter, heated to 980°C, is quenched in an oil bath at 40°C in which the heat transfer coefficient is 565 W/m²-°C. How long will it take for the centerline of the cylinder to reach 260°C? The bar is made of 19% Cr, 8% Ni, stainless steel.

Solution. Table A.1 gives

$$\rho = 7817 \text{ kg/m}^3, \qquad c_p = 0.46 \text{ kJ/kg-°C},$$

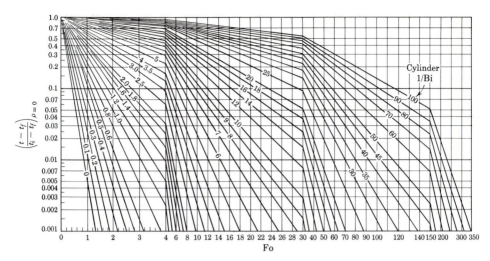

FIGURE 4.15. Heisler's chart for the temperature–time history at the center of an infinitely long cylinder initially at temperature t_i and placed in a medium at t_f. (From M. P. Heisler, *Trans. ASME*, Vol. 69, Apr. 1947, pp. 227–236. Used by permission.)

$$k = 16.3 \text{ W/m}^2\text{-}°\text{C}, \qquad \alpha = 0.444 \times 10^{-5} \text{ m}^2/\text{s}.$$

Thus, $\text{Bi} = (565 \times 0.1)/16.3 = 3.47$. The dimensionless temperature at the centerline is

$$\left(\frac{t - t_f}{t_i - t_f}\right)_{\rho=0} = \frac{260 - 40}{980 - 40} = 0.234.$$

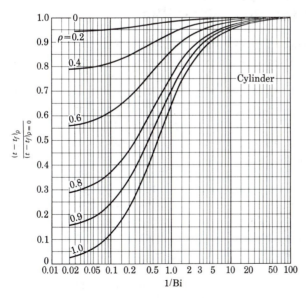

FIGURE 4.16. Heisler's position–correction chart for the infinitely long cylinder. (From M. P. Heisler, *Trans. ASME*, Vol. 69, Apr. 1947, pp. 227–236. Used by permission.)

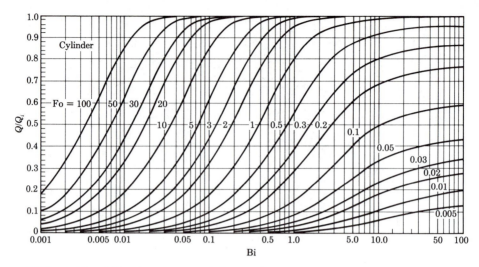

FIGURE 4.17. Heat flow from an infinitely long cylinder as a function of time and thermal resistance.

Figure 4.15 shows that Fo $= 0.53$. Thus, the time is

$$\tau = \frac{0.53(0.1)^2}{0.444 \times 10^{-5}} = 1194 \text{ s},$$

$$= 0.332 \text{ h.} \quad \blacksquare$$

4.7

TRANSIENT CONDUCTION IN MORE THAN ONE DIMENSION

The transient conduction problems discussed in Secs. 4.5 and 4.6 were ones limited to the very special and perhaps somewhat unrealistic configurations of the infinite plane slab and the infinite solid cylinder. These idealized shapes were chosen to ensure that the temperature of the solid would depend on only one space coordinate—in addition to time. In certain applications, neglecting end or edge effects (to which the above simplification of one-dimensional conduction is equivalent) may not greatly affect the desired results. However, in many instances such a simplification may not be possible, and transient conduction in more than one space dimension must be considered.

Under certain special conditions, the solution to problems of transient conduction in two or three space dimensions may be obtained by a simple product superposition of the solutions of one-dimensional problems. As an example of this method of superposition, consider the problem of transient conduction in a long rectangular bar—as depicted in Fig. 4.18(a). The bar has a thickness D in the x direction and a thickness W in the y direction. It is infinite in extent in the z direction so that the conduction will occur only in the x and y directions. Thus, a two-dimensional, transient problem results if the temperatures are nonsteady. For instance, let the bar be heated so that, initially, the temperature distribution depends only on x and y. Let this initial distribution be denoted $f(x, y)$. Then

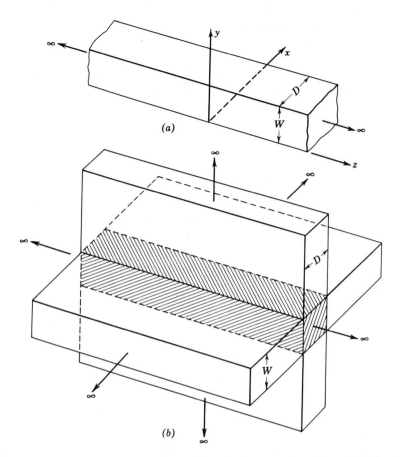

FIGURE 4.18. Rectangular bar formed by the intersection of two infinite slabs.

at time $\tau = 0$ assume that the bar is placed in contact with a convecting fluid at a constant temperature t_f. The heat transfer coefficient, h, is assumed to be constant over the entire surface. If the temperature difference variable is defined as

$$\theta = t - t_f, \tag{4.68}$$

then the differential equation to be solved is

$$\frac{1}{\alpha}\frac{\partial \theta}{\partial \tau} = \frac{\partial^2 \theta}{\partial x^2} + \frac{\partial^2 \theta}{\partial y^2} \tag{4.69}$$

and the boundary conditions are

For $\tau = 0$: $\theta = f(x, y) - t_f = F(x, y)$.

For $\tau \geq 0$:   at $x = 0$ and $x = D$, $\dfrac{\partial \theta}{\partial x} = \pm\dfrac{h}{k}\theta$,   (4.70)

at $y = 0$ and $y = W$, $\dfrac{\partial \theta}{\partial y} = \pm\dfrac{h}{k}\theta$.

[In the boundary conditions given in Eq. (4.70) the $+$ signs apply at $x = 0$ and $y = 0$, whereas the $-$ signs apply at $x = D$ and $y = W$.]

Now, if the initial temperature-*difference* distribution function, $F(x, y)$, is of such a form that it can be factored into a product of two other functions, each of which involves only one of the independent space variables, the initial condition may be replaced by

$$\text{At } \tau = 0: \quad \theta = F(x, y) \tag{4.71}$$

$$= F_1(x) \cdot F_2(y).$$

Thus, $F_1(x)$ represents a function of only x and $F_2(y)$ represents a function of only y. Under these conditions, the solution to Eq. (4.69), subjected to the conditions of Eq. (4.70), may be expressed as the *product* of two one-dimensional, transient solutions, as will now be shown.

Let the sought-after solution, $\theta(x, y, \tau)$ be represented as the following product:

$$\theta = \theta_x(x, \tau) \cdot \theta_y(y, \tau). \tag{4.72}$$

Here, $\theta_x(x, \tau)$ is a function of only x and time, τ, and $\theta_y(y, \tau)$ is a function of only y and τ. Substitution of Eq. (4.72) into Eq. (4.69) gives

$$\frac{1}{\alpha}\left(\theta_y \frac{\partial \theta_x}{\partial \tau} + \theta_x \frac{\partial \theta_y}{\partial \tau}\right) = \theta_y \frac{\partial^2 \theta_x}{\partial x^2} + \theta_x \frac{\partial^2 \theta_y}{\partial y^2}.$$

Upon rearrangement, this becomes

$$\theta_y\left(\frac{1}{\alpha} \frac{\partial \theta_x}{\partial \tau} - \frac{\partial^2 \theta_x}{\partial x^2}\right) + \theta_x\left(\frac{1}{\alpha} \frac{\partial \theta_y}{\partial \tau} - \frac{\partial^2 \theta_y}{\partial y^2}\right) = 0. \tag{4.73}$$

The boundary conditions, Eq. (4.70), and the initial condition, Eq. (4.71), become

$$\text{At } \tau = 0: \quad \theta = \theta_x \cdot \theta_y = F_1(x) \cdot F_2(y).$$

$$\text{For } \tau \geq 0: \quad \text{at } x = 0 \text{ and } x = D, \theta_y \frac{\partial \theta_x}{\partial x} = \pm\frac{h}{k} \theta_x\theta_y, \tag{4.74}$$

$$\text{at } y = 0 \text{ and } y = W, \theta_x \frac{\partial \theta_y}{\partial y} = \pm\frac{h}{k} \theta_x\theta_y.$$

Examination of Eqs. (4.73) and (4.74) will quickly show that they are satisfied if $\theta_x(x, \tau)$ and $\theta_y(y, \tau)$ are each obtained as the solutions of the two following one-dimensional problems:

$$\frac{1}{\alpha} \frac{\partial \theta_x}{\partial \tau} = \frac{\partial^2 \theta_x}{\partial x^2}.$$

$$\text{At } \tau = 0: \quad \theta_x = F_1(x). \tag{4.75}$$

$$\text{For } \tau \geq 0: \quad \text{at } x = 0, \frac{\partial \theta_x}{\partial x} = \frac{h}{k} \theta_x,$$

$$\text{at } x = D, \frac{\partial \theta_x}{\partial x} = -\frac{h}{k} \theta_x.$$

$$\frac{1}{\alpha} \frac{\partial \theta_y}{\partial \tau} = \frac{\partial^2 \theta_y}{\partial y^2}.$$

$$\text{At } \tau = 0: \quad \theta_y = F_2(y). \tag{4.76}$$

$$\text{For } \tau \geq 0: \quad \text{at } y = 0, \frac{\partial \theta_y}{\partial y} = \frac{h}{k} \theta_y,$$

$$\text{at } y = W, \frac{\partial \theta_y}{\partial y} = -\frac{h}{k} \theta_y.$$

The above shows that the solution to the two-dimensional, transient conduction problem set forth in Eqs. (4.69) and (4.70) may be obtained by taking the product of the two simpler one-dimensional problems given in Eqs. (4.75) and (4.76) when the initial temperature distribution is expressible in the factored form shown in Eq. (4.71). The case of a constant initial temperature distribution, as discussed in the two preceding sections, is of this form.

The last two sets of equations, Eqs. (4.75) and (4.76), are seen to be identical in form to the equations which govern the transient conduction of heat in the infinite plane slab discussed in Sec. 4.5. Thus, the solution to the problem of transient heat conduction in the rectangular bar depicted in Fig. 4.18(a) is obtained by taking the product of the two solutions for the two infinite slabs whose intersection forms the bar in question. This is illustrated in Fig. 4.18(b). For the instance of the rectangular bar heated initially to a uniform temperature, the solutions (and graphical results) obtained in Sec. 4.5 for plane slab initially at a uniform temperature may be utilized directly. It should be noted that the Biot and Fourier numbers for each of the two slabs making up the bar will be based on different nondimensionalizing factors unless the bar is square. It should also be remembered that the length L used in the formulas and charts of Sec. 4.5 is the half-width of the slab, so that L would have to be replaced by $D/2$ or $W/2$ before the results are applied to the problem set forth in this section.

The product superposition principle illustrated above for two-dimensional transient conduction in a rectangular bar can be extended to other configurations. Of particular practical value are the two configurations illustrated in Figs. 4.19 and 4.20. The solution for a parallelepiped of infinite dimensions may be obtained as the product of three infinite slabs. The solution for the finite solid circular cylinder may be obtained as the product of the solution for an infinite slab (Sec. 4.5) and the solution for a solid circular cylinder or infinite length (Sec. 4.6).

The product superposition principle described in this section is applicable only to cases in which the initial, dimensionless, temperature distribution can be factored into a form which is the product of functions, each involving only one of the independent space variables. The practical case of a uniform initial temperature is of this form and is the most frequently encountered case in which the product superposition principle is applied. The discussion of this chapter

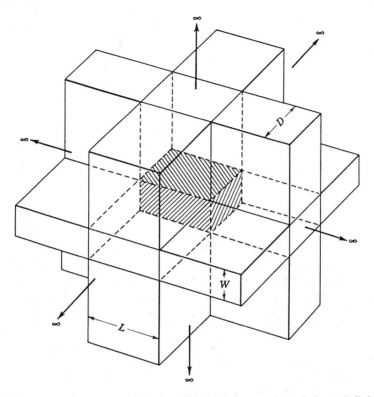

FIGURE 4.19. Parallelepiped formed by the intersection of three infinite slabs.

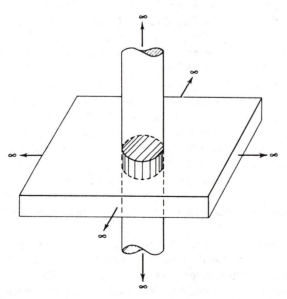

FIGURE 4.20. Finite cylinder formed by the intersection of an infinite cylinder and an infinite slab.

assumed that convection into a medium at constant temperature was taking place at the surface. The same superposition principle may be illustrated for cases in which the body surface temperature is suddenly changed to a constant value. In either instance the product superposition is valid only in the dimensionless form.

EXAMPLE 4.4

An aluminum cylinder has a diameter of 5 cm and a length of 10 cm. It is initially at a uniform temperature of 200°C. It is plunged into a quenching bath at 10°C with a heat transfer coefficient of 530 W/m²-°C. What is the temperature on the centerline of the cylinder, 2.5 cm from one end, after 1 min has passed?

Solution. For aluminum, Table A.1 gives

$$\alpha = 8.42 \times 10^{-5} \text{ m}^2/\text{s}, \quad k = 215 \text{ W/m-°C}.$$

The product superposition principle is applicable here, applied as the product of the solution of a slab of half-width $L = 10/2$ cm and a cylinder of $R = 5/2$ cm. The two solutions needed are

Slab: $L = 5$ cm, $x = 2.5$ cm.

$$\text{Thus, } \xi = \frac{x}{L} = 0.5,$$

$$\text{Fo} = \frac{8.42 \times 10^{-5} \times 1 \times 60}{(0.05)^2} = 2.02,$$

$$\text{Bi} = \frac{530 \times 0.05}{215} = 0.123.$$

Fig. 4.11: $\left(\dfrac{t - t_f}{t_i - t_f}\right)_{\xi=0} = 0.80.$

Fig. 4.12: $\dfrac{(t - t_f)_{\xi=0.25}}{(t - t_f)_{\xi=0}} = 0.98.$

$$\text{Thus, } \left[\frac{(t - t_f)}{(t_i - t_f)}\right]_{\text{slab}} = 0.80 \times 0.98 = 0.78.$$

Cylinder: $R = 2.5$ cm, $r = 0$.

$$\text{Thus, } \rho = \frac{r}{R} = 0,$$

$$\text{Fo} = \frac{8.42 \times 10^{-5} \times 1 \times 60}{(0.025)^2} = 8.08$$

$$\text{Bi} = \frac{530 \times 0.025}{215} = 0.0615$$

Fig. 4.15: $\left(\dfrac{t - t_f}{t_i - t_f}\right)_{\rho=0} = 0.381.$

4.7 TRANSIENT CONDUCTION IN MORE THAN ONE DIMENSION

So, for the finite cylinder at $\xi = 0.5$, $\rho = 0$,

$$\frac{t - t_f}{t_i - t_f} = 0.78 \times 0.381 = 0.297.$$

$$t = 101.6°C \ (214.9°F). \qquad \blacksquare$$

THE TRANSIENT RESPONSE OF BODIES WITH NEGLIGIBLE INTERNAL RESISTANCE

As pointed out in Sec 4.4 when the Biot number was introduced, it could be interpreted as the ratio of the internal resistance of a body, l/k, to the external surface resistance, $1/h$. The cases of specified surface temperature discussed in Secs. 4.5 and 4.6 may be taken as limiting cases for $\text{Bi} \rightarrow \infty$. The other solutions given in these sections were applicable for the finite Bi. This section will consider cases in which the internal resistance is negligible and $\text{Bi} \rightarrow 0$. Such a case may arise if the body has a high-enough thermal conductivity, compared with the surface film coefficient, that the interior temperature of the body may be taken as uniform at any instant of time. Thus, the entire temperature–time history of the body is regulated by the surface resistance. Under these circumstances, a heat balance on the body yields

$$hA_s(t - t_f) = -\rho V c_p \frac{dt}{d\tau}. \qquad (4.77)$$

In Eq. (4.77) A_s and V represent the exposed surface area and volume of the body, respectively. The density and specific heat of the body are denoted by ρ and c_p, while h represents the surface heat transfer coefficient between the body and an ambient fluid at temperature t_f.

Equation (4.77) may be rewritten as

$$\frac{d(t - t_f)}{t - t_f} = -\frac{hA_s}{\rho c_p V} d\tau.$$

Let it be presumed that the body is initially at a temperature t_i before it is thrust into the fluid at temperature t_f. Integration of the equation above and application of the initial condition that $t = t_i$ at $\tau = 0$ yields the following expression for the subsequent temperature of the body t, as a function of time:

$$\frac{t - t_f}{t_i - t_f} = e^{-(hA_s/\rho c_p V)\tau}$$

$$= e^{-\tau/\varphi}. \qquad (4.78)$$

The parameter

$$\varphi = \frac{\rho c_p V}{hA_s} \qquad (4.79)$$

is termed the *time constant* of the body—the larger the value of φ, the slower the body is to respond to the change in temperature. For given thermal properties, the time constant φ is proportional to the ratio V/A_s; thus, the smaller the surface area of a body, compared with its volume, the slower it will respond.

The instantaneous rate of heat flow between the body and the ambient fluid is given by

$$q = hA_s(t - t_f),$$

or, in dimensionless form,

$$\frac{q}{hA_s(t_i - t_f)} = e^{-\tau/\varphi}. \tag{4.80}$$

The cumulative heat flow between time $\tau = 0$ and $\tau = \tau$ is

$$Q = \int_0^\tau q \, d\tau,$$

so Eq. (4.80) leads to

$$\frac{Q}{\rho V c_p(t_i - t_f)} = 1 - e^{-\tau/\varphi}. \tag{4.81}$$

Equations (4.78), (4.80), and (4.81) afford simple means for the rapid determination of the temperature response of a body as long as the basic assumption of a uniform interior temperature may be made.

EXAMPLE 4.5

A 1-in.-diameter pure copper sphere and a 1-in. pure copper cube are each heated to 1200°F. They are then placed in air at 200°F with a surface heat transfer coefficient of 12 Btu/h-ft²-°F. Find the temperature of each body after a lapse of 5 min.

Solution.   Table A.1 gives, for copper, $\rho = 8954$ kg/m³, $c_p = 0.383$ kJ/kg-°C. According to Appendix F,

$$\rho = \frac{8954}{16.018} = 559 \text{ lb}_m/\text{ft}^3,$$

$$c_p = \frac{0.383}{4.1868} = 0.0915 \text{ Btu/lb}_m\text{-°F}.$$

For the sphere $V/A_s = \frac{4}{3}\pi r^3/4\pi r^2 = r/3 = \frac{1}{3}$ in. For the cube $V/A_s = d^3/6d^2 = d/6 = \frac{1}{6}$ in. Thus, the time constants are

$$\text{Sphere: } \varphi = \frac{559 \times 0.0915}{12 \times 3 \times 12} = 0.118 \text{ h.}$$

$$\text{Cube: } \varphi = \frac{559 \times 0.0915}{12 \times 6 \times 12} = 0.059 \text{ h.}$$

Thus, at $\tau = 5$ min $= \frac{1}{12}$ h, for the sphere,

$$\frac{\tau}{\varphi} = \frac{\frac{1}{12}}{0.118} = 0.706$$

$$\frac{t - 200}{1200 - 200} = e^{-0.706} = 0.492$$

$$t = 692°F \ (368°C).$$

Similarly, for the cube,

$$\frac{\tau}{\varphi} = 1.412$$

$$\frac{t - 200}{1200 - 200} = 0.244$$

$$t = 444°F \ (229°C). \qquad \blacksquare$$

REFERENCES

1. JAHNKE, E., and F. EMDE, *Tables of Functions*, 4th ed., New York, Dover, 1945.
2. CARSLAW, H. S., and J. C. JAEGER, *Conduction of Heat in Solids*, New York, Oxford U.P., 1957.
3. ÖZISIK, M. N., *Heat Conduction*, New York, Wiley, 1980.
4. ROHSENOW, W. M., and J. P. HARTNET, eds., *Handbook of Heat Transfer*, New York, McGraw-Hill, 1973.
5. HEISLER, M. P., "Temperature Charts for Induction and Constant-Temperature Heating," *Trans. ASME*, Vol. 69, No. 3, 1947, pp. 227–236.

PROBLEMS

4.1 Write the solution for the steady state temperature distributions in a rectangular plate which is infinite in extent in the y direction, of width L in the x direction, has temperature 0 imposed at $y = \infty$, $x = 0$ and $x = L$, and has a temperature distribution imposed on its edge at $y = 0$ as given by

$$t_{y=0} = T \sin\left(\frac{\pi x}{L}\right), \qquad 0 \leq x \leq L.$$

T is a constant.

4.2 A rectangular plate has the dimensions of 15 cm $\times$ 25 cm and has its edges maintained at 0°C except at one of the 15 cm edges. On this edge the temperature is maintained at

$$t = (100°C) \sin\left(\frac{\pi x}{15}\right), \qquad 0 \leq x \leq 15 \text{ cm}.$$

Draw the isothermal lines within the plate for 100°C, 90°C, 80°C, . . . , 10°C.

4.3　A rectangular plate 15 cm × 20 cm has its two 20 cm sides maintained at 100°C, one of its 15 cm sides maintained at 300°C, and its other 15 cm side maintained at 500°C. Find the temperature at the center of the plate.

4.4　A solid circular cylinder (outside radius R, finite length L) has its two circular ends maintained at 0°C. On the cylindrical surface the temperature is maintained at a distribution that depends *only* on the axial coordinate z. Denote this distribution by $f(z)$. Show that the steady state solution for the temperature distribution is

$$t = \frac{2}{L} \sum_{n=1}^{\infty} \frac{I_0(\lambda_n r)}{I_0(\lambda_n R)} \sin \lambda_n z \int_0^L f(z) \sin \lambda_n z \, dz, \qquad \lambda_n = \frac{n\pi}{L}.$$

4.5　A slab, 25 cm thick and infinite in extent in the other directions, is initially heated to a uniform temperature of 300°C. It then has its surface temperature suddenly reduced to and maintained at 50°C. If the material has a diffusivity of 5.7×10^{-5} m²/s, plot the variation of the centerline temperature and the heat removed, per unit area, as functions of time from 0 to 10 min. Take $k = 85$ W/m-°C.

4.6　A slab of 1.5% carbon steel, 2.5 cm thick and infinite in extent in the other directions, is heated to 980°C and then quenched in an oil bath at 90°C. If the convective heat transfer coefficient is 570 W/m²-°C, find the time required to reduce the temperature at the center of the slab to 425°C. What is the temperature at a depth of 0.5 cm from the surface at this time? How much heat, per unit area, has been removed from the slab up to this time?

4.7　Plot the time variation of the surface temperature of the slab in Prob. 4.6 from 0 to 200 s.

4.8　Large slabs of steel ($\rho = 8000$ kg/m³, $k = 17.3$ W/m-°C, $c_p = 0.194$ kJ/kg-°C) are to be annealed by placing them in an oven in which the ambient air temperature is 980°C and the heat transfer coefficient is 115 W/m²-°C. The slabs are at a uniform temperature of 150°C before being placed in the oven. It is desired to raise the *average* temperature of the slabs to 780°C, but the surface temperature must not be allowed to rise above 900°C. Find the maximum thickness of slabs that can be processed in this manner. How long does the annealing process take?

4.9　A large sheet of material ($\rho = 3200$ kg/m³, $c_p = 0.837$ kJ/kg-°C, $k = 19$ W/m-°C) is 5.0 cm thick and has an initial uniform temperature of -30°C. The temperature of the surrounding air is suddenly raised to 20°C, with a surface heat transfer coefficient of 8.5 W/m²-°C. How much time will elapse before the centerline temperature becomes 10°C?

4.10　A concrete wall 5 in. thick and insulated on the inside is initially at equilibrium with the surrounding ambient air at 80°F on the outside. The outside air temperature suddenly drops to 60°F. A heat transfer coefficient of 1.5 Btu/h-ft²-°F exists at the outside surface. Find the temperature of the inside and outside surfaces after 8 h and 24 h have elapsed.

4.11 An infinite slab of thickness L in the x direction is initially heated to a known temperature distribution, $f(x)$. If both surfaces normal to the x direction are then insulated so that they are impervious to heat flow, write the equation for the distribution of the temperature in the slab as a function of time. If the initial distribution $f(x)$ is a constant t_i, to what does the solution just obtained reduce?

4.12 A cylindrical bar of stainless steel (18% Cr, 8% Ni) 10 cm in diameter is removed from an annealing furnace where it has been maintained at 980°C. It is placed in air at 30°C to cool. If the surface heat transfer coefficient is 11 W/m²-°C, what is its centerline temperature after 2 h has passed?

4.13 A cylindrical bar of Duralumin is chilled to -100°C and then heated in an atmosphere at 38°C with a heat transfer coefficient of 140 W/m²-°C. If the bar is 2.5 cm in diameter, when does the surface temperature reach 15°C?

4.14 A 12-cm-diameter bar ($k = 16$ W/m-°C, $\alpha = 3.35 \times 10^{-5}$m²/s) is initially at 650°C. It is placed to cool in air at 40°C through a heat transfer coefficient of 68 W/m²-°C. How long will it be until the surface reaches 340°C? What will the center temperature be at this time?

4.15 A solid cylinder of radius R and infinite length is initially heated to a known temperature distribution which depends on the radial coordinate, r, only. Find the temperature in the cylinder as a function of r and time, if the surface is insulated.

4.16 Plot the temperature–time history for the geometric center of a steel bar ($k = 43$ W/m-°C, $\alpha = 1.172 \times 10^{-5}$ m²/s) that is 7.5 cm in diameter and 7.5 cm long. The cylinder is initially at 820°C and allowed to cool in air at 40°C with a heat transfer coefficient of 280 W/m²-°C. Let time range from 0 to 1200 s.

4.17 For the cylinder described in Prob. 4.16, find, after 4 min has passed, the temperature at the geometric center of the cylinder, at the center of the circular ends, and at the midpoint of the lateral cylindrical side.

4.18 A common brick (5.7 cm × 8.9 cm × 20 cm) is fired in a kiln at 1425°C. It is allowed to cool in air at 40°C through a heat transfer coefficient of 30 W/m²-°C. How long does it take for the surface temperature at the center of one of the 8.9 × 20 sides to reach 65°C?

4.19 A 4-in. × 4-in. wood timber is initially at 75°F. It is suddenly exposed to flames at 1000°F through a heat transfer coefficient of 3.0 Btu/h-ft²-°F. If the ignition temperature of the wood is 900°F, how much time will elapse before any portion of the timber starts burning? For the wood use $\rho = 50$ lb$_m$/ft³, $c_p = 0.6$ Btu/lb$_m$-°F, $k = 0.2$ Btu/h-ft-°F.

4.20 A very long copper cylinder, 5 cm in diameter, is heated to 30°C and then plunged into a liquid bath at 0°C. If, after 3 min, the cylinder center temperature is 4°C, determine the average heat transfer coefficient at the surface if the internal thermal resistance is neglected. Then find the Biot number and justify the neglection of the internal resistance.

4.21 A fine wire (0.07 cm diameter) is initially heated to 175°C. It is suddenly exposed to an environment at 40°C through a heat transfer coefficient of 60 W/m²-°C. After first calculating the Biot number to establish if the internal resistance can be neglected, find the wire temperature and the heat loss (per unit length) after 10 s if the wire is made of (a) copper, or (b) aluminum.

4.22 A plate 1 in. thick is made of 10% nickel steel. It is heated to some initial temperature and then exposed to air at 100°F through a heat transfer coefficient of $h = 2$ Btu/h-ft²-°F. After first justifying neglection of internal resistance, find the initial temperature of the plate if its temperature after 10 min is 1000°F.

CHAPTER 5

Numerical Methods for Heat Conduction

INTRODUCTORY REMARKS

The problems of heat conduction treated in Chapters 3 and 4 have stressed the use of mathematical methods. This approach, which treats the conducting body as a continuum, yields much information of a general nature concerning the particular problem being treated. By the successful application of analysis one can ascertain the temperature at *any* point, at *any* time, within the given body. Since the results are given in closed, analytical form, it is possible to deduce much useful information—the effect of various parameters, the effect of altering the body geometry, etc.

Although a great number of problems have been solved analytically, only a limited number of relatively simple geometrical shapes (e.g., cylinders, spheres, infinite slabs, etc.) can be handled. Also, only those boundary conditions which can be easily expressed mathematically may usually be applied. There are many heat conduction problems of considerable practical value for which no analytical solution is feasible. These problems usually involve geometrical shapes of a mathematically inconvenient sort. The absence of an analytical solution does not remove the need of an answer, and some other approach must be sought.

Numerical techniques exist which are able to handle almost any problem of any degree of complexity. The detail and accuracy of the answer obtained depends mainly upon the amount of effort one wishes to expend. The various numerical methods all yield *numerical* values for temperatures at selected, *discrete,* points within the body being considered and only at *discrete* time intervals. Thus answers are obtained only for a given set of conditions, a given set of discrete points and discrete time intervals. One must give up the generality of the analytical solution in order to obtain an answer.

Several different techniques of numerical analysis of heat conduction problems exist. The references at the end of the chapter may be consulted in this regard. Some of the techniques developed have been based on the use of hand calculations or desk calculators. The availability of high speed digital computers has considerably reduced the value of these methods. The formulations presented and the techniques recommended are not the most efficient or concise from the standpoint of hand or desk calculator solution; they are presented in forms most suitable for computer programming.

Both steady state and transient methods are treated. Formulations of steady state numerical methods are not presented in the most concise fashion for desk calculation, but rather are formulated in a way that carries over most to the transient methods.

The numerical methods presented here are all based on the representation of the derivatives in the heat conduction equation by finite difference approximations. In recent years a new and particularly powerful technique called the "finite element method" has been developed for the numerical analysis of heat conduction problems. A useful explanation of this technique is too lengthy for this text, and the reader is referred to Ref. 1 for a detailed treatment.

Attention is first directed toward the development of finite difference approximations to the heat conduction equation and the degree of accuracy of these approximations. On the basis of the finite difference formulation, an electrical network analogy becomes apparent. Generalizations to complex problems are then made on the basis of this analogy because of its greater physical appeal.

5.2

FINITE DIFFERENCE APPROXIMATIONS

The basic principle of the numerical approach to a heat conduction problem is the replacement of the differential equation for the continuous temperature distribution in a heat conducting solid by a finite difference equation which must be satisfied at only certain points in the solid. The relation of the finite difference expression to the differential equation can be understood best by deriving the latter from the former, via the use of a Taylor's expansion. Consider, then, a function of two independent variables:

$$f = f(\xi, \eta). \tag{5.1}$$

The general variables ξ and η have been used since the results will be applied to spatial variable (x, y, and z) as well as time (τ).

Let h_1 represent an increment in the variable ξ. A *forward* expansion of the function at $\xi = \xi + h_1$, $\eta = \eta$, in terms of its value at $\xi = \xi$, $\eta = \eta$, is

$$f(\xi + h_1, \eta) = f(\xi, \eta) + h_1\left(\frac{\partial f}{\partial \xi}\right)_{\xi,\eta} + \frac{h_1^2}{2}\left(\frac{\partial^2 f}{\partial \xi^2}\right)_{\xi,\eta}$$

$$+ \frac{h_1^3}{6}\left(\frac{\partial^3 f}{\partial \xi^3}\right)_{\xi,\eta} + \mathcal{O}(h_1^4). \tag{5.2}$$

The notation $\mathcal{O}(h_1^4)$ is used to indicate that subsequent terms are of the order of h_1^4, and *higher*. If terms of the order of h_1^2, and greater, are neglected, the following *forward* finite difference approximation to the first derivative is obtained:

$$\left[\left(\frac{\partial f}{\partial \xi}\right)_{\xi,\eta}\right]_{fwd} \approx \frac{1}{h_1}[f(\xi + h_1, \eta) - f(\xi, \eta)]. \tag{5.3}$$

Similarly, for a forward displacement of h_2 in η, the forward finite difference approximation for the first derivative is

$$\left[\left(\frac{\partial f}{\partial \eta}\right)_{\xi,\eta}\right]_{fwd} \approx \frac{1}{h_2}[f(\xi, \eta + h_2) - f(\xi, \eta)]. \tag{5.4}$$

In like fashion, *backward* finite difference expressions for the first derivatives may be obtained by writing Taylor's expansions for an increment in ξ of $-h_1$ or an increment in η of $-h_2$:

$$f(\xi - h_1, \eta) = f(\xi, \eta) - h_1\left(\frac{\partial f}{\partial \xi}\right)_{\xi,\eta} + \frac{h_1^2}{2}\left(\frac{\partial^2 f}{\partial \xi^2}\right)_{\xi,\eta}$$

$$- \frac{h_1^3}{6}\left(\frac{\partial^3 f}{\partial \xi^3}\right)_{\xi,\eta} + \mathcal{O}(h_1^4). \tag{5.5}$$

Again neglecting terms of order h_1^2, and greater, one obtains

$$\left[\left(\frac{\partial f}{\partial \xi}\right)_{\xi,\eta}\right]_{bkwd} \approx \frac{1}{h_1}[f(\xi, \eta) - f(\xi - h_1, \eta)]. \tag{5.6}$$

Also,

$$\left[\left(\frac{\partial f}{\partial \eta}\right)_{\xi,\eta}\right]_{bkwd} \approx \frac{1}{h_2}[f(\xi, \eta) - f(\xi, \eta - h_2)]. \tag{5.7}$$

The *central* finite difference approximation to the second derivatives may be found by adding Eqs. (5.2) and (5.5):

$$f(\xi + h_1, \eta) + f(\xi - h_1, \eta) = 2f(\xi, \eta) + h_1^2\left(\frac{\partial^2 f}{\partial \xi^2}\right)_{\xi,\eta} + \mathcal{O}(h_1^4). \tag{5.8}$$

When terms of the order of h_1^4, and greater, are neglected,

$$\left[\left(\frac{\partial^2 f}{\partial \xi^2}\right)_{\xi,\eta}\right]_{cent} \approx \frac{1}{h_1^2}[f(\xi + h_1, \eta) - 2f(\xi, \eta) + f(\xi - h_1, \eta)]. \tag{5.9}$$

Also,

$$\left[\left(\frac{\partial^2 f}{\partial \eta^2}\right)_{\xi,\eta}\right]_{cent} \approx \frac{1}{h_2^2}[f(\xi, \eta + h_2) - 2f(\xi, \eta) + f(\xi, \eta - h_2)]. \quad (5.10)$$

It is worth noting that the first derivative finite difference approximations neglect terms of the order of h^2, while the second derivative approximations neglect terms of the order of h^4. It should be pointed out that other approximations to the second derivative may be written (e.g., forward, backward, etc.). However, the central difference given here is most commonly used for heat conduction analyses.

Section 5.3 applies these approximations to steady state heat conduction, and subsequent sections illustrate their solutions. Section 5.7 and subsequent sections apply the difference equations to the case of transient conduction.

5.3

STEADY STATE NUMERICAL METHODS

From Chapter 1 the steady state conduction equation is known to be

$$k\left(\frac{\partial^2 t}{\partial x^2} + \frac{\partial^2 t}{\partial y^2} + \frac{\partial^2 t}{\partial z^2}\right) + q^* = 0. \quad (5.11)$$

The thermal conductivity, k, is left as a factor in Eq. (5.11) since this will be a useful form when extensions are made to nonsteady conduction. Also, reference to the derivation of the heat conduction equation in Sec. 1.4 reveals that the expression given on the left side of Eq. (5.11) represents the net rate, *per unit volume,* at which heat is stored at a point in the conducting body, and q^* represents the internal generated heat, *per unit volume.*

Consider, first, the one-dimensional case:

$$k\frac{d^2 t}{dx^2} + q^* = 0. \quad (5.12)$$

Figure 5.1 depicts a slender bar whose lateral faces are insulated. Thus, all heat conduction occurs in the x direction and Eq. (5.12) applies. If, however, the

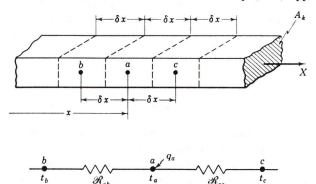

FIGURE 5.1. Finite difference approximation and corresponding nodal network for one-dimensional conduction.

second derivative in Eq. (5.12) is replaced by the finite difference approximation in Eq. (5.9) (with the coordinate x replacing ξ, the space increment δx replacing h_1, and the temperature t replacing f), the following expression results, which relates the temperatures at x, $x + \delta x$, and $x - \delta x$:

$$\frac{k}{(\delta x)^2}[t(x + \delta x) - 2t(x) + t(x - \delta x)] + q_x^* = 0.$$

In keeping with the central difference approximation used, the internally generated heat is evaluated at the central point x.

If, as suggested in Fig. 5.1, the points at x, $x - \delta x$, and $x + \delta x$ are denoted by a, b, c, respectively, the equation above is expressed more simply as

$$\frac{k}{(\delta x)^2}(t_b - 2t_a + t_c) + q_a^* = 0. \tag{5.13}$$

Equation (5.13) is the result of a Taylor's expansion of the heat conduction equation around the point a, and should thus be interpreted as a relation for the temperature t_a in terms of the temperatures at the neighboring points, t_b and t_c. A similar expression may be written for the point b—in terms of t_a and the temperature of the point δx to the left of b. Likewise, an expression may be written for point c, involving t_c, t_a, and the point δx to the right of c. Thus, if the body is subdivided into n distinct points, n such equations may be written. This set of n equations may then be solved for the temperatures at the n points presuming the q's are known. Consequently, one obtains the temperatures at these specific points without any consideration of the temperatures between them. The development of Eq. (5.9), upon which Eq. (5.13) is based, showed that the accuracy of the results obtained will increase with decreasing δx (i.e., by increasing the number of points), and, in fact, the error will be of the order of $(\delta x)^4$. Methods of solving the set of equations resulting from the application of Eq. (5.13) at each body point will be discussed in Sec. 5.6.

The representation just developed for the one-dimensional case may be extended to the two-dimensional and three-dimensional cases. Each term in Eq. (5.11) may be approximated in identical form to Eq. (5.13). The two-dimensional case is illustrated in Fig. 5.2. A plate of unit thickness, which is insulated on its plane faces, is subdivided in the x direction by increments of δx and in the y direction by increments of δy. The resulting finite difference approximation for the temperature at point a in terms of the temperatures at the neighboring points b, c, d, and e is

$$\frac{k}{(\delta x)^2}(t_b - 2t_a + t_c) + \frac{k}{(\delta y)^2}(t_d - 2t_a + t_e) + q_a^* = 0. \tag{5.14}$$

As before, such an equation is to be satisfied at each point in the plate—in terms of its four neighboring points. A body divided into an $n \times m$ network will require the solution of $(n \times m)$ such equations. Quite apparently, considerable simplification will result if one chooses $\delta x = \delta y$. This could prove to be valuable if hand calculations were to be employed; however, it is of little consequence when digital computers are used.

The three-dimensional case is not illustrated, but it should be apparent that

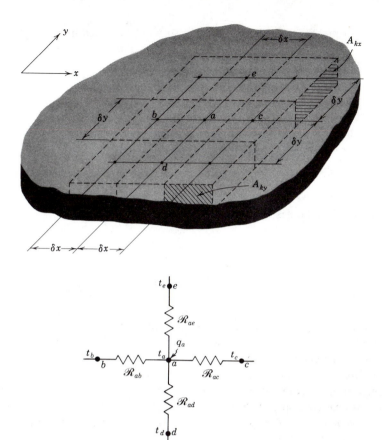

FIGURE 5.2. Finite difference approximation and corresponding nodal network for two-dimensional conduction.

at each point in a three-dimensional body one must satisfy an equation of the form.

$$\frac{k}{(\delta x)^2}(t_b - 2t_a + t_c) + \frac{k}{(\delta y)^2}(t_d - 2t_a + t_e)$$

$$+ \frac{k}{(\delta z)^2}(t_f - 2t_a + t_g) + q_a^* = 0. \tag{5.15}$$

The points f and g are the neighboring points in the z direction—at a spacing of δz.

5.4

NODAL NETWORK REPRESENTATIONS FOR THE STEADY STATE

The finite difference equations presented in the preceding section may be interpreted in a physical way that will permit rather convenient generalizations.

Consider first the one-dimensional case. It will be recalled from Chapter 3 that for conduction through a slab of thickness Δx, a *unit* thermal resistance could be defined,

$$R = \frac{\Delta x}{k},$$

so the heat *flux* through the slab would be

$$\frac{q}{A} = \frac{t_1 - t_2}{R_{12}}.$$

A *total* thermal resistance, $\mathcal{R}$, may be defined which gives the total heat flow, q, rather than the heat flux:

$$\mathcal{R} = \frac{\Delta x}{kA_k}, \tag{5.16}$$

$$q = \frac{t_1 - t_2}{\mathcal{R}_{12}}.$$

The subscript on the area symbol A_k is used to denote the fact that the area considered is the *conduction* area normal to the direction of heat flow.

Now, by reference to Fig. 5.1, imagine the one-dimensional bar to be divided into *lumps* of length δx so located that the nodal points a, b, and c lie at the center of these lumps. The area A_k is the cross-sectional area normal to the one-dimensional coordinate x. Since the left side of Eq. (5.13) represents the heat flow *per unit volume*, multiply it by the volume of the lump surrounding point a, namely $(A_k \times \delta x)$:

$$\frac{kA_k}{\delta x}(t_b - 2t_a + t_c) + q_a = 0,$$

where

$$q_a = q_a^*(A_k \delta x)$$

is the total heat generated, per unit time, in the lump volume around point a. The term $\delta x/kA_k$ is seen to be the thermal resistance between point a and either b or c, so the above expression may be written

$$\frac{t_b - t_a}{\mathcal{R}} + \frac{t_c - t_a}{\mathcal{R}} + q_a = 0, \tag{5.17}$$

$$\mathcal{R} = \frac{\delta x}{kA_k}.$$

This latter expression suggests the electrical network analogy identified earlier and illustrated in Fig. 5.1. The lumps into which the conducting body is divided are presumed to be of uniform temperature—equal to the temperature of the node it surrounds. In the network representation, these isothermal lumps are symbolized by the nodal points a, b, and c, and the thermal resistances are symbolized by the electrical resistors shown connecting the nodes. The expression given in Eq. (5.17) is simply a statement of the total heat flow into the node a through the resistances connecting this node to all neighboring nodes. The lump temperatures applied to the nodes are analogous to electrical potentials in the network representation, and the heat balance of Eq. (5.17) is the sum of the currents flowing into a node if q_a is interpreted as a current flow into node a from some external potential source. In the steady state this summation must be zero.

In a similar fashion, the two-dimensional plate, of unit thickness, pictured in Fig. 5.2 may be subdivided into lumps $\delta x \times \delta y \times 1$ in volume, centered on the nodal points $a, b, \ldots, e$. The finite difference expression given in Eq. (5.14) reduces to

$$\frac{t_b - t_a}{\mathscr{R}_x} + \frac{t_c - t_a}{\mathscr{R}_x} + \frac{t_d - t_a}{\mathscr{R}_y} + \frac{t_e - t_a}{\mathscr{R}_y} + q_a = 0,$$

$$\mathscr{R}_x = \frac{\delta x}{kA_{kx}}, \qquad A_{kx} = \delta y \times 1, \qquad (5.18)$$

$$\mathscr{R}_y = \frac{\delta y}{kA_{ky}}, \qquad A_{ky} = \delta x \times 1.$$

The equivalent nodal network is shown in Fig. 5.2. The three-dimensional case can be treated similarly, and the resultant expression should be apparent.

All the above results may be generalized into a single expression. For a nodal network representation of any steady state problem, if a thermal resistance is provided between a given node and each of its neighbors with which it conducts heat, each node i must satisfy the equations

$$\sum_j \frac{t_j - t_i}{\mathscr{R}_{ij}} + q_i = 0,$$

$$\mathscr{R}_{ij} = \frac{\delta_{ij}}{kA_{kij}}. \qquad (5.19)$$

In Eq. (5.19) j denotes all neighboring nodes connected to node i, δ_{ij} denotes the conduction distance between node i and node j, A_{kij} is the cross-sectional area for heat conduction normal to δ_{ij}, and q_i is the heat generated in or added to the volume lump at i. This representation permits the inclusion of unequal nodal spacings in any particular coordinate direction. Particular attention must then be paid to the evaluation of the various resistors, since a different conduction distance, δ_{ij}, would be involved in each case. Similarly, the volume of the lump and the conduction area associated with each node might be different. As will be seen later, the volume of the body lump associated with each node will become of considerable significance in the nonsteady case. In general, it is desirable to use equal net spacings, particularly if hand calculations are used, since all resistances become identical and can be eliminated from Eq. (5.19) altogether. In the instance of equal net spacings and zero heat addition, the two-dimensional case given by Eq. (5.18) reduces to

$$t_b + t_c + t_d + t_e - 4t_a = 0. \qquad (5.20)$$

With the use of digital computers, the advantage of equal net spacings becomes of less value.

The existence of convective losses from body lumps bounded by a free surface exposed to an ambient fluid may be accounted for by adding to the network another node representing the fluid. The node representing the body lump is connected to the node representing the fluid by a resistor whose value is found from the heat transfer coefficient (a unit conductance) according to

$$\mathcal{R} = \frac{1}{hA_c}. \tag{5.21}$$

Here A_c denotes the surface area of the body lump exposed to the convecting ambient fluid. The expression which must be satisfied at each point then takes the form: At each node i,

$$\sum_j \frac{t_j - t_i}{\mathcal{R}_{ij}} + q_i = 0, \tag{5.22}$$

$$\mathcal{R}_{ij} = \begin{cases} \dfrac{\delta_{ij}}{kA_{k_{ij}}} & \text{for conduction,} \\[2em] \dfrac{1}{h_{ij}A_{c_{ij}}} & \text{for convection.} \end{cases}$$

In Eq. (5.22) q_i represents the heat added at a node by means other than surface convection. In cases in which internal heat generation is present, the q_i's are known. In the event that the node under consideration is one subjected to a boundary condition of specified temperature, t_i is then known and q_i becomes an unknown—namely the heat flux necessary to maintain the node at the desired temperature. In the remaining discussions of this chapter no internal heat generation will be considered, and thus the latter instances of specified boundary temperature will be the only cases in which the q_i term must be included.

The convective resistance was included in order to write the general form of the equation as it is shown in Eq. (5.22). Actually, this convective condition is just one of several boundary conditions that could be encountered. Before illustrating the solution of the set of equations resulting from the application of Eq. (5.22) to actual networks, these boundary conditions must be discussed. A sample nodal network for a complex system is shown in Fig. 5.3 as an illustration of the application of the concepts discussed here.

5.5

BOUNDARY CONDITIONS

The discussion so far has been restricted mainly to points interior to the body in which the heat conduction is taking place. In order to apply the heat balances represented by Eq. (5.19) or (5.22), the imposed boundary conditions must also be satisfied. One such boundary condition, that of convection, has already been

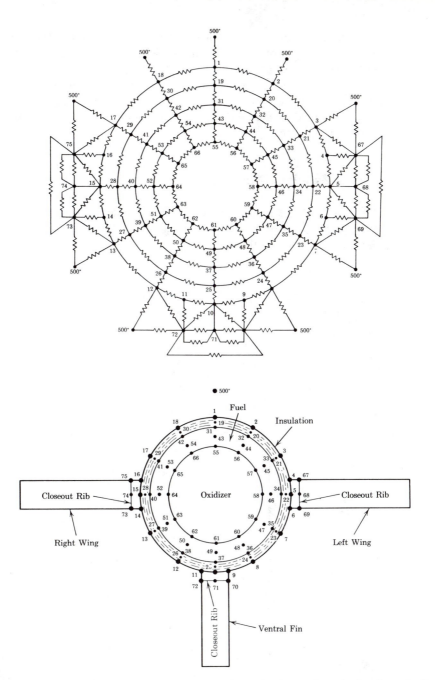

FIGURE 5.3. Complex nodal representation for the thermal analysis of a missile fuel tank. (From "Temperature Control Systems for Space Vehicles," *ASD Rep. TRD-62-493, Part II,* Air Force Flight Dynamics Laboratory, Wright-Patterson Air Force Base, Ohio, 1963.)

mentioned. Other typical conditions will be discussed in connection with two-dimensional conduction; however, the special form of these conditions for the one-dimensional case and their generalization to the three-dimensional case are sufficiently apparent to be left as exercises for the reader. Attention will first be directed to the instance in which regular network spacings (i.e., equal spacings in any one direction) are used, and the boundary coincides with a nodal point in this equal spacing.

Regular Spacing, Plane Boundary

Figure 5.4 illustrates a two-dimensional case in which nodal spacings of δx in the x direction and δy in the y direction are used. The conducting solid is presumed to have a unit thickness in the plane of the paper. A plane boundary

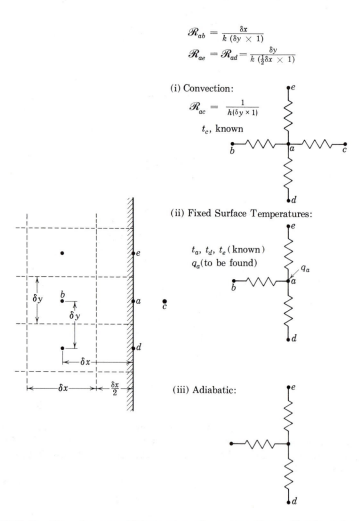

$$\mathscr{R}_{ab} = \frac{\delta x}{k\,(\delta y \times 1)}$$

$$\mathscr{R}_{ae} = \mathscr{R}_{ad} = \frac{\delta y}{k\,(\tfrac{1}{2}\delta x \times 1)}$$

(i) Convection:

$$\mathscr{R}_{ac} = \frac{1}{h(\delta y \times 1)}$$

t_c, known

(ii) Fixed Surface Temperatures:

t_a, t_d, t_e (known)
q_a (to be found)

(iii) Adiabatic:

FIGURE 5.4.   Boundary conditions at a plane, two-dimensional surface.

parallel to the y direction coincides with one column of nodal points. The limits of the volume lumps surrounding each node are shown. It is apparent that the lump volumes associated with the boundary nodes are one-half of the volume of a regular, interior node. Evaluation of the heat balances on node a will be considered.

The value of the conduction resistor connecting a and b is the same as for any other resistor connecting two interior points spaced in the x direction:

$$\mathcal{R}_{ab} = \frac{\delta_{ij}}{kA_{k_{ij}}} = \frac{\delta x}{k(\delta y \times 1)}. \tag{5.23}$$

For resistors $\mathcal{R}_{ad}$ and $\mathcal{R}_{ae}$, the conduction distance is δy, as for an interior resistance in the same direction; however, the conduction area between e and a or d and a is one-half of what it would be otherwise:

$$\mathcal{R}_{ad} = \mathcal{R}_{ae} = \frac{\delta y}{k(\frac{1}{2}\delta x \times 1)}. \tag{5.24}$$

The three resistances just discussed will remain unchanged regardless of the boundary condition to be imposed. The resistance $\mathcal{R}_{ac}$ (if any), the heat input at node a, etc., *will* depend on the imposed condition. Some conditions are listed below.

Convection to a Fluid of Known Temperature. If a fluid of known temperature is in contact with the body surface, and if the heat transfer coefficient, h, for convective heat transfer between the solid surface and the fluid is known, the fluid may be represented by a node—call it c—maintained at the known fluid temperature. Then the resistance connecting a and c is

$$\mathcal{R}_{ac} = \frac{1}{hA_c} = \frac{1}{h(\delta y \times 1)}. \tag{5.25}$$

The application of the heat balance in Eq. (5.22) is, at node a:

$$\frac{t_b - t_a}{\mathcal{R}_{ab}} + \frac{t_d - t_a}{\mathcal{R}_{ad}} + \frac{t_e - t_a}{\mathcal{R}_{ae}} + \frac{t_c - t_a}{\mathcal{R}_{ac}} = 0. \tag{5.26}$$

The $\mathcal{R}$'s are given in Eqs. (5.23) through (5.25). The applicable network is shown in Fig. 5.4. For a square network ($\delta x = \delta y = \delta$) this equation reduces to

$$t_b + \tfrac{1}{2}(t_d + t_e) + \frac{h\delta}{k} t_c - \left(2 + \frac{h\delta}{k}\right) t_a = 0. \tag{5.27}$$

No heat balance is necessary at node c, since its temperature is known.

Specified Surface Temperature. A temperature distribution may be specified and maintained at the boundary. If so, the three temperatures t_a, t_d, and t_e are known. The resistance between these nodes was given above. In order that the temperature at node a remain fixed, an unknown amount of heat must flow between the node and its surroundings. The network representation is shown and Eq. (5.22) yields, at node a:

$$\frac{t_b - t_a}{\mathcal{R}_{ab}} + \frac{t_d - t_a}{\mathcal{R}_{ad}} + \frac{t_e - t_a}{\mathcal{R}_{ae}} + q_a = 0. \tag{5.28}$$

Since t_a is known, the unknown quantities in the above equation are q_a and t_b. For a square network, Eq. (5.28) becomes

$$t_b + \tfrac{1}{2}(t_d + t_e) - 2t_a + \frac{q_a}{k} = 0. \tag{5.29}$$

If it is not desired to find the heat flux at a, this nodal equation can be eliminated completely from consideration.

If the surface is *isothermal*, $t_a = t_d = t_e$, so the nodal balance reduces to

$$\frac{t_b - t_a}{\mathcal{R}_{ab}} + q_a = 0, \tag{5.30}$$

or

$$t_b - t_a + \frac{q_a}{k} = 0 \tag{5.31}$$

for square networks.

An Adiabatic Boundary.   If the boundary is insulated, no interaction with the surroundings occurs, and node c is unnecessary. Thus,

$$\frac{t_b - t_a}{\mathcal{R}_{ab}} + \frac{t_d - t_a}{\mathcal{R}_{ad}} + \frac{t_e - t_a}{\mathcal{R}_{ae}} = 0, \tag{5.32}$$

where the $\mathcal{R}$'s are given above. For the square network

$$t_b + \tfrac{1}{2}(t_d + t_e) - 2t_a = 0. \tag{5.33}$$

Regular Spacing, Exterior Corner

Figure 5.5 illustrates a two-dimensional case in which the boundary of the solid is an exterior corner, and the regular nodal grid results in a node being located at the corner. In this instance the lump volume associated with nodes b and d is one-half the regular volume and that of node a is one-fourth that of an interior point. Without detailed discussion, the results are

$$\mathcal{R}_{ab} = \frac{\delta x}{k(\tfrac{1}{2}\delta y \times 1)}, \tag{5.34}$$

$$\mathcal{R}_{ad} = \frac{\delta y}{k(\tfrac{1}{2}\delta x \times 1)}. $$

Convection.   Using nodes c and e to represent the fluid (with $t_e = t_c$), one finds that

$$\mathcal{R}_{ae} = \frac{1}{h(\tfrac{1}{2}\delta x \times 1)}, \tag{5.35}$$

$$\mathscr{R}_{ab} = \frac{\delta x}{k\left(\frac{1}{2}\delta y \times 1\right)}$$

$$\mathscr{R}_{ad} = \frac{\delta y}{k\left(\frac{1}{2}\delta x \times 1\right)}$$

(i) Convection:

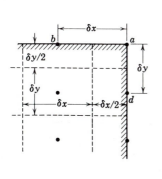

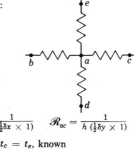

$$\mathscr{R}_{ae} = \frac{1}{h\left(\frac{1}{2}\delta x \times 1\right)} \qquad \mathscr{R}_{ac} = \frac{1}{h\left(\frac{1}{2}\delta y \times 1\right)}$$

$t_c = t_e$, known

(ii) Fixed Surface Temperatures:

t_a, t_b, t_d (known).
q_a (to be found) b

(iii) Adiabatic:

FIGURE 5.5.   Boundary conditions at a two-dimensional exterior corner.

$$\mathscr{R}_{ac} = \frac{1}{h\left(\frac{1}{2}\delta y \times 1\right)}.$$

Equation (5.26) still applies, with the $\mathscr{R}$'s now given by Eqs. (5.34) and (5.35), and $t_e = t_c$. For the square network,

$$t_b + t_d + 2\frac{h\delta}{k}t_c - 2\left(1 + \frac{h\delta}{k}\right)t_a = 0. \tag{5.36}$$

Specified Surface Temperature.   With t_a, t_b, and t_d known, Eq. (5.22) gives

$$\frac{t_b - t_a}{\mathscr{R}_{ab}} + \frac{t_d - t_a}{\mathscr{R}_{ad}} + q_a = 0, \tag{5.37}$$

and for the square network,

$$t_b + t_d - 2t_a + \frac{2q_a}{k} = 0. \tag{5.38}$$

If the surface is isothermal, either Eq. (5.37) or (5.38) yields $q_a = 0$, showing that no heat balance is necessary in this instance.

An Adiabatic Boundary. One has, simply,

$$\frac{t_b - t_a}{\mathscr{R}_{ab}} + \frac{t_d - t_a}{\mathscr{R}_{ad}} = 0 \tag{5.39}$$

which is, for the square network,

$$t_b + t_d - 2t_a = 0. \tag{5.40}$$

Irregular Boundary Points

In many practical applications of the numerical methods discussed here, the boundaries of the solid are often shaped such that it is either inconvenient or impossible to arrange net spacings so that the boundary points coincide with regular points of the net. In such cases special relations must be developed for the resistances connecting the boundary points to interior points. The case of a two-dimensional network is shown in Fig. 5.6. For simplicity of discussion, a square network, of spacing δ, is considered. A curved boundary is shown which passes between nodes in the square network. Points b and e represent boundary points on the network connected to the interior point a. The points b' and e' represent points spaced the regular distance, δ, from point a, and lying outside the boundary of the solid. The symbol δ' will denote the distance from node a

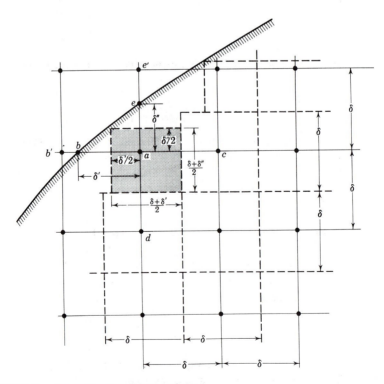

FIGURE 5.6. Irregular boundary points.

to node b, the real boundary point, while δ'' denotes the spacing from a to e. Apparently, both δ' and δ'' are less than δ.

In order to apply the heat balance of Eq. (5.22) to nodes a, b, c, d, and e, the resistances $\mathscr{R}_{ab}$, $\mathscr{R}_{ac}$, $\mathscr{R}_{ad}$, and $\mathscr{R}_{ae}$ must be evaluated. The method described here is developed in full in Ref. 5 wherein it is shown that the proposed representations neglect terms of the order of δ^3, and greater. It will be recalled from the discussion of Secs. 5.2 and 5.3 that the representation given by Eq. (5.22) for interior points neglects terms of the order of δ^4, and greater.

The recommended procedure consists of defining the limits of the solid lump surrounding node a as that rectangle which cuts the network lengths a-b, a-c, a-d, and a-e into equal parts. The results may be observed in Fig. 5.6—the lump volume associated with node a being shaded. The nodal point a is not located at the geometric center of the lump. With this representation, the conduction areas between node a and its neighbors are seen to be

$$A_{k_{ab}} = A_{k_{ac}} = \frac{\delta + \delta''}{2},$$

$$A_{k_{ad}} = A_{k_{ae}} = \frac{\delta + \delta'}{2}.$$

Thus, the resistances needed become

$$\mathscr{R}_{ab} = \frac{\delta'}{k(\delta + \delta'')/2},$$

$$\mathscr{R}_{ac} = \frac{\delta}{k(\delta + \delta'')/2}, \qquad (5.41)$$

$$\mathscr{R}_{ad} = \frac{\delta}{k(\delta + \delta')/2},$$

$$\mathscr{R}_{ae} = \frac{\delta''}{k(\delta + \delta')/2}.$$

Application of these relations is made in the examples of following sections.

5.6

NUMERICAL SOLUTION OF STEADY STATE NETWORK EQUATIONS

The foregoing sections have been devoted to the development of finite difference approximations for the heat conduction equation, network representations, boundary conditions, etc. The result of applying these concepts to a given conduction situation is that the *differential* equation for heat conduction in a body is replaced by an *algebraic* expression at each nodal subdivision of the body at which the temperature (or the heat flux in the case of an isothermal node) is desired. The expression for the temperature at a node involves the temperatures of the neighboring nodes. Thus, for a case in which N nodes of unknown temperature are involved, N algebraic equations are obtained which involve these unknowns. It remains, then, to be shown how such a system of N equations may be solved.

Several methods are available for the solution of a system of simultaneous algebraic equations (e.g., Refs. 2, 3, and 4). Certain methods particularly suited to the form of the equations involved here (usually linear) have been developed to a high degree and may be found discussed in detail in Refs. 4 through 7. Those methods particularly adaptable to digital computer use will be described here. These methods will be briefly discussed and then illustrated by example. The examples will be chosen to illustrate as many of the points developed in the foregoing sections as possible.

Solution by Iteration

The iteration technique is a method of successive approximations and follows a fixed sequence of operations. As such, no opportunity exists for modification of the sequence of the operations at the discretion of the user. Hence, the method is particularly adaptable to digital computer use.

Equation (5.22), the nodal heat balance, may be written for each node i:

$$t_i = \frac{q_i + \sum_j \dfrac{t_j}{\mathcal{R}_{ij}}}{\sum_j 1/\mathcal{R}_{ij}}. \tag{5.42}$$

The symbol j, it is recalled, represents all neighboring nodes with which node i exchanges heat. In general, the j nodes will number from one to, perhaps, six, depending on the dimensionality of the problem and the boundary conditions. The iteration method, properly called the Gauss–Seidel method, consists of the following sequence of operations:

1. An initial set of values of the nodal temperatures, t_i, is assumed.
2. New values of the nodal temperatures are calculated from Eq. (5.42), always using the *most recent* values of the t_j's. That is, as a new value of a given node's temperature is found, it is used in any subsequent calculation in which it appears.
3. This process is repeated, cycling through the N nodes, over and over, until

$$\left| [t_i]_{n+1} - [t_i]_n \right| \le \epsilon$$

for all nodes. The subscripts n and $n + 1$ indicate the values at a given node on two successive iterations. The tolerance ϵ is a selected convergence criterion.

An alternative method, the Jacoby method, calculates new temperatures at each node based on all the temperatures of the previous iteration instead of always using the most recent value. However, the Gauss-Seidel method is generally preferred because of its more rapid convergence. The above discussion is directed toward the instances in which the heat fluxes, q_i, are known and the temperatures are unknown. If a node is one of specified temperature, t_i is known and q_i becomes the sought-for unknown.

The above process will converge since Eq. (5.42) represents a weighted averaging process. Somewhere in the network, known, fixed, temperatures (or q's) are specified, and the iteration procedure continually averages these fixed values with the incorrect unknowns. Slowness of convergence should not be confused with accuracy of convergence. The accuracy of the final results depends upon the network spacing parameters and the convergence criterion ϵ. Improved accuracy may be obtained by using finer nets or smaller values of ϵ at the expense of increasing the number of computations and the required computing time, and no definite rule can be formulated for the choice of these parameters.

Quite apparently, the closeness of the initial assumptions for the t_i's to the correct solution has a profound effect upon the computing time. In some instances it may be advisable to make an initial calculation using a rather coarse net and then to use the results obtained to interpolate initial values for a much finer net.

The example which follows illustrates the iteration method.

EXAMPLE 5.1

Solve, by numerical means, the problem stated in Example 3.7 for the case of the iron rod. It is desired to find the temperature distribution and the total heat loss from a 1.25-cm-diameter iron rod ($k = 57$ W/m-°C) which is 30 cm long, heated to 120°C on one end, exposed to a convecting fluid at 20°C through a heat transfer coefficient of 9.0 W/m²-°C, and insulated on its free end.

Solution. Figure 5.7 illustrates the situation at hand. In keeping with the one-dimensional treatment employed in Chapter 3 for extended surfaces, it will be assumed that conduction occurs in the longitudinal direction only. The rod is, then, divided into five equal 5-cm lumps with 2.5-cm lumps at each end. This division results in seven equally spaced nodal points, with one at each end of the rod. Node 1 is an isothermal node maintained at $t_1 = 120$°C, and node 7 is a boundary node on an adiabatic surface. All other nodes are regular interior

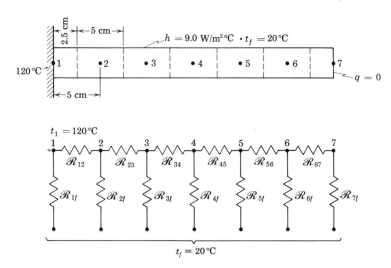

FIGURE 5.7. Nodal representation of a one-dimensional spine.

points. Each lump exchanges heat by convection with the ambient fluid at $t_f = 20°C$. The resultant nodal network is also shown in Fig. 5.7.

The conduction area for the conduction resistors $\mathcal{R}_{12} = \mathcal{R}_{23} = \cdots \mathcal{R}_{67}$ is the same in all cases:

$$A_k = \frac{\pi(0.0125)^2}{4} = 1.227 \times 10^{-4} \text{ m}^2.$$

Thus,

$$\mathcal{R}_{12} = \mathcal{R}_{23} = \cdots \mathcal{R}_{67} = \frac{\delta x}{kA_k} = \frac{0.05}{57 \times 1.227 \times 10^{-4}}$$

$$= 7.148 \frac{°C}{W}.$$

The convection area for the five interior nodes is $A_c = \pi(0.0125) \times 0.05 = 1.964 \times 10^{-3} \text{ m}^2$, so

$$\mathcal{R}_{2f} = \mathcal{R}_{3f} = \cdots = \mathcal{R}_{6f} = \frac{1}{hA_c} = \frac{1}{9.0 \times 1.964 \times 10^{-3}}$$

$$= 56.59 \frac{°C}{W}.$$

Since the convection area for nodes 1 and 7 is half that of the other nodes, the resistance is doubled:

$$\mathcal{R}_{1f} = \mathcal{R}_{7f} = 113.18 \frac{°C}{W}.$$

Recognizing the fact that the heat flow, q_1, exists at node 1 and is unknown and that no other node has an associated heat flow, application of Eq. (5.42) at each node gives

$$q_1 = 17.67 - 0.1399t_2,$$

$$t_2 = 0.4703t_3 + 57.62,$$

$$t_3 = 0.4703(t_2 + t_4) + 1.188,$$

$$t_4 = 0.4703(t_3 + t_5) + 1.188,$$

$$t_5 = 0.4703(t_4 + t_6) + 1.188,$$

$$t_6 = 0.4703(t_5 + t_7) + 1.188,$$

$$t_7 = 0.9406t_6 + 1.188.$$

Table 5.1 illustrates the results of the application of the Gauss–Seidel iteration technique, starting with an assumed linear temperature distribution and using a convergence criterion of $\epsilon = 0.05°C$. Eighteen iterations were necessary to achieve this degree of convergence. Note that the most recent values of the t's

TABLE 5.1

Number of Iterations	q_1 W	t_2 °C	t_3 °C	t_4 °C	t_5 °C	t_6 °C	t_7 °C
0	(t_1 = 120)	103.33	86.67	70.00	53.33	36.67	20.00
1	3.214	98.38	80.38	64.07	48.57	33.44	32.64
2	3.907	95.42	76.20	59.86	45.07	37.73	36.68
3	4.321	93.46	73.29	56.85	45.67	39.92	38.73
4	4.596	92.09	71.24	56.17	46.38	41.22	39.96
5	4.787	91.12	70.46	56.14	46.97	42.07	40.76
6	4.922	90.76	70.27	56.33	47.47	42.68	41.33
7	4.973	90.67	70.32	56.58	47.87	43.14	41.77
8	4.985	90.69	70.45	56.84	48.21	43.50	42.11
9	4.982	90.75	70.60	57.06	48.48	43.79	42.38
10	4.974	90.82	70.74	57.26	48.71	44.03	42.60
.	.	.	.	.	.	.	.
.	.	.	.	.	.	.	.
.	.	.	.	.	.	.	.
17	4.925	91.12	71.28	57.97	49.51	44.84	43.36
18 (ϵ = 0.05°C)	4.925	91.14	71.32	58.01	49.56	44.89	43.41
ϵ = 0.005°C, 30 iterations	4.908	91.22	71.46	58.20	49.77	45.10	43.61
ϵ = 0.5°C, 7 iterations	4.973	90.67	70.32	56.58	47.87	43.14	41.77
Analytic solution	4.834	91.095	71.265	57.980	49.541	44.382	43.382

are always used. For example, in iteration number 1, the value of t_2 = 98.38 results from the assumed initial t_3 = 86.67 while the calculated t_3 = 80.38 uses t_2 = 98.38 from the first iteration and t_4 = 70.00 from the initial assumption. The final results for ϵ = 0.5°C and 0.005°C are also shown, and the effect of the convergence criterion is apparent. Also shown are the results of the analytical methods of Example 3.7 for comparison purposes. ∎

Solution by Matrix Inversion

As was noted in the introductory discussion of this section, the problem at hand is that of solving N algebraic (usually linear) equations in N unknowns, where N is the number of nodal points at which either an unknown temperature or an unknown heat flux must be found. The application of Eq. (5.22) at each node generates a set of N equations of the form

$$a_{11}t_1 + a_{12}t_2 + a_{13}t_3 + \cdots + a_{1N}t_N = C_1$$
$$a_{21}t_1 + a_{22}t_2 + \cdots \qquad\qquad = C_2$$
$$a_{31}t_1 + \cdots \qquad\qquad\qquad\qquad = C_3 \qquad (5.43)$$
$$\vdots \qquad\qquad\qquad\qquad\qquad \vdots$$
$$a_{N1}t_1 + a_{N2}t_2 \qquad\quad + \cdots + a_{NN}t_N = C_N.$$

The coefficients, $a_{11}, a_{12}, \cdots a_{NN}$ involve the resistances between the nodes and will be different from zero only when the nodes indicated by the subscripts are in thermal communication. The C's are constant terms resulting from specified boundary conditions or heat fluxes.

If one defines the following matrix representations:

$$[A] = \begin{bmatrix} a_{11} & a_{12} & \cdots & a_{1N} \\ a_{21} & a_{22} & \cdots & \\ a_{31} & & & \\ \cdot & & & \\ \cdot & & & \\ \cdot & & & \\ a_{N1} & a_{N2} & \cdots & a_{NN} \end{bmatrix}, \quad [C] = \begin{bmatrix} C_1 \\ C_2 \\ \cdot \\ \cdot \\ \cdot \\ C_N \end{bmatrix}, \quad [t] = \begin{bmatrix} t_1 \\ t_2 \\ \cdot \\ \cdot \\ \cdot \\ t_N \end{bmatrix},$$

the above set of equations may be written

$$[A][t] = [C].$$

Reference 3 may be consulted for information on the rules of matrix algebra. If $[A]^{-1}$ represents the *inverse* of $[A]$, the solution for the temperatures is given by

$$[t] = [A]^{-1}[C]. \qquad (5.44)$$

If $[A]^{-1}$ has the elements

$$[A]^{-1} = \begin{bmatrix} b_{11} & b_{12} & \cdots & b_{1N} \\ b_{21} & \cdots & & \\ \cdot & & & \\ \cdot & & & \\ \cdot & & & \\ b_{N1} & b_{N2} & \cdots & b_{NN} \end{bmatrix},$$

the required t's are

$$t_1 = b_{11}C_1 + b_{12}C_2 + b_{13}C_3 + \cdots + b_{1N}C_N,$$
$$t_2 = b_{21}C_1 + \cdots$$
$$\vdots \qquad\qquad\qquad\qquad\qquad\qquad (5.45)$$
$$t_N = b_{N1}C_1 + b_{N2}C_2 + \cdots \qquad + b_{NN}C_N.$$

The problem, then, reduces to the inversion of a matrix. The process of matrix inversion (Ref. 3) is usually a laborious one when done by hand, and this method is not recommended in that instance, although the matrix [A] usually contains a large number of zero elements. Most digital computer installations, however, have matrix inversion and matrix multiplication routines available as parts of the system library. In such cases the above operations may be carried out rather simply. Even so, if the number of nodes, and hence the rank of the matrix [A], is large, the inversion process may become lengthy. Generally speaking, the matrix inversion method will usually be speedier for a modest number of nodes, but as the number of nodes increases, the iteration technique may prove to be faster.

EXAMPLE 5.2

Find the temperature distribution for Example 5.1 by matrix inversion.

Solution.   Since the heat flow is not desired in this case, the heat balance at node 1 may be eliminated and only the temperature t_2 through t_7 need to be found. The number of equations is, then, six, and the calculations of Example 5.1 give the coefficient matrix and the matrix of the nonhomogeneous terms to be

$$[A] = \begin{bmatrix} 1 & -0.4703 & 0 & 0 & 0 & 0 \\ -0.4703 & 1 & -0.4703 & 0 & 0 & 0 \\ 0 & -0.4703 & 1 & -0.4703 & 0 & 0 \\ 0 & 0 & -0.4703 & 1 & -0.4703 & 0 \\ 0 & 0 & 0 & -0.4703 & 1 & -0.4703 \\ 0 & 0 & 0 & 0 & -0.9406 & 1 \end{bmatrix}$$

$$[C] = \begin{bmatrix} 57.62 \\ 1.188 \\ 1.188 \\ 1.188 \\ 1.188 \\ 1.188 \end{bmatrix}.$$

Inversion of the matrix [A] and multiplication by the matrix [C] gives

$$t_2 = 91.24°C, \qquad t_5 = 49.79°C,$$

$$t_3 = 71.48°C, \qquad t_6 = 45.13°C,$$

$$t_4 = 58.22°C, \qquad t_7 = 43.63°C.$$

These results compare favorably with those found by iteration. ■

EXAMPLE 5.3

A long circular cylinder 8 cm in diameter has a 2-cm × 2-cm square hole located on its axis. The shaft is so long that conduction down its length may be neglected. The surface of the square hole is maintained at 100°C and the external cylindrical surface is maintained at 0°C. The thermal conductivity of the material is 1W/m-°C. Considering two-dimensional conduction only, establish a nodal network, determine the thermal resistance, and establish the nodal heat balance equations needed to determine the temperature distribution.

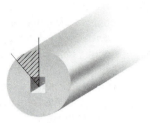

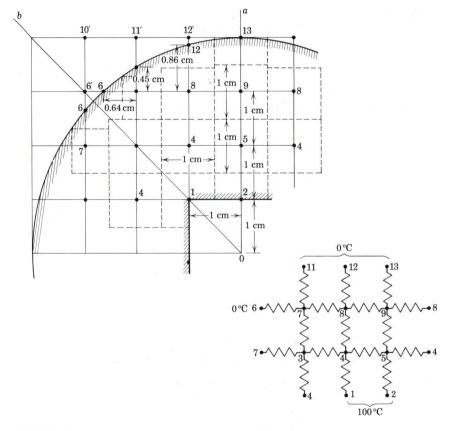

FIGURE 5.8.   Nodal representation of a hollow shaft.

Solution.   As may be noted in Fig. 5.8, symmetry considerations show that only one-eighth of the cross section need be considered. A $\delta x = \delta y = 1$-cm-square nodal network was chosen as shown in the figure. The symmetry requirement along radial lines 0-*a* and 0-*b* is handled by including in the network symmetrical points for all nodes which are thermally connected to nodes located on the lines of symmetry. Specifically, image points for nodes 4 (twice), 7 and 8 are included. The resulting resistor network is also shown. No resistors are shown for connections along the isothermal surfaces (i.e., points 1-2 and points 6-11-12-13) since no conduction will take place in these directions. Similarly, no heat balance expression will be written for these nodes since their temperatures are known, and there is no desire to find the associated heat fluxes.

The resulting equivalent network is also shown. Heat balance expressions need to be written only for the six interior nodes: 3, 4, 5, 7, 8, 9. Seventeen resistances must be found; however, many are identical for the regular interior nodes. By use of Eq. (5.23), per meter of cylinder length,

$$\mathscr{R}_{14} = \mathscr{R}_{25} = \mathscr{R}_{45} = \mathscr{R}_{59} = \mathscr{R}_{9\text{-}13} = \mathscr{R}_{34} = 1.0 \frac{°C}{W}.$$

Nodes 7 and 8 have irregularly spaced nodes connected with them (nodes 6, 11, 12) and the lengths of the unequal spacings are shown in the figure. By application of the relations given in Eq. (5.41):

$$\mathscr{R}_{67} = 0.882, \quad \mathscr{R}_{7\text{-}11} = 0.548, \quad \mathscr{R}_{78} = 1.38, \quad \mathscr{R}_{37} = 1.22,$$

$$\mathscr{R}_{89} = 1.075, \quad \mathscr{R}_{8\text{-}12} = 0.86, \quad \mathscr{R}_{48} = 1.0 \ \frac{°C}{W}.$$

Application of Eq. (5.22) and use of the facts that $t_1 = t_2 = 100$, $t_6 = t_{11} = t_{12} = t_{13} = 0$ gives the following nodal heat balances:

$$\text{Node 3:} \quad -2.22t_3 + 1.22t_4 + t_7 = 0,$$

$$\text{Node 4:} \quad t_3 - 4t_4 + t_5 + t_8 + 100 = 0,$$

$$\text{Node 5:} \quad 2t_4 - 4t_5 + t_9 + 100 = 0,$$

$$\text{Node 7:} \quad 1.13t_3 - 6.32t_7 + t_8 = 0,$$

$$\text{Node 8:} \quad 1.38t_4 + t_7 - 5.26t_8 + 1.28t_9 = 0$$

$$\text{Node 9:} \quad t_5 + 1.86t_8 - 3.86t_9 = 0.$$

These expressions may be solved for the unknown temperatures t_3, t_4, t_5, t_7, t_8, and t_9, by using either the iterative technique (after making an initial assumption) or the matrix inversion technique, to yield

$$t_3 = 33.68°C, \quad t_7 = 9.52°C,$$

$$t_4 = 53.49°C, \quad t_8 = 22.10°C,$$

$$t_5 = 58.18°C, \quad t_9 = 25.72°C. \quad \blacksquare$$

5.7

NONSTEADY NUMERICAL METHODS

The material presented thus far has been limited to the case of steady state conduction. Of considerable practical importance is the application of numerical techniques to the analysis of nonsteady conduction problems. The finite difference representation of the second derivatives in the space variables, the nodal resistance network representations, the boundary conditions, etc., developed in

Secs. 5.3 through 5.5 for the steady state will be equally applicable for nonsteady conduction analysis. The significant difference between the steady and nonsteady cases lies in finite difference representations which may be written for the partial derivative of temperature with respect to time. As will be recalled from Sec. 5.1, two possible finite difference expressions may be written for a first derivative, the forward difference and the backward difference. These representations lead to two different numerical methods for treatment of nonsteady problems.

In either instance, it is important to note that in addition to calculating temperatures at points spaced discrete intervals apart in space, in the nonsteady case the temperatures at these points are calculated at discrete intervals of time. That is, after the temperatures are known at all the spatial points, a finite increment of time is selected and all the temperatures are recalculated at the end of this time. Time is then progressively incremented, the spatial temperatures being calculated at each increment.

The nonsteady methods to be described here may be applied to the solution of steady state problems. With time-independent boundary conditions, an assumed temperature distribution may be regarded as an initial distribution of a nonsteady problem. As the nonsteady solution is carried forward in time, the desired steady state solution is eventually approached to any desired degree of accuracy.

Difference Equations

The nonsteady conduction equation is

$$k\left(\frac{\partial^2 t}{\partial x^2} + \frac{\partial^2 t}{\partial y^2} + \frac{\partial t}{\partial z^2}\right) = \rho c_p \frac{\partial t}{\partial \tau}. \tag{5.46}$$

Written in this form, each side of the conduction equation represents the time rate of heat storage, per unit volume, at a point. Finite difference approximations will now be written for this expression, but as noted above, two possible representations are possible. For conciseness of expression, the representations for the one-dimensional case will be written first. In this instance,

$$k \frac{\partial^2 t}{\partial x^2} = \rho c_p \frac{\partial t}{\partial \tau}. \tag{5.47}$$

Explicit Formulation. The explicit formulation is obtained by using the forward difference expression for the first derivative in place of the time derivative on the right side of Eq. (5.47). The central difference for the second derivative is used for the left side. In other words, to expand Eq. (5.47) about $x = x$ and $\tau = \tau$, use Eqs. (5.4) and (5.9) with t replacing f, τ replacing η, x replacing ξ, $\delta\tau$ replacing h_2, and δx replacing h_1. The result is

$$\frac{k}{(\delta x)^2}[t(x + \delta x, \tau) - 2t(x, \tau) + t(x - \delta x, \tau)]$$

$$= \frac{\rho c_p}{\delta \tau}[t(x, \tau + \delta \tau) - t(x, \tau)]. \tag{5.48}$$

If, as used in the steady state case and as suggested in Fig. 5.9, the subscripts a, b, and c are used to denote nodal locations at x, $x - \delta x$, and $x + \delta x$, respectively, and if t' is used to denote a temperature at time $\tau + \delta\tau$ while t is used simply to denote a temperature at time τ, Eq. (5.48) is more concisely stated as

$$\frac{k}{(\delta x)^2}(t_b - 2t_a + t_c) = \frac{\rho c_p}{\delta\tau}(t'_a - t_a). \tag{5.49}$$

This expression gives the *future* temperature at node a, t'_a, in terms of the *current* temperatures at node a and its surrounding nodes. This forward difference approximation for the nonsteady conduction equation is also termed the "explicit" formulation.

The equivalent two-dimensional, explicit, formulation is easily shown to be (see Fig. 5.9)

$$\frac{k}{(\delta x)^2}(t_b - 2t_a + t_c) + \frac{k}{(\delta y)^2}(t_d - 2t_a + t_e) = \frac{\rho c_p}{\delta\tau}(t'_a - t_a). \tag{5.50}$$

FIGURE 5.9.
Nodal representations of nonsteady conduction: (a) one-dimensional; (b) two-dimensional.

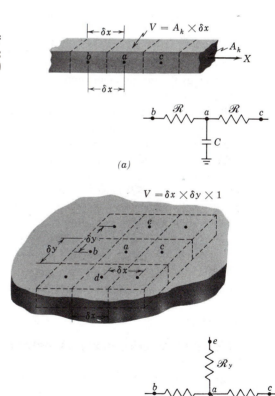

(a)

(b)

Implicit Formulation. The implicit representation is obtained by expanding Eq. (5.47) about $x = x$ and $\tau = \tau + \delta\tau$. This is done by use of Eqs. (5.7) and (5.9) with t replacing f, $\tau + \delta\tau$ replacing η, x replacing ξ, $\delta\tau$ replacing h_2, and δx replacing h_1. The result is

$$\frac{k}{(\delta x)^2}[t(x + \delta x, \tau + \delta\tau) - 2t(x, \tau + \delta\tau) + t(x - \delta x, \tau + \delta\tau)]$$

$$= \frac{\rho c_p}{\delta\tau}[t(x, \tau + \delta\tau) - t(x, \tau)].$$

Using the same abbreviated notation as employed above, one obtains

$$\frac{k}{(\delta x)^2}(t_b' - 2t_a' + t_c') = \frac{\rho c_p}{\delta\tau}(t_a' - t_a). \tag{5.51}$$

This implicit formulation gives the *future* temperature of point a, t_a', in terms of the current temperature at a, t_a, and the *future* temperatures of the neighboring points. The two-dimensional case reduces to

$$\frac{k}{(\delta x)^2}(t_b' - 2t_a' + t_c') + \frac{k}{(\delta y)^2}(t_d' - 2t_a' + t_e') = \frac{\rho c_p}{\delta\tau}(t_a' - t_a). \tag{5.52}$$

Explicit Versus Implicit Formulation. A rather obvious advantage of the explicit representation over the implicit is the fact the forward difference equation gives the *future* temperature of a single node in terms of *current* temperatures of that node and its neighbors. Thus, if at the end of a certain time period, all the nodal temperatures are known, then each of the nodal temperatures at the end of the next moment, $\delta\tau$, may be explicitly found, node by node. The implicit equation, however, expresses a *future* nodal temperature in terms of its current value and the *future* values of its neighbors' temperatures. Thus, to progress from one time step to the next, a system of equations of the form of Eq. (5.51) or (5.52) must be solved.

It would appear, then, that the explicit representation would be preferred to the implicit representation. However, as will be seen later, in the explicit case a serious restriction must be placed upon the magnitude of the time step $\delta\tau$ in relation to the spatial increment δx. Thus, although more direct, the explicit formulation may actually involve more calculation time than the less-direct implicit method.

Nodal Networks in the Nonsteady State

Directing attention for the moment to the one-dimensional case shown in Fig. 5.9, one may divide the conducting solid into lumps centered on the nodal points, as in the steady state cases discussed earlier. The total heat flow at a node may be had by multiplying Eq. (5.49), or (5.51), by the lump volume, $V_a = A_k \, \delta x$. Thus, in the explicit case:

$$\frac{t_b - t_a}{\frac{\delta x}{kA_k}} + \frac{t_c - t_a}{\frac{\delta x}{kA_k}} = \frac{V_a \rho c_p}{\delta \tau}(t_a' - t_a).$$

The resistances $\mathcal{R}_{ab} = \mathcal{R}_{ac} = \delta x / kA_k$ are defined as before. The term

$$C_a = V_a \rho c_p$$

is recognized as the thermal capacity of the lump volume surrounding the node a. Thus, the above expression becomes

$$\frac{t_b - t_a}{\mathcal{R}_{ab}} + \frac{t_c - t_a}{\mathcal{R}_{ac}} = C_a \frac{t_a' - t_a}{\delta \tau}. \tag{5.53}$$

Similarly, for the implicit case

$$\frac{t_b' - t_a'}{\mathcal{R}_{ab}} + \frac{t_c' - t_a'}{\mathcal{R}_{ac}} = C_a \frac{t_a' - t_a}{\delta \tau}. \tag{5.54}$$

Either Eq. (5.53) or (5.54) suggests the analogous electrical network illustrated in Fig. 5.9, wherein a capacitance is associated with each node—representing the thermal capacitance of the associated lump.

Generalizing upon the concepts just discussed, and drawing upon the concepts developed in Sec. 5.4, one may conclude that if a complex geometry (i.e., two or more dimensions) is subdivided into a nodal network of, perhaps unequal spacings, the following expression must be satisfied at each nodal point i:

$$\text{Explicit:} \quad \sum_j \frac{t_j - t_i}{\mathcal{R}_{ij}} + q_i = \frac{C_i}{\delta \tau}(t_i' - t_i). \tag{5.55}$$

$$\text{Implicit:} \quad \sum_j \frac{t_j' - t_i'}{\mathcal{R}_{ij}} + q_i = \frac{C_i}{\delta \tau}(t_i' - t_i). \tag{5.56}$$

The term q_i again denotes the rate of heat addition (external or internal) to the node. Thus, the future temperature of node i is determined by

$$\text{Explicit:} \quad t_i' = t_i + \delta \tau \left(\sum_j \frac{t_j - t_i}{\mathcal{R}_{ij} C_i} + \frac{q_i}{C_i} \right). \tag{5.57}$$

$$\text{Implicit:} \quad t_i' = t_i + \delta \tau \left(\sum_j \frac{t_j' - t_i'}{\mathcal{R}_{ij} C_i} + \frac{q_i}{C_i} \right), \tag{5.58}$$

where

$$\mathcal{R}_{ij} = \frac{\delta_{ij}}{kA_{k_{ij}}}, \qquad \text{for conduction,}$$

$$\mathcal{R}_{ij} = \frac{1}{h_{ij} A_{c_{ij}}}, \qquad \text{for convection,} \tag{5.59}$$

$$C_i = V_i \rho c_p.$$

These equations just given express the temperatures at a given node in terms of discrete spatial steps (i.e., δ_{ij}) and discrete time steps (i.e., $\delta\tau$). The method of solution depends on which formulation (explicit or implicit) is used but the important point to be noted is that the continuous solution of the original differential equation is replaced by the numerical solution of these difference equations at particular points in space and at particular intervals of time.

Boundary Conditions in the Nonsteady State

Equations of the form of Eq. (5.57) or (5.58) must be satisfied at each nodal point into which a conducting body is subdivided. As in the steady state case, certain special precautions must be taken at boundary points. In general, this amounts to calculating the resistances (conduction and convection) convected to the boundary nodes by use of special relations rather than those given in Eq. (5.59). The boundary conditions given in Sec. 5.5 for the steady case are equally applicable to the nonsteady case. Thus, no repetition will be made here for those conditions.

The case of time-dependent boundary temperatures enters as a possibility in the nonsteady case, however. In such cases, it is usual practice to treat such a case as that of a fixed temperature boundary condition [see Eqs. (5.28) through (5.30)]. During any time increment, the fixed boundary temperatures are chosen as the mean of the temperatures occurring at the boundary points at the beginning and the end of that time increment. As the calculations proceed in time, these boundary point temperatures must be continually revised according to the prescribed temperature function imposed at the surface. One of the examples given later will illustrate this procedure.

The Stability of the Solution

As has been pointed out previously, the numerical methods presented in this chapter represent approximate solutions to the original differential equations since derivatives are replaced by finite differences. Terms of the order of the fourth power of the spatial step size are neglected and, in the transient case, terms of the order of the square of the time increment are neglected. Errors introduced by these approximations are termed *truncation* errors, and the degree to which the approximate solution approaches the exact solution is termed the *convergence* of the finite difference representation. *Numerical* errors are also introduced into a solution by virtue of the inability, or impracticality, of performing the numerical computations with a sufficient number of significant figures.

In nonsteady numerical problems, the *stability* of the difference equations must also be considered. This matter is related to the way in which numerical and truncation errors introduced at one point in time either damp out or propagate and amplify in succeeding time steps. Detailed analyses of the stability properties of the equations used in this chapter are quite complex, but simple stability criteria may be developed on an elementary, intuitive, basis.

Rewrite Eqs. (5.57) and (5.58) in the following forms—expressing the sought-for future temperature at node i, t_i', in terms of the other quantities:

$$\text{Explicit:} \quad t_i' = t_i\left(1 - \sum_j \frac{\delta\tau}{\mathcal{R}_{ij}C_i}\right) + \sum_j \frac{t_j\delta\tau}{\mathcal{R}_{ij}C_i}. \tag{5.60}$$

$$\text{Implicit:} \quad t_i' = \frac{t_i + \sum_j t_j' \dfrac{\delta\tau}{\mathcal{R}_{ij}C_i}}{1 + \sum_j \dfrac{\delta\tau}{\mathcal{R}_{ij}C_i}}. \tag{5.61}$$

For simplicity of discussion, any internal heat generation, q_i, has been considered zero.

It may be noted that in the case of the explicit formulation, the coefficient of the t_i term might become negative—particularly if the time increment, $\delta\tau$, is chosen large enough. If the coefficient of t_i *does* become negative, t_i' could conceivably be less than t_i. This implies that for certain values of $\sum_j \delta\tau/\mathcal{R}_{ij}C_i$, the greater is the temperature t_i at the time τ, the *smaller* it will be at time $\tau + \delta\tau$. This fact does not make sense, thermodynamically, and it can be seen as a trend which will cause the temperature t_i to oscillate wildly from one time period to the next. Although this is not a precise analysis of stability, one would be certain of a stable procedure so long as

$$\sum_j \frac{\delta\tau}{\mathcal{R}_{ij}C_i} \leq 1. \tag{5.62}$$

More sophisticated analyses (Refs. 4 and 5) lead to less stringent stability criteria; however, the above limiting relation is generally used in practical cases.

The implication of Eq. (5.62) is that the choice of the time step $\delta\tau$ is intimately connected with the choice of the spatial increment, δ_{ij}, which is involved in the resistance $\mathcal{R}_{ij}$. As smaller spatial increments are chosen to reduce the associated truncation error, smaller time increments must also be used in order to satisfy Eq. (5.62). Thus, increased accuracy in the spatial network is obtained at the cost of smaller time increments—perhaps leading to prohibitive computation times. This is the principal disadvantage of the explicit method, and it will be examined in more detail in Sec. 5.9.

Examination of Eq. (5.61) for the implicit formulation reveals that no stability limitation exists as a result of possible negative coefficients. Thus, the time increment is not restricted by the choice of spatial increments. The only limitation on $\delta\tau$ in this instance is that imposed by the minimization of truncation errors in time.

5.8

SOLUTION OF NETWORK EQUATIONS FOR THE IMPLICIT CASE

The nodal heat balance in the implicit case is given by

$$t_i' = \frac{t_i + \sum_j t_j' \dfrac{\delta\tau}{\mathcal{R}_{ij}C_i}}{1 + \sum_j \dfrac{\delta\tau}{\mathcal{R}_{ij}C_i}}. \tag{5.61}$$

If all the resistances of a network are known, and if the capacitances of all the associated lumps are known, then with a specified *initial* temperature distribution, equations of the type of Eq. (5.61) may be written for each nodal point—care being taken to satisfy the established boundary conditions. The net result is that a set of equations are obtained for the *future* temperatures, t_i', of the nodal points—in terms of the *future* temperatures of neighboring points—once a time step, $\delta\tau$, is chosen. There will be as many equations as there are unknown future temperatures. Once this set of equations is solved, the resulting temperatures become the initial temperatures for the next time step calculation.

Thus, the implicit technique reduces to the solution of a set of simultaneous algebraic equations at each time increment. Consequently, the methods discussed in Secs. 5.3 through 5.6 for the solution of the steady state equations are applicable here—iteration or matrix inversion. The implicit formulation is seen, then, to reduce to a series of steady state calculations at each time step. It has the advantage of not having a restricted time step—in fact, the time step may be varied during the progression of the calculation. It has the disadvantage of requiring a *set* of calculations (i.e., iteration or matrix inversion) at each step in time, leading to increased calculation time and storage requirements as the number of nodes becomes large. Since the calculations do not differ, in principle, from those already shown for the steady state, no examples will be given.

5.9

SOLUTION OF NETWORK EQUATIONS FOR THE EXPLICIT CASE

As discussed earlier, the explicit formulation avoids the need of iterative or matrix inversion techniques, since each future nodal temperature can be individually calculated for a time increment $\delta\tau$ from only the current nodal temperatures. Thus, from an equation of the form

$$t_i' = t_i\left(1 - \sum_j \frac{\delta\tau}{\mathcal{R}_{ij}C_i}\right) + \sum_j t_j \frac{\delta\tau}{\mathcal{R}_{ij}C_i}, \tag{5.60}$$

new temperatures are successively calculated at each node, starting with the given initial temperature distribution in a network for a given $\delta\tau$. Time is then incremented and the calculations are repeated. No iterations or matrix inversions are required. Only the stability requirement stated in Eq. (5.62) need be satisfied:

$$\sum_j \frac{\delta\tau}{\mathcal{R}_{ij}C_i} \leq 1. \tag{5.62}$$

Special Stability Criteria

Examination of the stability requirement for special instances is appropriate. The relations to be developed emphasize the *upper* limit of the allowable time increment. In practice it is wise to use a time increment smaller than the maximum.

The reason for this practice is a combination of the desires to improve accuracy by reducing truncation error in time and to avoid instabilities by staying safely below the upper limit of $\delta\tau$.

Irregular Networks. For the most general instance in which the network contains irregular net spacings, convective resistors as well as conductive resistors, etc., the criterion given in Eq. (5.62) must be evaluated at each nodal point. Since it is necessary at any one time step to carry *each* node forward in time by the same $\delta\tau$, the time increment for the entire calculation is controlled by the node for which stability criterion yields the smallest time increment. That is,

$$\delta\tau \leq \left(\frac{1}{\sum_j \dfrac{1}{\mathcal{R}_{ij} C_i}} \right)_{min} \tag{5.63}$$

In general, nets that include lumps which are significantly smaller than the others should be avoided. Such small lumps usually have small capacitances and consequently control the time increment for the entire network.

Equally Spaced Networks. In the event that a net is one of equal spacings, special forms result for Eq. (5.63). Figure 5.10 illustrates one- and two-dimensional cases in which regular spacings are used. In the case of the two-dimensional network, regular spacing implies the use of a square grid. In either the one- or two-dimensional cases a half-lump must be provided at the boundary.

Consider first the one-dimensional case. The body dimensions in the two directions, other than the one in which conduction takes place, are taken as l_1 and l_2, respectively. (Actually the body may be of infinite extent in these directions, in which case one would carry out the calculation on a unit flow area basis and simply take $l_1 = l_2 = 1$.) For the case in which there is conduction only (i.e., no convection at the boundary) the equal spacing of δx results in identical resistors—two connected to each interior node and one at the surface node. The resistors all have the value

$$\mathcal{R} = \frac{\delta x}{k(l_1 l_2)}.$$

The capacitors of the interior nodes are all equal to

$$C = \rho V c_p = \rho l_1 l_2 \, \delta x c_p,$$

while the surface node has a capacitance of $C/2$. Since the nodes with capacitance C have two resistors, $\mathcal{R}$, connected to them, and that with capacitance $C/2$ has only one resistor connected to it, the stability criterion of Eq. (5.62) yields the same result for *all* nodes:

$$2 \frac{\delta\tau}{\mathcal{R}C} \leq 1.$$

Rewritten, the above expression becomes

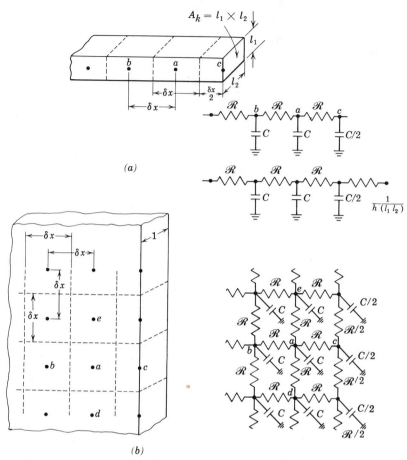

FIGURE 5.10. Regular boundary points in nonsteady conduction: (a) one-dimensional; (b) two-dimensional.

$$\delta\tau \le \tfrac{1}{2}\mathcal{R}C$$

$$\le \frac{1}{2}\frac{\rho c_p}{k}(\delta x)^2 \tag{5.64}$$

$$\le \frac{1}{2}\frac{(\delta x)^2}{\alpha}.$$

The latter form shows more clearly the restriction placed on the time increment by the choice of spatial increment.

For the two-dimensional case shown in Fig. 5.10, and for the three-dimensional case with an equally spaced net, arguments similar to that just given yield

Two dimensions: $\delta\tau \le \tfrac{1}{4}\mathcal{R}C$

$$\le \frac{1}{4}\frac{(\delta x)^2}{\alpha}. \tag{5.65}$$

Three dimensions: $\delta\tau \le \tfrac{1}{6}\mathcal{R}C$

$$\le \frac{1}{6}\frac{(\delta x)^2}{\alpha}. \tag{5.66}$$

If the upper limit of the time increment is used, the nodal heat balance, Eq. (5.60), takes on particularly simple forms:

$$\text{One dimension:} \quad t_a' = \frac{t_b + t_c}{2}. \tag{5.67}$$

$$\text{Two dimensions:} \quad t_a' = \frac{t_b + t_c + t_d + t_e}{4}. \tag{5.68}$$

The future temperature at a node is seen to be a simple mean of the surrounding nodes. A similar relation results for the three-dimensional case. These relations prove to be most useful for hand calculations and also form the basis of certain graphical methods (Ref. 5).

In the event that convection occurs at the boundary, the above stability criteria must be altered. In the one-dimensional case, also pictured in Fig. 5.10, the stability criterion yields the same result as given in Eq. (5.64) when applied at the interior nodes. At the surface node, however, there is a conduction resistance and a convection resistance:

$$\mathcal{R}_{\text{cond}} = \frac{\delta x}{k(l_1 l_2)},$$

$$\mathcal{R}_{\text{conv}} = \frac{1}{h(l_1 l_2)},$$

$$C = \tfrac{1}{2}\rho l_1 l_2 \, \delta x c_p.$$

Thus, Eq. (5.62) yields

$$\frac{\delta\tau}{\tfrac{1}{2}\rho l_1 l_2 \, \delta x c_p} \left[\frac{k l_1 l_2}{\delta x} + h l_1 l_2 \right] \le 1,$$

or

$$\delta\tau \le \frac{1}{2} \frac{(\delta x)^2}{\alpha} \left[\frac{1}{1 + (h \, \delta x/k)} \right]. \tag{5.69}$$

Since the term in brackets in Eq. (5.69) is always less than 1, it is apparent that the stability criterion applied at the surface node yields a smaller time increment than that resulting from consideration of the interior nodes. Thus, the surface node becomes the controlling factor as far as the selection of the time increment is concerned.

A similar analysis when applied to the two-dimensional case shown in Fig. 5.10 and in the three-dimensional case gives

$$\text{Two-dimensional:} \quad \delta\tau \le \frac{1}{4} \frac{(\delta x)^2}{\alpha} \left[\frac{1}{1 + \tfrac{1}{2}(h \, \delta x/k)} \right]. \tag{5.70}$$

$$\text{Three-dimensional:} \quad \delta\tau \le \frac{1}{6} \frac{(\delta x)^2}{\alpha} \left[\frac{1}{1 + \tfrac{1}{3}(h \, \delta x/k)} \right].$$

Relations similar to those given in Eq. (5.70) may be developed for other special configurations, such as an external corner, etc. These may be found in the references at the end of the chapter. When the surface film coefficient becomes quite large the permissible time increment allowed by the boundary nodes may become considerably less than that allowed by the interior nodes. In such an instance, the computation of the temperature history may become prohibitively long. Certain schemes have been developed whereby such severe limitations may be circumvented.

The implicit and explicit techniques just described are only two, rather direct, of many possible numerical methods available for solution of the heat conduction equation. For certain problems, accelerating techniques are available. These techniques, although more complex to carry out, are hybrid methods which attempt to combine the inherent stability of the implicit approach with the matrix-inversion-free advantages of the explicit approach. Reference 4 may be consulted for information on such techniques.

EXAMPLE 5.4

An infinite slab 30 cm thick is initially at a uniform temperature of 40°C. The temperature of both surfaces is suddenly raised to 260°C and maintained at that value. The slab is composed of a material with the following properties: $k = 1.84$ W/m-°C, $\rho = 2300$ kg/m³, $c_p = 0.8$ kJ/kg-°C, $\alpha = 1.0 \times 10^{-6}$ m²/s. Determine, numerically, the temperature history in the slab during the first 1.25 h after the surface temperature increased.

Solution. Because of symmetry considerations only one-half of the slab need be considered. Six nodal points, spaced 3 cm apart, are used, with the associated lumps as shown in Fig. 5.11. The network subdivision satisfies the requirements for Eq. (5.64), so the sum $\Sigma_j 1/\mathcal{R}_{ij}C_i$ is the same for all nodes. Specifically,

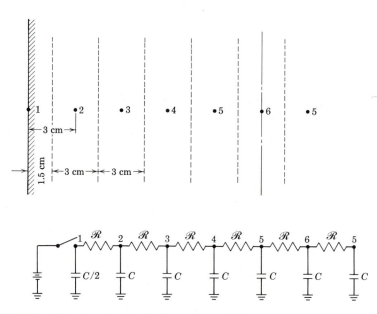

FIGURE 5.11. Nonsteady circuit for a plane slab.

$$\sum_j \frac{1}{\mathcal{R}_{ij} C_i} = \frac{2}{\mathcal{R}C},$$

$$\mathcal{R}C = \frac{(\delta x)^2}{\alpha} = \frac{(0.03)^2}{1 \times 10^{-6}} = 900 \text{ s}.$$

According to Eq. (5.62) or (5.64), the maximum permissible time increment is

$$\delta\tau \leq \left(\sum_j \frac{1}{\mathcal{R}_{ij} C_i} \right)^{-1} = \frac{\mathcal{R}C}{2}$$

$$\leq 450 \text{ s}.$$

For this time increment the nodal equations are all of the form of Eq. (5.67). Table 5.2 shows the results of the calculations.

For node 1, the temperatures are specified. In keeping with the principle mentioned in association with the boundary conditions, the temperature of node 1 during the first time increment is taken as the mean of its value at the beginning (40°C) and the end (260°C) of the increment, namely 150°C. At $\tau = 0$, the initial temperatures of nodes 2 through 6 are known to be 40°C. Then, at $\tau = 450$ s the new temperatures at nodes 2 through 6 are found by use of Eq. (5.67), or, equivalently, Eq. (5.60), upon the completion of each row of the table, the next row is constructed in a similar manner.

Figure 5.12 compares the results of the numerical calculations at selected time intervals with the results of the exact solution available from Sec. 4.5 and Fig. 4.8. Also compared in Fig. 5.13 are the temperature-time histories at the slab center (node 6) which result from using the maximum time increment, as above, $\delta\tau = (\delta\tau)_{max} = \frac{1}{2}(\delta x)^2/\alpha$, and a time increment one-half of that, $\delta\tau = \frac{1}{2}(\delta\tau)_{max} = \frac{1}{4}(\delta x)^2/\alpha$. The improvement in accuracy with decreased time increment is apparent. Also shown is the unstable character of the results

TABLE 5.2 Numerical Calculations for Infinite Slab

Time s	Node h	t_1 °C	t_2 °C	t_3 °C	t_4 °C	t_5 °C	t_6 °C
0	0	150.00	40.00	40.00	40.00	40.00	40.00
450	0.125	260.00	95.00	40.00	40.00	40.00	40.00
900	0.250	260.00	150.00	67.50	40.00	40.00	40.00
1350	0.375	260.00	163.75	95.00	53.75	40.00	40.00
1800	0.500	260.00	177.50	108.75	67.50	46.87	40.00
2250	0.625	260.00	184.38	122.50	77.81	53.75	46.87
2700	0.750	260.00	191.25	131.09	88.13	62.34	53.75
3150	0.875	260.00	195.55	139.69	96.72	70.94	62.34
3600	1.000	260.00	199.84	146.13	105.31	79.53	70.94
4050	1.125	260.00	203.07	152.58	112.83	88.13	79.53
4500	1.250	260.00	206.29	157.95	120.35	96.18	88.13

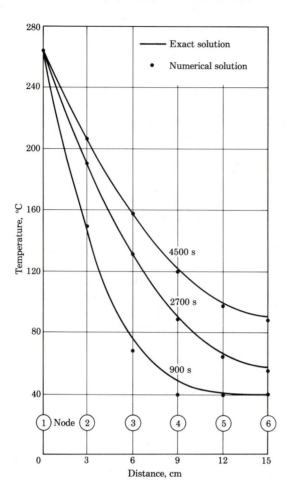

FIGURE 5.12. Comparison of analytical and numerical results for a plane slab, showing the temperature distribution at various times when using the maximum time step.

which are obtained when one uses a time increment which exceeds the maximum: $\delta\tau = 1.2(\delta\tau)_{\text{max}} = 0.60(\delta x)^2/\alpha$. ∎

EXAMPLE 5.5

Determine for the geometry given in Example 5.3, the maximum allowable time step that stability limitations will permit in a transient calculation. Assume that the boundary conditions do not involve convection and that the material has the following properties: $k = 1.0$ W/m-°C, $\rho = 2000$ kg/m³, $c_p = 0.5$ kJ/kg-°C.

Solution. As in Example 5.3, use a cylinder length of 1 m and the same nodal network. In that case, the resistances found in Example 5.3 apply here. Using the dimensions shown in Fig. 5.8, the lump capacitances may be found as noted below. The values shown for $\Sigma\, 1/\mathcal{R}_{ij}$ result from the data of Example 5.3:

$$\text{Node 3:} \quad \frac{1}{C_3} = 0.010\ \frac{°C}{W\text{-s}}; \quad \sum_j \frac{1}{\mathcal{R}_{3j}} = 3.64\ \frac{W}{°C}; \quad \sum_j \frac{1}{\mathcal{R}_{3j}C_3} = 0.036\ \frac{1}{s}.$$

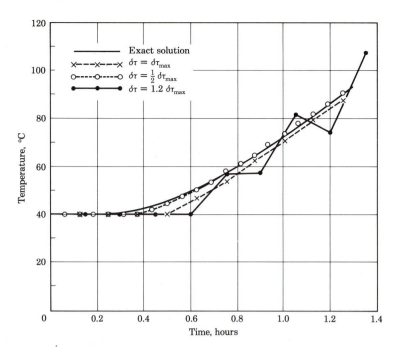

FIGURE 5.13. Comparison of analytical and numerical results for a plane slab, showing the temperature–time history of the centerline for different time increments.

Node 4: $\dfrac{1}{C_4} = 0.010 \dfrac{°C}{W\text{-}s}$; $\sum\limits_j \dfrac{1}{\mathcal{R}_{4j}} = 4.0 \dfrac{W}{°C}$; $\sum\limits_j \dfrac{1}{\mathcal{R}_{4j}C_4} = 0.040 \dfrac{1}{s}$.

Node 5: $\dfrac{1}{C_5} = 0.010$; $\sum\limits_j \dfrac{1}{\mathcal{R}_{5j}} = 4.0$; $\sum\limits_j \dfrac{1}{\mathcal{R}_{5j}C_5} = 0.040$.

Node 7: $\dfrac{1}{C_7} = 0.017$; $\sum\limits_j \dfrac{1}{\mathcal{R}_{7j}} = 4.41$; $\sum\limits_j \dfrac{1}{\mathcal{R}_{7j}C_7} = 0.075$.

Node 8: $\dfrac{1}{C_8} = 0.011$; $\sum\limits_j \dfrac{1}{\mathcal{R}_{8j}} = 3.83$; $\sum\limits_j \dfrac{1}{\mathcal{R}_{8j}C_8} = 0.042$

Node 9: $\dfrac{1}{C_9} = 0.010$; $\sum\limits_j \dfrac{1}{\mathcal{R}_{9j}} = 4.162$; $\sum\limits_j \dfrac{1}{\mathcal{R}_{9j}C_9} = 0.042$.

From the largest value of $\sum_j 1/\mathcal{R}_{ij}C_i = 0.075$, the associated maximum time increment is found to be

$$\delta\tau = \frac{1}{0.075} = 13.3 \text{ s}.$$

All other nodes will yield larger maxima increments, so the above value must be used—or any smaller value. ∎

REFERENCES

1. DESAI, C. S., *Elementary Finite Element Methods,* Englewood Cliffs, N.J., Prentice-Hall, 1979.

2. SOUTHWELL, R. V., *Relaxation Methods in Theoretical Physics,* New York, Oxford U.P., 1946.

3. HILDEBRAND, F. B., *Methods of Applied Mathematics,* Englewood Cliffs, N.J., Prentice-Hall, 1952.

4. PATANKAR, S. V., *Numerical Heat Transfer and Fluid Flow,* New York, McGraw-Hill, 1980.

5. SCHNEIDER, P. J., *Conduction Heat Transfer,* Reading, Mass., Addison-Wesley, 1955.

6. DUSINBERRE, G. M., *Heat Transfer Calculations by Finite Differences,* Scranton, Pa., International Textbook, 1961.

7. ROHSENOW, W. M., and J. P. HARTNET, eds., *Handbook of Heat Transfer,* New York, McGraw-Hill, 1973.

PROBLEMS

5.1   A 0.6-cm-diameter pure copper rod is 48 cm long and is heated to 90°C at *each* end. The rod surface is exposed to an ambient fluid at 25°C through a surface heat transfer coefficient of 25 W/m²-°C. By numerical means, find the temperature distribution in the rod using a nodal spacing of 4 cm. How much heat is dissipated by the rod? Compare the results with an analytic solution.

5.2   A straight fin of uniform thickness is attached to a surface at 150°C. The fin is 5 cm thick and 15 cm long and is composed of 20% nickel steel. The surface and the end of the fin are exposed to an ambient fluid at 10°C through a heat transfer coefficient of 30 W/m²-°C. Find, numerically, the temperature distribution in the fin, recognizing that the thickness length ratio is too large to treat the fin as one-dimensional.

5.3   The accompanying figure depicts a section of a chimney made of common brick. The inside surface is maintained at 350°F while the outside surface is maintained

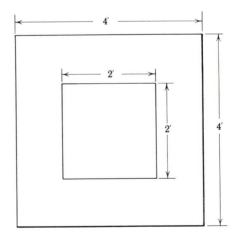

PROBLEM 5.3

at 100°F. Assuming that the chimney is quite tall so that the heat conduction through it may be considered as two-dimensional, find by numerical means the temperature distribution in the chimney. Use a network spacing of 4 in.

5.4 Repeat Prob 5.3 if a gas at 400°F flows inside the chimney with a heat transfer coefficient of 10 Btu/h-ft²-°F at the surface. Air at 70°F surrounds the outside of the chimney and the heat transfer coefficient there is 2 Btu/h-ft²-°F. Use a 6-in. net spacing.

5.5 A concrete block, 30 cm square, has a 15-cm-O.D. steam pipe buried concentrically inside. The steam pipe is maintained at 150°C. The exterior surfaces of the concrete are maintained at 10°C. Find, numerically, the temperature distribution in the concrete. Presume that the block is long enough to treat the heat flow as two-dimensional.

5.6 Imagine that the chimney in Prob. 5.3 is initially at 350°F throughout and that the outside surface temperature is subsequently lowered suddenly to 100°F. Determine, numerically, the temperature–time history in the chimney for a time span of 10 h.

5.7 A large plate of pure aluminum is 4 cm thick. It is in contact with convecting fluids on each side. On one side the heat transfer coefficient is 30 W/m²-°C, and on the other side it is 70 W/m²-°C. Initially both fluids are at 20°C, and the plate is in equilibrium. Suddenly the fluid temperatures are raised to 90°C. Determine, numerically, how long it takes the center of the plate to reach 80°C.

5.8 The accompanying figure depicts the cross section of a very long Duralumin I beam. All surfaces, other than the upper and lower faces, are insulated. Determine, numerically, the steady state rate of heat flow (per foot of beam length) which results when the upper surface is maintained at 40°C and the lower surface at 15°C.

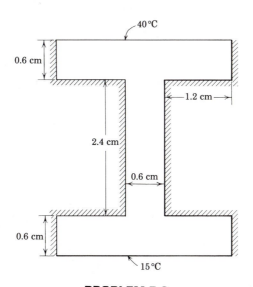

PROBLEM 5.8

5.9 For the network chosen in Prob. 5.8, determine the maximum time increment that could be used to numerically analyze the transient response of the I beam to a sudden change in one of the surface temperatures.

5.10 The rod described in Example 5.1 is initially at equilibrium with the ambient fluid at 20°C, the base of the rod being maintained at that temperature. For time greater than zero, the base temperature of the rod is raised, linearly, to 120°C over a time span of 20 min. Find the temperature history for the rod for a total time span of 1 h. Use $\rho = 7840 \text{ kg/m}^3$, $c_p = 0.46 \text{ kJ/kg-°C}$.

5.11 Derive relations analogous to Eqs. (5.34) through (5.40) for the applicable boundary conditions at an *internal* corner.

5.12 Derive the facts given in Eqs. (5.65), (5.66), and (5.70).

CHAPTER 6

The Fundamental Principles of Viscous Fluid Motion and Boundary Layer Motion

THE FLUID MECHANICAL ASPECTS OF CONVECTION

In the foregoing chapters, primary attention has been directed to the problems of heat conduction in solids. The mode of heat transfer known as convection has been considered only as one type of boundary condition to be applied at the surface of a conducting solid. This boundary condition has been treated in terms of a gross parameter, the surface heat transfer coefficient, h, defined by Newton's law of cooling in Eq. (1.16). In these applications h has been presumed known. The purpose of the next few chapters is to focus attention on the process of heat convection in a fluid and to develop methods of predicting the value of the heat transfer coefficient that will likely result under a given set of conditions.

As discussed earlier in Secs. 1.3 and 1.5, convection is the term applied to the energy transfer process which is observed to occur in fluids mainly because of the transport of energy by means of the motion of the fluid itself. The process of conduction of energy by molecular interchange is, of course, still present, but high energy (or hot) portions of the fluid are brought into contact with the lower energy regions (cooler regions) by virtue of the fact that there are gross displacements of the fluid particles. When the fluid motion is caused by the imposition of external forces in the form of pressure differences, this mechanism is called *forced convection*. The pumping of a fluid through or past solid surfaces

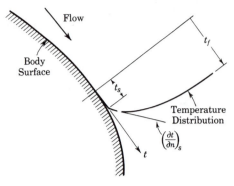

FIGURE 6.1.

of a temperature different from the fluid is an example of forced convection. When no external forces are applied to a fluid it may be set into motion by differences in density that would be caused by an immersed solid body whose temperature is different from that of the fluid. Such a heat exchange situation is termed *free convection* and may be observed in a heated pan of boiling water, in the air surrounding a room heater, etc.

In either case, an analytical approach to the determination of the heat transfer coefficient would involve the finding of the temperature distribution in the fluid surrounding the body. If, as is usually the case, the fluid motion in the region immediately adjacent to the surface is laminar, then the heat flux from the surface may be evaluated in terms of the fluid temperature gradient at the surface as illustrated in Fig. 6.1. Then the definition of the heat transfer coefficient as the ratio of the heat flux to the difference between the surface temperature and the fluid temperature enables one to write

$$h = \frac{-k_f (dt/dn)_s}{t_s - t_f},$$

where

k_f = thermal conductivity of the fluid,

t_s = temperature of surface,

t_f = temperature of fluid far removed from surface,

$\left(\dfrac{dt}{dn}\right)_s$ = fluid temperature gradient, measured at the surface in a direction normal to the surface.

In order to find the temperature distribution in the fluid so that the above formulation can be used to determine h, it is then necessary to solve the complete fluid mechanics problem in the region near the surface. For the general case of the motion of a convecting fluid in three dimensions a complete description of the fluid motion requires the determination of the three velocity components, the fluid pressure, the fluid temperature, and fluid density—all as functions of position and time. Six equations are required to find these six dependent variables. Newton's law of motion in each of the three coordinate directions yields three partial differential equations, and two additional partial differential equations are obtained from applying the principles of conservation of mass and conservation of energy. The equation of state of the fluid furnishes the sixth, algebraic, equation.

The general problem requires the simultaneous solution of these six equations, under the appropriate boundary conditions, so that the fluid temperature gradient at the wall may be found to yield h. Such general solutions have been obtained in only a few cases of practical significance. Thus, certain approximate methods of solution must be sought. Thus, in sections which follow, the above-mentioned equations for viscous fluid motion will be discussed. These discussions are not meant to be rigorous derivations of these equations, but, rather, are meant to illustrate the physical significance of the various terms so that intelligent approximations may be applied. A complete discussion and derivation of these equations may be found in Refs. 1 and 2.

For the sake of simplicity, this chapter will consider fluid motion in only *two space dimensions,* thus eliminating one unknown velocity component and one statement of Newton's law. Also, it will be presumed that the fluid may be treated as incompressible, eliminating density as a dependent variable and the need for an equation of state. Thus, only four equations involving two velocity components and the pressure and temperature will be needed. Also, in most instances, it will be assumed that steady fluid motion, independent of time, exists.

6.2

THE CONTINUITY EQUATION— THE CONSERVATION OF MASS

Since motion in only two space dimensions will be considered, it is convenient to think of the fluid flowing in the x-y plane (as illustrated in Fig. 6.2) with unit depth in the z direction (normal to the plane of the figure). All flow properties would then be constant over this unit depth.

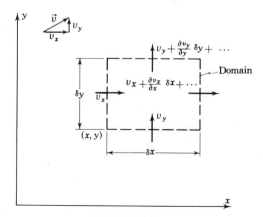

FIGURE 6.2.

Generally, then, the x-y plane represents a field in which the fluid velocity varies throughout. Representing the velocity vector, $\vec{v}$, by the x and y components, v_x and v_y, one has, in general, a variation of these components throughout the x-y plane, each component being a function of x and y.

Selecting arbitrarily a volume element of space having dimensions δx, δy, and unit depth, one sees that the rate of mass flow into the element must equal the rate of outflow if the principle of conservation of mass is to hold for the case in which no sources or sinks of mass exist within the fluid domain. This statement is true whether the flow is steady or not since no mass may be stored in the element if the fluid is incompressible.

Considering first flow in the x direction only, mass enters and leaves the element at the faces located at x and $x + \delta x$. The mass rate entering at x is

$$(\rho v_x)\ \delta y,$$

and that leaving at $x + \delta x$ may be written as an expansion of that at x:

$$(\rho v_x)\ \delta y + \frac{\partial}{\partial x}\ [(\rho v_x)\ \delta y]\ \delta x + \cdots.$$

Thus, the excess rate at which mass leaves the element over that which enters, in the x direction, is

$$\frac{\partial}{\partial x}(\rho v_x)\ \delta y\ \delta x + \cdots.$$

Similarly, the excess rate of mass flow out of the element for flow in the y direction is

$$\frac{\partial}{\partial y}(\rho v_y)\ \delta x\ \delta y + \cdots.$$

Since it is assumed that there are no sources or sinks of mass within the element, the principle of mass conservation requires that the sum of the above two expressions be zero:

$$\frac{\partial}{\partial x}(\rho v_x)\ \delta x\ \delta y + \frac{\partial}{\partial y}(\rho v_y)\ \delta x\ \delta y + \cdots = 0,$$

or, per unit volume of the element

$$\frac{\partial}{\partial x}(\rho v_x) + \frac{\partial}{\partial y}(\rho v_y) + \cdots = 0.$$

The above conservation principle must be satisfied for *all* elements, so letting $\delta x \to \delta y \to 0$ and using the fact that the density is constant, one obtains the *continuity equation* for incompressible flow:

$$\frac{\partial v_x}{\partial x} + \frac{\partial v_y}{\partial y} = 0. \tag{6.1}$$

As noted in the foregoing, this relation is valid for steady or nonsteady flow, even through time does not appear explicitly, as long as the density is constant.

VISCOUS RESISTANCE
FOR PLANE LAMINAR FLUID MOTION

Before discussing the equations of motion of a viscous fluid it will be necessary to extend the definition of viscosity made in Chapter 2. The present discussion and the resulting equations of motion will be restricted to *laminar* flow conditions. The concept of turbulence and the interpretation of the equations of motion for turbulent flow will be reserved for discussion in later sections.

In Sec. 2.7 the coefficient of dynamic viscosity, μ, was defined for laminar motion in one direction. That is, for motion all in one direction, the dynamic viscosity was defined as the proportionality factor between the shear stress in the fluid and the velocity gradient produced by that shearing resistance. Figure 6.3(a) shows an infinitesimal element of a fluid moving in only one direction. In this case, viscous shear acts only in that direction—the x direction in Fig. 6.3—and deforms the element as shown. The initially rectangular element deforms into a parallelogram if it is assumed infinitesimal in size and if only first order effects are considered.

Now, μ is defined by Newton's relation [Eq. (2.4)],

$$\tau_l = \mu \frac{\partial v_x}{\partial y}. \tag{6.2}$$

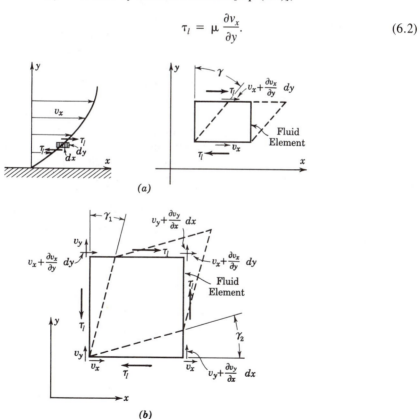

FIGURE 6.3. Angular deformation of fluid elements by shear stresses.

The symbol τ_l is used to denote the *laminar* shear stress. The subscript l is used to emphasize the fact that the definition applies only to the laminar case and to offer a distinction from the symbol τ, which is used to denote time. This use of the same symbol to denote shear stress and time may appear to be confusing; however, the context should make it clear which quantity is meant, and rarely will the two be involved in the same equation.

The above interrelation between the shear stress and the velocity gradient may be given a geometrical interpretation. If, as suggested in Fig. 6.3(a), the variation of the x velocity component, v_x, in the y direction is represented as linear (i.e., if only two terms in a Taylor's expansion of v_x are used), then the relative motion between the two sides of the element is $(\partial v_x/\partial y)\,dy$, so that the rate of *angular* deformation is

$$\frac{d\gamma}{d\tau} = \frac{\partial v_x}{\partial y},\tag{6.3}$$

γ being the angle shown in Fig. 6.3(a). Thus, the dynamic viscosity may be interpreted as a proportionality constant between the shear stress and the rate of angular deformation

$$\tau_l = \mu\,\frac{\partial v_x}{\partial y} = \mu\,\frac{d\gamma}{d\tau}.$$

If, now, one is concerned with motion in two directions, one type of deformation that may occur is that illustrated in Fig. 6.3(b). Here a fluid element (assumed initially square for simplicity) is acted on by shear stresses in two directions. A general two-dimensional motion would involve, perhaps, rotation and translation of the element as well as the shearing action shown. Also, linear deformation of the sides of the element might also be present. However, for simplicity of discussion assume that only the shearing actions, as represented by the cross derivatives $(\partial v_x/\partial y)$ and $(\partial v_y/\partial x)$, are present and that any rotation or translation be removed by superposition, resulting in the relative motion shown in the figure. Thus, the relative motion of the sides of the originally square fluid element deforms it into the suggested diamond shape. The absence of any rotation means that the shear stresses in the two directions must be equal and oriented as shown. If γ_1 and γ_2 represent the relative angular deformation of the two sides of the element, it is apparent that the *total rate* of angular deformation is

$$\frac{d\gamma_1}{d\tau} + \frac{d\gamma_2}{d\tau} = \frac{\partial v_x}{\partial y} + \frac{\partial v_y}{\partial x}.$$

Extending the above-mentioned concept of dynamic viscosity to this case, define μ such that

$$\tau_l = \mu\left(\frac{d\gamma_1}{d\tau} + \frac{d\gamma_2}{d\tau}\right),$$

$$= \mu\left(\frac{\partial v_x}{\partial y} + \frac{\partial v_y}{\partial x}\right).\tag{6.4}$$

This extended definition of μ includes the earlier one as a special case.

The deformations considered in the foregoing have been only angular deformations. These angular deformations are produced by the cross derivatives of the velocity components, $\partial v_x/\partial y$ and $\partial v_y/\partial x$, and give rise to the shear stress τ_l. It is also possible to produce a *linear* deformation of the fluid through derivatives of the form $\partial v_x/\partial x$ and $\partial v_y/\partial y$. Such relative motion of a fluid element produces linear stretching (or compression) deformation, giving rise to *normal* stresses. Figure 6.4 illustrates how a normal stress in the x direction (all other deformations being omitted for clarity), call it σ_x, would deform an initially square fluid particle. One face of the fluid element normal to the x direction will move relative to the other by an amount given by $\partial v_x/\partial x$, to a first-order approximation. It is reasonable to expect that the normal stress, σ_x, is related to the linear deformation $\partial v_x/\partial x$. This relation may be shown to be (Ref. 1), for an incompressible fluid,

$$\sigma_x = 2\mu \frac{\partial v_x}{\partial x}.$$

The fact that the same proportionality constant, μ, is involved in this definition as that for the shear stress arises from the fact that the linear deformation, $\partial v_x/\partial x$, induces a shear stress within the fluid element since it produces a rotation (or angular deformation) of the diagonal of the element, as suggested in Fig. 6.4. Incompressibility requires that the area of the fluid element remain constant, and thus the linear deformation may be related to the angular deformation of the diagonal.

The foregoing discussion was oversimplified in order to illustrate the principles involved. In general, fluid motion in two dimensions involves both linear deformations, as given by the normal derivatives $\partial v_x/\partial x$ and $\partial v_y/\partial y$, and angular deformations, as given by the cross derivatives $\partial v_x/\partial y$ and $\partial v_y/\partial x$. These deformations are related to the normal stresses, σ_x and σ_y, and the shear stress τ_l. The stresses and the deformations are interrelated by Stokes' hypothesis:

FIGURE 6.4. Deformation of a fluid element by normal stress.

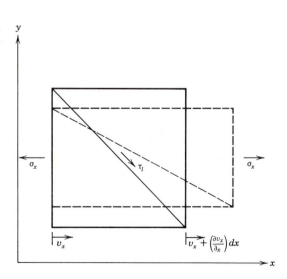

$$\tau_l = \mu \left(\frac{\partial v_x}{\partial y} + \frac{\partial v_y}{\partial x} \right),$$

$$\sigma_x = 2\mu \frac{\partial v_x}{\partial x}, \qquad (6.5)$$

$$\sigma_y = 2\mu \frac{\partial v_y}{\partial y}.$$

In the event the fluid is compressible, more complicated stress–strain relations result (Ref. 1).

The above relations between the fluid deformation and the viscous stresses will be used in the application of Newton's second law of motion to the analysis of viscous fluid motion.

6.4

THE SUBSTANTIAL DERIVATIVE

The fundamental laws of physics (e.g., Newton's law of motion, the laws of conservation of energy, etc.) are usually expressed as applying to a constant collection of matter. In the analysis of fluid motion, then, it is necessary to apply these principles to a given fluid element or "particle" which always consists of the same mass of fluid. The fluid is moving with respect to a selected, fixed coordinate system; and it is desirable, for practical reasons, to obtain expressions of these fundamental laws in terms of this coordinate system.

Since the fluid is moving relative to the coordinate system, a given fluid element will experience changes in the flow properties because of the fact that the element has moved from one point in the fluid domain to another point where the flow properties (pressure, temperature, velocity, etc.) have different values. Since it takes a certain length of time for the fluid element to move from one point to another, the above-described changes in the flow properties may be interpreted as changes with respect to time. These changes may also be accompanied by changes caused by a local dependence of the flow properties on time. In the case of steady flow this latter time dependence is absent, but the fluid element will still undergo time changes in its properties due to the mechanism described above. For this reason, care must be taken to distinguish between partial derivatives and total derivatives taken with respect to time in a moving fluid field.

To present this argument in a more formal way, consider a fluid element for which φ is used to symbolize any scalar flow parameter (e.g., pressure, temperature, velocity component, etc.), the element experiences a change $d\varphi$ as it moves from one point, (x, y), to another point, $(x + dx, y + dy)$. This change, $d\varphi$, is composed of the two changes described above. In general, since φ is dependent on time and the coordinate location of the fluid element, one may write

$$\varphi = \varphi(x, y, \tau),$$

where τ represents time. Then, the change in φ as the element moves from (x, y) to $(x + dx, y + dy)$ is given by the total differential of φ:

$$d\varphi = \frac{\partial \varphi}{\partial \tau} \, d\tau + \frac{\partial \varphi}{\partial x} \, dx + \frac{\partial \varphi}{\partial y} \, dy.$$

If $d\varphi$ is the change, then dx and dy represent the coordinate displacements of the fluid element, and $d\tau$ represents the lapse of time during this displacement. Then the total time rate of change of φ is

$$\frac{d\varphi}{d\tau} = \frac{\partial\varphi}{\partial\tau} + \frac{\partial\varphi}{\partial x}\frac{dx}{d\tau} + \frac{\partial\varphi}{\partial y}\frac{dy}{d\tau}.$$

Since dx and dy are the x and y displacements of the fluid element, $dx/d\tau$ and $dy/d\tau$ are identified as the x and y components of the velocity of the fluid element—i.e., $v_x = dx/d\tau$ and $v_y = dy/d\tau$. Thus,

$$\frac{d\varphi}{d\tau} = \frac{\partial\varphi}{\partial\tau} + v_x\frac{\partial\varphi}{\partial x} + v_y\frac{\partial\varphi}{\partial y}. \tag{6.6}$$

This last expression is usually referred to as the *substantial derivative* of φ with respect to time. It expresses the total time rate of change of φ for a fluid element moving in a time-dependent velocity field. The first term on the right side of Eq. (6.6), $\partial\varphi/\partial\tau$, is called the *local derivative* of φ with respect to time and is the rate of change of φ due to local variations. The second term, $v_x(\partial\varphi/\partial x) + v_y(\partial\varphi/\partial y)$, is called the *convective derivative* of φ and is due to the fact that the element is moving in a field in which φ varies from point to point.

The term *steady flow* is used to denote the flow condition in which there is no *local* dependence on time of any of the flow properties. That is, "steady flow" means that $\partial\varphi/\partial\tau = 0$, *but* $d\varphi/d\tau \neq 0$, necessarily. In the case of steady flow the substantial derivative consists solely of the convective part:

$$\frac{d\varphi}{d\tau} = v_x\frac{\partial\varphi}{\partial x} + v_y\frac{\partial\varphi}{\partial y}. \tag{6.7}$$

Thus, a time rate of change of a flow parameter may occur in steady flow when attention is directed to a particular fluid element.

6.5

THE EQUATION OF MOTION

Newton's second law may be applied to the motion of a fluid element to relate its acceleration to the forces acting on it. The forces acting on the fluid element include the viscous forces discussed in Sec. 6.3 but also involve forces resulting from gradients in the fluid pressure and "body forces" which act throughout the fluid domain. The force of gravity is the most frequently encountered example of a body force.

For a body of given mass, Newton's second law states that the product of the mass and the body's acceleration is equal to the sum of the forces acting on the body. As just stated, this law of motion applies to a body of constant mass; thus its application to the analysis of fluid motion must consider a particular fluid element and move with that element as discussed in Sec. 6.4. Figure 6.5 depicts such a fluid element—of dimensions δx, δy, and unit depth. Since Newton's law is a vector statement, its application to the fluid element yields two scalar equations, one in each coordinate direction. For the sake of clarity, only

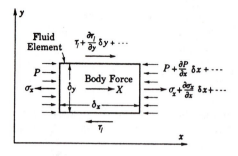

FIGURE 6.5

the motion in the coordinate direction x is depicted in Fig. 6.5. Acting in this direction, then, are the following forces:

Static pressure on the δy faces: $\quad P$ and $P + \dfrac{\partial P}{\partial x} \delta x + \cdots$.

Normal stress on the δy faces: $\quad \sigma_x$ and $\sigma_x + \dfrac{\partial \sigma_x}{\partial x} \delta x + \cdots$.

Shear stress on the δx faces: $\quad \tau_l$ and $\tau_l + \dfrac{\partial \tau_l}{\partial y} \delta y + \cdots$.

A body force (per unit mass): $\quad X$.

The body force per unit mass in the x direction, X, is included for generality. If this force is that of gravity and if the coordinate directions are chosen so that gravity acts parallel to the y direction, X may be zero.

Noting that the static pressure P and the normal stress σ_x act on areas equal to $\delta y \cdot 1$, whereas the shear stresses act on areas equal to $\delta x \cdot 1$, Newton's law applied to the fluid element in the x direction gives

$$
\begin{aligned}
a_x(\delta x \, \delta y)\, \rho = {} & \left[P - \left(P + \frac{\partial P}{\partial x} \delta x + \cdots \right) \right] \delta y \\
& - \left[\sigma_x - \left(\sigma_x + \frac{\partial \sigma_x}{\partial x} \delta x + \cdots \right) \right] \delta y \\
& - \left[\tau_l - \left(\tau_l + \frac{\partial \tau_l}{\partial y} \delta y + \cdots \right) \right] \delta x \\
& + (\delta x \, \delta y)\rho X.
\end{aligned}
$$

In the above expression ρ denotes the fluid density and a_x is the acceleration of the element in the x direction. Dividing by the mass of the element $(\delta x \, \delta y)\rho$ and allowing the element size to vanish, $\delta x \to \delta y \to 0$, one obtains

$$
a_x = X - \frac{1}{\rho} \frac{\partial P}{\partial x} + \frac{1}{\rho}\left(\frac{\partial \sigma_x}{\partial x} + \frac{\partial \tau_l}{\partial y} \right).
$$

The acceleration a_x is $dv_x/d\tau$, the substantial derivative of the x component of the velocity. Thus, Eq. (6.6) gives

$$\frac{\partial v_x}{\partial \tau} + v_x \frac{\partial v_x}{\partial x} + v_y \frac{\partial v_x}{\partial y} = X - \frac{1}{\rho}\frac{\partial P}{\partial x} + \frac{1}{\rho}\left(\frac{\partial \sigma_x}{\partial x} + \frac{\partial \tau_l}{\partial y}\right). \qquad (6.8a)$$

Similarly, motion in the y direction must satisfy

$$\frac{\partial v_y}{\partial \tau} + v_x \frac{\partial v_y}{\partial x} + v_y \frac{\partial v_y}{\partial y} = Y - \frac{1}{\rho}\frac{\partial P}{\partial y} + \frac{1}{\rho}\left(\frac{\partial \sigma_y}{\partial y} + \frac{\partial \tau_l}{\partial x}\right), \qquad (6.8b)$$

where Y is the body force component in the y direction. The above equations relate the fluid motion to the stress state acting in the fluid. As discussed in Sec. 6.3, the stresses may be related to the velocity field by use of Stokes' hypothesis given in Eq. (6.5). Substitution, then, of Eq. (6.5) into Eqs. (6.8) and the subsequent use of the continuity equation, given in Eq. (6.1), to simplify terms, yield finally

$$\frac{\partial v_x}{\partial \tau} + v_x \frac{\partial v_x}{\partial x} + v_y \frac{\partial v_x}{\partial y} = X - \frac{1}{\rho}\frac{\partial P}{\partial x} + \nu\left(\frac{\partial^2 v_x}{\partial x^2} + \frac{\partial^2 v_x}{\partial y^2}\right).$$

$$\frac{\partial v_y}{\partial \tau} + v_x \frac{\partial v_y}{\partial x} + v_y \frac{\partial v_y}{\partial y} = Y - \frac{1}{\rho}\frac{\partial P}{\partial y} + \nu\left(\frac{\partial^2 v_y}{\partial x^2} + \frac{\partial^2 v_y}{\partial y^2}\right). \qquad (6.9)$$

The last two equations govern the laminar, nonsteady motion of an incompressible fluid in two dimensions and are known as the *Navier–Stokes equations of motion*. It should be noted that in Eqs. (6.9) the ratio of the dynamic viscosity to the density has been replaced by the kinematic viscosity, ν, defined as μ/ρ in Sec. 2.7. In certain fluids and under the appropriate conditions, the effect of the viscous forces may be neglected in comparison to the inertia forces in the above equations. If such is the case, setting $\nu = 0$ in the Navier–Stokes equations yields the simpler *Euler equations of motion* of an inviscid fluid:

$$\frac{\partial v_x}{\partial \tau} + v_x \frac{\partial v_x}{\partial x} + v_y \frac{\partial v_x}{\partial y} = X - \frac{1}{\rho}\frac{\partial P}{\partial x},$$

$$\frac{\partial v_y}{\partial \tau} + v_x \frac{\partial v_y}{\partial x} + v_y \frac{\partial v_y}{\partial y} = Y - \frac{1}{\rho}\frac{\partial P}{\partial y}. \qquad (6.10)$$

For steady motion $\partial v_x/\partial \tau = \partial v_y/\partial \tau = 0$, and the Navier–Stokes equations become

$$v_x \frac{\partial v_x}{\partial x} + v_y \frac{\partial v_x}{\partial y} = X - \frac{1}{\rho}\frac{\partial P}{\partial x} + \nu\left(\frac{\partial^2 v_x}{\partial x^2} + \frac{\partial^2 v_x}{\partial y^2}\right),$$

$$v_x \frac{\partial v_y}{\partial x} + v_y \frac{\partial v_y}{\partial y} = Y - \frac{1}{\rho}\frac{\partial P}{\partial y} + \nu\left(\frac{\partial^2 v_y}{\partial x^2} + \frac{\partial^2 v_y}{\partial y^2}\right). \qquad (6.11)$$

THE ENERGY EQUATION—
THE FIRST LAW OF THERMODYNAMICS

If only the hydrodynamic picture of a given flow condition is desired, the quantities required to describe completely the flow (for the two-dimensional, incompressible case) are the velocity components, v_x and v_y, and the fluid pressure P. Thus, for a given set of boundary conditions one needs to determine how these quantities depend on time and on the coordinates x and y. The continuity equation, Eq. (6.1), and the two Navier–Stokes equations, Eqs. (6.9), are the three equations needed to find the three unknowns.

If heat transfer by convection is being considered, it is also necessary to determine the fluid temperature, t, at each point in the field of flow. This fourth dependent variable requires the use of a fourth fundamental equation. This additional relation, known as the *energy equation,* is obtained from the first law of thermodynamics.

Since, as in the derivation of the Navier–Stokes equations, a particular fluid element (always consisting of the same mass of fluid) is being considered, it is thermodynamically a "closed system." For a closed system, the first law of thermodynamics may be written in the following way:

$$\frac{dQ}{d\tau} = \frac{dU}{d\tau} + \frac{dW}{d\tau}.$$

Here $dQ/d\tau$ is used to symbolize the rate at which heat is crossing the boundary of the fluid element (presumably by conduction), $dU/d\tau$ is the rate of change of the internal energy of the element, and $dW/d\tau$ is the rate at which work is done on the element. Since only incompressible fluids are being considered, no work may be done in the form of compression or expansion of the element. Thus, the only form in which work will be present is that of dissipation through the action of the viscous forces. If $dW_d/d\tau$ is used to represent this rate of dissipatively done work,

$$\frac{dQ}{d\tau} = \frac{dU}{d\tau} + \frac{dW_d}{d\tau}.$$

If, now, $dQ'/d\tau$ and $dW_d'/d\tau$ are used to denote the rates of heat addition and viscous dissipative work *per unit volume* and if u denotes the *specific* internal energy,

$$\frac{dQ'}{d\tau} = \rho \frac{du}{d\tau} + \frac{dW_d'}{d\tau}. \tag{6.12}$$

By assuming that the thermal internal energy, u, is dependent on temperature only, this equation may be written

$$\frac{dQ'}{d\tau} = \rho c \frac{dt}{d\tau} + \frac{dW_d'}{d\tau},$$

where c is the specific heat of the fluid. The specific heat which is customarily used is that at constant pressure, although this actually involves the making of additional assumptions.* Thus, for the incompressible fluid,

$$\frac{dQ'}{d\tau} = \rho c_p \frac{dt}{d\tau} + \frac{dW'_d}{d\tau}.$$

Now, $dt/d\tau$ is the substantial derivative of the fluid element temperature so that Eq. (6.6) leads to

$$\frac{dQ'}{d\tau} = \rho c_p \left(\frac{\partial t}{\partial \tau} + v_x \frac{\partial t}{\partial x} + v_y \frac{\partial t}{\partial y} \right) + \frac{dW'_d}{d\tau}. \tag{6.13}$$

The heat that is added to the fluid element from its surroundings is added by the mechanism of *conduction* arising from the temperature difference between the element and the surrounding fluid. The derivation of the heat conduction equation in Sec. 1.4 showed that the heat conducted per *unit time* per *unit volume* into an element is (for two dimensions)

$$\frac{dQ'}{d\tau} = k \left(\frac{\partial^2 t}{\partial x^2} + \frac{\partial^2 t}{\partial y^2} \right). \tag{6.14}$$

The expression for $dW'_d/d\tau$, the rate at which dissipative work is done on the fluid element by the viscous forces per unit volume is difficult to obtain in a concise fashion. Hence, only the result is given here—the reader being referred to texts on fluid dynamics, such as Ref. 1, for a derivation of the fact that for two-dimensional, incompressible flow

$$\frac{dW'_d}{d\tau} = - \left[\sigma_x \frac{\partial v_x}{\partial x} + \sigma_y \frac{\partial v_y}{\partial y} + \tau_l \left(\frac{\partial v_x}{\partial y} + \frac{\partial v_y}{\partial x} \right) \right].$$

*A general thermodynamic relation (see, for instance, Ref. 3) between the specific heats at constant volume is

$$c_p - c_v = T \frac{\beta^2 B}{\rho}.$$

In this expression β is the coefficient of volume expansion (see Sec. 2.6), B is the isothermal bulk modulus, T is the absolute temperature, and ρ is the density. Since the bulk modulus is defined as $B = \rho(\partial p/\partial \rho)_T$, then

$$c_p - c_v = \frac{T\beta^2}{(\partial \rho/\partial p)_T}.$$

If by "incompressible" it is meant that the density is truly a constant, this means that ρ does not change with *pressure or temperature;* i.e., no thermal expansion is allowed. If this is taken as the definition of an incompressible fluid, then $c_p = c_v$ since β^2 goes to zero faster than $(\partial \rho/\partial p)_T$. Thus, the use of c_p in the energy equation, Eq. (6.13), is justified. In some cases, however, this equation is used even though $\beta \neq 0$, although the fluid may be treated as incompressible as far as continuity principles are concerned. In such cases (e.g., low-speed air flow) the use of Eq. (6.13) is only an approximation.

The same result is often obtained a second way by introducing the enthalpy, h, into Eq. (6.12). Since $h = u + p/\rho$, Eq. (6.12) is

$$\frac{dQ'}{d\tau} = \rho \frac{dh}{d\tau} + \frac{dW'_d}{d\tau} - \rho \frac{d}{d\tau} \left(\frac{p}{\rho} \right).$$

Since ρ is constant $d(p/\rho)/d\tau = 1/\rho(dp/d\tau)$. Then, by using the fact that $dh = c_p\, dt$, Eq. (6.13) is obtained *if* it is assumed that the substantial derivative of p is negligible—i.e., if $dp/d\tau = \partial p/\partial \tau + v_x(\partial p/\partial x) + v_y(\partial p/\partial y) = 0$. Neglecting $dp/d\tau$ is equivalent to assuming that the fluid is incompressible and $\beta = 0$.

Introducing the definitions of the viscous stress given by Eq. (6.5), one obtains

$$\frac{dW_d'}{d\tau} = -\mu \left[2\left(\frac{\partial v_x}{\partial x}\right)^2 + 2\left(\frac{\partial v_y}{\partial y}\right)^2 + \left(\frac{\partial v_x}{\partial y} + \frac{\partial v_y}{\partial x}\right)^2 \right]. \qquad (6.15)$$

This term represents the rate at which mechanical energy is being dissipated into thermal energy by the action of the viscous forces. It must, because of the requirements of the second law of thermodynamics, be negative; i.e., there must be a loss of mechanical energy and a corresponding gain in thermal energy.

Substitution of Eqs. (6.14) and (6.15) into Eq. (6.13) gives, finally, the accepted form of the *energy equation* for two-dimensional, incompressible, laminar flow of a viscous fluid:

$$k\left(\frac{\partial^2 t}{\partial x^2} + \frac{\partial^2 t}{\partial y^2}\right) + \mu\left[2\left(\frac{\partial v_x}{\partial x}\right)^2 + 2\left(\frac{\partial v_y}{\partial y}\right)^2 + \left(\frac{\partial v_x}{\partial y} + \frac{\partial v_y}{\partial x}\right)^2\right]$$
$$= c_p\rho\left(v_x\frac{\partial t}{\partial x} + v_y\frac{\partial t}{\partial y} + \frac{\partial t}{\partial \tau}\right), \qquad (6.16)$$

or

$$\alpha\left(\frac{\partial^2 t}{\partial x^2} + \frac{\partial^2 t}{\partial y^2}\right) + \frac{\mu}{\rho c_p}\left[2\left(\frac{\partial v_x}{\partial x}\right)^2 + 2\left(\frac{\partial v_y}{\partial y}\right)^2 + \left(\frac{\partial v_x}{\partial y} + \frac{\partial v_y}{\partial x}\right)^2\right]$$
$$= \left(v_x\frac{\partial t}{\partial x} + v_y\frac{\partial t}{\partial y} + \frac{\partial t}{\partial \tau}\right). \qquad (6.17)$$

In the latter form the thermal diffusivity defined in Eq. (1.9) has been introduced.

In problems of heat convection in a two-dimensional incompressible laminar fluid motion this equation, Eq. (6.16), must be solved simultaneously with the two Navier–Stokes equations, Eqs. (6.9), and the continuity equation, Eq. (6.1), to find the unknown quantities v_x, v_y, P, and t for a given set of boundary conditions (such as the shape and temperature of the heated body about which the fluid flows, the fluid velocity and temperature far removed from the body, etc.). Once the temperature distribution of the fluid is known, the convective heat transfer coefficient could be determined at a particular point on the body by the method outlined in the introductory paragraphs of this chapter.

Obviously, the actual execution of such an analytical procedure would be an extremely complex and laborious, if not impossible, undertaking—even for the simplest of geometrical bodies and boundary conditions.

The main reason that these basic equations governing the convection mechanism are presented here is not to impress the student with a complex system of differential equations merely for the reason of mathematical elegance. The complexity of the equations and the inability to obtain analytical solutions quickly to practical convection problems do not eliminate the need of the engineer to have answers to *real* problems. These basic equations are presented to justify making some rather far-reaching, simplifying assumptions for the very practical reason that there is no other way to obtain a useful engineering answer. These basic laws will also provide a rational basis for such simplifications. They also aid the engineer in being logical when an empirical attack on a problem is necessary. Thus, in this respect the study of heat transfer involves the effective combination of powerful analytical methods and the empirical results of experiment.

SIMILARITY PARAMETERS IN HEAT TRANSFER

As noted in the foregoing sections, the analytic solution to a heat transfer problem involves the simultaneous solution of the two Navier–Stokes equations, Eq. (6.9), the continuity equation, Eq. (6.1), and the energy equation, Eq. (6.17), for the four dependent variables v_x, v_y, P, and t. Once t is known as a function of position, the heat transfer coefficient h may be found from its definition. Some solutions will be demonstrated in the next chapter, but it should be apparent that the equations involved are sufficiently complex that analytic solutions can be obtained for only a few cases involving rather simple boundary conditions and geometries. However, certain information concerning the form of such solutions can be obtained without solving the equations themselves by examining the conditions under which the solution of two different problems have similarity. Thus, in much the same way as was done in Sec. 4.4 to show that the Fourier and Biot numbers are important similarity parameters for transient conduction, one may use the above-named basic equations to deduce the significant parameters for convection. This is not only useful in predicting the form of expected solutions, it will also be valuable in correlating data when experimental studies are conducted.

The question then is: Given the flow of a fluid of known velocity and temperature about a body of known temperature, under what conditions will the resultant velocity, pressure, and temperature fields be similar to that of some other fluid at other conditions flowing about another body having *geometric* similarity? Or, as depicted in Fig. 6.6, given two bodies that are geometrically similar, but subjected to different hydrodynamic and thermal conditions, when can the heat transfer results in one case yield useful information about the other? By appropriately nondimensionalizing the governing equations, one may ascertain the conditions under which geometrically similar situations also have *dynamic* and *energetic* similarity.

For simplicity of discussion, the case of *steady* flow of a fluid at a temperature t_f past a surface maintained at t_s will be considered, such as depicted in Fig. 6.6. To maintain some generality, let a body force due to bouyancy exist. By selecting the x coordinate in the direction of the gravity vector, the body forces in the Navier–Stokes equations become $Y = 0$, $X = g\beta(t - t_f)$, where β is the coefficient of thermal expansion. (This expression for the bouyant force involves some assumptions that are discussed in more detail in Chapter 9.) Under these stated conditions, Eqs. (6.1), (6.9), and (6.17) become

$$\frac{\partial v_x}{\partial x} + \frac{\partial v_y}{\partial y} = 0,$$

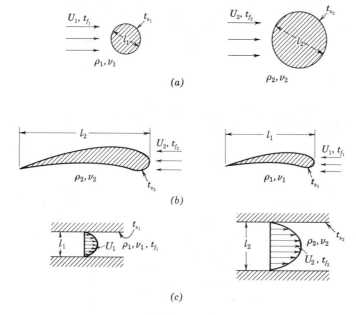

FIGURE 6.6

$$v_x \frac{\partial v_x}{\partial x} + v_y \frac{\partial v_x}{\partial y} = -\frac{1}{\rho} \frac{\partial P}{\partial x} + \nu \left(\frac{\partial^2 v_x}{\partial x^2} + \frac{\partial^2 v_x}{\partial y^2} \right) + g\beta(t - t_f),$$

$$v_x \frac{\partial v_y}{\partial x} + v_y \frac{\partial v_y}{\partial y} = -\frac{1}{\rho} \frac{\partial P}{\partial y} + \nu \left(\frac{\partial^2 v_y}{\partial x^2} + \frac{\partial^2 v_y}{\partial y^2} \right),$$

$$k \left(\frac{\partial^2 t}{\partial x^2} + \frac{\partial^2 t}{\partial y^2} \right) + \mu \left[2 \left(\frac{\partial v_x}{\partial x} \right)^2 + 2 \left(\frac{\partial v_y}{\partial y} \right)^2 + \left(\frac{\partial v_x}{\partial y} + \frac{\partial v_y}{\partial x} \right)^2 \right]$$

$$= \rho c_p \left(v_x \frac{\partial t}{\partial x} + v_y \frac{\partial t}{\partial y} \right).$$

The nondimensionalization of these equations is performed by first selecting certain characteristic quantities. For instance, one selects a characteristic length, call it l, which characterizes the size of the geometrically similar bodies—say, the diameter of the cylinders in Fig. 6.6(a), or the airfoil chord in Fig. 6.6(b). Similarly, one chooses a characteristic velocity, U (usually the free stream velocity in external flow cases, or the average or maximum velocity in internal flows). Since the density is constant, a characteristic pressure can be defined in terms of the already chosen characteristic velocity by simply defining it as $\frac{1}{2}\rho U^2$. To avoid difficulties with a temperature scale, one needs to work in terms of the temperature difference, say $t - t_f$. A natural characteristic temperature difference is then $\Delta t = t_s - t_f$ where t_s is the surface temperature. Thus, in summary, the characteristic parameters chosen are:

Characteristic length: l.

Characteristic velocity: U.

Characteristic pressure: $\frac{1}{2}\rho U^2$.

Characteristic temperature difference: $\Delta t = t_s - t_f$.

With these characteristic parameters chosen, the independent and dependent variables may be nondimensionalized according to the following definitions:

Dimensionless coordinates: $\xi = \dfrac{x}{l}$, $\eta = \dfrac{y}{l}$.

Dimensionless velocities: $v_\xi = \dfrac{v_x}{U}$, $v_\eta = \dfrac{v_y}{U}$.

Dimensionless pressure: $\Pi = P/\frac{1}{2}\rho U^2$.

Dimensionless temperature: $\varphi = \dfrac{t - t_f}{\Delta t} = \dfrac{t - t_f}{t_s - t_f}$.

Then, in two geometrically similar cases, similar spatial points result for equal values of ξ and η; the velocities are similar when v_ξ and v_η are equal in the two cases; the pressures are similar when the Π's are the same; the temperatures are similar when the φ's are equal.

If the dimensionless variables just defined are introduced into the above equations, the following results are obtained:

$$\frac{\partial v_\xi}{\partial \xi} + \frac{\partial v_\eta}{\partial \eta} = 0,$$

$$v_\xi \frac{\partial v_\xi}{\partial \xi} + v_\eta \frac{\partial v_\xi}{\partial v} = -\frac{1}{2}\frac{\partial \Pi}{\partial \xi} + \frac{v}{Ul}\left(\frac{\partial^2 v_\xi}{\partial \xi^2} + \frac{\partial^2 v_\xi}{\partial \eta^2}\right) + \frac{g\beta l\,\Delta t}{U^2}\,\varphi,$$

$$v_\xi \frac{\partial v_\eta}{\partial \xi} + v_\eta \frac{\partial v_\eta}{\partial \eta} = -\frac{1}{2}\frac{\partial \Pi}{\partial \eta} + \frac{v}{Ul}\left(\frac{\partial^2 v_\eta}{\partial \xi^2} + \frac{\partial^2 v_\eta}{\partial \eta^2}\right),$$

$$\left(\frac{\partial^2 \varphi}{\partial \xi^2} + \frac{\partial^2 \varphi}{\partial \eta^2}\right) + \frac{\mu U^2}{k\Delta t}\left[2\left(\frac{\partial v_\xi}{\partial \xi}\right)^2 + 2\left(\frac{\partial v_\eta}{\partial \eta}\right)^2 + \left(\frac{\partial v_\xi}{\partial \eta} + \frac{\partial v_\eta}{\partial \xi}\right)^2\right]$$

$$= \frac{\rho c_p Ul}{k}\left(v_\xi \frac{\partial \varphi}{\partial \xi} + v_\eta \frac{\partial \varphi}{\partial \eta}\right).$$

Examination of these equations reveals that identical solutions for the dimensionless variables will be obtained when the various coefficients involving groupings of the characteristic parameters (l, U, Δt) and fluid properties (β, v, μ, ρ, c_p, k) are the same. Thus, these groupings identify the relevant similarity parameters. These similarity parameters may be identified by remembering that the object of the analysis is the determination of the temperature distribution in the fluid for subsequent evaluation of the heat transfer coefficient. Imagine, then, that the four equations above are solved for the dimensionless temperature, φ. This solution may be represented by the functional relation

$$\varphi = f\left(\xi, \eta, \frac{v}{Ul}, \frac{g\beta l\,\Delta t}{U^2}, \frac{\mu U^2}{k\,\Delta t}, \frac{\rho c_p Ul}{k}\right). \tag{6.18}$$

The above solution for φ is seen to depend on, in addition to the dimensionless space variables, four dimensionless groupings of the physical parameters of the problem. These four groupings could be used as similarity parameters as they stand; however, it is customary to rearrange them in the following way:

$$\varphi = f\left(\xi,\ \eta,\ \frac{v}{Ul},\ \frac{gl^3\beta\,\Delta t}{v^2}\frac{v^2}{U^2l^2},\ \frac{U^2}{c_p\,\Delta t}\frac{\mu c_p}{k},\ \frac{Ul}{v}\frac{\mu c_p}{k}\right).$$

Certain groups of parameters are seen to occur more than once in the above formulation, and since only a functional dependence is being examined, the above also implies the following:

$$\varphi = f\left(\xi,\ \eta,\ \frac{Ul}{v},\ \frac{gl^3\beta\,\Delta t}{v^2},\ \frac{U^2}{c_p\,\Delta t},\ \frac{\mu c_p}{k}\right). \tag{6.19}$$

The four groups of physical parameters in Eq. (6.19) are the sought-for similarity parameters for the general case of the flow of a viscous fluid past a heated surface. The first of these, Ul/v, is recognized as the Reynolds number so frequently encountered in fluid mechanics. The others also bear special symbols and names:

$$\mathrm{Re} = \frac{Ul}{v} = \text{the Reynolds number,}$$

$$\mathrm{Gr} = \frac{gl^3\beta\,\Delta t}{v^2} = \text{the Grashof number,}$$

$$\mathrm{Ec} = \frac{U^2}{c_p\,\Delta t} = \text{the Eckert number,}$$

$$\mathrm{Pr} = \frac{\mu c_p}{k} = \frac{v}{\alpha} = \text{the Prandtl number.}$$

$$\tag{6.20}$$

The Reynolds number, Re, represents a measure of the magnitude of the inertia forces in the fluid to the viscous forces. The Grashof number, Gr, is a measure of the ratio of the buoyant forces to the viscous forces and is absent if buoyancy is neglected. The Eckert number, Ec, is a measure of the thermal equivalent of the kinetic energy of the flow to the imposed temperature difference and arises from the inclusion of the viscous dissipation term in the energy equation. Hence, Ec, is absent when dissipation is neglected. The Prandtl number, Pr, was discussed in Sec. 2.8. Unlike the others, it is composed only of physical fluid properties and represents one of the most significant heat transfer parameters— being a measure of the relative magnitude of the diffusion of momentum, through viscosity, and the diffusion of heat, through conduction, in the fluid.

Remembering now that the ultimate result desired is the heat transfer coefficient, one would obtain it from the definition

$$h = -\frac{k(\partial t/\partial n)_s}{t_s - t_f},$$

where $(\partial t/\partial n)_s$ represents the fluid temperature gradient at the body surface, measured in a direction normal to the surface. If the dimensionless variables $\varphi = (t - t_s)/\Delta t$ and $\bar{n} = n/l$ are introduced, the above definition of h becomes

$$\text{Nu} = \frac{hl}{k} = -\left(\frac{\partial \varphi}{\partial \bar{n}}\right)_s. \tag{6.21}$$

Thus, a dimensionless heat transfer coefficient has been defined:

$$\text{Nu} = \frac{hl}{k} = \text{the Nusselt number.} \tag{6.22}$$

By use of this definition, Eqs. (6.20) and (6.21) yield

$$\text{Nu} = f(\xi_s, \eta_s, \text{Re}, \text{Gr}, \text{Ec}, \text{Pr}). \tag{6.23}$$

Thus, one expects that the local Nusselt number (i.e., local heat transfer coefficient) will depend on the position on the body surface and Re, Gr, Ec, and Pr. Often it will be desirable to evaluate an average h, or an average Nu, by appropriately integrating over the body surface. Such average coefficients will be discussed in greater detail later, but it is sufficient here to note that the resultant average Nusselt number will have the following functional dependence:

$$\text{Nu} = f(\text{Re}, \text{Gr}, \text{Ec}, \text{Pr}). \tag{6.24}$$

The implications of Eqs. (6.23) and (6.24) are twofold. First, in the event that an experimental investigation is being carried out, instead of seeking h as a function of U, l, ν, g, β, Δt, c_p, ρ, and k, the above shows that the data should correlate according to Eq. (6.24)—reducing the number of variables from nine to four. Thus, a considerable saving of effort in the taking of data, and the subsequent analysis and representation of these data, may be obtained. Second, this similarity analysis gives advance notice of the form any analytic solution would be expected to take.

Certain special forms of the above results will also be encountered. In pure *forced convection*, one often neglects any effects of buoyant forces so that

$$\text{Nu} = f(\text{Re}, \text{Ec}, \text{Pr}) \tag{6.25}$$

when viscous dissipation is included, and

$$\text{Nu} = f(\text{Re}, \text{Pr}) \tag{6.26}$$

when viscous dissipation is neglected. In the case of pure free convection, the main fluid stream is absent, so that Re is no longer significant. In addition, the flow is generally slow enough to ignore viscous dissipation. Thus, in free convection, one expects

$$\text{Nu} = f(\text{Gr}, \text{Pr}). \tag{6.27}$$

6.8

THE CONCEPT OF THE BOUNDARY LAYER

The Navier–Stokes equations of motion of a viscous fluid along with the energy equations are, mathematically speaking, very complex. Exact solutions to these equations are known for only a few cases, most of which have very specialized and often impractical boundary conditions. The complicated nature of these equations does not, however, eliminate the *need* for answers to problems of practical importance. For this reason an engineer must be content with the compromise of accepting either an approximate solution to these fundamental equations, an exact solution to simplified or approximate versions of the equations, or, sometimes, an approximate solution of approximate equations. This does not destroy the value of the exact fundamental equations, for they are necessary in order to understand the physical implications of making any approximation or simplification.

Since the Navier–Stokes equations express a balance among inertia forces, viscous forces, pressure forces, and body forces, one such simplification may come from neglecting certain of the forces as being small in comparison to others when the conditions of the flow and the relative magnitude of the terms will permit. A very important case of this is found in instances in which one decides that *all* the viscous forces are negligible with respect to the inertia and pressure forces. If this is so, one can assume that the fluid viscosity is negligibly small and the Navier–Stokes equations reduce to the simpler Euler equations [as stated in Eqs. (6.10)]. This simplification has many applications, the general theories of aerodynamics being an example.

However, when one considers the flow of a fluid past a solid surface the assumption of a vanishingly small viscosity may lead to results that are not verified in experiment. If the basic assumption is made that the fluid particles adjacent to the surface adhere to it and have zero velocity, there must exist velocity gradients in the fluid motion since the velocity must change from zero at the surface to some finite value at points removed from the surface. Immediately in the vicinity of the surface the velocity gradient may be so large that even if the fluid viscosity is small, the product of the velocity gradient and the viscosity (i.e., the viscous stress) may not be negligible. The extent of the region near the wall in which the viscous stresses may not be negligible, even in fluids of small viscosity, will depend on the properties of the fluid, the shape of the wall, the velocity of the main stream of the fluid, etc.

Even in such a region it may be possible that not all the viscous stresses are of a sizable nature, nor may all the components of the inertia forces be significant. It is on this basis that Ludwig Prandtl, in 1904, proposed his boundary layer theory. Prandtl's fundamental concept was that the motion of a fluid of small viscosity about a solid surface could be divided into two regions. One region, termed the *boundary layer,* near the body surface is defined as a region in which the velocity gradients are large enough so that the influence of viscosity cannot be neglected. The other region, termed the *potential region* (or *potential core* if flow inside a body is considered), is defined as the region in which the influence of the presence of the body has died out enough so that the velocity gradients are so small that the fluid viscosity can be ignored.

In actual flows the influence of the body extends to all regions of the fluid, but this influence and the associated changes in the fluid velocity decrease rapidly with the distance from the body—particularly if the viscosity is small and the main fluid velocity is large (i.e., at large Reynolds numbers). Since this decrease is continuous, it is obviously not possible to define a precise locus of the limit of the boundary layer and the beginning of the potential region. In practice, the limit of the boundary layer, the *boundary layer thickness,* is taken to be the distance away from the surface at which the flow velocity has achieved some arbitrary percentage of the undisturbed, free stream flow—say 90 or 99%.

6.9

THE EQUATION OF MOTION AND THE ENERGY EQUATION OF THE LAMINAR BOUNDARY LAYER

The application of Prandtl's boundary layer theory to the solution of real problems can be carried out in two ways. The first is based on differential equations of motion and energy that are obtained by simplifying the Navier–Stokes equations given above. These simplifications arise out of Prandtl's hypothesis that the boundary layer will be thin compared to the dimensions of the body about which it flows. This section will consider this approach.

The second method is based on integral equations of momentum and energy. This method attempts to describe, approximately, the over-all behavior of the boundary layer. Although the results obtained by this integral approach are not as complete and detailed as the results that may be obtained by the application of the differential equations, a greater variety of problems may be handled by this method. The integral equations of the boundary layer will be derived in Sec. 6.10.

Chapter 7 illustrates the solution of problems by these two methods.

The Equation of Motion

Visualize now the fluid motion depicted in Fig. 6.7(a), in which the fluid is moving past a solid wall, shown to be plane for simplicity. The x coordinate is measured parallel to the surface, and the y coordinate is measured normal to it. At any station, x, along the surface the x-component velocity distribution in the y direction varies somewhat as shown—from a zero value at the wall to a uniform value away from the wall.

Defining the *boundary layer thickness* as the distance from the wall beyond which the velocity gradient is small enough that the viscous stresses may be neglected, one finds that this thickness increases in some manner with x as more and more of the fluid is influenced by the retarding effect of the wall—i.e., as the flow proceeds along the surface, more and more momentum must be fed into the boundary layer to overcome the viscous stresses in the region of the high velocity gradients. Thus, the boundary layer thickness is a function of the coordinate x and will be denoted by $\delta(x)$.

The velocity of the fluid in the region external to the boundary layer—the potential region—is indicated in Fig. 6.7(a) to be a function of x, namely $U(x)$. This is done for the sake of generality since there could be other bodies or surfaces present in the flow to cause a change in the nonviscous region. For example, another wall placed above the one shown (parallel or at an angle to the one drawn) would cause an acceleration or deceleration in the potential core.

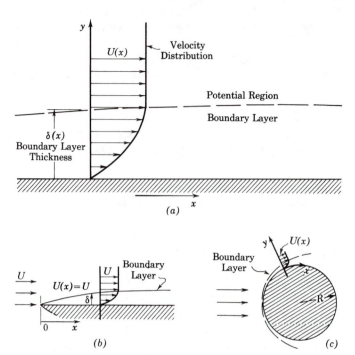

FIGURE 6.7. Boundary layer flow past solid surfaces.

If the flow depicted is that resulting from a uniform stream approaching a flat plate placed parallel to the approaching flow [such as in Fig. 6.7(b)], then $U(x)$ would be a constant. If, however, the plate were oriented at even a slight angle with the oncoming flow, the potential region would have an x-direction velocity component which would not be constant. Flow about a curved surface, such as in Fig. 6.7(c), is another case in which a nonconstant $U(x)$ would result.

This analysis will consider only the steady flow case. If one assumes that the body forces, X and Y, are negligible compared to the other terms, the Navier–Stokes equations to be applied *in* the boundary layer are, from Eqs. (6.11),

$$v_x \frac{\partial v_x}{\partial x} + v_y \frac{\partial v_x}{\partial y} = -\frac{1}{\rho}\frac{\partial P}{\partial x} + \nu\left(\frac{\partial^2 v_x}{\partial x^2} + \frac{\partial^2 v_x}{\partial y^2}\right),$$

$$v_x \frac{\partial v_y}{\partial x} + v_y \frac{\partial v_y}{\partial y} = -\frac{1}{\rho}\frac{\partial P}{\partial y} + \nu\left(\frac{\partial^2 v_y}{\partial x^2} + \frac{\partial^2 v_y}{\partial y^2}\right).$$

(6.28)

Now, from the definition of the boundary layer, one must account for the viscous stresses only when $y < \delta(x)$. Prandtl's fundamental hypothesis for the boundary layer was that the thickness must be quite small in comparison to the linear dimensions of the body [i.e., $\delta(x) << x$]. With this fact, an examination of the above two equations may be made to determine the relative order of magnitude of the various terms. The order-of-magnitude analysis is tedious, so only the results will be quoted here, although certain general justifications of the results may be made. For a detailed discussion of the analysis the reader is referred to Refs. 1 and 4.

Consider first the equation of motion for the x direction. If the layer is quite thin, the v_x component of the velocity must change very rapidly from 0 at $y = 0$ to $U(x)$ at $y = \delta$. Thus, the gradient $\partial v_x/\partial y$ may be expected to be of significant magnitude as will $\partial^2 v_x/\partial y^2$. If one takes the inertia force in the x direction, $[v_x(\partial v_x/\partial x) + v_y(\partial v_x/\partial y)]$, to be the standard order of magnitude, it is not surprising that the end results show that the viscous force $\nu(\partial^2 v_x/\partial y^2)$ may be of the same order, whereas the viscous force in the x direction due to $\nu(\partial^2 v_x/\partial x^2)$ will be of a considerably lower order of magnitude. Upon comparing the terms of the second equation of motion to the same standard order of magnitude, all the terms are found to be of a much smaller magnitude. This is due principally to the fact that the v_y component will be small if the boundary layer thickness is likewise small. If these terms of smaller orders of magnitude are neglected, the equations reduce to

$$v_x \frac{\partial v_x}{\partial x} + v_y \frac{\partial v_x}{\partial y} = -\frac{1}{\rho}\frac{\partial P}{\partial x} + \nu \frac{\partial^2 v_x}{\partial y^2},$$

(6.29)

$$\frac{\partial P}{\partial y} = 0.$$

An important fact to be observed here is, under the assumptions noted above, that the pressure gradient normal to the surface is zero throughout the boundary layer! This result implies the very useful fact that the longitudinal pressure gradient, $\partial P/\partial x$, in the boundary layer, taken in a direction parallel to the wall, is equal to the pressure gradient in the external potential flow region taken in the same direction. Thus, a basic knowledge of the laws of the flow of an inviscid fluid, as governed by the Euler equations [(6.10)], is necessary for a complete understanding of boundary layer theory.

The boundary layer equations of motion then reduce to the following single equation with a total derivative replacing the partial derivative of pressure in the x direction.*

$$v_x \frac{\partial v_x}{\partial x} + v_y \frac{\partial v_x}{\partial y} = -\frac{1}{\rho}\frac{dP}{dx} + \nu \frac{\partial^2 v_x}{\partial y^2}.$$

(6.30)

It must be remembered that this equation applies only *inside* the boundary layer. It may appear that the single equation (together with the continuity equation) is not enough to determine the dependent quantities v_x, v_y, and P. However, as just noted, the gradient dP/dx is determined from the external potential flow by first solving the inviscid Euler equations for the given geometry and then using this solution to provide dP/dx as input information to the boundary layer equation. For example, if the flow is that past a flat surface, the inviscid solution gives $\partial P/\partial x = 0$, so that $dP/dx = 0$ in Eq. (6.30).

An important point to note is that the basic assumption leading to the simplified boundary layer equations shown above is that the layer is quite thin compared to the other linear dimensions. This assumption implies a fluid of low

*In the case of free convection where the fluid is caused to move by density differences, the body force is the driving force and hence cannot be neglected as was done here. In such a case the body force must be related to the fluid temperature and the coefficient of thermal expansion. This is discussed in Sec. 6.7 and Chapter 9.

viscosity or a high external flow velocity—or both. Whether or not this condition is met in practice depends on the particular situation at hand, and the accuracy obtained by replacing the Navier–Stokes equations with these simpler expressions depends on how well this assumption is satisfied in reality.

In cases in which the fluid is flowing past nonplane surfaces (but still in two space dimensions), e.g., in the case of flow about airfoils or, say, cylinders as shown in Fig. 6.7(c), the same equations may be applied to the boundary layer formed along the curved surface if certain additional conditions are met.

Let the x coordinate be interpreted as the arc length measured along the curved surface, and let y be the coordinate measured normal to the local x direction. Thus x-y coordinates are an intrinsic system attached to the flow. If, in this system of coordinates, the boundary layer thickness is presumed to be small in comparison with the radius of curvature ($\delta << R$), the same boundary layer equations are obtained if the following additional (but not unreasonable) conditions are met: (a) There must be no *sudden* changes in the radius of curvature, and (b) the radius of curvature must be large compared to the linear displacement along the surface. In such cases of nonplane surfaces the external potential flow velocity $U(x)$ will be a function of the arc coordinate x.

The Energy Equation

Application of the same order-of-magnitude considerations to the energy equations discussed in Sec. 6.6 and shown in Eq. (6.16) will also lead to a simplified version of that equation. Equation (6.16) is repeated here for the steady state:

$$k\left(\frac{\partial^2 t}{\partial x^2} + \frac{\partial^2 t}{\partial y^2}\right) + \mu\left[2\left(\frac{\partial v_x}{\partial x}\right)^2 + 2\left(\frac{\partial v_y}{\partial y}\right)^2 + \left(\frac{\partial v_x}{\partial y} + \frac{\partial v_y}{\partial x}\right)^2\right]$$

$$= c_p\rho\left(v_x \frac{\partial t}{\partial x} + v_y \frac{\partial t}{\partial y}\right).$$

If a difference exists between the temperature at the surface and the temperature of the main fluid stream, one presumes the presence of a *thermal boundary layer* of thickness δ_t. Rather than being related to the exchange of momentum as in the case of the velocity boundary layer, the thermal boundary layer results from an exchange of thermal energy between the surface and the potential region. This thermal boundary layer thickness is also assumed to be quite small with respect to the linear dimensions of the surface, so that it is of the same order of magnitude as (but not necessarily equal to) the velocity boundary layer thickness defined above. Thus, the distribution of the temperature in the thermal boundary layer would be similar to that of velocity in the velocity boundary layer, i.e., a rapid change from some value at the surface to the temperature of the potential core, when moving in a direction normal to the surface, but with a relatively gradual gradient in the direction parallel to the surface if the rate of dissipation of mechanical energy is not too large.

Figure 6.8 depicts the typical variation of the temperature through a thermal boundary layer on a surface which is cooler than the main stream. As indicated, the thermal boundary layer may not have the same thickness as the velocity

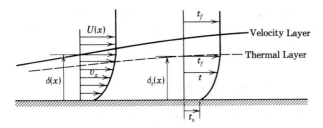

FIGURE 6.8. Velocity and thermal boundary layers.

layer. As will be shown later, the thermal layer is thinner than the velocity layer when Pr > 1 and thicker when Pr < 1. By taking the internal energy term (the right side of the above equation) as the standard order of magnitude, then, owing to the large possible gradient in the y direction, the first part of the first term of the left side of the equation may be expected to be negligible. In keeping with the assumptions made for the equations of motion, the viscous dissipation terms (i.e., those terms preceded by μ in the above equation) should again be negligible, excepting the term involving the velocity gradient normal to the wall, $\mu(\partial v_x/\partial y)^2$. The resulting simplified, approximate form of the energy equation would then be

$$v_x \frac{\partial t}{\partial x} + v_y \frac{\partial t}{\partial y} = \frac{k}{\rho c_p} \frac{\partial^2 t}{\partial y^2} + \frac{\mu}{\rho c_p} \left(\frac{\partial v_x}{\partial y} \right)^2. \tag{6.31}$$

The equations of the laminar, two-dimensional boundary layer just discussed are naturally simpler than the full Navier–Stokes and energy equations, but the analytical solution is still difficult because of the nonlinearity of the equations. A brief discussion of a so-called "exact" solution to these equations for flow parallel to a flat plate will be presented in Chapter 7. Before turning to that example, it is appropriate to discuss another approach to the problem of boundary layer flow, developed by Theodore von Kármán. Von Kármán's method consists of an integral approach to the gross behavior of the boundary layer without examining in detail the equations governing the motion of each fluid particle in the layer. This approach is considered in the next section.

6.10

THE INTEGRAL EQUATIONS OF THE LAMINAR BOUNDARY LAYER

The *integral methods* of analyzing boundary layers consist of fixing one's attention on the over-all behavior of the boundary layer (i.e., as far as the conservation of energy and momentum principles are concerned) rather than on the detailed motion of an individual fluid particle. With this approach rather crude approximations may be made to the distribution of velocity and temperature in the boundary layer without appreciable error in the resultant overall behavior. The necessary governing equations for the integral treatment of laminar incompressible boundary layers can be obtained directly by integration of the approximate equation of motion and energy equations [Eqs. (6.30) and (6.31)]. However, the equations may also be obtained by a direct derivation that will more clearly illustrate the physical processes taking place.

The Momentum Integral Equation

Turning first to the velocity boundary layer, consider the flow picture shown in Fig. 6.9. Here boundary layer flow along a surface is shown. (The surface is assumed to be plane, although x may, in reality, represent the arc-length displacement along a curved surface as discussed earlier.) It is assumed, in this analysis, that the boundary layer has a definite thickness, $\delta(x)$, which is a function of the coordinate x. It is this thickness and the way in which it varies with x, among other things, that one wishes to determine in order to analyze the convection of heat away from the wall.

At some station, say $x = x$, the distribution of the x component of the velocity is somewhat as shown in Fig. 6.9—a variation from 0, at $y = 0$, to a uniform value for all values of $y > \delta(x)$. At some station farther downstream, say $x = x + dx$, the variation is of the same nature, except that the boundary layer thickness may be greater, and the uniform potential value reached at the edge of the boundary layer may also be different. If a reference value of y, say y_r, is chosen so that it lies outside the boundary layer between x and $x + dx$, the domain enclosed between x and $x + dx$ and between $y = 0$ and y_r is defined, and the conservation laws may be applied thereto. This domain is denoted as the rectangle $ABCD$ in Fig. 6.9, and unit depth in the plane of the paper is assumed.

The mass flow rates, denoted by $\dot{m}$, across each of the sides of the rectangle may be deduced. The results, for constant density, are summarized below:

$$\dot{m}_{AD} = \left[\rho \int_0^{y_r} v_x \, dy \right]_x,$$

$$\dot{m}_{BC} = \left[\rho \int_0^{y_r} v_x \, dy \right]_{x+dx} = \left[\rho \int_0^{y_r} v_x \, dy \right]_x + \frac{d}{dx} \left[\rho \int_0^{y_r} v_x \, dy \right]_x dx, \qquad (6.32)$$

$$\dot{m}_{AB} = 0 \ (\text{solid wall}),$$

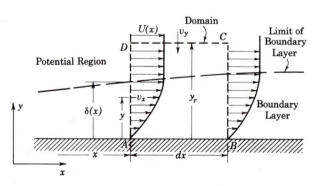

FIGURE 6.9. Laminar boundary layer flow past a solid surface.

$$\dot{m}_{DC} = \dot{m}_{BC} - \dot{m}_{AD} - \dot{m}_{AB} = \frac{d}{dx}\left[\rho \int_0^{y_r} v_x \, dy \right]_x dx.$$

A Taylor expansion has been written above, to express the mass flow at $x + dx$ in terms of the mass flow at x, with only the first two terms considered since dx will subsequently be allowed to approach 0. The necessity for the existence of a flow velocity in the y direction across the line DC is shown by the existence of $\dot{m}_{DC}$, which must be present in order to account for the difference in the flows across AD and BC.

Denoting by M the momentum flux in the x *direction* across each of the sides of the rectangle, one may write the following expressions noting that the y direction flow carries x direction momentum into the domain due to the existence of $U(x)$:

$$\dot{M}_{AD} = \left[\int_0^{y_r} (\rho v_x) v_x \, dy \right]_x = \left[\rho \int_0^{y_r} v_x^2 \, dy \right]_x,$$

$$\dot{M}_{BC} = \left[\rho \int_0^{y_r} v_x^2 \, dy \right]_x + \frac{d}{dx}\left[\rho \int_0^{y_r} v_x^2 \, dy \right]_x dx,$$

$$\dot{M}_{AB} = 0,$$

$$\dot{M}_{DC} = \dot{m}_{DC} U(x) = U(x)\frac{d}{dx}\left[\rho \int_0^{y_r} v_x \, dy \right]_x dx.$$

(6.33)

The forces acting on the domain will consist of a shear force at $y = 0$ (denote the shear stress there by τ_0), and pressure and normal stresses on the AD and BC faces. Since the dimension y_r is chosen so that DC lies outside the boundary layer, there will be no viscous shear on that face. If the dominant velocity gradient is that of v_x in the y direction, the normal stresses on AD and BC may be ignored, and the following equations may be written for the x-direction forces, F, acting on each face:

$$F_{AD} = [P]_x y_r,$$

$$F_{BC} = \left[P + \frac{dP}{dx} dx \right]_x y_r,$$

$$F_{AB} = \tau_0 \, dx = \mu\left(\frac{\partial v_x}{\partial y}\right)_{y=0} dx,$$

(6.34)

$$F_{DC} = 0.$$

The momentum principle states that for steady flow the net flux of momentum over the influx of momentum (in a selected direction) from a domain must equal the algebraic sum of the forces acting on the domain in that direction. This principle may be applied to the x-direction motion for the domain $ABCD$ of Fig. 6.9, resulting in the following expression after substitution of Eqs. (6.33) and (6.34):

$$\left[\rho \int_0^{y_r} v_x^2 \, dy + \frac{d}{dx}\left(\rho \int_0^{y_r} v_x^2 \, dy\right) dx\right]_x - \left[\rho \int_0^{y_r} v_x^2 \, dy\right]_x$$

$$- \left[U(x)\frac{d}{dx}\left(\rho \int_0^{y_r} v_x \, dy\right)\right]_x dx = [P \cdot y_r]_x - \left[P + \frac{dP}{dx}\,dx\right]_x y_r - \tau_0 \, dx.$$

Canceling some terms and noting that ρ is constant, one finds that

$$\frac{d}{dx}\int_0^{y_r} v_x^2 \, dy - U(x)\frac{d}{dx}\int_0^{y_r} v_x \, dy = -\frac{1}{\rho}\frac{dP}{dx}y_r - \frac{\tau_0}{\rho}.$$

If more of the Taylor expansion had been included, then division by dx and subsequent progression to the limit, $dx \to 0$, would, more rigorously, have achieved the same result. With dx thus interpreted as an infinitesimal, y_r may be set equal to the boundary layer thickness, $\delta(x)$, since the two integrals on the left of the above equation would yield identical results for $y > \delta$ [i.e., $v_x = U(x)$]. Thus, the following equation is obtained:

$$\frac{d}{dx}\int_0^\delta v_x^2 \, dy - U(x)\frac{d}{dx}\int_0^\delta v_x \, dy = -\frac{\delta}{\rho}\frac{dP}{dx} - \frac{\tau_0}{\rho}. \tag{6.35}$$

This equation constitutes the *integral momentum equation* of the steady laminar incompressible boundary layer. For laminar flow the wall shear, τ_0, may be replaced with $\tau_0 = \mu(\partial v_x/\partial y)_{y=0}$, so

$$\frac{d}{dx}\int_0^\delta v_x^2 \, dy - U(x)\frac{d}{dx}\int_0^\delta v_x \, dy = -\frac{\delta}{\rho}\frac{dp}{dx} - v\left(\frac{\partial v_x}{\partial y}\right)_0. \tag{6.36}$$

While turbulent flow and the associated turbulent boundary layer equations are not treated until the next two sections, it is worth noting here that the form of the integral momentum equation given in Eq. (6.35) will apply for turbulent layers if the time average values of the velocity components are used. The laminar law relating τ_0 to $(\partial v_x/\partial y)_{y=0}$ as used in Eq. (6.36) may not be used in this instance, and some other appropriate representation will have to be employed.

If the velocity distribution $v_x(y)$ in Eq. (6.35) or Eq. (6.36) is known, or assumed, the two integrands are known and these equations then yield a differential equation for δ, the boundary layer thickness, as a function of x. This approach is investigated in Chapter 7.

The Energy Integral Equation

The very same approach as applied above may be used to arrive at an integral energy relation for the thermal boundary layer. As mentioned above, the thermal boundary layer thickness is defined as the distance away from the surface at which the fluid temperature approaches the uniform temperature of the free stream potential flow. This thickness, $\delta_t(x)$, will be of the same order of magnitude as, but not necessarily equal to, the velocity boundary layer thickness $\delta(x)$. As in the case of the momentum integral analysis, it will be assumed that

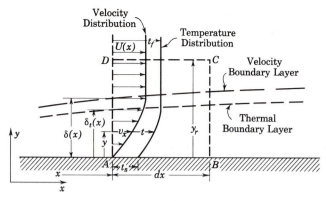

FIGURE 6.10. Laminar velocity and thermal boundary layers on a heated surface.

the thermal layer thickness is a definite quantity, although in reality it would be difficult to define such a precise limit.

Figure 6.10 shows a fluid flowing past a surface heated to a constant temperature t_s. The velocity boundary layer, δ, is shown along with the thermal boundary layer, δ_t, which is assumed to be thinner than the velocity layer. The symbol t_f denotes the uniform temperature of the potential flow and, for purposes of discussion, it is assumed that the fluid is hotter than the surface ($t_f > t_s$), although the reverse might be true. Once more a control domain, of length dx along the surface, is chosen with the ordinate y_r taken to be outside the velocity layer. The energy transported across the boundaries of the domain may be expressed in terms of the flux of enthalpy across those boundaries since kinetic energy effects will be ignored. Again using a linear approximation in the incremental distance dx, the excess of the enthalpy flux out the face BC (for unit depth) over the enthalpy flux in face AD is

$$\frac{d}{dx}\left[\int_0^{y_r} c_p(\rho v_x)t\, dy\right]dx.$$

The symbols t and v_x denote the velocity and temperature of the fluid at an ordinate y. As shown in Eqs. (6.32), the mass crossing face DC is

$$\dot{m}_{DC} = \frac{d}{dx}\left[\rho\int_0^{y_r} v_x\, dy\right]dx.$$

This mass carries into the domain an amount of energy equal to

$$c_p t_f \frac{d}{dx}\left[\rho\int_0^{y_r} v_x\, dy\right]dx.$$

Finally, the heat entering or leaving the domain by conduction at the surface may be expressed in terms of the wall heat flux:

$$\left(\frac{q}{A}\right)_0 dx.$$

The principle of conservation of energy gives then, neglecting the energy dissipated by friction and neglecting kinetic energy effects,

$$\rho c_p \frac{d}{dx} \int_0^{y_r} t_f v_x \, dy - \rho c_p \frac{d}{dx} \int_0^{y_r} t v_x \, dy = -\left(\frac{q}{A}\right)_0.$$

For values of $y > \delta_t$ the two integrals give identical results since t will be the same as t_f, so the integration needs to be carried only to $y = \delta_t(x)$, the thermal boundary layer thickness. The result is the *integral energy equation* of the laminar, incompressible boundary layer:

$$\frac{d}{dx} \int_0^{\delta_t} (t_f - t)v_x \, dy = -\frac{(q/A)_0}{\rho c_p}. \tag{6.37}$$

For laminar flow the wall heat flux may be written in terms of the wall temperature gradient:

$$\frac{d}{dx} \int_0^{\delta_t} (t_f - t)v_x \, dy = \alpha \left(\frac{\partial t}{\partial y}\right)_0. \tag{6.38}$$

As in the case of the integral momentum equation, the integral energy equation in the form of Eq. (6.37) may be applied in turbulent cases if an appropriate rate law for $(q/A)_0$ can be obtained.

6.11

TURBULENT FLOW
AND TURBULENT BOUNDARY LAYERS

In their nonsteady forms, the equations of continuity, motion, and energy developed in the earlier portions of this chapter may, conceptually, be applied to all forms of fluid motion. However, in their steady state forms and in the boundary layer simplifications, one must consider these relations as applying only to laminar motion. Steady laminar motion is characterized by streamlines that run in a well-ordered manner with adjacent layers of fluid sliding relative to one another and without any motion or change taking place normal to the streamlines of the main flow. Under certain conditions such a flow may undergo instabilities and change into a flow state known as *turbulent*. In such instances close examination of the flow will reveal that what appears to be steady motion is actually a time-dependent flow oscillating around an *apparent* steady condition. The oscillations are observed to occur in directions which are both parallel and normal to the main flow. Turbulent motion, then is inherently unsteady in that the velocity at any point in the fluid varies in time; however in "steady" turbulent flow this variation occurs around a constant average value.

Considering a two-dimensional motion, a steady turbulent flow may be represented in the following way:

$$v_x = \bar{v}_x + v_x',$$

$$v_y = \bar{v}_y + v_y',$$

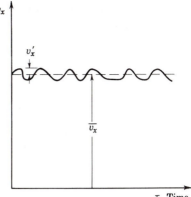

FIGURE 6.11. Turbulent velocity fluctuations about a time average.

in which v_x and v_y are the local *time-dependent* velocity components, $\bar{v}_x$ and $\bar{v}_y$ are the *time-average* values of the components, and v'_x and v'_y are the time-dependent fluctuations about the average. Figure 6.11 graphically illustrates the meaning of v_x, $\bar{v}_x$, and v'_x. A similar representation exists for the y component of the velocity. Thus, the true fluid velocity at a point is a nonsteady one, and the paths of the individual fluid particles form an interlacing network. This means that a certain amount of mixing and energy exchange occurs in a direction transverse to the main, time-average, flow. Such a situation is depicted in Fig. 6.12 for the simplified case in which the main flow is parallel to a surface. In this picture, mixing occurs in the y direction due to fluctuations v'_y even though the time-average flow has no y component. This transverse mixing is a characteristic of turbulent motion that is missing in laminar flow and is of great importance in heat transfer, since such motion may profoundly alter the rate of heat and momentum exchange within the fluid because of the associated movement of large, macroscopic fluid "lumps."

To make the discussion more formal, let the following representations be used to characterize a turbulent motion—including now the other two dependent variables, pressure and temperature:

$$v_x = \bar{v}_x + v'_x,$$

$$v_y = \bar{v}_y + v'_y,$$

$$P = \bar{P} + P',$$

$$t = \bar{t} + t'.$$

(6.39)

FIGURE 6.12. Mass transport due to transverse turbulent fluctuations.

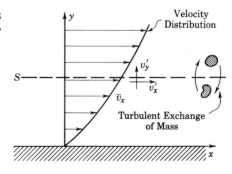

In Eq. (6.39) the quantities bearing the bar represent time-averaged values, which over a time interval of $\Delta\tau$ may be written in the following way for *steady* turbulent flow:

$$\bar{v}_x = \frac{1}{\Delta\tau} \int_{\tau}^{\tau+\Delta\tau} v_x \, d\tau = \text{constant in time,}$$

(6.40)

similarly for $\bar{v}_y$, $\bar{P}$, $\bar{t}$.

Also, for steady turbulent motion, the time average of the fluctuating components (v_x', v_y', P', t') must vanish:

$$\overline{v_x'} = \frac{1}{\Delta\tau} \int_{\tau}^{\tau+\Delta\tau} v_x' \, d\tau = 0,$$

(6.41)

similarly for $\overline{v_y'}$, $\overline{P'}$, $\overline{t'}$.

Clearly, the above representation is dependent on the time interval, $\Delta\tau$, over which the average is taken. This time interval must be large enough that the fluctuations can be observed but small enough that the time averages are constant. Thus, exactly what constitutes a steady turbulent flow is highly dependent upon the physical situation being examined. What constitutes steady turbulence for an engineer studying water flow in a pipe may not satisfy the needs of a meteorologist studying motion in the earth's atmosphere.

It must also be pointed out that while the time average of the fluctuations is zero, the time average of *products* of these fluctuations is not necessarily zero. That is,

$$\overline{v_x' v_y'} = \frac{1}{\Delta\tau} \int_{\tau}^{\tau+\Delta\tau} v_x' v_y' \, d\tau \neq 0,$$

$$\overline{v_y' t'} = \frac{1}{\Delta\tau} \int_{\tau}^{\tau+\Delta\tau} v_y' t' \, d\tau \neq 0.$$

(6.42)

In fact, the above two quantities are of extreme importance in the study of turbulence, as will be seen later.

The foregoing representation of steady turbulent motion, as given in Eqs. (6.39) through (6.42), may in theory be applied to the nonsteady continuity, motion, and energy equations to study the transport of momentum and energy in such flows. However, such analyses are extremely complex, due mainly to the apparent random nature of the fluctuations, and have resulted in an enormous literature on the subject. For the engineering applications to heat transfer to be considered in this text it will be adequate to treat the simpler case in which the effects of turbulence on the boundary layer equations are studied—as in the next section.

Turbulent Boundary Layers

The effects of turbulence on the exchange of momentum and energy in a boundary layer is best illustrated by observing the differences between the governing equations for the laminar case and the turbulent case. For these purposes it is useful to rewrite the laminar boundary layer equation in an alternative form.

Recalling that the fundamental concept leading to the laminar boundary layer simplifications is that the dominant variations of the velocity and temperature are in the direction normal to the surface so that the derivatives $\partial v_x/\partial y$ and $\partial t/\partial y$ are much larger than $\partial v_y/\partial x$ and $\partial t/\partial x$. Thus in the laminar boundary layer the shear stress given by Eq. (6.4) and the heat flux given by Fourier's law are approximated by

$$\tau_l = \mu \frac{\partial v_x}{\partial y} = \rho \nu \frac{\partial v_x}{\partial y},$$

$$\left(\frac{q}{A}\right)_l = -k \frac{\partial t}{\partial y} = -\rho c_p \alpha \frac{\partial t}{\partial y},$$

(6.43)

where the subscript l is used to emphasize that the shear stress and heat flux are those for laminar flow. It should be remembered that these laminar quantities represent the flux of momentum and energy, in a direction normal to the main flow in the boundary layer, due to *molecular* scale activity. With these representations, the laminar boundary layer equations of continuity, motion, and energy may be rewritten:

$$\frac{\partial v_x}{\partial x} + \frac{\partial v_y}{\partial y} = 0,$$

(6.44a)

$$v_x \frac{\partial v_x}{\partial x} + v_y \frac{\partial v_x}{\partial y} = -\frac{1}{\rho}\frac{dP}{dx} + \nu \frac{\partial^2 v_x}{\partial y^2},$$

$$= -\frac{1}{\rho}\frac{dP}{dx} + \frac{1}{\rho}\frac{\partial}{\partial y}(\tau_l),$$

(6.44b)

$$v_x \frac{\partial t}{\partial x} + v_y \frac{\partial t}{\partial y} = \frac{\nu}{c_p}\left(\frac{\partial v_x}{\partial y}\right)^2 + \alpha \frac{\partial^2 t}{\partial y^2},$$

$$= \frac{1}{\rho c_p}\left(\tau_l \frac{\partial v_x}{\partial y}\right) + \frac{1}{\rho c_p}\frac{\partial}{\partial y}\left(-\frac{q}{A}\right)_l.$$

(6.44c)

Counterparts of these expressions for turbulent motion will now be developed.

Since the representation of steady turbulent flow given in Eqs. (6.39) through (6.41) is that for a nonsteady flow (even if its time average is steady), nonsteady versions of the boundary layer equations must be used. The continuity equation in Eq. (6.44a) is valid for steady or nonsteady motion. Had the nonsteady terms been retained in the development of the boundary layer equations of motion and energy [Eqs. (6.30) and (6.31)] the following forms would result:

$$\frac{\partial v_x}{\partial x} + \frac{\partial v_y}{\partial y} = 0,$$

$$\frac{\partial v_x}{\partial \tau} + v_x \frac{\partial v_x}{\partial x} + v_y \frac{\partial v_x}{\partial y} = -\frac{1}{\rho}\frac{dP}{dx} + \nu \frac{\partial^2 v_x}{\partial y^2},$$

(6.45)

$$\frac{\partial t}{\partial \tau} + v_x \frac{\partial t}{\partial x} + v_y \frac{\partial t}{\partial y} = \frac{1}{\rho c_p}\left(\frac{\partial v_x}{\partial y}\right)^2 + \alpha \frac{\partial^2 t}{\partial y^2}.$$

If, now, these equations are applied to a turbulent boundary layer by introducing the nonsteady representations for v_x, v_y, P, and t given in Eqs. (6.39), equations of considerable complexity will result. However, if each term of these resultant equations is time-averaged, and the fact that the time averages for *steady* turbulent motion are subject to the conditions of Eqs. (6.40) through (6.42), considerable algebraic manipulation will give (Ref. 1):

$$\frac{\partial \bar{v}_x}{\partial x} + \frac{\partial \bar{v}_y}{\partial y} = 0,$$

$$\bar{v}_x \frac{\partial \bar{v}_x}{\partial x} + \bar{v}_y \frac{\partial \bar{v}_x}{\partial y} = -\frac{1}{\rho}\frac{d\bar{P}}{dx} + \nu \frac{\partial^2 \bar{v}_x}{\partial y^2} - \frac{\partial}{\partial y}(\overline{v_x' v_y'}), \qquad (6.46)$$

$$\bar{v}_x \frac{\partial \bar{t}}{\partial x} + \bar{v}_y \frac{\partial \bar{t}}{\partial y} = \frac{\nu}{c_p}\left(\frac{\partial \bar{v}_x}{\partial y}\right)^2 - \frac{1}{c_p}(\overline{v_x' v_y'})\frac{\partial \bar{v}_x}{\partial y} + \alpha \frac{\partial^2 \bar{t}}{\partial y^2} - \frac{\partial}{\partial y}(\overline{v_y' t'}).$$

Comparison of Eqs. (6.46) with Eqs. (6.44) reveals that while the time-average velocities in steady turbulent flow satisfy the steady laminar continuity equation, they do not satisfy the same forms of the equations of motion and energy as does steady laminar flow because of the terms $\overline{v_x' v_y'}$ and $\overline{v_y' t'}$ are not zero.

A physical interpretation of the additional terms in Eqs. (6.46) may be found by referring to the simplified situation depicted in Fig. 6.12. As discussed in Sec. 2.7, a physical interpretation of the laminar viscosity may be obtained from the molecular theory of gases and liquids. An exchange of molecules between the fluid layers on either side of a plane, say, S in Fig. 6.12, will produce an exchange of x-directed momentum across this plane which is proportional to the gradient in the time-average velocity $\bar{v}_x$. This momentum exchange is accompanied by a force in the x direction which is interpreted as the laminar shear stress τ_l. If, in addition, turbulent fluctuations v_x' and v_y' occur about the mean values $\bar{v}_x$ and $\bar{v}_y$, fluid masses which are large compared to molecular sizes may also be transported across S. If one considers masses which move upward in Fig. 6.12 (i.e., v_y' positive), such motion will be associated with a negative v_x' since they are moving from a region where, on the whole, a smaller mean velocity $\bar{v}_x$ prevails. This will tend to produce a negative v_x' in the layer above S. The momentum change, per unit area, experienced by such masses will then be $(\rho v_y')(-v_x')$. By similar reasoning, a negative v_y' will be most often associated with a positive v_x' with an x-directed momentum change of $(-\rho v_y')(v_x')$. Thus the time-average value of the x-directed momentum flux through s will be $-\rho \overline{v_x' v_y'}$, an amount different from zero. This momentum exchange, on macroscopic scale, gives rise to a "turbulent" shear stress:

$$\tau_t = -\rho \overline{v_x' v_y'}. \qquad (6.47)$$

Thus, the additional terms involving $-\rho \overline{v_x' v_y'}$ in Eqs. (6.46) may be thought of as additional contributions resulting from this turbulent shear stress.

Similar reasoning may be applied to interpret the term involving $\overline{v_y' t'}$ as a *turbulent heat flux* arising from the fluctuating components:

$$\left(\frac{q}{A}\right)_t = \rho c_p (\overline{v_y' t'}). \qquad (6.48)$$

The formal substitution of the concept of a turbulent shear stress and a turbulent heat flux into Eqs. (6.46) results in

$$\frac{\partial \bar{v}_x}{\partial x} + \frac{\partial \bar{v}_y}{\partial y} = 0,$$

$$\bar{v}_x \frac{\partial \bar{v}_x}{\partial x} + \bar{v}_y \frac{\partial \bar{v}_x}{\partial y} = \frac{1}{\rho}\frac{d\bar{P}}{dx} + \frac{1}{\rho}\frac{\partial}{\partial y}(\tau_l + \tau_t), \tag{6.49}$$

$$\bar{v}_x \frac{\partial \bar{t}}{\partial x} + \bar{v}_y \frac{\partial \bar{t}}{\partial y} = \frac{1}{\rho c_p}(\tau_l + \tau_t)\frac{\partial \bar{v}_x}{\partial y} + \frac{1}{\rho c_p}\left[-\left(\frac{q}{A}\right)_l - \left(\frac{q}{A}\right)_t\right].$$

In the form of Eqs. (6.49), the governing equations for steady turbulent flow are seen to correspond formally to those for steady laminar flow, Eqs. (6.44), as long as one intreprets the contribution of the turbulent fluctuations to be macroscopic augmentations to the laminar, microscopic shear stress and heat flux.

The turbulent equations given in Eqs. (6.49) are merely alternative forms of the basic equations given in Eqs. (6.46). Solution of either set requires the finding of the turbulent fluctuations v_x', v_y', t'. This constitutes the main problem of turbulent flow analysis—the relation of these fluctuations to the mean flow properties, the boundary conditions, etc. The interpretation of v_x', v_y', t' as being related to a turbulent shear or a turbulent heat flux is convenient for physical interpretations, but neither the fluctuations nor these turbulent fluxes are *fluid properties*—they are dependent on the solution of the time-dependent flow velocities and temperatures themselves. It should not be surprising then, to realize that the analysis of turbulent flow is one of the most complex fields of modern mathematical physics. Consequently, for engineering applications in cases where turbulence is present, certain simplified models must be adopted.

Eddy Viscosity and Eddy Diffusivity

One model often used to represent the turbulent boundary layers is that involving the concepts of *eddy viscosity* and *eddy diffusivity*. In an attempt to make the turbulent boundary layer equations look as much like the laminar ones, an eddy viscosity ϵ and an eddy diffusivity ϵ_H are defined:

$$\epsilon = \frac{\overline{v_x'v_y'}}{\partial \bar{v}_x/\partial y},$$

$$\epsilon_H = \frac{\overline{v_y't'}}{\partial \bar{t}/\partial y},$$

so that the turbulent shear and heat flux may be written

$$\tau_t = \rho\epsilon\left(\frac{\partial \bar{v}_x}{\partial y}\right), \tag{6.50}$$

$$\left(\frac{q}{A}\right)_t = -\rho c_p \epsilon_H\left(\frac{\partial \bar{t}}{\partial y}\right). \tag{6.51}$$

Thus, an *apparent*, or total, shear and an apparent heat flux may be defined:

$$\tau_{app} = \tau_l + \tau_t = \rho(\nu + \epsilon)\frac{\partial \bar{v}_x}{\partial y}, \tag{6.52}$$

$$\left(\frac{q}{A}\right)_{app} = \left(\frac{q}{A}\right)_l + \left(\frac{q}{A}\right)_t = -\rho c_p(\alpha + \epsilon_H)\frac{\partial \bar{t}}{\partial y}, \tag{6.53}$$

so that the turbulent boundary layer equations, Eqs. (6.46) or Eqs. (6.49), may be written

$$\frac{\partial \bar{v}_x}{\partial x} + \frac{\partial \bar{v}_y}{\partial y} = 0,$$

$$\bar{v}_x\frac{\partial \bar{v}_x}{\partial x} + \bar{v}_y\frac{\partial \bar{v}_x}{\partial y} = -\frac{1}{\rho}\frac{d\bar{P}}{dx} + (\nu + \epsilon)\frac{\partial^2 \bar{v}_x}{\partial y^2}, \tag{6.54}$$

$$\bar{v}_x\frac{\partial \bar{t}}{\partial x} + \bar{v}_y\frac{\partial \bar{t}}{\partial y} = \frac{\nu + \epsilon}{c_p}\left(\frac{\partial \bar{v}_x}{\partial y}\right)^2 + (\alpha + \epsilon_H)\frac{\partial \bar{t}}{\partial y}.$$

Comparison of these latter equations with those for laminar flow in Eqs. (6.44) shows the two sets to be formally the same in terms of the time average velocities and temperatures—with $(\nu + \epsilon)$ replacing ν and $(\alpha + \epsilon_H)$ replacing α.

The introduction of the eddy coefficients ϵ and ϵ_H are convenient formal representations to make the turbulent shear stress and heat flux rate laws in Eqs. (6.52) and (6.53) analogous to the laminar ones in Eqs. (6.43) and to make the turbulent governing equations in Eqs. (6.54) analogous to the laminar ones in Eqs. (6.44). However, the procedure is somewhat like trying to bury a pile of dirt—while convenient in a formal sense, the introduction of ϵ and ϵ_H does not provide any insight or understanding of the basic complexities of turbulent motion. The parameters ϵ and ϵ_H are *not* fluid properties as are the molecular quantities ν and α. The eddy coefficients depend on the nature of the fluctuating quantities v_x', v_y', and t' and are known to vary from point to point in the flow field itself since, for example, they must vanish at or near a solid surface where the transverse fluctuations must disappear. At any rate, one may remember that while ν and α are related to *molecular* scale transport of momentum and energy, ϵ and ϵ_H are functions of the larger scale, macroscopic fluctuations and may be much greater than ν and α. Thus, again for engineering applications, simplified models, semiempirical in nature, are applied, and one of these, "Prandtl's mixing length" model, is discussed in the next section.

The general characteristics of turbulent boundary layers are similar to those of laminar layers—thin regions exist near the surface of a body where the time-average velocity and temperature vary rapidly from uniform values in the potential core to either zero or a fixed value at the surface. As a result of the transverse turbulent fluctuations, there exists a more uniform velocity or temperature distribution near the outer edge of the boundary layer than in the corresponding laminar case. Figure 6.13 illustrates, in a pictorial way, the typical distribution for the velocity layer; an analogous picture would result for the thermal layer.

Often the turbulent layer is further divided into subregions as indicated in

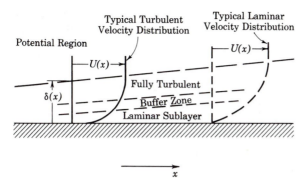

FIGURE 6.13. Turbulent boundary layer on a flat surface.

Fig. 6.13. Near the wall one may imagine a region in which the transverse fluctuations must be suppressed by the wall. In this region one would expect that the eddy viscosity and diffusivity are dominated by the laminar ones, or $\epsilon \ll \nu$ and $\epsilon_H \ll \alpha$. Thus, the apparent shear stress and heat flux given in Eqs. (6.52) and (6.53) follow the laminar laws and the region is sometimes called the *laminar sublayer*. Near the outer edge of the turbulent layer a region is observed to exist in which the opposite is true: $\epsilon \gg \nu$, $\epsilon_H \gg \alpha$. Such a region is referred to as a "fully turbulent" zone, in which the eddy quantities dominate. There may be a region between the laminar sublayer and the fully turbulent zone in which $\epsilon \approx \nu$, $\epsilon_H \approx \alpha$, and it is called the *buffer zone*. The three regions just discussed (laminar sublayer, buffer zone, and fully turbulent region) are not distinctly identifiable, of course, and actually there must be a continuous variation from one to the other.

One additional point concerning turbulent boundary layers may be made in conjunction with the integral equations developed in Sec. 6.10. Those equations in the forms given in Eqs. (6.35) and (6.37) may be applied to turbulent layers as long as the time-averaged velocity and temperature are used. The extension to the forms given in Eqs. (6.36) and (6.38) may be made only if ν is replaced by $(\nu + \epsilon)$ and α by $(\alpha + \epsilon_H)$.

Prandtl's Mixing Length

As was pointed out above, the principal difficulty associated with the analysis of turbulent boundary layers is the determination of the turbulent shear stress and turbulent heat flux either by modeling the distribution of the fluctuations v'_x, v'_y, t' or the eddy coefficients. That is, one is seeking representations for

$$\tau_t = -\rho \overline{v'_x v'_y} = \rho \epsilon \frac{\partial \overline{v}_x}{\partial y},$$

$$\left(\frac{q}{A}\right)_t = \rho c_p (\overline{v'_y t'}) = -\rho c_p \epsilon_H \frac{\partial \overline{t}}{\partial y}.$$

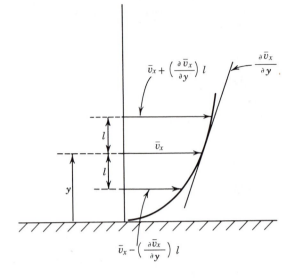

FIGURE 6.14. Turbulent mixing length.

$$\bar{v}_x + \left(\frac{\partial \bar{v}_x}{\partial y}\right) l$$

$$\frac{\partial \bar{v}_x}{\partial y}$$

$$\bar{v}_x$$

$$y$$

$$\bar{v}_x - \left(\frac{\partial \bar{v}_x}{\partial y}\right) l$$

Since laminar shear and heat flux arise from microscopic momentum and energy transport by molecules moving over a mean free path between collisions, Prandtl hypothesized that the macroscopic turbulent eddies, or fluid "lumps," associated with the fluctuations v'_x, v'_y, and t' could similarly be envisioned as traveling over some finite distance before losing their identities. He called this macroscopic distance the *mixing length, l.* Figure 6.14 illustrates the concepts involved. A fluid lump, or eddy, coming from $(y + l)$ would have, on the average, a higher $\bar{v}_x$ than the fluid at y; a lump coming from $(y - l)$ would have a lower $\bar{v}_x$. The averages of these differences would be, from the definition of the mixing length, the fluctuation v'_x:

$$v'_x \propto l \left|\frac{\partial \bar{v}_x}{\partial y}\right|.$$

Then, by applying continuity principles, Prandtl concluded that v'_y should be of the same order of magnitude as v'_x so that the turbulent shear stress should be

$$\tau_t = -\rho \overline{v'_x v'_y} \propto \rho l^2 \left(\frac{\partial \bar{v}_x}{\partial y}\right)^2,$$

$$\tau_t = \rho l^2 \left(\frac{\partial \bar{v}_x}{\partial y}\right)^2,$$

(6.55)

in which the constant of proportionality has been absorbed into the definition of the mixing length. Similar reasoning leads to the following representation for the turbulent heat flux

$$\left(\frac{q}{A}\right)_t = -\rho l_H^2 \left(\frac{\partial \bar{t}}{\partial y}\right)^2$$

(6.56)

in which l_H is a thermal mixing length.

Prandtl's representation amounts to the assumption that the eddy viscosity and eddy diffusivity are of the form

$$\epsilon = l^2 \left(\frac{\partial \bar{v}_x}{\partial y} \right),$$

$$\epsilon_H = l_H^2 \left(\frac{\partial \bar{t}}{\partial y} \right). \tag{6.57}$$

The momentum and thermal mixing lengths need not be the same; however, the similarity of the two turbulent mixing processes leads one to expect that l and l_H will not be very different. Application of Prandtl's mixing length theory to the solution of real problems becomes that of finding suitable representations of how the mixing length varies in the flow field. An example of this is presented in the next chapter.

The Turbulent Prandtl Number

It will be recalled that the definition of the molecular Prandtl number is

$$\text{Pr} = \frac{\nu}{\alpha}.$$

Thus, Pr is clearly a fluid property. For turbulent flow one also defines a turbulent Prandtl number Pr_t:

$$\text{Pr}_t = \frac{\epsilon}{\epsilon_H}.$$

Quite obviously the turbulent Prandtl number is *not* a fluid property, but, instead, depends on the distribution of ϵ and ϵ_H through the flow field. Because of the similarity of the turbulent momentum and energy exchange, one simplifying assumption often made, particularly in the fully turbulent zone, is that $\text{Pr}_t \approx 1$.

REFERENCES

1. SCHLICTING, H., *Boundary Layer Theory,* 7th ed., New York, McGraw-Hill, 1979.
2. ECKERT, E. R. G., and R. M. DRAKE, *Analysis of Heat and Mass Transfer,* New York, McGraw-Hill, 1972.
3. VAN WYLEN, G. J., and R. E. SONNTAG, *Fundamentals of Classical Thermodynamics,* 2nd ed., New York, Wiley, 1978.
4. WHITE, F. M., *Viscous Fluid Flow,* New York, McGraw-Hill, 1974.

Examples of Analytic Solutions to Problems of Forced Convection

INTRODUCTORY REMARKS

The application of the principles discussed in Chapter 6 to the analytical prediction of problems in forced convection can become quite complex and can involve much computational labor. It is not the purpose of this chapter to equip the reader with the ability to apply the boundary layer theory to the solution of all subsequent heat convection problems that may be encountered. Rather it is meant to illustrate, by the consideration of the simplest of examples, the basic methods involved and to develop an understanding of these methods. Also, these simple examples will familiarize the reader with the important physical parameters involved. Subsequently, in Chapter 8, the results of similar analyses of more complex problems will be presented with a minimum of explanation, as will certain empirical results whose bases lie in these elementary cases.

The examples to be presented in this chapter will all be limited to cases of "pure" forced convection. Thus, the body force term will be omitted from all expressions of the equations of motion. If one were to examine a problem in terms of the differential equations of the boundary layer, then a solution of

$$\frac{\partial v_x}{\partial x} + \frac{\partial v_y}{\partial y} = 0,$$

$$v_x \frac{\partial v_x}{\partial x} + v_y \frac{\partial v_x}{\partial y} = -\frac{1}{\rho}\frac{dP}{dx} + \nu \frac{\partial^2 v_x}{\partial y^2}, \tag{7.1}$$

$$v_x \frac{\partial t}{\partial x} + v_y \frac{\partial t}{\partial y} = \frac{\mu}{\rho c_p}\left(\frac{\partial v_x}{\partial y}\right)^2 + \alpha \frac{\partial^2 t}{\partial y^2},$$

for v_x, v_y, and t is required. One notes from Eqs. (7.1) that if the fluid properties ρ, μ, c_p, ν, and α are assumed to be independent of temperature, then the first two equations may be solved independently for the velocity boundary layer (i.e., v_x and v_y). Then the fluid temperature may be determined by solving the energy equation. Thus, the problem becomes "uncoupled" and considerably simpler algebraically. Cases in which free convection is present (and hence a body force term involving the fluid temperature appears in the momentum equation), or those in which property dependence on temperature is included, are thus more complex algebraically. Free convection problems are discussed in Chapter 9.

The same conclusion regarding the uncoupling of the velocity and temperature problems may be reached by examining the integral equations of the boundary layer given in Eqs. (6.36) and (6.38). Solutions for forced convection using both the differential boundary layer equations, above, or the integral counterparts will be demonstrated in the following sections.

7.2

LAMINAR FORCED CONVECTION PAST A FLAT SURFACE, DISSIPATION NEGLECTED

Perhaps the simplest example of the application of boundary layer theory to the solution of forced convection is that of laminar flow of a fluid of uniform velocity U and uniform temperature t_f past a flat plate maintained at a uniform temperature t_s and placed parallel to the incident flow. Such a situation is depicted in Fig. 7.1. It will be recalled that the term dP/dx in Eq. (7.1) is determined by first solving the inviscid Euler equations, Eq. (6.10), for the same geometry. For the present case this solution is obtained by inspection—giving a constant $v_x = U$ and $dP/dx = 0$. While the effects of viscous dissipation will be considered in some detail later, this first solution will be limited to those cases in which the dissipated energy, proportional to $(\partial v_x/\partial y)^2$, is negligible compared

FIGURE 7.1. Velocity and thermal boundary layers for laminar flow past a flat plate.

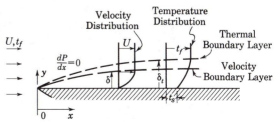

to the other terms in the energy equation. Thus, for the flat surface with negligible dissipation one has to solve

$$\frac{\partial v_x}{\partial x} + \frac{\partial v_y}{\partial y} = 0, \tag{7.2}$$

$$v_x \frac{\partial v_x}{\partial x} + v_y \frac{\partial v_x}{\partial y} = \nu\left(\frac{\partial^2 v_x}{\partial y^2}\right), \tag{7.3}$$

$$v_x \frac{\partial t}{\partial x} + v_y \frac{\partial t}{\partial y} = \alpha\left(\frac{\partial^2 t}{\partial y^2}\right), \tag{7.4}$$

with the boundary conditions:

$$\text{Velocity:} \quad \text{at } y = 0, \; v_x = v_y = 0, \tag{7.5}$$

$$\text{as } y \to \infty, \; v_x \to U.$$

$$\text{Temperature:} \quad \text{at } y = 0, \; t = t_s, \tag{7.6}$$

$$\text{as } y \to \infty, \; t \to t_f.$$

Before proceeding with the solution of the above set of equations, it is useful to note the similarity of the forms of the momentum equation, Eq. (7.3), and the energy equation, Eq. (7.4). Equation (7.4) is identical in form to Eq. (7.3) if t is replaced by v_x and if $\alpha = \nu$. The boundary conditions are of similar form if t is replaced by $t - t_s$. Thus, one would expect similar forms for the velocity and temperature distributions when $\alpha = \nu$. Since the Prandtl number is defined as $\text{Pr} = \nu/\alpha$, this similarity exists when $\text{Pr} = 1$. While examination of the tables in Appendix A reveals that Pr may take on a wide range of values, it is noteworthy that for many fluids, particularly gases, it is not too different from 1. This similarity between heat and momentum exchange will be examined more fully in later portions of this chapter.

The Velocity Boundary Layer
and the Skin Friction Coefficient

As mentioned in the foregoing, the assumption of constant fluid properties enables one to solve the velocity problem separately from the thermal problem. That is, Eqs. (7.2) and (7.3) may be solved under the conditions of Eq. (7.5) for v_x and v_y, independently of t.

Considerable insight into the problem can be had by first making a rather crude, order-of-magnitude analysis. The terms on the left of Eq. (7.3) represent the fluid inertia forces, and the first term, $v_x(\partial v_x/\partial x)$, is typical. Since v_x can be as great as U, the free stream velocity, this term may, at some downstream location x, be as great as $U(U/x)$. Since in the boundary layer y varies from 0 at the wall to δ at the boundary layer limit (δ being the layer thickness at x), then the viscous term on the right is of the order of $\nu U/\delta^2$. Thus in the boundary layer, since these two terms are of the same order of magnitude, one would expect

$$\frac{U^2}{x} \approx \frac{\nu U}{\delta^2}, \qquad \text{or} \qquad \delta \approx \sqrt{\frac{\nu x}{U}}. \tag{7.7}$$

Thus, the boundary layer may be expected to grow as $\sqrt{x}$ as it proceeds down the plate. More generally, Eq. (7.7) shows that the ratio of the *local* boundary layer thickness to the local distance x is

$$\frac{\delta}{x} \approx \sqrt{\frac{\nu}{Ux}} = \text{Re}_x^{-1/2}. \tag{7.8}$$

That is, the ratio δ/x is inversely proportional to the square root of the *local length Reynolds number*, Re_x:

$$\text{Re}_x = \frac{Ux}{\nu},$$

where the subscript x is used to emphasize that the Reynolds number involved is based on the local distance from the leading edge of the plate. Equation (7.8) shows that Re_x must be large in order that the boundary layer be thin, as assumed.

Based on the above, Blasius (Refs. 1 and 2) greatly simplified the solution to the problem by noting that, in reference to Fig. 7.1, the absence of any characteristic dimension in the problem and the constancy of U implies that the dimensionless velocity v_x/U when plotted against the local dimensionless distance y/δ must be the same for all locations on the plate. That is, since $\delta \sim \sqrt{\nu x/U}$, Blasius assumed that if a new parameter η is defined

$$\eta = y\sqrt{\frac{U}{\nu x}}, \tag{7.9}$$

then v_x/U is dependent on x and y only in the combination suggested in Eq. (7.4), or that v_x/U is a function only of η. For simplicity of later algebra, Blasius assumed a solution to Eqs. (7.2) and (7.3) of the form

$$\frac{v_x}{U} = \frac{df}{d\eta} = f'(\eta), \tag{7.10}$$

in which $f(\eta)$ is a function of η only.

Substitution of Eq. (7.10) into (7.2) and subsequent integration shows that continuity is satisfied if v_y/U is of the form

$$\frac{v_y}{U} = \frac{1}{2}\sqrt{\frac{\nu}{Ux}}\,(\eta f' - f). \tag{7.11}$$

The velocity components v_x and v_y may be found if the function $f(\eta)$ can be found.

The function $f(\eta)$ is found by substituting Eqs. (7.10) and (7.11) into Eqs. (7.3) and (7.5), which, after considerable algebraic manipulation, gives

$$f''' + \tfrac{1}{2}ff'' = 0,$$

$$f(0) = f'(0) = 0, \tag{7.12}$$

$$f'(\infty) = 1.$$

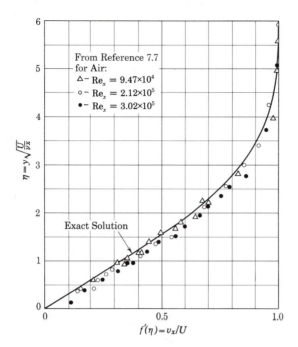

FIGURE 7.2 Dimensionless results for the velocity distribution in the laminar boundary layer on a flat surface as given by the exact solution of the boundary layer equations. Experimental values are shown for comparison.

The absence of any quantities other than f and η in Eq. (7.12) justifies the similarity assumption implied by Eq. (7.10).

The nonlinear equation given in Eq. (7.12) has been solved by series techniques (Ref. 2) and the resulting functions $f(\eta)$, $f'(\eta)$, and $f''(\eta)$ have been tabulated. Since the velocity layer is not of primary concern these series representations and tabulations are not given here, but the significant results are shown graphically in Figs. 7.2 and 7.3.

Figure 7.2 displays $f'(\eta)$ as a function of η, and since $v_x/U = f'(\eta)$ this figure amounts to a dimensionless velocity profile in the boundary layer. The figure compares the prediction of the Blasius solution with experimental results and the good agreement is apparent. The nature of the boundary condition that $v_x/U = f'(\infty) = 1$ implies that the full free stream velocity is not achieved anywhere in the fluid domain except in the limit as $y \rightarrow \infty$. However, for practical purposes it is desirable to know in what region the viscous effects are dominant. For this reason it is customary to define the boundary layer thickness, δ, as the point at which v_x/U is about 0.99. As Fig. 7.2 indicates, this occurs at a value of $\eta \approx 5.0$. Thus, it is customary to stipulate that $y = \delta$ at $\eta = 5.0$, or

$$\delta = 5.0 \sqrt{\frac{\nu x}{U}},$$

$$\frac{\delta}{x} = 5.0 \, \text{Re}_x^{-1/2}.$$

(7.13)

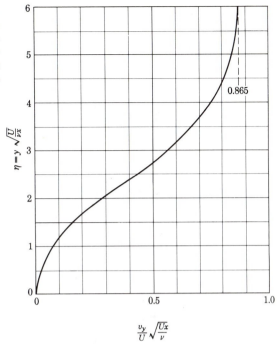

FIGURE 7.3. Dimensionless results for the normal velocity component in the laminar boundary layer on a flat plate as given by the exact solution.

Incidentally, the form of Eq. (7.13) supports the order of magnitude determination given in Eq. (7.7) which led to the definition of η.

Another important result of the solution to Eq. (7.12) given in Fig. 7.2 is the value of $f''(0)$, or the slope of the f' versus η curve at $\eta = 0$. The numerical value of this quantity, obtained from the series solution, is known to be (Ref. 1)

$$f''(0) = 0.33206. \tag{7.14}$$

The significance of this parameter lies in its relation to the shear stress at the plate surface:

$$\tau_0 = \mu \left(\frac{\partial v_x}{\partial y} \right)_{y=0},$$

$$= \mu U \left[\frac{d}{d\eta} \left(\frac{v_x}{U} \right) \right] \frac{\partial \eta}{\partial y},$$

$$= \mu U \sqrt{\frac{U}{\nu x}} f''(0).$$

The surface shear stress is obviously related to the viscous drag force on the plate. Usually, the wall shear stress is written nondimensionally by defining a *skin friction coefficient*

$$C_f = \frac{\tau_0}{\frac{1}{2}\rho U^2} \tag{7.15}$$

**7.2 LAMINAR FORCED CONVECTION
PAST A FLAT SURFACE**

so that

$$C_f = 2f''(0) \sqrt{\frac{v}{Ux}} \tag{7.16}$$

$$= 0.664\text{Re}_x^{-1/2}.$$

If the plate has a total length L and width b, the total drag force due to viscous shear on one side of the plate is

$$D = b \int_0^L \tau_0 \, dx.$$

A drag coefficient is often used, defined as

$$C_D = \frac{D}{(\frac{1}{2}\rho U^2)(bL)} = \frac{1}{L} \int_0^L \frac{\tau_0}{\frac{1}{2}\rho U^2} \, dx, \tag{7.17}$$

$$C_D = \frac{1}{L} \int_0^L C_f \, dx.$$

Thus, the drag coefficient is the integrated average of C_f over the plate length. Apparently, in the case considered here

$$C_D = \frac{1}{L} \int_0^L 2f''(0) \sqrt{\frac{v}{Ux}} \, dx, \tag{7.18}$$

$$= 4f''(0) \sqrt{\frac{v}{UL}}, \tag{7.19}$$

$$= 1.328\text{Re}_L^{-1/2}. \tag{7.20}$$

In this last equation $\text{Re}_L = UL/v$ is the *plate length Reynolds number* using the total plate length as the characteristic dimension.

The Thermal Boundary Layer

The ultimate aim of the analysis, the determination of the surface heat transfer coefficient, may now be obtained by solving the energy equation, Eq. (7.4), under the boundary conditions of Eq. (7.6), since the quantities v_x and v_y are now known.

It is reasonable to assume that the temperature distribution in the boundary layer should exhibit the same similarity property with respect to the parameter $\eta = y \sqrt{U/vx}$ as did the velocity distribution since the thermal layer, while not identical to the velocity layer, should be of the same order of magnitude. Thus, a dimensionless temperature, φ, may be defined

$$\varphi = \frac{t - t_s}{t_f - t_s}, \tag{7.21}$$

and is assumed that φ is a function of η alone,

$$\varphi = \varphi(\eta). \tag{7.22}$$

Remembering that v_x and v_y are now known functions of the previously determined function $f(\eta)$, as given in Eqs. (7.10) and (7.11), substitution of Eqs. (7.10), (7.11), and (7.22) into Eqs. (7.4) and (7.6) yields

$$\varphi'' + \frac{\text{Pr}}{2} f(\eta)\varphi' = 0,$$

$$\varphi(0) = 0, \qquad (7.23)$$

$$\varphi(\infty) = 1.$$

The solution to Eqs. (7.23), due to Pohlhausen (Ref. 3), may be obtained by noting that it is a linear first-order differential equation in φ' and by using the fact that, from Eq. (7.12), $f = -2f'''/f''$. The solution is

$$\varphi(\eta) = \frac{t - t_s}{t_f - t_s} = \frac{\displaystyle\int_0^\eta (f'')^{\text{Pr}}\, d\eta}{\displaystyle\int_0^\infty (f'')^{\text{Pr}}\, d\eta}. \qquad (7.24)$$

With $f(\eta)$, and hence $f''(\eta)$, known from the solution of Eq. (7.12), at least in numerical form, Eq. (7.24) may be solved (perhaps numerically) for the temperature distribution. The results are, of course, dependent on the Prandtl number Pr, and results are shown in Fig. 7.4 for various Pr. The curve for Pr = 0.7 is typical for air.

For the special case in which Pr = 1, the above solution for $\varphi(\eta)$ becomes

$$\varphi(\eta) = \frac{\displaystyle\int_0^\eta f''\, d\eta}{\displaystyle\int_0^\infty f''\, d\eta} = \frac{f'(\eta) - f'(0)}{f'(\infty) - f'(0)} = f'(\eta). \qquad (7.25)$$

Thus, as noted earlier, for the case in which Pr = 1, the dimensionless velocity and temperature distributions are the same:

$$\frac{t - t_s}{t_f - t_s} = \frac{v_x}{U}.$$

Further examination of Fig. 7.4 shows that for Pr > 1, the thermal layer is thinner than the velocity layer, whereas for Pr < 1, the reverse is true.

The Heat Transfer Coefficient and the Nusselt Number

For the determination of the heat transfer between the plate and the fluid, the gradient of the temperature at the surface is of more significance than the temperature distribution itself. From Eq. (7.24), the surface temperature gradient is

$$\varphi'(0) = \frac{\sqrt{\nu x/U}}{t_f - t_s} \left(\frac{\partial t}{\partial y}\right)_{y=0} = \frac{[f''(0)]^{\text{Pr}}}{\displaystyle\int_0^\infty (f'')^{\text{Pr}}\, d\eta},$$

$$= \varphi_0'(\text{Pr}). \qquad (7.26)$$

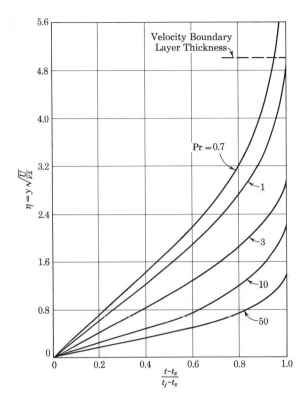

FIGURE 7.4. Temperature distributions in the laminar boundary layer on a heated flat plate, viscous dissipation neglected.

The notation $\varphi_0'(\text{Pr})$ has been introduced to symbolize the dimensionless wall temperature gradient in order to emphasize the fact that it is a function of only the fluid Prandtl number, as given by Eq. (7.26). It is also worth noting that the parameter $f''(0) = 0.33206$ found in the velocity solution is again of importance. Table 7.1 gives numerical values for $\varphi_0'(\text{Pr})$ for various Pr.

Now, by definition, the surface heat transfer coefficient is

$$h_x = -\frac{k\left(\dfrac{\partial t}{\partial y}\right)_{y=0}}{t_s - t_f} = k\varphi_0'(\text{Pr})\sqrt{\frac{U}{\nu x}}. \tag{7.27}$$

Equation (7.27) shows that the heat transfer coefficient varies with the distance along the plate. Thus, it becomes necessary to identify the coefficient in Eq. (7.27) as the *local* heat transfer coefficient—hence the notation h_x. In many applications it is useful to use an average heat transfer coefficient h over the whole plate length L:

$$\begin{aligned} h &= \frac{1}{L}\int_0^L h_x\, dx, \\ &= \frac{k}{L}\, 2\,\varphi_0'(\text{Pr})\sqrt{\frac{UL}{\nu}}. \end{aligned} \tag{7.28}$$

TABLE 7.1 Laminar Forced Convection Functions*

Pr	$\varphi_0'(Pr)$	$0.332(Pr)^{1/3}$	$\Psi_0(Pr)$	$Pr^{1/2}$
0.6	0.276	0.280	0.770	0.775
0.7	0.293	0.295	0.835	0.837
0.8	0.307	0.308	0.895	0.894
0.9	0.320	0.321	0.950	0.949
1.0	0.332	0.332	1.000	1.000
1.1	0.344	0.343	1.050	1.049
7.0	0.645	0.635	2.515	2.646
10.0	0.730	0.715	2.865	3.162
15.0	0.835	0.819	3.535	3.873

*From H. SCHLICHTING, *Boundary Layer Theory*, 6th ed., New York, McGraw-Hill, 1968.

It is convenient to nondimensionalize Eqs. (7.27) and (7.28) by introducing the Nusselt number defined in Sec. 6.7:

$$\text{Nu} = \frac{hl}{k}, \tag{7.29}$$

in which l is the characteristic dimension. By using Nu_x to denote the *local* Nusselt number based on the local coefficient h_x and the local coordinate x, and using Nu_L to denote an *average* plate length Nusselt number based on the average h and the total plate length L, Eqs. (7.27) and (7.28) are

$$\text{Nu}_x = \frac{h_x x}{k} = \varphi_0'(Pr)\sqrt{\frac{Ux}{\nu}},$$

$$\text{Nu}_L = \frac{hL}{k} = 2\varphi_0'(Pr)\sqrt{\frac{UL}{\nu}}. \tag{7.30}$$

Equations (7.30) may be used, together with the tabulations in Table 7.1, for the determination of the local and average heat transfer coefficients; however, for typical engineering calculations the functional dependence of φ_0' on Pr may be replaced by the rather accurate approximation

$$\varphi_0'(Pr) \approx f''(0)Pr^{1/3} = 0.332Pr^{1/3}. \tag{7.31}$$

Table 7.1 compares the predictions of Eq. (7.31) with the true values of φ_0'. Thus, Eqs. (7.30) may be written

$$\text{Nu}_x = 0.332Pr^{1/3}\text{Re}_x^{1/2},$$

$$\text{Nu}_L = 0.664Pr^{1/3}\text{Re}_L^{1/2}, \tag{7.32}$$

where the local length Reynolds number Re_x and the plate length Reynolds number have been introduced. Equations (7.32) are of the form predicted in Sec. 6.7—that in the absence of buoyancy and viscous dissipation, forced convection solutions are of the form $\text{Nu} = f(Pr, Re)$.

7.2 LAMINAR FORCED CONVECTION PAST A FLAT SURFACE

SOLUTION OF LAMINAR FORCED CONVECTION ON A FLAT PLATE BY USE OF THE INTEGRAL EQUATIONS OF THE BOUNDARY LAYER

The example just considered in Sec. 7.2 showed the methodology used when applying the differential equations of the boundary layer to the solution of a convection problem. This section will present an *alternative* technique which employs the integral boundary layer equations developed in Sec. 6.10. The same example will be considered—that of the flow of a stream at uniform velocity and temperature past a surface of fixed temperature. Clearly, if one already has a solution, as given in Sec. 7.2, then solving the problem again by another means is pointless. The same example is used to show how the integral method works so that results from its application to more complex problems may be used with confidence. As this example will show, the integral technique will give less detailed information as to the structure of the velocity and thermal layers, but it will yield virtually the same overall information as the differential equations, with considerably less algebra.

For flow past a flat plate in which $U(x) = U$ (a constant) and $dP/dx = 0$, the integral momentum and energy equations [Eqs. (6.36) and (6.38)] become

$$\frac{d}{dx} \int_0^\delta v_x(v_x - U)\, dy = -\frac{\tau_0}{\rho} = -\nu\left(\frac{\partial v_x}{\partial y}\right)_{y=0}, \tag{7.33}$$

$$\frac{d}{dx} \int_0^{\delta_t} v_x(t - t_f)\, dy = -\alpha\left(\frac{\partial t}{\partial y}\right)_{y=0}.$$

These equations will now be applied to illustrate this alternative approach to the determination of the velocity and temperature distribution (and the resultant heat transfer coefficient) in the laminar boundary layer on a flat plate.

The Velocity Boundary Layer

The method of application of the momentum integral equation involves, usually, the assumption that the velocity profiles are similar at different stations, x. That is, as in the above analysis, a similarity parameter, η, is defined as

$$\eta = \frac{y}{\delta}. \tag{7.34}$$

It should be noted that the η defined in Eq. (7.34) is *not* the same similarity parameter that was used in the analysis of Sec. 7.2. The boundary layer thickness, δ, is presumed to be a finite quantity—although it is a function of x, the distance along the plate. Also it is assumed that the dimensionless velocity profile, v_x/U, is a function of η only:

$$\frac{v_x}{U} = g(\eta). \tag{7.35}$$

Introducing this definition of $g(\eta)$ into the first equation of Eq. (7.33), the integral momentum equation, one obtains

$$\frac{d}{dx} \int_0^\delta \frac{v_x}{U}\left(\frac{v_x}{U} - 1\right) dy = -\frac{\nu}{U}\left[\frac{\partial(v_x/U)}{\partial y}\right]_{y=0},$$

$$\frac{d}{dx}\left\{\delta \int_0^1 g(\eta)[g(\eta) - 1]\, d\eta\right\} = -\frac{\nu}{U\delta}\left(\frac{dg}{d\eta}\right)_{\eta=0}.$$

(7.36)

If the shape of the velocity distribution profile, $g(\eta)$, is known, Eq. (7.36) results in a differential equation for δ as a function of x. The main advantage of this integral approach is that integration tends to suppress errors. That is, even if a velocity profile shape, $g(\eta)$, is *assumed* which differs from the correct result, the resultant integral will not differ as much from the correct integral. Experience shows that even rather crude guesses for $g(\eta)$ may yield very reasonable and usable results for δ and the associated film coefficient h. Customarily one represents the unknown velocity distribution as a polynomial in η,

$$\frac{v_x}{U} = g(\eta) = a + b\eta + c\eta^2 + d\eta^3 + \cdots. \qquad (7.37)$$

As many terms may be taken as one desires—depending on what accuracy is wanted and how much algebraic labor can be tolerated. If, for instance, a third-degree polynomial is used, one has four constants to be determined. These may be fixed by application of conditions at the wall ($\eta = 0$) and at the outer limit of the boundary layer ($\eta = 1$).

At the outer limit of the boundary layer, $\eta = 1$, it is known that $v_x = U$, or $g(\eta) = 1$. Also for a smooth transition from the boundary layer to the potential region, all the derivatives of v_x with respect to y must vanish. This means that all the derivatives of $g(\eta)$ with respect to η are zero at $\eta = 1$ [i.e., $g'(1) = 0$, $g''(1) = 0, \ldots$].

At the solid surface $v_x = 0$, so $g(0) = 0$. The first derivative of v_x with respect to y at $y = 0$ determines the viscous shear stress there. Thus, $g'(0)$ may not be specified. Now, the boundary layer equation of motion, for zero pressure gradient, is [Eq. (7.3)]

$$v_x \frac{\partial v_x}{\partial x} + v_y \frac{\partial v_x}{\partial y} = \nu \frac{\partial^2 v_x}{\partial y^2}.$$

At the surface $v_x = v_y = 0$, and thus

$$\left(\frac{\partial^2 v_x}{\partial y^2}\right)_{y=0} = 0, \quad \text{or} \quad g''(0) = 0.$$

Differentiation of the equation of motion and evaluation at $y = 0$ will show that all the higher derivatives of $g(\eta)$ are also zero for $\eta = 0$.

These conditions at $\eta = 0$ and $\eta = 1$ may be employed to evaluate the constants of the polynomial for $g(\eta)$. As many will be employed as are required by the degree of the polynomial. The use of a polynomial as an *approximation* for $g(\eta)$ means, of course, that *all* the boundary conditions cannot be satisfied. It is best to satisfy as many conditions at the plate surface as at the outer limit of the boundary layer—alternating between one and the other as higher degree polynomials are used. Summarizing, the available conditions are

$$
\left.
\begin{array}{l}
\left.
\begin{array}{l}
\left.
\begin{array}{l}
g(1) = 1 \\
g(0) = 0
\end{array}
\right\} \text{first degree} \\
g'(1) = 0
\end{array}
\right\} \text{second degree} \\
g''(0) = 0
\end{array}
\right\} \text{third degree.} \tag{7.38}
$$

$$
g''(1) = 0
$$
$$
g'''(0) = 0
$$
$$
\vdots
$$

Considering, as an example, a third-degree polynomial

$$
\frac{v_x}{U} = a + b\eta + c\eta^2 + d\eta^3,
$$

one finds that the application of the first four conditions of Eq. (7.38) gives

$$
1 = a + b + c + d,
$$
$$
0 = a,
$$
$$
0 = b + 2c + 3d,
$$
$$
0 = 2c.
$$

Thus, $a = 0$, $b = \frac{3}{2}$, $c = 0$, and $d = -\frac{1}{2}$, so

$$
\frac{v_x}{U} = g(\eta) = \frac{3}{2}\eta - \frac{1}{2}\eta^3. \tag{7.39}
$$

This equation is, then, the assumed form of the velocity distribution through the boundary layer. It is, of course, approximate, but comparison with the exact solution of Blasius (Sec. 7.2) will show a close correspondence. Figure 7.5 plots

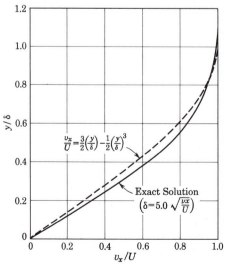

FIGURE 7.5. Comparison of the velocity distribution for a laminar boundary layer on a flat plate as given by the exact solution and the momentum integral method.

ANALYTIC SOLUTIONS OF FORCED CONVECTION

the velocity distribution given by the exact solution with that given by Eq. (7.39).

Substitution of Eq. (7.39) into the momentum integral equation, Eq. (7.36), gives the following differential equation in δ:

$$\frac{d}{dx}\left[\delta \int_0^1 \left(\frac{1}{4}\eta^6 - \frac{6}{4}\eta^4 + \frac{1}{2}\eta^3 + \frac{9}{4}\eta^2 - \frac{3}{2}\eta\right) d\eta\right] = -\frac{\nu}{U\delta}\frac{3}{2},$$

$$\frac{d}{dx}\left(\frac{39}{280}\delta\right) = \frac{\nu}{U\delta}\frac{3}{2},$$

$$\delta\,d\delta = \frac{140}{13}\frac{\nu}{U}dx.$$

Since the boundary layer has zero thickness at the leading edge of the flat plate—where $x = 0$—the integration of the above equation gives

$$\delta^2 = \frac{280}{13}\frac{\nu x}{U}.$$

Thus, the boundary layer thickness is given as the following function of x:

$$\delta = \sqrt{\frac{280}{13}}\sqrt{\frac{\nu x}{U}} = 4.64\sqrt{\frac{\nu x}{U}},$$

(7.40)

$$\frac{\delta}{x} = \sqrt{\frac{280}{13}}\sqrt{\frac{\nu}{Ux}} = \frac{4.64}{\sqrt{Re_x}}.$$

Comparison of these results with Eqs. (7.12) and (7.13)—the "exact" solution of the problem—again shows a close correspondence between the two methods.

The Thermal Boundary Layer and the Heat Transfer Coefficient

Turning now, as before, to the thermal boundary layer, one can apply the results just obtained to the integral energy equation to determine the temperature distribution in the fluid and, eventually, the heat transfer coefficient. If δ_t is the thermal boundary layer thickness, then let a similarity parameter η_t be defined as

$$\eta_t = \frac{y}{\delta_t}.$$

(7.41)

Similarity of the dimensionless temperature distribution implies the assumption that

$$\varphi = \frac{t - t_s}{t_f - t_s} = \varphi(\eta_t)$$

is a function of η_t only. The energy equation [the second equation of Eqs. (7.33)] may now be rewritten in terms of these new variables, η_t and φ:

$$\frac{d}{dx}\left[\delta_t \int_0^1 g(\eta)[1 - \varphi(\eta_t)] \, d\eta_t\right] = \frac{\alpha}{U\delta_t}\left(\frac{d\varphi}{d\eta_t}\right)_{\eta_t=0}. \tag{7.42}$$

In Eq. (7.42) the velocity distribution $v_x/U = g(\eta)$, from the above analysis of the velocity boundary layer, has been introduced. It will be remembered that in Sec. 6.10, in which the integral energy equation was derived, it was assumed that the thermal boundary layer was thinner than the velocity boundary layer (i.e., $\delta_t < \delta$). This implies that $\mathrm{Pr} > 1$, if the plate is heated to t_s over its entire length from the leading edge.

It is reasonable to approximate the temperature distribution in the boundary layer with a polynomial of the same degree as was used for the velocity layer. In this case this is a third-degree polynomial. Hence, let

$$\varphi(\eta_t) = a + b\eta_t + c\eta_t^2 + d\eta_t^3.$$

By the same reasoning used for the velocity boundary layer, the following conditions may be applied to determine the constants a, b, c, and d:

At $y = \delta_t$, $t = t_f$,   or at $\eta_t = 1$, $\varphi = 1$.

At $y = 0$, $t = t_s$,   or at $\eta_t = 0$, $\varphi = 0$.

At $y = \delta_t$, $\dfrac{\partial t}{\partial y} = 0$,   or at $\eta_t = 1$, $\dfrac{d\varphi}{d\eta} = 0$.

At $y = 0$, $\dfrac{\partial^2 t}{\partial y^2} = 0$,   or at $\eta_t = 0$, $\dfrac{d^2\varphi}{d\eta^2} = 0$.

These conditions determine the constants to be $a = 0$, $b = \frac{3}{2}$, $c = 0$, $d = -\frac{1}{2}$, so

$$\varphi(\eta_t) = \frac{3}{2}\eta_t - \frac{1}{2}\eta_t^3. \tag{7.43}$$

This expression for $\varphi(\eta_t)$ and the previously found expression for $g(\eta)$ in the velocity boundary layer may now be introduced into Eq. (7.42) to give

$$\frac{d}{dx}\left[\delta_t \int_0^1 \left(\frac{3}{2}\eta - \frac{1}{2}\eta^3\right)\left(1 - \frac{3}{2}\eta_t + \frac{1}{2}\eta_t^3\right) d\eta_t\right] = \frac{\alpha}{U\delta_t}\frac{3}{2}.$$

It is convenient to define a variable (a function of x only) ζ as the ratio of the thermal layer thickness to the velocity layer thickness:

$$\zeta = \frac{\delta_t}{\delta} = \frac{\eta}{\eta_t}. \tag{7.44}$$

Introducing this fact into the equation above, one obtains

$$\frac{d}{dx}\left[\delta_t\zeta \int_0^1 \left(\frac{3}{2}\eta_t - \frac{1}{2}\zeta^2\eta_t^3\right)\left(1 - \frac{3}{2}\eta_t + \frac{1}{2}\eta_t^3\right) d\eta_t\right] = \frac{\alpha}{U\delta_t}\frac{3}{2}.$$

Upon evaluation of the integral this expression becomes

$$\frac{d}{dx}\left[\delta_t\zeta\left(\frac{3}{20}-\frac{3}{280}\zeta^2\right)\right] = \frac{\alpha}{U\delta_t}\frac{3}{2}.$$

A rather far-reaching simplification can be made at this point if one accepts the fact that ζ, the ratio of the boundary layer thicknesses, will be near 1. This will be true, as seen in the previous sections, if the Prandtl number, Pr, is close to 1—a fact which is substantially met for a great number of fluids. If this is assumed, then $\frac{3}{280}$ can be neglected when compared with $\frac{3}{20}$, so the equation above becomes, approximately,

$$\frac{d}{dx}\left(\delta_t\zeta\frac{3}{20}\right) = \frac{3}{2}\frac{\alpha}{U\delta_t}.$$

Since $\delta_t = \delta\zeta$, this equation may be written as

$$\frac{d}{dx}\delta\zeta^2 = 10\frac{\alpha}{U\zeta\delta}.$$

Now, δ is a function of only x, and this functional relation has already been determined in the analysis of the velocity boundary layer, Eq. (7.40), as

$$\delta = \sqrt{\frac{280}{13}}\sqrt{\frac{\nu x}{U}}.$$

Thus, the above differential equation for δ is expressible as

$$\frac{d}{dx}(x^{1/2}\zeta^2) = \frac{13}{28}\frac{\alpha}{\nu}\frac{1}{\zeta x^{1/2}},$$

or

$$\zeta^3 + \frac{4}{3}x\frac{d}{dx}(\zeta^3) = \frac{13}{14}\frac{\alpha}{\nu}.$$

This latter expression is a first-order linear equation in ζ^3 which has the solution

$$\zeta^3 = \frac{13}{14}\frac{\alpha}{\nu} + Cx^{-3/4}.$$

The constant C in this equation must be zero to avoid an indeterminate solution at the leading edge, $x = 0$, so

$$\zeta = \left(\frac{13}{14}\frac{\alpha}{\nu}\right)^{1/3} \approx \left(\frac{\alpha}{\nu}\right)^{1/3}.$$

This is recognized as involving the Prandtl number, $\mathrm{Pr} = \nu/\alpha$, so

$$\zeta = \frac{1}{(\mathrm{Pr})^{1/3}}. \tag{7.45}$$

If the Prandtl number is close to 1, then $\zeta \approx 1$, as was assumed above when $\frac{3}{280}\zeta^2$ was neglected in comparison with $\frac{3}{20}$.

It is now possible to evaluate the heat transfer coefficient since

$$\left(\frac{\partial t}{\partial y}\right)_{y=0} = (t_f - t_s)\left(\frac{d\varphi}{d\eta}\right)_{\eta_t=0}\frac{1}{\delta_t}$$

$$= (t_f - t_s)\frac{3}{2}\frac{1}{\delta_t}.$$

By using the same notation as before for the local heat transfer coefficient, h_x Eq. (7.27) gives

$$h_x = -\frac{k\left(\dfrac{\partial t}{\partial y}\right)_{y=0}}{t_s - t_f} = \frac{3}{2}\frac{k}{\delta_t} = \frac{3}{2}\frac{k}{\delta\zeta}.$$

But δ and ζ are known from Eqs. (7.40) and (7.45), so

$$h_x = \frac{3}{2}k\sqrt{\frac{13}{280}}\sqrt{\frac{U}{\nu x}}\,\mathrm{Pr}^{1/3}$$

$$h_x = 0.331k\sqrt{\frac{U}{\nu x}}\,\mathrm{Pr}^{1/3} \tag{7.46}$$

$$\mathrm{Nu}_x = 0.331\mathrm{Re}_x^{1/2}\mathrm{Pr}^{1/3}.$$

The average film coefficient, h, and Nusselt number for a plate of length L are, then,

$$h = 0.662k\sqrt{\frac{U}{\nu L}}\,\mathrm{Pr}^{1/3},$$

$$\mathrm{Nu}_L = 0.662\mathrm{Re}_L^{1/2}\mathrm{Pr}^{1/3}. \tag{7.47}$$

The close correspondence between these relations and those obtained in Sec. 7.2 using the differential boundary layer equations of motion [see Eqs. (7.27) through (7.32)] is rather striking. Figure 7.6 shows a plot of Eq. (7.32) or (7.47) for the Nusselt number. This plot is compared with the experimentally measured values of Ref. 4 for air. The close agreement between theory and experiment is apparent.

The discussions of this last section were presented mainly to illustrate the basic thoughts behind the methods of determining suitable descriptions of the behavior of boundary layer flow and the associated heat convection mechanism. Each of the two methods has various advantages over the other—the first approach attempts to satisfy exactly the boundary layer governing equations, whereas the second approach attempts to give an over-all picture of the processes and to satisfy conditions only at the limits of the boundary layer. The integral method may be made more exact by using polynomials of higher degree, but the labor of computation increases therewith. For cases other than flow past a flat plate the "exact" approach becomes very complex—and often impossible—whereas the integral method will usually yield a usable result. For certain very complex geometrical configurations neither method may be of value, and some other attack must be made to obtain suitable engineering answers. This is usually done by experimentation—using the dimensionless parameters developed in Sec. 6.7.

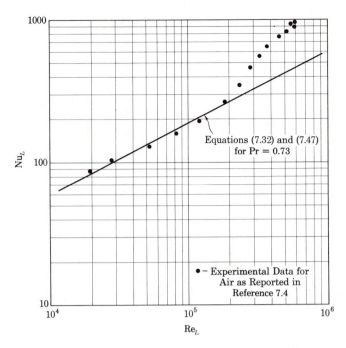

FIGURE 7.6. Comparison of theory and experiment for heat transfer in laminar flow of air past a flat plate.

AERODYNAMIC HEATING—LAMINAR FORCED CONVECTION, WITH VISCOUS DISSIPATION, ON A FLAT SURFACE

The examples treated in Secs. 7.2 and 7.3 have all been instances in which it was presumed that the rate of viscous dissipation was small enough, when compared with the other terms of the energy equation, that it could be neglected. Such a situation could be expected to occur at relatively low values of the free stream velocity U. However, situations occur in which the velocity U is large enough that the dissipation term must be included even though the flow is still laminar. This section will consider such a case—exactly the same problem as that posed in Sec. 7.2, except that the dissipation term will be retained in the energy equation. The solution to this problem is best described and understood if a simpler, subproblem, is considered first.

Laminar Flow past an Adiabatic Surface, the Recovery Temperature

First, rather than treat the case in which the surface temperature is fixed and the surface temperature gradient is sought, consider instead the case in which the surface temperature gradient is fixed as zero and the resultant surface temperature is sought.

Consider, then, a stream at velocity U, temperature t_f, flowing past a flat plate which is otherwise thermally isolated. After sufficient time has elapsed, the plate will reach an equilibrium temperature, and at the surface $(\partial t/\partial y)_{y=0}$ will be zero, since no heat will flow into or out of the plate when such equilibrium is reached. This equilibrium temperature, called the "recovery temperature," t_r, is to be found. Under these conditions, the energy equation, given in Eq. (7.1), must be solved under the conditions given below:

$$v_x \frac{\partial t}{\partial x} + v_y \frac{\partial t}{\partial y} = \alpha \frac{\partial^2 t}{\partial y^2} + \frac{\mu}{\rho c_p}\left(\frac{\partial v_x}{\partial y}\right)^2.$$

$$\text{At } y = 0: \quad \frac{\partial t}{\partial y} = 0. \tag{7.48}$$

$$\text{As } y \to \infty: \quad t \to t_f.$$

As before, introduce the similarity parameter

$$\eta = y\sqrt{\frac{U}{\nu x}},$$

so that

$$\frac{v_x}{U} = f'(\eta),$$

$$\frac{v_y}{U} = \frac{1}{2}\sqrt{\frac{\nu}{Ux}}(\eta f' - f),$$

as in the Blasius solution of Sec. 7.2. If, in addition, a dimensionless temperature parameter, Ψ is defined

$$\Psi = \frac{t - t_f}{U^2/2c_p}, \tag{7.49}$$

Eq. (7.48) may be transformed into the following form if Ψ is assumed to depend on η only:

$$\frac{d^2\Psi}{d\eta} + \frac{\text{Pr}}{2} f(\eta) \frac{d\Psi}{d\eta} = -2\text{Pr}[f''(\eta)]^2,$$

$$\Psi(\infty) = 0, \tag{7.50}$$

$$\Psi'(0) = 0.$$

The solution to the system given in Eq. (7.50) may be shown to be

$$\Psi(\eta) = 2\text{Pr} \int_{\eta}^{\infty} (f'')^{\text{Pr}}\left\{\int_{0}^{\eta} [f''(\zeta)]^{2-\text{Pr}} \, d\zeta\right\} d\eta, \tag{7.51}$$

where ζ is a dummy integration variable. Since it is the equilibrium surface temperature, t_r, which is sought,

TABLE 7.2   Recovery Temperature in Air Due to Viscous Dissipation

Air Velocity m/s	$t_r - t_f$ °C
10	0.04
25	0.3
50	1.0
100	4.1
150	9.2
200	16.4
300	36.8
500	102.3

$$\frac{t_r - t_f}{U^2/2c_p} = \Psi(0) = 2\Pr \int_0^\infty (f'')^{\Pr}\left[\int_0^\eta [f''(\zeta)]^{2-\Pr}\,d\zeta\right]d\eta \tag{7.52}$$

$$= \Psi_0(\Pr).$$

The notation $\Psi_0(\Pr)$ is introduced to point out the fact that $\Psi(0)$ is a function of *only* the Prandtl number. $\Psi_0(\Pr)$ is a constant for a given fluid, and values of it are tabulated for various Pr in Table 7.1.

Since $\Psi_0(\Pr) \geq 0$, the recovery temperature is always greater than the ambient free stream. The difference $(t_r - t_f)$ is a measure of the rate of viscous dissipation in the boundary layer. The rise of t_r above t_f depends on only the fluid and its velocity. For air, $\Pr \cong 0.7$ and $c_p \cong 1.02$ kJ/kg-°C. Thus, Table 7.1 gives $\Psi_0(\Pr) = 0.835$, so that the values of the temperature rise, $t_r - t_f$, shown in Table 7.2, result. The dissipative effect of the viscosity at high velocities is readily apparent.

The quantity $\Psi_0(\Pr)$ is often referred to as the *recovery factor*. As an approximate relation, it may be observed that the dependence of Ψ_0 on Pr is rather accurately given by

$$\Psi_0(\Pr) \approx \Pr^{1/2}. \tag{7.53}$$

Laminar, Viscous Flow past a Heated or Cooled Flat Plate

Return now to the original problem of this section—what would be the expected heat transfer (i.e., temperature gradient) at the surface of a flat plate when viscous dissipation is accounted for and when the plate is maintained at a temperature, t_s? The system of equations to be solved is

$$v_x \frac{\partial t}{\partial x} + v_y \frac{\partial t}{\partial y} = \alpha \frac{\partial^2 t}{\partial y^2} + \frac{\mu}{\rho c_p}\left(\frac{\partial v_x}{\partial y}\right)^2.$$

$$\text{At } y = 0: \quad t = t_s.$$
$$\text{As } y \to \infty: \quad t \to t_f. \tag{7.54}$$

Once again, the Blasius parameter $\eta = y\sqrt{U/vx}$ may be introduced so that v_x and v_y are known functions—according to Eqs. (7.10) and (7.11). If the dimensionless temperature ratio $(t - t_f)/(t_s - t_f)$ is assumed to depend on η, only, the solution of Eq. (7.54) may be shown (Ref. 2) to be

$$\frac{t - t_f}{t_s - t_f} = [1 - \varphi(\eta)]\left[1 - \frac{\Psi_0(\text{Pr})}{2}\frac{U^2}{c_p(t_s - t_f)}\right] + \frac{1}{2}\Psi(\eta)\frac{U^2}{c_p(t_s - t_f)}.$$

(7.55)

In Eq. (7.55) the functions $\varphi(\eta)$ and $\Psi(\eta)$ are the functions given in Eqs. (7.24) and (7.51). Thus, the temperature distribution may be calculated for any given set of conditions. However, for heat transfer calculations, the temperature gradient at the surface is desired.

Equation (7.55) yields

$$\frac{dt/d\eta}{t_s - t_f} = -\varphi'(\eta)\left[1 - \frac{1}{2}\Psi_0(\text{Pr})\frac{U^2}{c_p(t_s - t_f)}\right] + \frac{\Psi'(\eta)}{2}\frac{U^2}{c_p(t_s - t_f)}.$$

Since at $\eta = 0$, $\varphi'(0) = \varphi_0'(\text{Pr})$ and $\Psi'(0) = 0$,

$$\left(\frac{dt}{d\eta}\right)_0 = -(t_s - t_f)\varphi_0'(\text{Pr})\left[1 - \frac{1}{2}\Psi_0(\text{Pr})\frac{U^2}{c_p(t_s - t_f)}\right]. \qquad (7.56)$$

Equation (7.56) shows an interesting fact. When the plate is cooler than the ambient fluid, $t_s < t_f$, $(dt/d\eta)_0$ is always positive and heat flows *into* the surface. However, when the plate is hotter than the fluid, $t_s > t_f$, $(dt/d\eta)_0$ may be positive or negative depending on the relative sizes of $\Psi_0(\text{Pr})$ and $U^2/c_p(t_s - t_f)$. Thus, even though the surface is hotter than the fluid, heat may still flow *into* it!

These facts are better illustrated in terms of the recovery temperature, t_r, defined in Eq. (7.52):

$$\frac{t_r - t_f}{U^2/2c_p} = \Psi_0(\text{Pr}).$$

In terms of t_r, Eq. (7.56) becomes

$$\left(\frac{dt}{d\eta}\right)_0 = -(t_s - t_f)\varphi_0'(\text{Pr})\left(1 - \frac{t_r - t_f}{t_s - t_f}\right)$$

$$= -(t_s - t_r)\varphi_0'(\text{Pr}).$$

(7.57)

In this formulation, the sign of $(dt/d\eta)_0$, and the direction of the heat flow, are determined by the temperature difference $(t_s - t_r)$ rather than the customary difference $(t_s - t_f)$. The recovery temperature is that temperature which the surface would achieve if it were not maintained at t_s but allowed to come to equilibrium with the fluid stream flowing past. Thus, the heat flux is dependent upon the difference between this equilibrium temperature and the imposed t_s. Figure 7.7 illustrates the possible situations. For $t_s > t_r$, heat flows *from* the wall; for $t_s = t_r$, no heat flows; for $t_s < t_r$ heat flows into the wall.

Since $t_r > t_f$, the case of heat flow *from* the wall is always associated with $t_s > t_f$. However, for heat flow into the wall with $t_s < t_r$, the surface may be

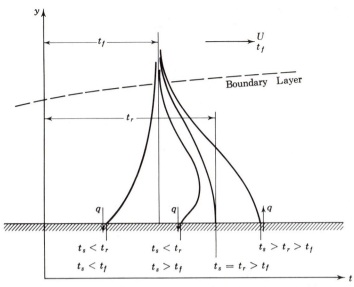

FIGURE 7.7. Temperature distributions in a laminar boundary layer when viscous dissipation is present.

hotter or cooler than the free stream temperature—depending on the relative sizes of $(t_r - t_s)$ and $(t_r - t_f)$. A case in which $t_s > t_f$ but in which heat flows *into* the wall is shown in Fig. 7.7. An inversion of the temperature profile is apparent, indicating that such a high rate of viscous dissipation occurs in the boundary layer that heat flows *both* into the fluid stream *and* the wall.

If the conventional definition of the heat transfer coefficient is used, based on $(t_s - t_f)$ as in Eq. (7.27), the above discussion indicates that a negative h could result even though $t_s > t_f$. A more rational basis for the definition of the heat transfer coefficient is the temperature difference $t_s - t_r$:

$$h_x = \frac{(q/A)_0}{t_s - t_r} = -\frac{k\left(\dfrac{\partial t}{\partial y}\right)_0}{t_s - t_r}. \tag{7.58}$$

Thus, Eq. (7.57) gives

$$h_x = k\varphi_0'(\text{Pr})\sqrt{\frac{U}{\nu x}}.$$

This latter expression is identical to that obtained in Eq. (7.27). Thus, in terms of the Nusselt numbers, the local and average film coefficients are given by Eqs. (7.30) and (7.32):

$$\text{Nu}_x = \frac{h_x x}{k} = \varphi_0'(\text{Pr})\text{Re}_x^{1/2}$$

$$\approx 0.332\text{Pr}^{1/3}\text{Re}_x^{1/2}, \tag{7.59}$$

$$\mathrm{Nu}_L = \frac{hL}{k} = 2\varphi_0'(\mathrm{Pr})\mathrm{Re}_L^{1/2}$$

$$\approx 0.664\mathrm{Pr}^{1/3}\mathrm{Re}_L^{1/2}.$$

In order to use these equations with Eq. (7.58) for the heat flux, the recovery temperature must be found from the known values of t_s, t_f, and U:

$$\frac{t_r - t_f}{U^2/2c_p} = \Psi_0(\mathrm{Pr}),$$

or

$$\frac{t_r - t_f}{t_s - t_f} = \frac{1}{2}\Psi_0(\mathrm{Pr})\frac{U^2}{c_p(t_s - t_f)}.$$

The last factor in this equation is the *Eckert number* identified in Eq. (6.20), so that

$$\frac{t_r - t_f}{t_s - t_f} = \frac{1}{2}\Psi_0(\mathrm{Pr})\mathrm{Ec}$$

$$\tag{7.60}$$

$$\approx \frac{1}{2}\mathrm{Pr}^{1/2}\mathrm{Ec}.$$

Equations (7.58), (7.59), and (7.60), when used together to find the heat flux when viscous dissipation is present, verify the predictions of Sec. 6.7, where it was found that the Nusselt, Reynolds, Prandtl, and Eckert numbers would be involved in any such analysis.

Since $\mathrm{Ec} \to 0$ as $U \to 0$, $t_r \to t_f$ under such conditions, and the definition of h given in Eq. (7.58) reduces to the definition used in the cases in which dissipation was neglected.

7.5

VELOCITY DISTRIBUTION AND SKIN FRICTION COEFFICIENT FOR TURBULENT FLOW PAST A FLAT SURFACE

The examples presented in the preceding sections have all been limited to the laminar motion. Even though rather simple examples were considered, it should be apparent that an analytical treatment of these cases is quite complex. In the case of turbulent motion, the additional complexities introduced by the turbulent fluctuations discussed in Sec. 6.11 make the analysis even more involved. The simplest case of a turbulent boundary layer is that of flow past a flat surface since, as noted earlier, there is no pressure gradient in the direction of the main flow. Turbulent flow past a flat surface has many practical applications, and flows past moderately curved surfaces, such as airfoils, turbine blades, etc., may be treated as approximately flat. This section will be devoted, then, to a discussion of turbulent flows without pressure gradients. Extensions to important cases with pressure gradients, as in pipes, will be made later.

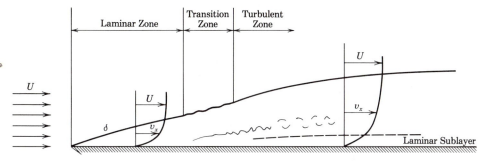

FIGURE 7.8. Laminar-turbulent transition of a boundary layer on a flat surface.

Consider, then, the situation depicted in Fig. 7.8. Here a uniform flow is shown incident upon a flat surface placed parallel to it. Beginning at the leading edge a region develops, the laminar region, in which the laminar viscous shear dominates, and the boundary layer grows, with an approximately parabolic velocity profile, according to the relations developed in Sec. 7.2. Any random transverse disturbances are rapidly damped out. After some distance down the surface the presence of transverse disturbances becomes amplified and a transition region is encountered in which turbulence begins to develop. Eventually, a region, the turbulent region, is reached in which the random disturbances become so amplified that the motion may be considered as fully turbulent steady motion as discussed in Sec. 6.11.

In the turbulent region the velocity distribution in the boundary layer is observed to be much flatter as a result of the enhanced transverse mixing. A typical turbulent velocity profile is depicted in Fig. 7.8. Near the surface the transverse turbulent fluctuations must be damped out, and the velocity variation is nearly linear—creating a region that is often characterized as the *laminar sublayer*.

The initial laminar region has already been treated in Sec. 7.2. It is the purpose of this section to describe the characteristics of the turbulent region. The intervening transition region is quite complex and is not well understood. The transition from the laminar state of flow to the turbulent state is, as noted, the result of the amplification of random disturbances in the flow. Very complex analyses exist for the prediction of the laminar–turbulent transition, but they are too involved to be presented here. As one would expect, this transition depends on the free stream fluid velocity, the distance from the leading edge, the roughness of the surface, and the properties of the fluid. For most engineering applications the laminar–turbulent transition may be taken to occur at local length Reynolds number, $Re_x = Ux/\nu$, between 5×10^5 and 10^6—although under special conditions laminar layers have been maintained up to values of $Re_x = 3 \times 10^6$, and turbulent layers have been observed for Re_x as low as 80,000. This text will adopt the frequently used criterion that transition will take place for a *critical Reynolds number*:

$$Re_{x,c} \cong 5 \times 10^5. \tag{7.61}$$

Only flow in the fully established turbulent zone will be discussed here. The total, or average, behavior of a surface with an initial laminar zone will be treated later. While a vast amount of data and theories exist for the description of turbulent boundary layers, only the simplest will be discussed since they yield sufficiently accurate results for the majority of engineering heat transfer calculations. Basically, one needs to know how the turbulent fluctuations (v_x', v_y', t') depend on the flow, the geometry, etc. Equivalently, one needs to know how the eddy viscosity and eddy diffusivity, ϵ and ϵ_H, depend on the flow. Rather than explore these directions, this work will simply quote two kinds of empirical information from which the above may be deduced. That is, on the basis of experimental observation, this text will use empirical data of two basic kinds: skin friction (or wall shear stress) and velocity distribution.

The Turbulent Skin Friction Coefficient

On the basis of extensive measurements in smooth tubes, Blasius (Ref. 5) deduced the following relation for the skin friction coefficient in turbulent flow on a smooth flat plate:

$$C_f = \frac{\tau_0}{\frac{1}{2}\rho U^2} = 0.045\left(\frac{\nu}{U\delta}\right)^{1/4}, \qquad 5 \times 10^5 < \text{Re}_x < 10^7. \quad (7.62)$$

In Eq. (7.61), δ is the local turbulent boundary layer thickness at the location x from the leading edge of the plate. Thus, the term $U\delta/\nu$ is recognized as the Reynolds number based on the boundary layer thickness and is sometimes called the *thickness Reynolds number*.

The above expression for the skin friction coefficient can be related to the length Reynolds number once one determines how the thickness δ varies with x. This is done in the next section, along with some further empirical adjustments.

Power Law Velocity Distribution

Velocity profiles in turbulent boundary layers have been obtained by a number of investigators. As would be expected, turbulent layers behave much in a laminar fashion near the surface (the laminar "sublayer" mentioned earlier) with an eventual transition to a fully turbulent condition farther away from the wall. A detailed description of this variation is complex, and one representation which gives this variation in some detail is described in the next section. However, for certain applications the gross, overall behavior of a turbulent boundary layer may be represented adequately by the following power law:*

$$\frac{v_x}{U} = \left(\frac{y}{\delta}\right)^{1/7} \quad (7.63)$$

*In this equation, and all others in this and subsequent sections relating to turbulent flow, the bar notation for the time-average velocity will be dropped for simplicity of expression—with the understanding that all turbulent velocities referred to are time-average velocities.

FIGURE 7.9. Velocity distribution and skin friction coefficient for turbulent flow of air past a flat surface.

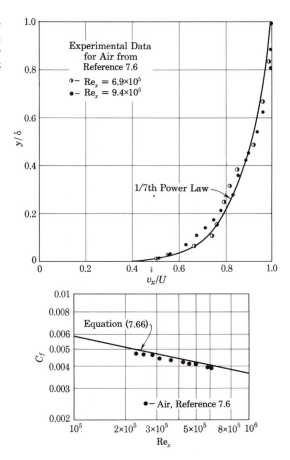

Figure 7.9 compares this one-seventh power law with measured data for air, and the good agreement is apparent. This power law represents well the observed behavior of a turbulent boundary layer for the range: $5 \times 10^5 < Re_x < 10^7$. However, Eq. (7.63) can be used only to describe the overall behavior of a turbulent boundary and not the detailed behavior near the wall since

$$\frac{\partial v_x}{\partial y} = U \frac{1}{7} \left(\frac{y}{\delta}\right)^{-6/7} \frac{1}{\delta}$$

is clearly indeterminate when $y \to 0$. Thus, some other representation, such as the "universal velocity distribution," described in the next section, must be applied near the surface.

As discussed in Secs. 6.10 and 6.11, the integral momentum equation for the boundary layer, Eq. (6.35), will apply equally well to turbulent layers as long as the wall shear term, τ_0, is properly represented for turbulent flow, i.e., as in Eq. (7.62). For flow without pressure gradient, so that $dP/dx = 0$ and $U(x) = U = $ constant, Eq. (6.35) becomes (with v_x denoting the time-average velocity)

$$\frac{d}{dx} \int_0^\delta v_x (v_x - U) \, dy = -\frac{\tau_0}{\rho}.$$

If Eqs. (7.62) and (7.63) are introduced to represent v_x and τ_0, one obtains

$$\frac{d}{dx} \int_0^\delta \left(\frac{y}{\delta}\right)^{1/7} \left[\left(\frac{y}{\delta}\right)^{1/7} - 1\right] dy = -\frac{0.045}{2} \left(\frac{v}{U\delta}\right)^{1/4},$$

or

$$\delta^{1/4} d\delta = \frac{0.045}{2} \frac{72}{7} \left(\frac{v}{U}\right)^{1/4} dx.$$

Integration of the above expression under the assumption of a zero layer thickness at $x = 0$ (an assumption to be discussed more fully in Sec. 7.7) gives

$$\frac{\delta}{x} = \left[\frac{0.045}{2} \frac{72}{7} \frac{5}{4}\right]^{4/5} \left(\frac{v}{Ux}\right)^{1/5} \quad (7.64)$$

$$= 0.371 \mathrm{Re}_x^{-1/5}$$

as the growth rate law for a turbulent boundary layer. Equation (7.64) shows that the turbulent layer grows in thickness at a greater rate than does the laminar layer for which Eq. (7.13) showed δ/x growing as $\mathrm{Re}_x^{1/2}$. Equation (7.64) may be combined with Eq. (7.62) to relate the skin friction coefficient to the local *length* Reynolds number rather than the *thickness* Reynolds number:

$$C_f = \frac{0.045}{(0.371)^{1/4}} \left(\frac{v}{Ux}\right)^{1/4} [\mathrm{Re}_x^{1/5}]^{1/4}, \quad (7.65)$$

$$= 0.0577 \mathrm{Re}_x^{-1/5}.$$

On the basis of experimental evidence, Schlicting (Ref. 2) suggests that inaccuracies related to the one-seventh power law be accounted for by empirically adjusting Eqs. (7.64) and (7.65) to

$$\frac{\delta}{x} = 0.381 \mathrm{Re}_x^{-1/5},$$

$$5 \times 10^5 < \mathrm{Re}_x < 10^7 \quad (7.66)$$

$$C_f = 0.0592 \mathrm{Re}_x^{-1/5}.$$

Figure 7.9 also compares experiment with Eq. (7.66) for C_f. For Re_x in excess of 10^7, Schlicting further recommends, on the basis of experiments reported in Ref. 7, that

$$C_f = 0.37(\log_{10} \mathrm{Re}_x)^{-2.584}, \quad 10^7 < \mathrm{Re}_x. \quad (7.67)$$

Universal Velocity Distribution

As noted in the foregoing, the one-seventh power law velocity distribution does not represent well the detailed velocity profile in a turbulent boundary layer, particularly near the surface. A more detailed representation is provided by the so-called "universal velocity distribution," based in part on Prandtl's mixing length concept described in Sec. 6.11.

It will be recalled from Sec. 6.11, that the main problem in describing the velocity in a turbulent boundary is that of having a suitable representation for the turbulent shear stress given in either Eq. (6.47) or (6.50):

$$\tau_t = -\rho \overline{v'_x v'_y} = \rho \epsilon \frac{\partial v_x}{\partial y}. \tag{7.68}$$

In other words, one seeks expressions relating either the fluctuating components v'_x, v'_y or the eddy viscosity ϵ to the flow field. Prandtl's mixing length concept represented the eddy diffusivity as [Eq. (6.57)]

$$\epsilon = l^2 \left(\frac{\partial v_x}{\partial y} \right), \tag{7.69}$$

in which l, the mixing length, represented a characteristic length over which the turbulent eddies moved before losing their identity with the mean fluid flow. Near the wall in a turbulent boundary layer on a flat surface, Prandtl theorized that l had to be zero at the surface where the turbulent fluctuations vanish and that it varied linearly with distance into the fully turbulent layer. Thus, he assumed

$$l = Cy, \tag{7.70}$$

in which C is a constant. Substitution of Eqs. (7.69) and (7.70) into (7.68) gives

$$\tau_t = \rho C^2 y^2 \left(\frac{\partial v_x}{\partial y} \right)^2.$$

Upon taking the square root of this equation and integrating, one obtains the following if the shear stress is assumed to be constant at its wall value, τ_0:

$$v_x = C_1 \sqrt{\frac{\tau_0}{\rho}} \ln y + C_2 \tag{7.71}$$

in which $C_1 = 1/C$ and C_2 is an integration constant. Thus, the assumption of the existence of a mixing length, that the mixing length varies linearly, and that the shear stress is constant results in a logarithmic velocity variation in the turbulent layer as given in Eq. (7.71). Experimental verification of this law will be shown presently.

The logarithmic velocity distribution of Eq. (7.71) is usually written in a dimensionless form by introducing a characteristic "shear velocity," v_x^*, defined as

$$v_x^* = \sqrt{\frac{\tau_0}{\rho}}. \tag{7.72}$$

Then, a nondimensional velocity and a nondimensional y coordinate are defined

$$v_x^+ = \frac{v_x}{v_x^*}, \tag{7.73}$$

$$y^+ = y \frac{v_x^*}{\nu}. \tag{7.74}$$

With these definitions, Eq. (7.71) becomes

$$v_x^+ = C_1 \ln y^+ + C_3. \tag{7.75}$$

Before proceeding, it is useful to note that the relation for the total, or apparent, shear stress given in Eq. (6.52)

$$\tau = \rho(\nu + \epsilon)\frac{\partial v_x}{\partial y}$$

becomes

$$\frac{dv_x^+}{dy^+} = \frac{1}{1 + (\epsilon/\nu)} \tag{7.76}$$

when v_x^+ and y^+ are introduced and when one again assumes the shear stress in the boundary layer to be constant at the wall value τ_0. Thus, a knowledge of the dimensionless velocity distribution $v_x^+(y^+)$ amounts to knowledge of how the eddy viscosity, ϵ, varies in the boundary layer.

The universal velocity distribution results from representing the turbulent boundary layer as separable into three sublayers as was suggested earlier in Sec. 6.11 and Fig. 6.13: a *laminar sublayer* in which the eddy viscosity is negligible ($\dot\epsilon \ll \nu$); a *buffer zone* in which the eddy viscosity and kinematic viscosity are comparable ($\epsilon \approx \nu$); and a *fully turbulent* zone in which the eddy viscosity dominates ($\epsilon \gg \nu$). For the laminar sublayer, the fact that $\epsilon \ll \nu$ allows one to put $\epsilon/\nu \approx 0$ in Eq. (7.76) so that a linear velocity distribution $v^+ = y^+$ results. For the buffer and fully turbulent zones the constants C_1 and C_3 in Eq. (7.75) must be determined experimentally. Then in these two zones the ratio ϵ/ν may be found from Eq. (7.76):

$$\frac{\epsilon}{\nu} = \frac{y^+}{C_1} - 1. \tag{7.77}$$

Figure 7.10 illustrates experimental data on turbulent boundary layers, and on the basis of data like these, the generally accepted representation for the universal velocity distribution is taken to be:

Laminar sublayer: $0 < y^+ < 5$: $v_x^+ = y^+$

$$\frac{\epsilon}{\nu} = 0.$$

Buffer zone: $5 < y^+ < 30$: $v_x^+ = 5.0 \ln y^+ - 3.05$

$$\frac{\epsilon}{\nu} = \frac{y^+}{5} - 1, \quad 0 < \frac{\epsilon}{\nu} < 5. \tag{7.78}$$

Fully turbulent zone: $y^+ > 30$: $v_x^+ = 2.5 \ln y^+ + 5.5$

$$\frac{\epsilon}{\nu} = \frac{y^+}{2.5} - 1, \quad \frac{\epsilon}{\nu} > 11.$$

It must be remembered that in the use of the universal distribution given in Eq. (7.78), it is assumed that the wall shear τ_0 is known in order that the shear

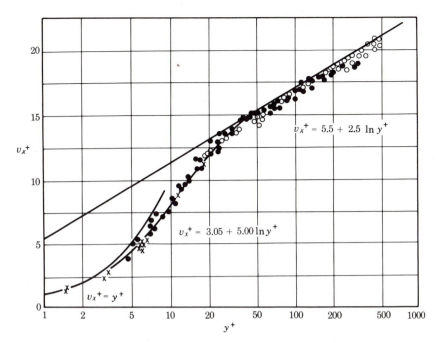

FIGURE 7.10. Universal velocity distribution. (From H. Reichardt, "Heat Transfer Through Turbulent Boundary Layers." *NACA Tech. Memo. 1047*, Washington, D.C., 1943.)

velocity $v_x^* = \sqrt{\tau_0/\rho}$ involved in the definitions of v_x^+ and y^+ can be found. Knowledge of τ_0 is found from the skin friction coefficient as given by Eq. (7.62), (7.66), or (7.67).

EXAMPLE 7.1

Atmospheric pressure air at 30°C flows with a velocity of 15 m/s past a flat surface. At a distance 0.6 m from the leading edge, find the thickness of the boundary layer, the laminar sublayer, and the buffer zone. At this location, find the ratio ϵ/ν at 0.0003 m from the surface, 0.003 m from the surface, and at the edge of the boundary layer.

Solution. At 80°C, Table A.6 gives $\rho = 1.1644$ kg/m³ and $\nu = 16.01 \times 10^{-6}$ m²/s. Thus, the local length Reynolds number is

$$\text{Re}_x = \frac{Ux}{\nu} = \frac{15.0 \times 0.6}{16.01 \times 10^{-6}} = 5.62 \times 10^5.$$

Consequently, Eq. (7.66) gives

$$\frac{\delta}{x} = 0.381\text{Re}_x^{-1/5} = 0.0270,$$

$$C_f = 0.0592\text{Re}_x^{-1/5} = 0.00419,$$

from which the boundary layer thickness at $x = 0.6$ m is

$$\delta = 0.0162 \text{ m (0.638 in.)}.$$

The shear velocity is

$$v_x^* = \sqrt{\frac{\tau_0}{\rho}} = \sqrt{\frac{C_f}{2} U^2}$$

$$= \sqrt{\frac{0.00419}{2}(15)^2} = 0.687 \text{ m/s}.$$

From Eq. (7.74),

$$y^+ = y\frac{v_x^*}{\nu} = y \times 4.291 \times 10^4, \, y \text{ in m}.$$

Thus:

Limit of laminar sublayer: $y^+ = 5$

$y = 0.000117$ m (0.0046 in.)

Limit of buffer zone: $y^+ = 30$

$y = 0.000699$ m (0.027 in.).

at $y = 0.0003$ m, $y^+ = 12.87$. This location is in the buffer zone, and Eq. (7.78) gives

$$\frac{\epsilon}{\nu} = \frac{y^+}{5} - 1 = 1.575.$$

At $y = 0.003$ m, $y^+ = 128.7$. This location is in the fully turbulent zone, and Eq. (7.78) gives

$$\frac{\epsilon}{\nu} = \frac{y^+}{2.5} - 1 = 50.49.$$

At $y = \delta = 0.0162$ m, $y^+ = 695.2$ and $\epsilon/\nu = 277$. ■

7.6

LOCAL HEAT TRANSFER IN TURBULENT BOUNDARY LAYERS ON A FLAT SURFACE

Now that the characteristics of the velocity boundary layer in turbulent flow have been established, the determination of the thermal boundary layer and the heat transfer coefficient may be determined. Again, much work on the heat transfer aspects of turbulent flow has been reported in the literature. Rather than examine these often rather complex analyses, useful engineering results may be obtained by using the similarity between heat and momentum transfer already alluded to in Secs. 6.11 and 7.2.

The Similarity Between Heat and Momentum Transfer

The similarity between heat and momentum transfer may best be seen by first examining the laminar case already discussed in Sec. 7.2. For laminar flow Eq.

(7.12) for the dimensionless velocity distribution $f' = v_x/U$ may be rewritten

$$\frac{d^2}{d\eta^2}\left(\frac{v_x}{U}\right) + \tfrac{1}{2}f\frac{d}{d\eta}\left(\frac{v_x}{U}\right) = 0,$$

(7.79)

$$\frac{v_x(0)}{U} = 0, \qquad \frac{v_x(\infty)}{U} = 1.$$

Equation (7.23) for the dimensionless temperature $\varphi = (t - t_s)/(t_f - t_s)$ may be written

$$\frac{d^2\varphi}{d\eta^2} + \tfrac{1}{2}f\,\mathrm{Pr}\,\frac{d\varphi}{d\eta} = 0,$$

(7.80)

$$\varphi(0) = 0, \qquad \varphi(\infty) = 1.$$

Since laminar motion is being considered, Pr in Eq. (7.80) is the molecular Prandtl number. If $\mathrm{Pr} = 1$ (a condition approached by many, but certainly not all, gases and some liquids), examination of Eqs. (7.79) and (7.80) shows that

$$\varphi = \frac{v_x}{U},$$

(7.81)

$$\frac{t - t_s}{t_f - t_s} = \frac{v_x}{U}.$$

Thus, for the special case of laminar flow when $\mathrm{Pr} = 1$, the dimensionless temperature and velocity profiles are identical.

More important, however, is the fact that Eq. (7.81) implies that

$$\frac{1}{t_f - t_s}\frac{\partial t}{\partial y} = \frac{1}{U}\frac{\partial v_x}{\partial y},$$

(7.82)

so that for laminar flow in which the shear stress and heat flux are given by

$$\tau = \rho\nu\frac{\partial v_x}{\partial y},$$

$$\frac{q}{A} = -\rho c_p \alpha \frac{\partial t}{\partial y},$$

one may deduce that

$$\frac{q/A}{\tau} = -c_p\frac{t_f - t_s}{U}.$$

(7.83)

The implication of Eq. (7.83) is that for $\mathrm{Pr} = 1$, the ratio of the heat flux to the shear stress in a laminar boundary layer is *constant!* This latter fact will be extended to turbulent layers in the next section in what is called the *Reynolds analogy,* but before doing so it is instructive to make another deduction for the laminar case first.

Equation (7.83) for laminar flow may be rewritten

$$\frac{q/A}{t_s - t_f} = \frac{\tau c_p}{U}.$$

When evaluated at the surface where the heat flux is $(q/A)_0$ and the shear stress is τ_0, the left side of the above equation is recognized as the surface heat transfer coefficient, h. Thus,

$$h = \frac{\tau_0 c_p}{U}. \tag{7.84}$$

Equation (7.84) may be interpreted as a statement of Reynolds analogy for laminar flow. It states that for fluids with $\mathrm{Pr} = 1$, the heat transfer coefficient and the surface shear stress are uniquely related. The implication of this similarity between heat and momentum transfer is quite important. High heat transfer rates are invariably accompanied by high viscous drag forces; in some circumstances, viscous drag measurements may be used to deduce heat transfer coefficients.

Equation (7.84) may be written in a dimensionless form as follows, recognizing that $\mathrm{Pr} = 1$ implies $\nu = \alpha$ or $\mu c_p = k$:

$$\frac{hx}{k} = \frac{\tau_0 c_p x}{kU},$$

$$= \frac{\tau_0 x}{\mu U}, \tag{7.85}$$

$$= \frac{C_f}{2} \frac{\rho U x}{\mu},$$

$$\mathrm{Nu}_x = \frac{C_f}{2} \mathrm{Re}_x.$$

Examination of Eqs. (7.16) and (7.32) when $\mathrm{Pr} = 1$ bears out the statement given in Eq. (7.85).

The similarity of heat and momentum transfer implied in Eqs. (7.83), (7.84), and (7.85) for laminar flow may be extended to the turbulent case when $\mathrm{Pr} \neq 1$, providing considerable simplifications, as is illustrated next.

The Reynolds Analogy for Turbulent Heat Transfer

In the case of turbulent flow past a flat surface, the total apparent shear stress and heat flux are given by

$$\tau = \rho(\nu + \epsilon) \frac{\partial v_x}{\partial y},$$

$$\frac{q}{A} = -\rho c_p(\alpha + \epsilon_H) \frac{\partial t}{\partial y}. \tag{7.86}$$

Reynolds assumed, as a simple model, that the turbulent boundary layer consisted of only the fully turbulent zone—that is, he presumed that the laminar sublayer and the buffer zone are negligible. Thus, in the boundary layer $\epsilon \gg \nu$ and $\epsilon_H \gg \alpha$, so that ν and α may be taken as negligible and the ratio of the two expressions in Eq. (7.86) yields

$$\frac{q/A}{\tau} = -c_p \frac{\epsilon_H}{\epsilon} \frac{\partial t}{\partial v_x} = -\frac{c_p}{\text{Pr}_t} \frac{\partial t}{\partial v_x},$$

in which $\text{Pr}_t = \epsilon/\epsilon_H$ is the *turbulent* Prandtl number. Reynolds further assumed that since the eddy viscosity and eddy diffusivity arise from the same mechanism of transverse fluctuation, then $\text{Pr}_t \approx 1$, i.e., $\epsilon \approx \epsilon_H$, so that

$$\frac{q/A}{\tau} = -c_p \frac{\partial t}{\partial v_x}. \tag{7.87}$$

The similarity between heat and momentum transfer discussed for laminar flow in the preceding section noted that the ratio $(q/A)/\tau$ was constant. Reynolds assumed that this same similarity exists for the fully turbulent boundary layer and integrated Eq. (7.87) across the boundary layer from $t = t_s$, $v_x = 0$ to $t = t_f$, $v_x = U$, while taking $(q/A)/\tau$ as constant and equal to the wall values, to obtain

$$\frac{(q/A)_0}{\tau_0} = -c_p \frac{t_f - t_s}{U},$$

or

$$h = \frac{\tau_0 c_p}{U}.$$

While this latter expression appears to be the same as that in Eq. (7.84) for laminar flow, one should note that although it has been presumed that the turbulent Prandtl number is unity, $\text{Pr}_t \approx 1$, it has *not* been assumed that the molecular Prandtl number $\text{Pr} = \nu/\alpha$ is 1. Thus, the above statement may also be written

$$h = \frac{\tau_0 k \text{Pr}}{\mu U}, \tag{7.88}$$

$$\text{Nu}_x = \frac{C_f}{2} \text{Re}_x \text{Pr}.$$

Equation (7.88) is known as the *Reynolds analogy* for turbulent heat transfer on a flat surface. When Eq. (7.88) is combined with Eqs. (7.66) and (7.67), Reynolds analogy for turbulent heat transfer on a flat surface is

$$\text{Nu}_x = 0.0296 \text{Re}_x^{0.8} \text{Pr}, \qquad 5 \times 10^5 < \text{Re}_x < 10^7, \tag{7.89}$$

$$\text{Nu}_x = \frac{0.185 \text{Re}_x \text{Pr}}{(\log_{10} \text{Re}_x)^{2.584}}, \qquad 10^7 < \text{Re}_x.$$

Equations (7.89) may be used to predict turbulent heat transfer coefficients with reasonable accuracy; however, comparison with experiments will be deferred until after other improvements have been discussed. Also, integrated forms of Eqs. (7.89) for the average plate length Nusselt number will be obtained later in Sec. 7.7.

The Colburn Analogy

On the basis of experimental evidence Colburn (Ref. 8) empirically modified the Reynolds analogy as

$$\mathrm{Nu}_x = \frac{C_f}{2}\,\mathrm{Re}_x\mathrm{Pr}^{1/3}, \tag{7.90}$$

or for the flat surface:

$$\mathrm{Nu}_x = 0.0296\mathrm{Re}_x^{0.8}\mathrm{Pr}^{1/3}, \quad 5 \times 10^5 < \mathrm{Re}_x < 10^7, \tag{7.91}$$

$$\mathrm{Nu}_x = \frac{0.185\mathrm{Re}_x\mathrm{Pr}^{1/3}}{(\log_{10}\mathrm{Re}_x)^{2.584}}, \quad 10^7 < \mathrm{Re}_x.$$

The Prandtl Analogy

Prandtl modified the Reynolds analogy by including the laminar sublayer while still excluding the buffer zone. That is, he modeled the turbulent boundary layer as noted in Fig. 7.11. In this model a laminar sublayer, in which the velocity distribution is taken as linear, is joined to a fully turbulent region at some distance, δ_s, from the surface at which the velocity and temperature are denoted as v_{xj} and t_j, respectively.

In the laminar sublayer it is assumed that $\epsilon \ll \nu$ and $\epsilon_H \ll \alpha$, so that

$$\tau = \rho(\nu + \epsilon)\frac{\partial v_x}{\partial y} \cong \rho\nu\frac{\partial v_x}{\partial y},$$

$$\frac{q}{A} = -\rho c_p(\alpha + \epsilon_H)\frac{\partial t}{\partial y} \cong -\rho c_p\alpha\frac{\partial t}{\partial y},$$

or

$$\frac{q/A}{\tau} = -\frac{c_p}{\mathrm{Pr}}\frac{\partial t}{\partial v_x}.$$

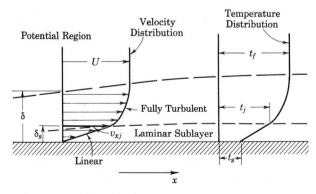

FIGURE 7.11. Simplified representation of a turbulent boundary layer on a flat surface.

Again, invoking the similarity of heat and momentum transfer by assuming $(q/A)/\tau = (q/A)_0/\tau_0$ through the laminar sublayer, integration of the above from the wall where $v_x = 0$, $t = t_s$ to the limit of the sublayer where $v_x = v_{xj}$, $t = t_j$, gives

$$\frac{(q/A)_0}{\tau_0} = -\frac{c_p}{\text{Pr}} \frac{t_j - t_s}{v_{xj}}. \tag{7.92}$$

The fully turbulent region from the limit of the sublayer to the free stream was characterized, as in Reynolds analogy, by $\epsilon \gg \nu$, $\epsilon_H \gg \alpha$, and $\epsilon \approx \epsilon_H$. Then the integration of Eq. (7.87) from δ_s to the free stream, with $(q/A)/\tau$ again being taken as constant, gives

$$\frac{(q/A)_0}{\tau_0} = -c_p \frac{t_f - t_j}{U - v_{xj}}.$$

Elimination of t_j between the above expression and Eq. (7.92) gives

$$h = \frac{(q/A)_0}{t_s - t_f} = \frac{c_p \tau_0}{U\left[1 + \dfrac{v_{xj}}{U}(\text{Pr} - 1)\right]},$$

or

$$\text{Nu}_x = \frac{\frac{1}{2}C_f \text{Re}_x \text{Pr}}{1 + \dfrac{v_{xj}}{U}(\text{Pr} - 1)}.$$

The ratio v_{xj}/U may be found by several methods (such as joining the linear sublayer velocity with a one-seventh power law at δ_s), but the simplest is to take the limit of the sublayer as that given in Eq. (7.78):

$$v_x^+ = y^+ = 5,$$

$$v_{xj} = 5\sqrt{\frac{\tau_0}{\rho}},$$

$$v_{xj} = 5\sqrt{\frac{C_f}{2}}\,U,$$

$$\frac{v_{xj}}{U} = 5\sqrt{\frac{C_f}{2}}.$$

Thus, the Prandtl analogy for turbulent heat transfer on a flat plate is

$$\text{Nu}_x = \frac{\frac{1}{2}C_f \text{Re}_x \text{Pr}}{1 + 5\sqrt{C_f/2}[\text{Pr} - 1]}, \tag{7.93}$$

or with Eq. (7.66),

$$\text{Nu}_x = \frac{0.0296\text{Re}_x^{0.8}\text{Pr}}{1 + 0.860\text{Re}_x^{-0.1}(\text{Pr} - 1)}, \qquad 5 \times 10^5 < \text{Re}_x < 10^7 \tag{7.94}$$

A similar expression for $\text{Re}_x > 10^7$ can be written using Eq. (7.67).

Prandtl's analogy, Eq. (7.94), is found to give reasonable predictions for turbulent heat transfer on a flat surface. Again, comparison with experiment is deferred until later; although the reader may wish to refer to Fig. 7.19 in which the counterpart of Eq. (7.94) for pipe flow is compared with experiment.

The Von Kármán Analogy

Von Kármán (Ref. 9) improved on Prandtl's analysis by including the buffer zone given in the universal velocity distribution of Eq. (7.78). Applying Eq. (7.87) for the fully turbulent zone between $y^+ = 30$ and the free stream, again with the ratio $(q/A)_0/\tau_0$ constant, gives

$$\frac{(q/A)_0}{\tau_0} = -c_p \frac{t_f - t_{30}}{U - v_{x_{30}}}. \tag{7.95}$$

The temperature at the limit of the buffer zone, t_{30}, is found by integrating

$$\left(\frac{q}{A}\right)_0 = -\rho c_p (\alpha + \epsilon_H) \frac{\partial t}{\partial y}$$

$$= -\rho c_p \left(\frac{1}{\Pr} + \frac{\epsilon_H}{\nu}\right) \sqrt{\frac{\tau_0}{\rho}} \frac{\partial t}{\partial y^+}$$

between $y^+ = 0$, $t = t_s$ and $y^+ = 30$, $t = t_{30}$. In this integration one uses $\epsilon_H/\nu = 0$ in the interval $0 < y^+ < 5$ (the laminar sublayer) and for $5 < y^+ < 30$ one uses

$$\frac{\epsilon_H}{\nu} = \frac{\epsilon}{\nu} = \frac{y^+}{5} - 1$$

as given in Eq. (7.78) for the buffer zone. Considerable algebra gives

$$t_s - t_{30} = \sqrt{\frac{\rho}{\tau_0}} \frac{(q/A)_0}{\rho c_p} \left[5\Pr + 5 \ln \frac{\left(\dfrac{1}{\Pr} - 1\right) + 6}{\left(\dfrac{1}{\Pr} - 1\right) + 1} \right]. \tag{7.96}$$

The velocity $v_{x_{30}}$ required in Eq. (7.95) is readily found from Eq. (7.78):

$$v_{x_{30}} = \sqrt{\frac{\tau_0}{\rho}} (5 \ln 30 - 3.05)$$

$$= \sqrt{\frac{\tau_0}{\rho}} (5 \ln 6 + 5).$$

If this latter expression and Eq. (7.96) are introduced into Eq. (7.95) more algebraic manipulation yields finally the von Kármán analogy:

$$h = \frac{(q/A)_0}{t_s - t_f} = \frac{\tau_0 c_p}{U + \sqrt{\tau_0/\rho} \{5(\Pr - 1) + 5 \ln [1 + \frac{5}{6}(\Pr - 1)]\}},$$

$$Nu_x = \frac{\frac{1}{2}C_f Re_x Pr}{1 + 5\sqrt{C_f/2}\{(Pr - 1) + \ln[1 + \frac{5}{6}(Pr - 1)]\}}. \quad (7.97)$$

With Eq. (7.66), one has

$$Nu_x = \frac{0.0296 Re_x^{0.8} Pr}{1 + 0.860 Re_x^{-0.1}\{(Pr - 1) + \ln[1 + \frac{5}{6}(Pr - 1)]\}} \quad (7.98)$$

for $5 \times 10^5 < Re_x < 10^7$. As in the case of the Prandtl analogy, the pipe flow counterpart of Eq. (7.98) is compared with experiment in Fig. 7.19.

As was mentioned earlier, the main purpose of this chapter was to illustrate the *method* of analysis in the prediction of heat transfer coefficients. The examples noted are just that, *examples*. More elaborate theories exist for the prediction of heat transfer and skin friction; however, the basic concepts involved are not different, just the amount of algebra. Chapter 8 will present a summary of the above analyses, and others, that are recommended for the solution of practical engineering problems.

7.7

AVERAGE SKIN FRICTION AND HEAT TRANSFER FOR TURBULENT FLOW ON A FLAT SURFACE

The analyses presented in Secs. 7.5 and 7.6 were confined to descriptions of the *local* turbulent skin friction coefficient and Nusselt number. In many instances one is equally interested in average values for a plate of finite length, L. As in the laminar cases examined in Sec. 7.2 such averages may be found by appropriately integrating over the total plate length. However, in such instances one must account for the laminar boundary layer which occurs on the surface between the leading edge and the point at which the laminar-turbulent transition is assumed to take place. If the presence of a transition region is ignored, one may apply the results of Sec. 7.2 for laminar flow up to the distance, x_c (at which transition takes place) and the results of Sec. 7.5 and 7.6 been x_c and the plate length L. The distance x_c is determined from

$$Re_{x,c} = \frac{Ux_c}{\nu},$$

where $Re_{x,c}$ is the critical length Reynolds number.

The drag coefficient for a plate of length $L > x_c$ is then found by integration of the local skin friction coefficients given in Eqs. (7.16) and (7.66):

$$
\begin{aligned}
C_D &= \frac{1}{L}\int_0^L C_f \, dx \\
&= \frac{1}{L}\left[\int_0^{x_c} 0.664\left(\frac{\nu}{Ux}\right)^{1/2} dx + \int_{x_c}^L 0.0592\left(\frac{\nu}{Ux}\right)^{1/5} dx\right], \\
&= 1.328\left(\frac{\nu}{Ux_c}\right)^{1/2}\frac{x_c}{L} + 0.074\left(\frac{\nu}{UL}\right)^{1/5}\left[1 - \left(\frac{x_c}{L}\right)^{4/5}\right],
\end{aligned}
$$

TABLE 7.3

$Re_{x,c}$	A
3×10^5	1055
5×10^5	1743
1×10^6	3341
3×10^6	8944

$$C_D = 0.074\left(\frac{\nu}{UL}\right)^{1/5} - \frac{x_c}{L}\left[0.074\left(\frac{\nu}{Ux_c}\right)^{1/5} - 1.328\left(\frac{\nu}{Ux_c}\right)^{1/2}\right],$$

$$= 0.074Re_L^{-1/5} - ARe_L^{-1}, \qquad Re_{x,c} < Re_L < 10^7, \qquad (7.99)$$

where

$$A = 0.074Re_{x,c}^{4/5} - 1.328Re_{x,c}^{1/2}, \qquad (7.100)$$

since $x_c/L = Re_{x,c}/Re_L$. The quantity A depends on the transition Reynolds number, $Re_{x,c}$, and is given in Table 7.3. For higher Reynolds numbers, Schlicting (Ref. 2) approximates the drag coefficient with the following empirical formula:

$$C_D = 0.455(\log_{10} Re_L)^{-2.584} - ARe_L^{-1}, \qquad Re_{x,c} < Re_L < 10^9. \quad (7.101)$$

In a similar fashion, one may deduce average heat transfer coefficients and average Nusselt numbers using

$$h = \frac{1}{L}\int_0^L h_x \, dx$$

and

$$Nu_L = \frac{hL}{k} = \int_0^L \frac{1}{x} Nu_x \, dx.$$

[Note that $Nu_L \neq (1/L)\int_0^L Nu_x \, dx$, as one might be inclined to write!] For the laminar portion of the plate, one uses for h_x the result given in Eq. (7.27) or (7.32). However in the turbulent portion, $x_c < x < L$, the result depends on which of the analogies of Sec 7.6 is used. Clearly, the Prandtl analogy, Eq. (7.94), or the von Kármán analogy, Eq. (7.97), present difficulties in carrying out the desired integration. For this reason, either the Reynolds analogy or the Colburn analogy is usually used. The Colburn analogy is preferred since, as will be shown later, it agrees better with experimental data for fluids with Prandtl numbers near 1 and since it involves $Pr^{1/3}$ as does the laminar law.

Thus, using the laminar heat transfer relation in Eq. (7.32) and the Colburn analogy in Eq. (7.91), one finds the average Nusselt number to be

$$Nu_L = \int_0^{x_c} 0.332Pr^{1/3}\left(\frac{Ux}{\nu}\right)^{1/2}\frac{1}{x} \, dx + \int_{x_c}^L 0.0296Pr^{1/3}\left(\frac{Ux}{\nu}\right)^{0.8}\frac{1}{x} \, dx,$$

$$= 0.664Pr^{1/3}Re_{x,c}^{1/2} + 0.037Pr^{1/3}[Re_L^{0.8} - Re_{x,c}^{0.8}],$$

$$= \mathrm{Pr}^{1/3}[0.037\mathrm{Re}_L^{0.8} - (0.037\mathrm{Re}_{x,c}^{0.8} - 0.664\mathrm{Re}_{x,c}^{1/2})].$$

As in the case of the drag coefficient, the value of the average plate length Nusselt number depends on the value chosen for the transition Reynolds number, $\mathrm{Re}_{x,c}$. In terms of the quantity A defined in Eq. (7.100), the above equation may be written

$$\mathrm{Nu}_L = \frac{hL}{k} = \mathrm{Pr}^{1/3}[0.037\mathrm{Re}_L^{0.8} - \tfrac{1}{2}A], \tag{7.102}$$

with A given in Table A.3. For the customarily used $\mathrm{Re}_{x,c} = 5 \times 10^5$, one has

$$\mathrm{Nu}_L = \frac{hL}{k} = \mathrm{Pr}^{1/3}[0.037\mathrm{Re}_L^{0.8} - 872], \qquad 5 \times 10^5 < \mathrm{Re}_L < 10^7. \tag{7.103}$$

Comparison of Eqs. (7.99) for C_D and (7.103) for Nu_L suggests that the Colburn analogy applies for the total plate length as well as locally. That is,

$$\mathrm{Nu}_L = \frac{C_D}{2} \mathrm{Re}_L \mathrm{Pr}^{1/3}.$$

Thus, for Reynolds numbers in excess of that specified for Eq. (7.103), one may use Eq. (7.101) to show

$$\mathrm{Nu}_L = \mathrm{Pr}^{1/3}[0.228\mathrm{Re}_L(\log_{10}\mathrm{Re}_L)^{-2.584} - \tfrac{1}{2}A], \qquad \mathrm{Re}_{x,c} < \mathrm{Re}_L < 10^9,$$

$$= \mathrm{Pr}^{1/3}[0.228\mathrm{Re}_L(\log_{10}\mathrm{Re}_L)^{-2.584} - 872], \qquad 5 \times 10^5 < \mathrm{Re}_L < 10^9. \tag{7.104}$$

7.8

VISCOUS FLOW IN PIPES OR TUBES— FULLY DEVELOPED FLOW

Of equal practical importance as the cases of flow past a flat surface just considered is that of flow through circular pipes or tubes. The remaining sections of this chapter will be devoted to the analysis of heat transfer for this important geometry.

As a viscous fluid, initially of uniform velocity, enters a pipe or tube a boundary layer builds up along the surface of the pipe. This is illustrated in Fig. 7.12. Near the entrance of the pipe this growth of the boundary layer is much

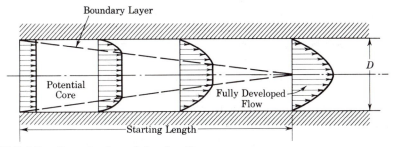

FIGURE 7.12. Starting length in pipe flow.

like the flow along the flat plate—particularly if the curvature of the pipe is not large. However, the flow cannot be the same as in the flat plate case due to the presence of the opposite wall—where a boundary layer is also developing. As the flow proceeds down the pipe, the boundary layer growing along the pipe wall gets thicker, eventually growing together from opposite sides and filling the pipe with "boundary layer" flow. The state of flow cannot actually be considered as boundary layer flow since the fundamental assumption that the viscous flow region is thin compared to the geometric dimensions of the pipe is no longer satisfied. However, experiment shows that many of the empirical laws quoted in the foregoing sections for flow along a flat surface may also be used in the case of pipe flow when appropriate definitions are made for the undisturbed stream velocity, boundary layer thickness, etc.

Pipe flow is termed *fully developed flow* once the viscous layers have grown together and filled the pipe cross section. The pipe length required to establish this fully developed flow is called the *starting* or *entrance* length. In the starting length the velocity distribution across a diameter consists of a potential core region near the center of the pipe (the velocity being uniform) which joins the boundary layer region at each surface where the velocity varies from the potential core value to zero at the wall. As one moves along the pipe in the starting region, the viscous or boundary layer portion of the velocity distribution curve increases while the potential core portion decreases. For constancy of mass flow rate, the mean velocity across the cross section (mass flow rate divided by density and cross-sectional area) must remain constant—for incompressible flow. Hence, the velocity of the potential core region must increase as the flow proceeds down the starting length. This increase is also shown in Fig. 7.12.

The transition from the laminar to turbulent state of flow may occur either in the boundary layer in the starting length or in the fully developed flow—or not at all—depending on the conditions of the flow (i.e., velocity, viscosity, etc.). Thus, it is possible to have a laminar or turbulent fully developed region. The transition in the starting region can be predicted reasonably well by use of the criterion based on the local length Reynolds number, Re_x, as noted in Sec. 7.5.

For fully developed pipe flow the length Reynolds number loses its significance, and it is customary to employ the *diameter Reynolds number, Re_D*, defined as

$$Re_D = \frac{U_m D \rho}{\mu} = \frac{U_m D}{\nu}. \tag{7.105}$$

Here D denotes the pipe diameter and U_m is the mean velocity of the flow in the pipe. Experiments indicate that a critical, or transition, diameter Reynolds number of 2300 may be used to predict reasonably well the existence of fully developed laminar or turbulent pipe flow. For values of $Re_D < 2300$, laminar flow may be expected, whereas greater values indicate the presence of turbulent flow. As discussed before in connection with transition in boundary layers, this critical value of $Re_D = 2300$ is not to be treated as a precise value since certain extraneous conditions can cause the transition to occur at other values.

A formula, developed by Langhaar (Ref. 10), which is useful for predicting the length of the starting section of laminar pipe flow is

$$\frac{L_s}{D} = 0.0575 \text{Re}_D. \tag{7.106}$$

This formula must be treated as approximate since the fully developed condition is reached asymptotically, and hence the starting length, L_s, is difficult to define. If the laminar–turbulent transition takes place in the starting region, the starting length decreases from the value indicated by Eq. (7.106). No adequate theory exists which will predict the starting length in turbulent flow since the nature of the pipe entrance, the pipe roughness, etc., will have a serious effect on the growth and transition of the boundary layer in the pipe. Starting lengths of 25 to 40 pipe diameters are typical for turbulent flow.

7.9

FULLY DEVELOPED PIPE FLOW FRICTION FACTORS AND VELOCITY DISTRIBUTIONS

In fully developed flow it is customary to represent the surface shear stress in a manner similar to that used for the skin friction coefficient in external flow. If τ_R is used to denote the fluid shear stress at the pipe wall (i.e., at the pipe radius R) the *Darcy friction factor*, f, is defined

$$f = \frac{4\tau_R}{\frac{1}{2}\rho U_m^2}. \tag{7.107}$$

Here U_m again represents the mean flow velocity over the pipe cross section. Some workers prefer a friction factor, the Fanning factor, which is one-fourth of that just defined, in analogy to the skin friction coefficient; however, the Darcy factor just defined is more often used.

Wall friction in a pipe must be balanced by a gradient in pressure along the pipe length. Thus, if dP is the pressure difference between two cross sections dx apart,

$$-\pi R^2 \, dP = 2\pi R \tau_R \, dx,$$

so that the pressure gradient along the pipe is

$$-\frac{dP}{dx} = \frac{2\tau_R}{R} = \frac{4\tau_R}{D},$$

$$= \frac{1}{2} \rho U_m^2 \frac{f}{D}.$$

For fully developed flow in which the fluid properties remain constant U_m and f are constant along the pipe, and the pressure drop due to friction over a pipe length L is, then,

$$-\Delta P = f \frac{L}{D} \frac{1}{2} \rho U_m^2. \tag{7.108}$$

Thus, the friction factor is related to the pressure gradient in the pipe and is used to calculate pressure losses due to friction. As will be seen later, application of the various analogies between heat and momentum transfer will relate f to the heat transfer coefficient in the pipe.

Laminar Flow

For fully developed laminar flow, the velocity distribution across a pipe cross section is given by the well-known parabolic form,

$$\frac{v}{U_m} = 2\left[1 - \left(\frac{r}{R}\right)^2\right] \tag{7.109}$$

as illustrated in Fig. 7.13. In Eq. (7.109), v is the local velocity at the radial location r, and R is the pipe radius. Since the surface shear stress is

$$\tau_R = -\mu\left(\frac{\partial v}{\partial r}\right)_{r=R},$$

the negative sign arising from the fact that r is measured positive away from the centerline, one has

$$\tau_R = 4\mu\frac{U_m}{R},$$

or by Eq. (7.107),

$$f = 64\frac{\mu}{U_m D\rho} = \frac{64}{Re_D}. \tag{7.110}$$

Thus, the friction factor is easily found for fully developed laminar flow from the pipe diameter Reynolds number.

Turbulent Flow

For fully developed turbulent flow, as in the case of flow past a flat surface, the velocity profile is much flatter than the parabolic laminar profile. Experiments have shown that while pipe flow is not a flow with a zero pressure gradient, as

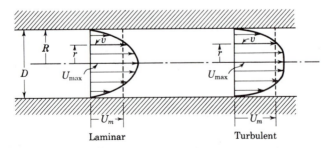

FIGURE 7.13. Fully developed profiles for laminar and turbulent pipe flow.

in the flat plate case, many of the relations for flat surfaces may be applied with reasonable accuracy to fully developed flow in a pipe with the boundary layer thickness, δ, replaced by R and the free stream velocity U replaced by U_{max}. The symbol U_{max} is used to denote the maximum tube axial velocity along the centerline.

Power Law Velocity Distribution. The power law velocity distribution given in Eq. (7.63) becomes, for pipe flow,

$$\frac{v}{U_{max}} = \left(\frac{R-r}{R}\right)^{1/7} \qquad (7.111)$$

As in the case of flat plate flow, this relation is only approximate and cannot be expected to describe the fully developed velocity profile near the pipe wall. It does give an adequate overall representation and yields the flat profile suggested in Fig. 7.13. Since it is customary to relate pipe flow parameters to the mean velocity, U_m, Eq. (7.111) may be integrated to yield

$$U_m = \frac{1}{\pi R^2} \int_0^R v(2\pi r) \, dr \qquad (7.112)$$

$$= \frac{98}{120} U_{max} = 0.817 U_{max} \approx 0.8 U_{max}.$$

Schlicting (Ref. 2) suggests the latter adjustment that $U_m \approx 0.8 U_{max}$ as being more physically realistic.

A friction factor for fully developed pipe flow based on the one-seventh power law may now be deduced. The Blasius relation for flat plate shear stress in terms of the boundary layer thickness may be translated into that for pipe flow by replacing in Eq. (7.62) τ_0 with τ_R, U with U_{max}, and δ with $R/2$:

$$\frac{\tau_R}{\frac{1}{2}\rho U_{max}^2} = 0.045(\nu/U_{max}R)^{1/4}.$$

With $U_{max} = U_m/0.8$, one finds

$$f = \frac{4\tau_R}{\frac{1}{2}\rho U_m^2} = \left[4 \times 0.045 \frac{(2 \times 0.8)^{1/4}}{(0.8)^2}\right]\left(\frac{\nu}{U_m D}\right)^{1/4}, \qquad (7.113)$$

$$f = 0.316 Re_D^{-1/4}, \qquad 10^4 < Re_D < 5 \times 10^4.$$

For higher Re_D, one may use the following empirical relation (Ref. 12):

$$f = 0.184 Re_D^{-1/5}, \qquad 3 \times 10^4 < Re_D < 10^6. \qquad (7.114)$$

Both Eqs. (7.113) and (7.114) are for smooth tubes and result principally from the one-seventh power law. Equivalent relations resulting from the universal distribution and some including the effects of surface roughness are noted next.

Universal Velocity Distribution. For fully developed pipe flow, the universal velocity distribution given in Eq. (7.78) may be used as long as the dimensionless variables are defined as

$$v^* = \sqrt{\frac{\tau_R}{\rho}},$$

$$y^+ = (R - r)\frac{v^*}{\nu},$$

$$v^+ = \frac{v}{v^*}.$$

Since

$$f = \frac{4\tau_R}{\frac{1}{2}\rho U_m^2}$$

$$= \frac{8}{(U_m^+)^2}$$

the friction factor can be predicted from the mean value of the dimensionless velocity distribution. Equations (7.78) may be integrated (Ref. 11) to show

$$U_m^+ = 2.5 \ln\left(\frac{R}{\nu}\sqrt{\frac{\tau_R}{\rho}}\right) + 1.75,$$

from which

$$\frac{1}{\sqrt{f}} = 0.88 \ln(\mathrm{Re}_D \sqrt{f}) - 0.91.$$

Reference 2 suggests that the above equation fits experimental data better if empirically adjusted to

$$\frac{1}{\sqrt{f}} = 0.87 \ln(\mathrm{Re}_D \sqrt{f}) - 0.8. \tag{7.115}$$

Equation (7.115) is probably the most reliable relation for smooth pipes; however, Eqs. (7.113) and (7.114) are easier to use and give comparable results.

All of Eqs. (7.113) through (7.115) are applicable only to "smooth" tubes—tubes in which the surface irregularities lie well within the laminar sublayer. More detailed analyses (Refs. 13 through 15) include the effects of surface roughness as measured by the relative roughness ϵ/D (ϵ being the absolute pipe surface roughness). Moody (Ref. 15) combined these studies into the diagram shown in Fig. 7.14. This diagram can be used to determine f with a fair degree of accuracy. The curve for $\epsilon/D = 0$ is equivalent to Eq. (7.113), (7.114), or (7.115).

7.10

HEAT TRANSFER IN FULLY DEVELOPED PIPE FLOW

In the case of flow in a pipe in which heat transfer is taking place, there is a difference between the pipe wall temperature and the fluid temperature. This difference produces a temperature profile in the fluid with some sort of variation

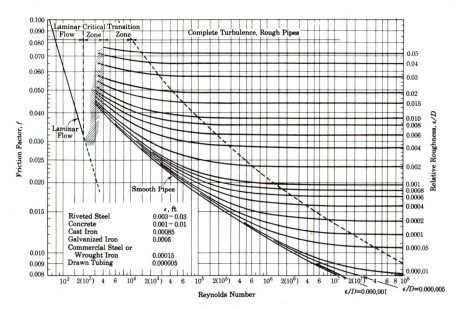

FIGURE 7.14. Moody diagram for pipe friction. (From L. F. Moody, "Friction Factors for Pipe Flow," *Trans. ASME*, Vol. 66, 1944, p. 671. Used by permission.)

such as suggested in Fig. 7.15. Here the case in which the surface is hotter than the fluid is shown, and some variation of the fluid temperature occurs between the value at the surface to that at the centerline. The exact form of this variation depends on the thermal boundary conditions imposed at the surface, whether the flow is laminar or turbulent, and possible entry length effects. Since there is no easily identified characteristic temperature of the flow, such as the free stream temperature in the case of external flow, some definition must be made to characterize the fluid temperature and the heat transfer coefficient. This is done by defining the *bulk* fluid temperature (or "mixing cup" temperature) t_b as the energy-average temperature across the cross section,

$$t_b = \frac{\int_0^R \rho c_p \, 2\pi r t \, dr}{\int_0^R \rho c_p \, 2\pi r \, dr}. \tag{7.116}$$

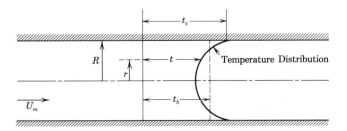

FIGURE 7.15. Temperature distribution in pipe flow.

Then, the local heat transfer coefficient is defined as

$$h = \frac{(q/A)_s}{t_s - t_b}.$$

(7.117)

Clearly, the bulk temperature is an integrated average for the cross section and is useful in expressing the heat transferred to the fluid between two different cross sections (where the bulk temperature is t_{b_1} and t_{b_2}) by writing an energy balance on the fluid:

$$q = \dot{m}c_p(t_{b_2} - t_{b_1}),$$

$\dot{m}$ being the mass flow rate. In some differential length of pipe, dx, the above two statements enable one to write

$$dq = \dot{m}c_p \, dt_b = h2\pi R \, dx(t_s - t_b).$$

(7.118)

As in the case of the development of the velocity profile discussed in the preceding section, there are also starting length considerations with respect to the development of the temperature profile in pipe flow. Figures 7.16 and 7.17 depict a fluid of initially uniform temperature entering a pipe, the wall of which is hotter than the fluid. As the flow proceeds down the pipe, a thermal boundary layer develops at the surface in which the fluid temperature is altered by heat transfer with the surface; however, a central core, in which the inlet temperature is retained, exists. Eventually, the thermal layer completely fills the pipe at the thermal starting length L_{s_t} and the temperature profile becomes fully developed. In this condition there is a continuous variation in the fluid temperature from

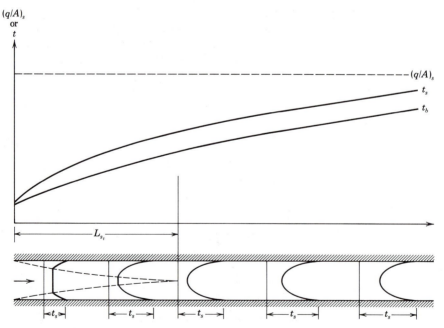

FIGURE 7.16. Developing and fully developed temperature profiles in pipe flow, constant wall heat flux.

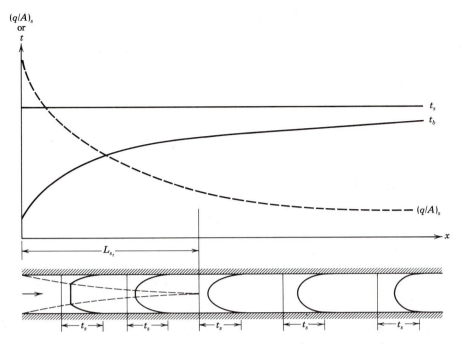

FIGURE 7.17. Developing and fully developed temperature profiles in pipe flow, constant wall temperature.

the surface value to some minimum value at the pipe centerline and no central core of uniform temperature exists. After the fully developed temperature profile has been established, the profile may or may not remain unchanged, depending on the nature of the imposed boundary conditions.

Figure 7.16 depicts the case in which the heat flux at the surface, $(q/A)_s$ is held uniform. Since heat is being transferred to the fluid, Eq. (7.118) requires that the bulk temperature increase but that the difference between the surface temperature, t_s, and the bulk temperature, t_b, remain fixed. Thus, both t_s and t_b increase with x, while the dimensionless profile $(t_s - t)/(t_s - t_b)$ remains unchanged.

Figure 7.17 depicts a different situation in which the surface temperature t_s is held fixed. The bulk temperature must still increase, and in the fully developed region the temperature profile appears to flatten as t_b increases. However, the dimensionless profile $(t_s - t)/(t_s - t_b)$ remains fixed and the rate of heat transfer, $(q/A)_s$, decreases in accord with Eq. (7.118).

Thus, a fully developed thermal profile is established when

$$\frac{\partial}{\partial x} \frac{t_s - t}{t_s - t_b} = 0.$$

Clearly, the dimensionless temperature gradient at the surface is also constant, with length, in thermally fully developed flow. Hence, for fluids with constant properties, the heat transfer coefficient and Nusselt number are constant in fully developed pipe flow.

The above comments have been made without reference to whether the flow involved is, hydrodynamically, laminar or turbulent since the thermal starting length concept is observed in both instances. The exact nature of the shape of the developing and fully developed temperature profiles is, of course, very dependent on whether the flow is laminar or turbulent.

In all subsequent discussions in this chapter, and those which follow, whenever the temperature of a fluid flowing in a pipe is referred to it is presumed that it is the *bulk* temperature unless otherwise specified.

Laminar Flow

In the instance of laminar pipe flow, the problem of the determination of the developing temperature profile and the fully developed profile has been subjected to analysis by a number of investigators. Such analyses are not necessarily so complex as to be beyond the scope of this text; however, as will be seen shortly, the magnitudes of typical entry lengths for fluids viscous enough to maintain fully laminar pipe flow are quite large. As a result many applications of practical interest involve pipes which never achieve fully developed flow. Also, while the analyses almost always assume constant fluid properties, real fluids of high viscosity usually exhibit strong temperature dependencies and considerable deviation from theoretical predictions result. For these reasons, most engineering calculations for laminar pipe flow are done using either empirical relations or empirical adjustments to the theoretical results; hence only the results of the analytical predictions are presented here. Their empirical counterparts will be presented in Chapter 8.

Actually, there are two basic types of thermal entry problems. The simpler is that which assumes the thermal conditions develop in the presence of a fully developed velocity profile—such as would be the case if the point at which heat transfer begins is preceded by an unheated hydrodynamic starting length. Highly viscous oils with large Prandtl numbers provide a reasonable approximation to this assumption. The more complex case is that called the *combined* entry length problem in which both the velocity and temperature profiles develop simultaneously. The results quoted here refer only to the first case.

The case in which one considers fully developed laminar flow, both hydrodynamic and thermal, can be rather simply analyzed (Ref. 16) for the condition of constant wall heat flux (so that $t_s - t_b$ is constant) to show that the diameter Nusselt number is constant and equal to

$$\text{Nu}_D = \frac{hD}{k} = \frac{48}{11} = 4.36. \tag{7.119}$$

Nusselt (Ref. 17), in one of the earliest heat transfer analyses, showed that the corresponding solution for the case of a fixed wall temperature is given by

$$\text{Nu}_D = 3.66. \tag{7.120}$$

The above deductions, then, are the asymptotic values approached, in either

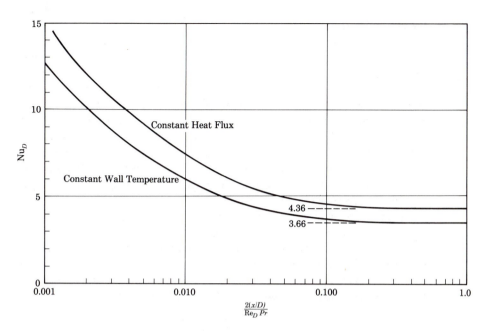

FIGURE 7.18. Local Nusselt number in the thermal starting region for laminar flow in a pipe.

case, in the thermal entry region. Analyses due to Nusselt as well as Sellars et al. (Ref. 18), summarized in Ref. 12, determine the local diameter Nusselt number in the developing region for the constant wall temperature and constant wall heat flux cases as functions of distance along the pipe. The results of these analyses are shown in Fig. 7.18. The approach to the fully developed values of Eqs. (7.119) and (7.120) is apparent. On the basis of the results shown in Fig. 7.18, Kays (Ref. 12) recommends that the thermal starting length for laminar pipe flow be defined:

$$\frac{2(x/D)}{\mathrm{Re}_D \, \mathrm{Pr}} \approx 0.1,$$

or

$$\frac{L_{S_t}}{D} = 0.05 \mathrm{Re}_D \mathrm{Pr}. \qquad (7.121)$$

When compared with the hydrodynamic starting length given by Eq. (7.106), the above equation shows that the thermal starting length differs by a factor of Pr. Thus, for cold water with $\mathrm{Pr} \approx 10$, the thermal length is about 10 times the hydrodynamic length, and for diameter Reynolds numbers near the transition value, L_{S_t} can be as great as 1000 pipe diameters. For viscous oils (see Table A.4) $\mathrm{Pr} \approx 10^4$ and it is apparent that fully developed thermal flow will rarely ever be achieved in practice.

Turbulent Flow

The thermal starting length in turbulent flow is a very difficult quantity to determine. Kays (Ref. 12) summarizes many calculations, and these results show that L_{st}/D is a complex function of Re and Pr that cannot be expressed in a simple way. In general, turbulent thermal starting lengths are shorter than laminar ones; the higher the value of Pr, the shorter is the starting length. Except for the liquid metals, for which Pr is quite small, one can expect thermal starting lengths of the order of 20 to 30 pipe diameters. Because of the complexity of determining turbulent thermal starting lengths, such effects are usually treated by empirical adjustments to the fully developed results. Hence, only the fully developed analyses will be presented here.

The treatment of fully developed turbulent heat transfer in pipes and tubes will be treated here by use of heat and momentum analogies discussed in Sec. 7.6 for flow past a flat surface. While it is possible to develop these analogies anew for the fully developed pipe flow case, it is more simply done by noting certain corresponding facts between the two cases. For flow on the flat surface, the local surface shear stress was related to the skin friction coefficient by

$$\tau_0 = \frac{C_f}{2} \rho U^2$$

and the surface heat transfer coefficient was defined:

$$h = \frac{(q/A)_0}{t_s - t_f}.$$

For fully developed pipe flow the wall shear and heat transfer coefficient are given by

$$\tau_R = \frac{f}{8} \rho U_m^2,$$

$$h = \frac{(q/A)_s}{t_s - t_b}.$$

In the flat plate flow the analogies were based on the similarity of the exchange of heat and momentum between the viscous layers and the reservoir of heat and momentum represented by the free stream at U and t_f. In the case of fully developed pipe flow, corresponding similarities may be drawn for exchanges between the viscous layers and the reservoir of heat and momentum represented by the bulk flow at U_m and t_b. Thus, the analogies developed in Sec. 7.6 may be translated into corresponding analogies for pipe flow by simply replacing $C_f/2$ by $f/8$, U by U_m, and t_f by t_b. The corresponding results are noted below.

The relations obtained below all give *local* Nusselt numbers (and, hence, local heat transfer coefficients) in a pipe. The method of calculating results which account for the changing bulk temperature as the fluid flows down the pipe will be treated in Chapter 8.

The Reynolds Analogy. For fully developed pipe flow, Reynolds analogy in Eq. (7.88) becomes

$$\text{Nu}_D = \frac{f}{8} \text{Re}_D \text{Pr}. \qquad (7.122)$$

Values for f may be found from Eqs. (7.113) through (7.115), or Fig. 7.14, as appropriate. However, this will not be done here as the Colburn analogy, given next, is found to provide better results.

The Colburn Analogy. For fully developed pipe flow the Colburn analogy of Eq. (7.90) becomes

$$\text{Nu}_D = \frac{f}{8} \text{Re}_D \text{Pr}^{1/3}. \qquad (7.123)$$

Again, f may be found from Fig. 7.14 or Eq. (7.115), as appropriate; however, for smooth pipes Eq (7.113) or (7.114) gives expressions which are easily evaluated for computational purposes:

$$\text{Nu}_D = 0.0395 \text{Re}_D^{0.75} \text{Pr}^{1/3}, \qquad 10^4 < \text{Re}_D < 5 \times 10^4,$$
$$\text{Nu}_D = 0.023 \text{Re}_D^{0.8} \text{Pr}^{1/3}, \qquad 3 \times 10^4 < \text{Re}_D < 10^6. \qquad (7.124)$$

Figure 7.19 compares the prediction of Eq. (7.124) with some measured values for air. Considering the influence of such factors as the variation of property values with temperature, etc., the agreement is not bad—typical of results obtained in convective heat transfer.

The Prandtl Analogy. The Prandtl analogy given in Eq. (7.93) similarly may be converted to fully developed turbulent pipe flow, and for smooth pipes Eqs. (7.113) and (7.114) provide:

$$\text{Nu}_D = \frac{(f/8)\text{Re}_D \text{Pr}}{1 + 5\sqrt{f/8}(\text{Pr} - 1)},$$
$$f = 0.316 \text{Re}_D^{-1/4}, \qquad 10^4 < \text{Re}_D < 5 \times 10^4, \qquad (7.125)$$
$$f = 0.184 \text{Re}_D^{-1/5}, \qquad 3 \times 10^4 < \text{Re}_D < 10^6.$$

As before, Eq. (7.115) may be used for f, or for rough pipes Fig. 7.14 may be used. Figure 7.19 also shows the predictions of Eq. (7.125) compared with experimental results.

The Von Kármán Analogy. The von Kármán analogy of Eq. (7.97) similarly has a counterpart for fully developed pipe flow. For smooth pipes, one has

$$\text{Nu}_D = \frac{(f/8)\,\text{Re}_D \text{Pr}}{1 + 5\sqrt{f/8}\{(\text{Pr} - 1) + \ln[1 + \frac{5}{6}(\text{Pr} - 1)]\}},$$
$$f = 0.316 \text{Re}_D^{-1/4}, \qquad 10^4 < \text{Re}_D < 5 \times 10^4, \qquad (7.126)$$
$$f = 0.184 \text{Re}_D^{-1/5}, \qquad 3 \times 10^4 < \text{Re}_D < 10^6.$$

The predictions of Eq. (7.126) are also shown in Fig. 7.19.

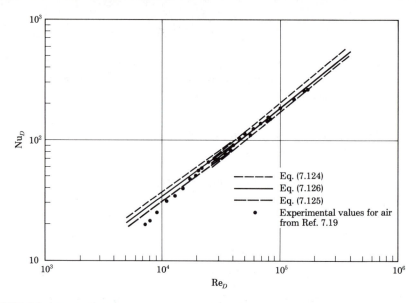

FIGURE 7.19. Turbulent heat transfer in pipes. Comparison between experimental values for air and the predictions of the Colburn, Prandtl, and von Kármán analogies.

7.11

CLOSURE

As mentioned at the outset of this chapter, its main purpose was to illustrate the methods employed to apply the principles of fluid dynamics and boundary layer theory to the *study* of forced convection problems. Even these analyses had to use certain empirical information, particularly for turbulent flow. The examples considered were not all-inclusive, but, rather, were chosen to illustrate method. Many, much more complex analyses are available in the published literature; however, the basic concepts remain the same. Chapter 8 presents a summary of some of the results of this chapter, and some other results, that may be used for typical engineering design calculations.

REFERENCES

1. BLASIUS, H., "Grenzschichten in Flüssigkeiten mit kleiner Reibung," *Z. Math. Phys.*, Vol. 56, 1908, p. 1. English translation in *NACA Tech. Memo. 1256*, Washington, D.C., 1950.

2. SCHLICTING, H., *Boundary Layer Theory*, 7th ed., New York, McGraw-Hill, 1979.

3. POHLHAUSEN, E., "Der Wärmeaustausch zwischen festen Körpern und Flüssigkeiten mit kleiner Wärmeleitung," *Z. Angew. Math. Mech.*, Vol. 1, 1921, p. 115.

4. PARMELEE, G. V., and R. G. HUEBSCHER, "Heat Transfer by Forced Convection Along a Smooth Flat Surface," *Heat. Piping Air Cond.*, Vol. 19, No. 8, 1947, p. 115.

5. Blasius, H., "Das Ahnlichkeitsgesetz bei Reibungvorgängen in Flüssigkeiten," *Forsch. Geb. Ingenieurwes.*, No. 131, 1913.

6. Dhawan, S., "Direct Measurements of Skin Friction," *NACA Tech. Note 2567*, Washington, D.C., 1952.

7. Schultz-Grunow, F., "Neues Widerstansgesetz für glatte Platten," *Luftfahrtforschung*, Vol. 17, 1940, p. 239.

8. Colburn, A. P., "A Method of Correlating Forced Convection Heat Transfer Data and a Comparison with Liquid Friction," *Trans. AIChE*, Vol 29, 1933, p. 174.

9. von Kármán, T., "The Analogy Between Fluid Friction and Heat Transfer," *Trans. ASME*, Vol. 61, 1939, p. 705.

10. Langhaar, H. L., "Steady Flow in the Transition Length of a Straight Tube," *Trans. ASME*, Vol. 64, 1942, p. A-55.

11. Özisik, M. N., *Basic Heat Transfer*, New York, McGraw-Hill, 1977.

12. Kays, W. M., and M. E. Crawford, *Convection Heat and Mass Transfer*, 2nd ed., New York, McGraw-Hill, 1980.

13. Nikaradse, J., "Strömungsgetze in rauhen Rohren," *Forsch. Geb. Ingenieurwes.*, No. 361, 1933.

14. von Kármán, T., "On Laminar and Turbulent Friction," *NACA Tech. Memo. 1092*, Washington, D.C., 1946.

15. Moody, L. F., "Friction Factors for Pipe Flow," *Trans. ASME*, Vol. 66, 1944, p. 671.

16. Holman, J. P., *Heat Transfer*, 5th ed., New York, McGraw-Hill, 1981.

17. Nusselt, W., "Die Abhängigkeit der Wärmeübergangszahl von der Rohrlänge," *Z.V.D.I.*, Vol. 54, 1910, p. 1154.

18. Sellars, J. R., M. Tribus, and J. S. Klein, "Heat Transfer to Laminar Flow in a Round Tube or Flat Conduit—The Graetz Problem Extended," *Trans. ASME*, Vol. 78, 1956, p. 441.

19. Deissler, R. G., and C. S. Eian, "Analytical and Experimental Investigation of Fully Developed Turbulent Flow of Air in a Smooth Tube with Heat Transfer with Variable Fluid Properties," *NACA Tech. Note 2629*, Washington, D.C., 1952.

PROBLEMS

7.1 Find the indicated dimensionless parameters for the conditions indicated.

    (a) Find Re_x for: $U = 20$ m/s

                                 $x = 0.6$ m

                                 air at 25°C, 1 atm

    (b) Find Re_D for: $\dot{m} = 0.5$ kg/s

                                 $D = 10$ cm

                                 water at 40°C

    (c) Find Nu_L for: $h = 10$ W/m²-°K

                                 $L = 1$ m

                                 CO_2 at 50°C, 1 atm

    (d) Find Nu_x for: $h = 110$ Btu/h-ft²-°F

                                 $x = 3$ ft

                                 steam at 100 psia, 400°F

    (e) Find Pr for: $\mu = 2.04 \times 10^{-5}$ kg/m-s

                                 $c_p = 0.958$ kJ/kg-°C

                                 $k = 0.02631$ W/m-°C

7.2 Verify the algebra leading from Eqs. (7.3) and (7.5) to Eq. (7.12).

7.3   Verify the algebra leading from Eqs. (7.4) and (7.6) to Eq. (7.23).

7.4   Show that the solution to Eq. (7.23) is Eq. (7.24).

7.5   Air at atmospheric pressure and 5°C flows past a flat surface at 45 m/s. Plot the laminar boundary layer thickness as a function of the distance from the leading edge up to a distance corresponding to a local length Reynolds number of 5×10^5.

7.6   Saturated water at 80°C flows with a velocity of 1.5 m/s past a flat surface at 40°C. Using the results of Sec. 7.2 as given in Figs. 7.2 and 7.4, plot the velocity and temperature profiles in the boundary layer at stations 2.5, 5.0, and 7.5 cm from the leading edge. For ease in reading Figs. 7.2 and 7.4, use even values of the parameter η and calculate the corresponding value of y at each x. Evaluate the fluid properties at the mean of the free stream and surface temperature.

7.7   Atmospheric air at 15°C flows with a velocity of 7.5 m/s over a flat plate 24 cm long and maintained at 45°C. Find (a) the local heat transfer coefficient at 6.0, 12.0, 18.0, and 24.0 cm from the leading edge; and (b) the average heat transfer coefficient for the entire plate.

7.8   Repeat Prob. 7.6 using the results of the integral methods in Sec. 7.3.

7.9   Using a fourth-degree polynomial, apply the integral momentum equation of the laminar boundary layer to find how δ/x varies with Re_x on a flat plate.

7.10   If a first-degree polynomial is used to represent the velocity and temperature distribution in a laminar boundary layer on a flat plate, apply the integral equations of the boundary layer to obtain expressions for the dependence of the local boundary layer thickness, the local skin friction coefficient, and the local Nusselt number on the local Reynolds number.

7.11   For a velocity distribution in a laminar boundary layer on a flat surface represented as a third-degree polynomial, find how the drag coefficient, C_D, depends on the total plate length Reynolds number.

7.12   Verify the algebra leading from Eq. (7.42) to Eq. (7.45).

7.13   Air at atmospheric pressure and at 15°C flows with a velocity of 185 m/s past a flat plate. Accounting for the effects of viscous dissipation, find the heat flux between the air and the plate at a location 2.5 cm from the leading edge if the plate is maintained at (a) 20°C, (b) 30°C, and (c) 40°C. Evaluate the air properties at the free stream temperature; however, Chapter 8 will suggest an alternative method to be used in practical applications.

7.14   Repeat Prob. 7.13 if the effects of viscous dissipation are neglected and compare the results.

7.15   The lubricating oil listed in Table A.4 flows at 100°F with a velocity of 100 ft/s past a flat surface. Evaluating the fluid properties at 100°F, find the recovery temperature. What is the heat transfer coefficient at a point 4 in. from the leading edge?

7.16   Verify the algebra leading from Eq. (7.62) to Eq. (7.65).

7.17   Water at 40°C flows past a flat surface with a free stream velocity of 0.75 m/s. Using the universal velocity distribution, find, for a distance 0.75 m from the leading edge, (a) the skin friction coefficient and (b) the velocity and eddy diffusivity at a distance (i) 0.5 cm away from the surface and (ii) 0.04 cm away from the surface.

7.18   Atmospheric pressure air flows past a flat surface at a velocity of 12 m/s and at a temperature of 90°C. Using the universal velocity distribution, find, for a distance 1.0 m from the leading edge, (a) the skin friction coefficient and (b) the velocity and eddy diffusivity at a distance (i) 0.06 cm away from the surface and (ii) 0.6 cm away from the surface.

7.19   Verify the algebra leading from Eq. (7.92) to Eq. (7.93).

7.20   Verify the algebra leading from Eq. (7.95) to Eq. (7.97).

7.21   Find the local heat transfer coefficient for the conditions specified in Prob. 7.17, using (a) the Reynolds analogy, (b) the Colburn analogy, and (c) the von Kármán analogy.

7.22   Find the local heat transfer coefficient for the conditions specified in Prob. 7.18 using (a) the Reynolds analogy, (b) the Colburn analogy, and (c) the von Kármán analogy.

7.23   Water at 60°C flows with a velocity of 0.8 m/s past a flat plate 1 m long. Find the average drag coefficient and average heat transfer coefficient for the entire plate.

7.24   Atmospheric air at a temperature of 50°C flows with a velocity of 15 m/s past a plate 1.25 m long. Find the average drag coefficient and the average heat transfer coefficient for the entire plate.

7.25   Water at 60°C flows at the rate of 10 kg/s through a pipe with a 7.5-cm inside diameter.
    (a) Find the heat transfer coefficient using (i) the Colburn analogy, (ii) the Prandtl analogy, and (iii) the von Kármán analogy.
    (b) Find the eddy viscosity and eddy diffusivity at a point 0.05 cm from the pipe wall.

7.26   Steam at 800°F, 1000 psia, flows in a 6-in. schedule 40 pipe with an average velocity of 20 ft/s. Find the local heat transfer coefficient using (a) the Colburn analogy, (b) the Prandtl analogy, and (c) the von Kármán analogy.

CHAPTER 8

Working Formulas for Forced Convection

INTRODUCTORY REMARKS

The analyses in Chapter 7 and their applications of boundary layer theory to the solution of forced heat convection problems were meant, as stated, to illustrate the methods of approach and to show typical results. A detailed exposition of the derivation and development of all the relationships having engineering importance for the prediction of forced convection heat transfer coefficients is beyond the scope of this book. In fact, some of the most useful engineering formulas for forced convection coefficients have no real theoretical basis. They are the results of extensive experimental analyses.

The purpose of this chapter is to present a collection of the most useful of the existing relations for the more frequently encountered cases of forced convection. Some of these will be relations having theoretical bases, and some will be empirical dimensionless correlations of experimental data. In some instances more than one relation will be given for a particular case. The relations considered to be the preferred—or "recommended"—ones will be marked by a ★. This presentation of working relations for forced convection will, of course, be limited. For sources of relations for cases not covered here, the reader is advised to consult Refs. 1 through 5.

Unless otherwise noted, all the relations to be quoted here will be limited (like those derived in Chapter 7) to the imposed condition of a constant surface temperature. That is, the thermal boundary condition imposed at the solid surface (either analytically or experimentally) is that of a uniform temperature. This is in contrast to other possible conditions—such as uniform heat flux, constant temperature difference, etc. Again, the reader is advised to consult the quoted references for relations based on those other conditions.

The theoretical analyses of Chapter 7 assumed that the physical properties (μ, c_p, k, and ρ) were constants. This is, of course, not true in reality, as the discussions of Chapter 2 pointed out. In almost every case there is a significant dependence of these properties on temperature, and in the case of ρ (and v) there is also a pressure effect for gases. The temperature dependence of the properties means that there may be a significant variation in the quantities through the thermal boundary layer. The accuracy of the results of these theoretical relations and of the dimensionless experimental correlations depends on the temperature chosen for the evaluation of the properties. In most cases one of two mean temperatures is used. One is the bulk fluid temperature, or mixing cup temperature, as defined in Eq. (7.116). This is usually applied in the case of forced convection inside a closed duct or pipe and is the mean fluid temperature at a cross section—the temperature that would result if the fluid flowing through a cross section were to be thoroughly mixed. The symbol t_b will be used to denote the *bulk temperature*. The second mean temperature that is often used is the *mean film temperature*. This is the arithmetic mean of the surface temperature, t_s, and (in the case of flow external to a body) the free stream or undisturbed fluid temperature, t_f. Using t_m to denote the mean film temperature,

$$t_m = \frac{t_s + t_f}{2}.$$

Sometimes in the case of internal flow a mean film temperature which is the average of the surface temperature and the bulk temperature will be used in place of the bulk temperature.

In the working formulas to be presented here, it will be stated in each case what temperature should be used for evaluation of the fluid properties. Likewise, the range of the various dimensionless parameters (e.g., Re, Pr, etc.) over which a particular formula is valid will also be stated.

As noted in Chapter 6, pure forced convection heat transfer results (i.e., buoyancy neglected) may be expected to be of the form

$$\text{Nu} = f(\text{Re, Pr, Ec}).$$

The Eckert number, Ec, arises from the inclusion of the effects of the dissipation of mechanical energy through viscous friction. When this dissipation is neglected, the expected heat transfer correlation is of the form

$$\text{Nu} = f(\text{Re, Pr}).$$

The solutions of Chapter 7 were of the above forms. In instances in which empirical methods are used, the resulting formulas are invariably presented in the above forms. However, certain effects such as the temperature dependence of properties, the influences of the starting length, etc., may necessitate the inclusion of additional parameters as some of the following examples will show.

Finally, it should be pointed out that uncertainties in the many fluid properties involved, experimental errors, unaccounted for effects, geometrical deviations, surface roughness, etc., may cause, at times, considerable deviation between theory and experiment or scatter in the correlation of experimental data. As a result such deviations or scatter of the order of magnitude of $\pm 10\%$ in predicted Nusselt numbers are not uncommon in forced convection—and even greater deviations may be encountered in some complex cases.

8.2

FORCED CONVECTION PAST PLANE SURFACES, DISSIPATION NEGLECTED

The case of forced convection of a fluid of known velocity and temperature past a flat surface of known temperature was discussed extensively in Chapter 7. Being the case of the simplest geometry and boundary conditions, the analytic expressions found in Chapter 7 for the flat surface are found to hold rather well with certain modifications as noted in the following.

Laminar Flow

For laminar flow past a flat plate, the analytical expressions of Eqs. (7.32) for the local and average Nusselt numbers have been verified experimentally and are recommended for use as noted:

$$\mathrm{Nu}_x = 0.332\mathrm{Re}_x^{1/2}\mathrm{Pr}^{1/3},$$

$$\mathrm{Nu}_L = 0.664\mathrm{Re}_L^{1/2}\mathrm{Pr}^{1/3}, \qquad (8.1)\bigstar$$

$$\begin{bmatrix} 0.6 \le \mathrm{Pr} \le 50 \\ \mathrm{Re} < \mathrm{Re}_{x,c} \approx 5 \times 10^5 \\ \text{properties at } t_m \end{bmatrix},$$

as long as the fluid properties are evaluated at the mean film temperature, t_m. Clearly, the length Reynolds number, based on the free stream velocity, must be less than that for turbulent transition, usually taken as 5×10^5.

The lower limit of 0.6 is established for the Prandtl number since that is the lowest possible for gases. The next lower values normally encountered for Pr are those for liquid metals (Table A.9), which are of the order of 0.01. In such instances the thermal boundary layer is very much thicker than the velocity layer and the basic assumption involved in the derivation of Eq. (8.1) that these layers are of the same order is violated. For fluids of such low Pr, Kays and Crawford (Ref. 1) derived the following:

$$\text{Nu}_x = 0.565(\text{Re}_x\text{Pr})^{1/2},$$

$$\text{Nu}_L = 1.130(\text{Re}_L\text{Pr})^{1/2}, \qquad (8.2)\bigstar$$

$$\begin{bmatrix} \text{Pr} < 0.05 \\ \text{Re} < \text{Re}_{x,c} \approx 5 \times 10^5 \\ \text{properties at } t_m \end{bmatrix}.$$

For the entire range of Prandtl numbers, Churchill and Ozoe (Ref. 6) recommend the following semiempirical relations:

$$\text{Nu}_x = \frac{0.3387\text{Re}_x^{1/2}\text{Pr}^{1/3}}{[1 + (0.0468/\text{Pr})^{2/3}]^{1/4}},$$

$$\text{Nu}_L = \frac{0.6774\text{Re}_L^{1/2}\text{Pr}^{1/3}}{[1 + (0.0468/\text{Pr})^{2/3}]^{1/4}}, \qquad (8.3)\bigstar$$

$$\begin{bmatrix} \text{RePr} > 100 \\ \text{properties at } t_m \end{bmatrix}.$$

Turbulent Flow—Local Heat Transfer

For turbulent flow past flat surfaces, only correlations for the local Nusselt number will be given since average values must account for an initial laminar section. Cases for such average, mixed flow are treated in the next section.

For local turbulent heat transfer, the analytic expression of von Kármán in Eq. (7.97) together with the skin friction coefficient in Eqs. (7.66) and (7.67) may be used as long as the Prandtl number is not too different from unity and the fluid properties are evaluated at t_m:

$$\text{Nu}_x = \frac{\frac{1}{2}C_f\text{Re}_x\text{Pr}}{1 + 5\sqrt{C_f/2}\{(\text{Pr} - 1) + \ln[1 + \frac{5}{6}(\text{Pr} - 1)]\}},$$

$$C_f = 0.0592\text{Re}_x^{-1/5}, \qquad 5 \times 10^5 < \text{Re}_x < 10^7, \qquad (8.4)$$

$$C_f = 0.37(\log_{10}\text{Re}_x)^{-2.584}, \qquad \text{Re}_x > 10^7.$$

However, the above equation is awkward to use and difficult to integrate for average values. Thus, the Colburn analogy, Eq. (7.91), is recommended:

$$\text{Nu}_x = 0.0296\text{Re}_x^{0.8}\text{Pr}^{1/3}, \qquad\qquad 5 \times 10^5 < \text{Re}_x < 10^7,$$

$$\text{Nu}_x = 0.185\text{Re}_x(\log_{10}\text{Re}_x)^{-2.584}\text{Pr}^{1/3}, \quad \text{Re}_x > 10^7, \qquad (8.5)\bigstar$$

$$\begin{bmatrix} 0.6 \le \text{Pr} \le 60 \\ \text{properties at } t_m \end{bmatrix}.$$

Mixed Flow—Average Heat Transfer

For a plate of total length L in which Re_L exceeds the critical value, the average heat transfer must account for the initial laminar section of the boundary layer as was done in Sec. 7.7. The relations derived there are found to give reliable results:

$$\mathrm{Nu}_L = [0.037\mathrm{Re}_L^{0.8} - 872]\mathrm{Pr}^{1/3}, \qquad\qquad 5 \times 10^5 < \mathrm{Re}_L < 10^7,$$

$$\mathrm{Nu}_L = [0.228\mathrm{Re}_L(\log_{10}\mathrm{Re}_L)^{-2.584} - 872]\mathrm{Pr}^{1/3}, \ 10^7 < \mathrm{Re}_L < 10^9, \ (8.6)\bigstar$$

$$\left[\begin{array}{l} 0.6 \le \mathrm{Pr} \le 60 \\ \mathrm{Re}_{x,c} = 5 \times 10^5 \\ \text{properties at } t_m \end{array}\right].$$

In Eq. (8.6) the customary transition Reynolds number $\mathrm{Re}_{x,c} = 5 \times 10^5$ has been used. For other values of $\mathrm{Re}_{x,c}$ the constant 872 must be replaced with $A/2$ from Table 7.3. If the presence of the laminar portion of the boundary layer is ignored, then the constant 872 is replaced with zero.

EXAMPLE 8.1

Atmospheric pressure air at 40°C flows with a velocity of 40 m/s past a flat plate 0.75 m long and 0.5 m wide. The surface is maintained at 260°C. Treat the air as incompressible and neglect viscous dissipation.

(a) Find the total heat transferred and the total drag force on one side of the plate.
(b) Find the fractions of the heat transfer and drag force for the laminar portion of the flow.
(c) What error in part (a) results if the boundary layer is assumed to be turbulent from the leading edge?
(d) Repeat parts (a) and (b) if the velocity is doubled.

Solution. The mean film temperature is $t_m = (40 + 260)/2 = 150°C$. At this temperature Table A.6 gives

$$\rho = 0.8342 \text{ kg/m}^3, \qquad \nu = 28.58 \times 10^{-6} \text{ m}^2/\text{s},$$
$$k = 34.59 \times 10^{-3} \text{ W/m-°C}, \qquad \mathrm{Pr} = 0.701.$$

(a) The plate length Reynolds number is

$$\mathrm{Re}_L = \frac{40 \times 0.75}{28.58 \times 10^{-6}} = 1.05 \times 10^6.$$

For transition at 5×10^5, Eqs. (8.6) and (7.99) give

$$\mathrm{Nu}_L = [0.037(1.05 \times 10^6)^{0.8} - 872](0.701)^{1/3} = 1381,$$

$$C_D = 0.074(1.05 \times 10^6)^{-0.2} - 1743(1.05 \times 10^{-6})^{-1} = 0.00296.$$

Thus, the average heat transfer coefficient is

$$h = Nu_L \frac{k}{L} = 1381 \times \frac{0.03459}{0.75} = 63.7 \text{ W/m}^2\text{-°C.}$$

The total heat transfer and drag force are then

$$q = Ah(t_s - t_f) = 0.75 \times 0.5 \times 63.7 \times (260 - 40)$$

$$= 5255 \text{ W (17,930 Btu/h)},$$

$$D = \tfrac{1}{2}\rho U^2 A \, C_D = \tfrac{1}{2} \times 0.8342 \times (40)^2 \times 0.75 \times 0.5 \times 0.00296$$

$$= 0.741 \text{ N (0.167 lb}_f\text{).}$$

(b) For the laminar portion it is necessary to locate the point of transition. For a critical Reynolds number of 5×10^5,

$$x_c = 5 \times 10^5 \frac{28.58 \times 10^{-6}}{40} = 0.357 \text{ m.}$$

For the laminar portion the average Nusselt number and drag coefficient are, from Eqs. (8.1) and (7.20),

$$Nu = 0.664 \times (5 \times 10^5)^{1/2} (0.701)^{1/3} = 404.5,$$

$$C_D = 1.328(5 \times 10^5)^{-1/2} = 0.00188.$$

Thus, the average heat transfer coefficient for the laminar portion is

$$h_{\text{lam}} = 404.5 \times \frac{0.03459}{0.357} = 39.2 \text{ W/m}^2\text{-°C},$$

and the heat transfer and drag force are

$$q_{\text{lam}} = 0.357 \times 0.5 \times 39.2 \times (260 - 40)$$

$$= 1539 \text{ W (5251 Btu/h), 30\% of the total.}$$

$$D_{\text{lam}} = \tfrac{1}{2} \times 0.8342 \times (40)^2 \times 0.357 \times 0.5 \times 0.00188$$

$$= 0.224 \text{ N (0.050 lb}_f\text{), 30\% of the total.}$$

(c) If the assumption is made that the boundary layer is turbulent from the leading edge, Eqs. (8.8) and (7.99) give

$$Nu_L = 0.037(1.05 \times 10^6)^{0.8}(0.701)^{1/3} = 2156,$$

$$C_D = 0.074(1.05 \times 10^6)^{-0.2} = 0.00462,$$

for which

$$h = 99.4 \text{ W/m}^2\text{-°C},$$

$$q = 8202 \text{ W (27,990 Btu/h)},$$

$$D = 1.157 \text{ N (0.260 lb}_f\text{).}$$

The heat flow and drag are 56% higher than that found in part (a).

(d) Doubling the flow velocity changes only the plate length Reynolds number and the position of the transition point. Performing identical calculations gives

$$\mathrm{Re}_L = 2.10 \times 10^6,$$

$$\mathrm{Nu}_L = 2979,$$

$$C_D = 0.00320,$$

$$h = 137.4 \ \mathrm{W/m^2\text{-}°C},$$

$$q = 11{,}338 \ \mathrm{W} \ (38{,}690 \ \mathrm{Btu/h}),$$

$$D = 3.199 \ \mathrm{N} \ (0.719 \ \mathrm{lb}_f),$$

$$x_c = 0.179 \ \mathrm{m},$$

$$h_{\mathrm{lam}} = 78.3 \ \mathrm{W/m^2\text{-}°C},$$

$$q_{\mathrm{lam}} = 1539 \ \mathrm{W} \ (5251 \ \mathrm{Btu/h}), \ 14\% \ \text{of the total},$$

$$D_{\mathrm{lam}} = 0.449 \ \mathrm{N} \ (0.050 \ \mathrm{lb}_f), \ 14\% \ \text{of the total.} \qquad ■$$

8.3

FORCED CONVECTION PAST PLANE SURFACES, DISSIPATION INCLUDED

The preceding correlations apply to cases in which the effects of the dissipation of mechanical energy into thermal energy through viscous friction were neglected. Many applications exist, particularly in connection with high speed aerodynamics, in which the fluid velocities are sufficiently high that these dissipative effects are significant. Such effects are present in both compressible and incompressible flow.

Viscous Dissipation in Incompressible Flow

The effect of viscous dissipation may be significant in certain cases of incompressible flow, such as might be encountered in the use of highly viscous oils in lubrication applications. Also, as will be seen shortly, the incompressible correlations form the basis for the most commonly used correlations in compressible flow.

The case in which the effects of viscous dissipation in incompressible flow are included was treated in Sec. 7.4 in some detail. Thus, only the principal results will be quoted here. In that section it was shown that the relations for laminar heat transfer when viscous dissipation is neglected apply equally well when dissipation is included if one alters the definition of the heat transfer coefficient so that it is based on the recovery temperature, t_r, instead of the free stream temperature, t_f:

$$h = \frac{(q/A)_s}{t_s - t_r}.$$

The recovery temperature is the temperature at which the surface would equilibrate if allowed to come to equilibrium with the viscous stream and was shown to be [Eq. (7.52)],

$$t_r - t_f = \Psi_0(\mathrm{Pr})\frac{U^2}{2c_p},$$

in which $\Psi_0(\mathrm{Pr})$ is a function of Pr determined in the solution of Eq. (7.51). For purposes here, it is convenient to change the notation and define the *recovery factor*, $r = \Psi_0(\mathrm{Pr})$, so that

$$r = \frac{t_r - t_f}{U^2/2c_p}. \tag{8.7}$$

This may be rewritten

$$\frac{t_r - t_f}{t_s - t_f} = r\frac{1}{2}\frac{U^2}{c_p(t_s - t_f)} \tag{8.8}$$

$$= r\frac{\mathrm{Ec}}{2},$$

where the Eckert number, Ec, has been introduced:

$$\mathrm{Ec} = \frac{U^2}{c_p(t_s - t_f)}. \tag{8.9}$$

Thus, given the free stream and surface conditions, Ec is readily found from Eq. (8.9), and the recovery temperature is found from Eq. (8.8) if r is known.

The analyses of Sec. (7.4) showed that for laminar flow the recovery factor is approximately [Eq. (7.53)],

$$r = \mathrm{Pr}^{1/2} \tag{8.10}$$

and the resulting expressions for the Nusselt number were identical to those when dissipation is neglected. Thus, in summary, for laminar flow one finds r from Eq. (8.10), the heat transfer coefficient from the usual formulas, and the heat flow from

$$\left(\frac{q}{A}\right)_s = h(t_s - t_r)$$

after finding t_r from

$$\frac{t_r - t_f}{t_s - t_f} = r\frac{\mathrm{Ec}}{2}.$$

The possibility exists, as noted in Sec. 7.4, that $(q/A) < 0$ even when $t_s > t_f$ if $\mathrm{Ec} > 2/r$ since then $t_r > t_s > t_f$.

Similar analyses for turbulent flow (Refs. 7 and 8) show that in this instance, the nondissipative relations also hold for the dissipative case, the only difference being that the recovery factor is given by

$$r = \text{Pr}^{1/3}. \tag{8.11}$$

Thus, for *incompressible* flow (notably liquids with large Pr) the following recommendations are made, using the mean film temperature for the fluid properties: Use

$$\left(\frac{q}{A}\right)_s = h(t_s - t_r),$$

$$\frac{t_r - t_f}{t_s - t_f} = r\frac{\text{Ec}}{2}, \tag{8.12}$$

$$\text{Ec} = \frac{U^2}{c_p(t_s - t_f)},$$

with:

Laminar flow on plates:

$$r = \text{Pr}^{1/2},$$

$$\text{Nu}_x = 0.332\text{Re}_x^{1/2}\text{Pr}^{1/3},$$

$$\text{Nu}_L = 0.664\text{Re}_L^{1/2}\text{Pr}^{1/3}, \tag{8.13}\bigstar$$

$$\begin{bmatrix} 0.6 \leq \text{Pr} \leq 50 \\ \text{Re} < \text{Re}_{x,c} \approx 5 \times 10^5 \\ \text{properties at } t_m \end{bmatrix}.$$

Turbulent flow on plates:

$$r = \text{Pr}^{1/3},$$

$$\text{Nu}_x = 0.0296\text{Re}_x^{0.8}\text{Pr}^{1/3}, \qquad\qquad 5 \times 10^5 < \text{Re}_x < 10^7,$$

$$\text{Nu}_x = 0.185\text{Re}_x(\log_{10}\text{Re}_x)^{-2.584}\text{Pr}^{1/3}, \qquad \text{Re}_x > 10^7, \tag{8.14}\bigstar$$

$$\begin{bmatrix} 0.6 \leq \text{Pr} \leq 60 \\ \text{properties at } t_m \end{bmatrix}.$$

Again, no correlation is shown for the average Nu in the turbulent case since account must be made for the initial laminar portion as in the preceding section. In this instance, however, integration of Eqs. (8.13) and (8.14) over a total plate length is pointless since the recovery temperature is different for the laminar and turbulent portions of the boundary layer and an average plate length h

is meaningless. Thus, each section has to be treated separately, as a later example will show.

Viscous Dissipation in Compressible Flow

In the consideration of heat transfer on the surface of bodies moving at high velocities in gases, consideration must be made for the effects of the fluid compressibility as well as viscous dissipation. Applications in modern aerospace technology are numerous. Aerodynamic heating effects on high velocity aircraft, missiles, re-entry of spacecraft, etc., result at times in very large recovery temperatures and heat flow rates.

Compressibility effects in gases are associated with the Mach number of the free stream flow, which for an ideal gas is defined as

$$M_f = \frac{U}{\sqrt{\gamma R T_f}}, \tag{8.15}$$

in which γ is the ratio of specific heats, R is the gas constant, and T_f is the absolute temperature of the free stream. For $M_f \lesssim 0.2$ to 0.3, the flow may be treated as incompressible and the correlations quoted above may be used. For higher Mach numbers, compressibility effects become important and the basic equations of continuity, momentum, and energy for the boundary layer must be altered.

Solutions have been carried out for the compressible case of flow past a flat surface and while the velocity and temperature distributions in the boundary layer differ markedly from the incompressible solutions, the vlaue of the velocity and temperature *gradients* at the surface are virtually unchanged. This result is, perhaps, not too surprising since even though the full stream may be of high Mach number, the flow just adjacent to the wall is usually of a rather low Mach number. The net result is that the basic relations for the skin friction coefficient and Nusselt number are not very different between the incompressible and compressible cases.

The main difference observed in the two cases is that which is produced by the rather extreme temperature variations which are produced in the compressible boundary layer. Studies reported in Refs. 9 and 10 which account for variable fluid properties indicate that the use of the mean film temperature, t_m, for the evaluation of fluid properties is inappropriate for accurate results. A relatively simple technique recommended by Eckert (Ref. 9) to account for fluid property dependence on temperature is that the properties be evaluated at a reference temperature, t^*, given by

$$t^* = 0.5(t_s + t_f) + 0.22(t_r - t_f), \tag{8.16}$$

where t_r is the recovery temperature defined earlier. For low speed flow, since $t_r \rightarrow t_f$, then t^* becomes t_m, the mean film temperature already recommended for use in such cases.

As a result of all the above discussion, the following recommendation is made for compressible dissipative flow past flat surfaces: Use

$$\left(\frac{q}{A}\right)_s = h(t_s - t_r),$$

$$\frac{t_r - t_f}{t_s - t_f} = r\frac{\text{Ec}}{2}, \qquad\qquad (8.17)$$

$$\text{Ec} = \frac{U^2}{c_p(t_s - t_f)},$$

with:

Laminar flow on plates:

$$r = \text{Pr}^{*1/2},$$

$$\text{Nu}_x^* = 0.332\text{Re}_x^{*1/2}\text{Pr}^{*1/3},$$

$$\text{Nu}_L^* = 0.664\text{Re}_L^{*1/2}\text{Pr}^{*1/3}, \qquad\qquad (8.18)\bigstar$$

$$\begin{bmatrix} 0.6 \le \text{Pr}^* \le 50 \\ \text{Re}^* < \text{Re}_{x,c} \approx 5 \times 10^5 \\ \text{properties at } t^* \end{bmatrix}.$$

Turbulent flow on plates:

$$r = \text{Pr}^{*1/3},$$

$$\text{Nu}_x^* = 0.0296\text{Re}_x^{*0.8}\text{Pr}^{*1/3}, \qquad\qquad 5 \times 10^5 < \text{Re}_x^* < 10^7,$$

$$\text{Nu}_x^* = 0.185\text{Re}_x^*(\log_{10}\text{Re}_x^*)^{-2.584}\text{Pr}^{*1/3}, \qquad \text{Re}_x^* > 10^7, \qquad (8.19)\bigstar$$

$$\begin{bmatrix} 0.6 \le \text{Pr}^* \le 60 \\ \text{properties at } t^* \end{bmatrix}.$$

Again, no plate length averages for turbulent flow are given here since the recovery temperature and t^* are different in the laminar and turbulent portions of the boundary layer.

It might be worthwhile noting an alternative formulation often used by aerodynamicists. The *stagnation* temperature of the free stream is defined as

$$T_0 = T_f + \frac{1}{2}\frac{U^2}{c_p},$$

which, for an ideal gas, becomes the following when the Mach number of the free stream, M_f in Eq. (8.15) is introduced:

$$T_0 = T_f\left(1 + \frac{\gamma - 1}{2}M_f^2\right).$$

The Eckert number, Ec, is replaced by the Mach number since

$$\text{Ec} = \frac{U^2}{c_p(T_s - T_f)} = (\gamma - 1)M_f^2\frac{T_f}{T_s - T_f}$$

$$= 2\frac{T_0 - T_f}{T_s - T_f}.$$

Then the defining relation for the recovery temperature becomes

$$\frac{t_r - t_f}{t_s - t_f} = \frac{T_r - T_f}{T_s - T_f} = r \frac{Ec}{2},$$

or

$$\frac{T_r - T_f}{T_0 - T_f} = r. \tag{8.20}$$

Thus, instead of Eq. (8.17), the following are used:

$$\left(\frac{q}{A}\right)_s = h(T_s - T_r),$$

$$\frac{T_r - T_f}{T_0 - T_f} = r, \tag{8.21}$$

$$T_0 = T_f \left(1 + \frac{\gamma - 1}{2} M_f^2\right),$$

$$M_f = \frac{U}{\sqrt{\gamma R T_f}}.$$

This formulation is obviously exactly equivalent to that given above.

EXAMPLE 8.2

In a wind tunnel test, air at $-40°C$ flows with a velocity of 900 m/s past a flat plate 0.6 m long and 0.3 m wide. The air pressure is 3.45 kN/m² (approximately 0.5 psia). Find the heat transferred to or from the plate if its surface is maintained at 40°C.

Solution. The given data suggest that part of the plate will be in laminar flow and part in turbulent flow. If this is so, then each part must be treated separately since t_r and t^* will be different in each.

Treating the laminar portion first, assume as tentative values $c_p = 1.009$ kJ/kg-°C and Pr = 0.706. These assumptions may have to be adjusted later after t^* is found. The Eckert number and the recovery factor are then

$$Ec = \frac{U^2}{c_p(t_s - t_f)} = \frac{(900)^2}{1.009 \times [40 - (-40)] \times 1000} = 10.03,$$

$$r = Pr^{1/2} = 0.840.$$

Thus Eq. (8.17) gives the recovery temperature:

$$\frac{t_r - t_f}{t_s - t_f} = r \frac{Ec}{2},$$

$$\frac{t_r - (-40)}{40 - (-40)} = 0.840 \times \frac{10.03}{2}; \qquad t_r = 297.3°C.$$

Then the reference temperature, Eq. (8.16), is

$$t^* = 0.5(t_s + t_f) + 0.22(t_r - t_f),$$

$$= 0.5[40 + (-40)] + 0.22[297.3 - (-40)] = 74.2°C.$$

At this temperature, Table A.6 shows $c_p^* = 1.009$ kJ/kg-°C, $Pr^* = 0.707$. These values are sufficiently close to the tentative values chosen above. Had they been significantly different, new values would have to be assumed and the above calculations repeated until a satisfactory check is obtained. The remaining fluid properties may then be found from Table A.6, noting that v^* and ρ^* may not be read from that table since the pressure is not atmospheric. Thus, μ^* must be used and the air density is found from the ideal gas law. Thus:

Laminar section:

$$Ec = 10.03, \quad r = 0.840,$$

$$t_r = 297.3°C, \quad t^* = 74.20°C,$$

$$c_p^* = 1.009 \text{ kJ/kg-°C}, \quad \mu^* = 20.66 \times 10^{-6} \text{ kg/m-s},$$

$$k^* = 29.57 \times 10^{-3} \text{ W/m-°C}, \quad Pr^* = 0.707,$$

$$\rho^* = \frac{P}{RT^*} = \frac{3.45}{(8.314/28.97)(74.2 + 273.15)}$$

$$= 0.0346 \text{ kg/m}^3.$$

Corresponding calculations must be carried out for the turbulent portion of the plate, the only difference being the use of $r = Pr^{1/3}$. Again, one must assume trial values of c_p and Pr which are then verified, or adjusted, and the final results are:

Turbulent section:

$$Ec = 10.03, \quad r = 0.890,$$

$$t_r = 317.4°C, \quad t^* = 78.6°C,$$

$$c_p^* = 1.009 \text{ kJ/kg-°C}, \quad \mu^* = 20.86 \times 10^{-6} \text{ kg/m-s},$$

$$k^* = 29.81 \times 10^{-3} \text{ W/m-°C}, \quad Pr^* = 0.706,$$

$$\rho^* = 0.0342 \text{ kg/m}^3.$$

Note that the recovery temperature is 20°C higher in the turbulent portion than in the laminar portion.

The transition point may now be located using the properties of the laminar portion:

$$Re_{x,c} = 5 \times 10^5 = \frac{Ux_c\rho^*}{\mu^*} = \frac{900 \times x_c \times 0.0346}{20.66 \times 10^{-6}},$$

$$x_c = 0.332 \text{ m}.$$

Thus, more than half the plate is in laminar flow.

The average heat transfer coefficient in the laminar section is found from Eq. (8.18):

$$\frac{hx_c}{k^*} = 0.664 \ \text{Re}_{x,c}^{*}{}^{1/2}\text{Pr}^{*1/3}$$

$$= 0.664(5 \times 10^5)^{1/2}(0.707)^{1/3} = 418.3,$$

$$h = 418 \times \frac{0.02951}{0.332} = 37.2 \ \text{W/m}^2\text{-}°\text{C},$$

so that the heat transferred in that section is

$$q_{\text{lam}} = hA(t_s - t_r)$$

$$= 37.2 \times (0.332 \times 0.3)(40 - 297.3)$$

$$= -952.6 \ \text{W} \ (3250 \ \text{Btu/h}).$$

The average heat transfer coefficient in the turbulent section has to be found by integration of Eq. (8.19) from x_c to L and division by $L - x_c$:

$$h = k^* \frac{0.0296}{0.8} \frac{1}{L - x_c}\left(\frac{U\rho^*}{\mu^*}\right)^{0.8}(L^{0.8} - x_c^{0.8})\text{Pr}^{*1/3}$$

$$= 0.02981 \times \frac{0.0296}{0.8} \frac{1}{0.6 - 0.332}\left(\frac{900 \times 0.0342}{20.86 \times 10^{-6}}\right)^{0.8}$$

$$\times (0.6^{0.8} - 0.332^{0.8})(0.706)^{1/3}$$

$$= 79.1 \ \text{W/m}^2\text{-}°\text{C}.$$

The turbulent heat transfer is then

$$q_{\text{turb}} = 79.1(0.6 - 0.332) \times 0.3 \times (40 - 317.4)$$

$$= -1764.2 \ \text{W} \ (6020 \ \text{Btu/h}).$$

Thus, the total heat transfer is

$$q = -952.6 - 1764.2$$

$$= -2716.8 \ \text{W} \ (9270 \ \text{Btu/h}).$$

Note that even though $t_s > t_f$, the high rate of dissipation in the boundary layer requires that the surface be cooled to maintain its temperature at 40°C. ■

8.4

FORCED CONVECTION INSIDE CYLINDRICAL PIPES AND TUBES

The engineering importance of forced flow through pipes and tubes for the purpose of transport, heating, cooling, etc., is apparent. Most heat exchangers involve the heating or cooling of fluids in tubes. Most ducts or pipes used to transport gases or liquids are invariably at temperatures different from the surroundings, producing a flow of heat to or from the fluids. Both laminar and turbulent conditions are encountered.

In the following sections both analytical and empirical correlations recommended for engineering use in this important geometry will be summarized. These correlations will yield methods of predicting the average value of the heat transfer coefficient on the inner tube surface applicable over the finite length of the tube. The manner in which this average value of h can be used to calculate the heat transferred to the fluid over a given length as the bulk temperature changes (or, equivalently, how the change in the bulk temperature can be calculated) will be discussed after the correlations are described.

Laminar Flow

As discussed in Sec. 7.10, the analytical prediction of heat transfer for forced laminar convection in a tube is quite complex, due mainly to the effects of the starting length (both thermal and hydrodynamic) and the temperature dependence of the fluid properties. As given in Eq. (7.120), Nusselt showed that for fully developed laminar flow with a constant tube surface temperature, the Nusselt number was given by

$$\text{Nu}_D = \frac{hD}{k} = 3.66. \tag{8.22}$$

For tubes that are so long that the starting length may be ignored, the average Nusselt number over the tube length is also given by Eq. (8.22)—with the fluid properties evaluated at the mean of the inlet and outlet bulk temperatures. What constitutes a tube long enough to use this relation will be noted shortly.

Hausen (Ref. 11) developed the following semi-empirical relation for thermally fully developed flow

$$\text{Nu}_D = 3.66 + \frac{0.0668(D/L)\text{Re}_D\text{Pr}}{1 + 0.4[(D/L)\text{Re}_D\text{Pr}]^{2/3}}, \tag{8.23}$$

in which the effect of the tube length is included. Clearly, as $L \to \infty$, the Nusselt value is approached. Equation (8.23), again, yields an average h over the tube length and the properties are to be evaluated at the mean temperature. However, Eq. (8.23) assumes the flow is hydrodynamically fully developed before entering the pipe and is, hence, generally not applicable.

The preferred correlation for "short" tubes which includes both the thermal and hydrodynamic starting length is the empirical relation of Sieder and Tate (Ref. 12):

$$\text{Nu}_D = 1.86 \left[\left(\frac{D}{L}\right)\text{Re}_D \text{Pr} \right]^{1/3} \left(\frac{\mu}{\mu_s}\right)^{0.14},$$

$$\begin{bmatrix} 0.48 < \text{Pr} < 16{,}700 \\ (D/L)\text{Re}_D\text{Pr} > 10 \\ \text{properties, except } \mu_s, \text{ at mean } t_b \\ \mu_s \text{ at } t_s \end{bmatrix}. \tag{8.24}★$$

Equation (8.24) gives the average h to be applied over the pipe length L. The quoted limit of $(D/L)\mathrm{Re}_D\mathrm{Pr} > 10$ establishes what is meant by a short tube for the use of Eq. (8.24). For values below this limit, Eq. (8.22) may be used.

Turbulent Flow

For fully developed turbulent flow in pipes the relation of von Kármán given in Eq. (7.126) may be used, with the properties evaluated at the bulk temperature, as long as the Prandtl number is not too different from 1:

$$\mathrm{Nu}_D = \frac{(f/8)\,\mathrm{Re}_D\mathrm{Pr}}{1 + 5\,\sqrt{f/8}\{(\mathrm{Pr} - 1) + \ln\,[1 + \tfrac{5}{6}(\mathrm{Pr} - 1)]\}},$$

$$f = 0.316\mathrm{Re}_D^{-1/4}, \qquad 10^4 < \mathrm{Re}_D < 5 \times 10^4, \tag{8.25}$$

$$f = 0.184\mathrm{Re}_D^{-1/5}, \qquad 3 \times 10^4 < \mathrm{Re}_D < 10^6.$$

However, this equation is awkward to use, and that of Colburn, Eq. (7.124), is much simpler:

$$\mathrm{Nu}_D = 0.0395\mathrm{Re}_D^{0.75}\mathrm{Pr}^{1/3}, \qquad 10^4 < \mathrm{Re}_D < 5 \times 10^4,$$

$$\mathrm{Nu}_D = 0.023\mathrm{Re}_D^{0.8}\mathrm{Pr}^{1/3}, \qquad 3 \times 10^4 < \mathrm{Re}_D < 10^6. \tag{8.26}$$

A variation on Eq. (8.26) which partially accounts for the effects of variable fluid properties is the widely used relation known as the Dittus–Boelter (Ref. 13) equation:

$$\mathrm{Nu}_D = 0.023\mathrm{Re}_D^{0.8}\mathrm{Pr}^n,$$

$$
\begin{bmatrix}
n = 0.4 \text{ for } t_s > t_b \\
n = 0.3 \text{ for } t_s < t_b \\
0.7 < \mathrm{Pr} < 160 \\
10^4 < \mathrm{Re}_D < 10^6 \\
|t_s - t_b| < 6°\mathrm{C} \text{ for liquids} \\
|t_s - t_b| < 60°\mathrm{C} \text{ for gases} \\
\text{properties at } t_b
\end{bmatrix}
\tag{8.27}\bigstar
$$

The Dittus–Boelter equation is compared in Fig. 8.1 with the same experimental data shown in Fig. 7.19.

For differences between t_s and t_b greater than those specified in Eq. (8.27), the following expression due to Sieder and Tate (Ref. 12) will give better results:

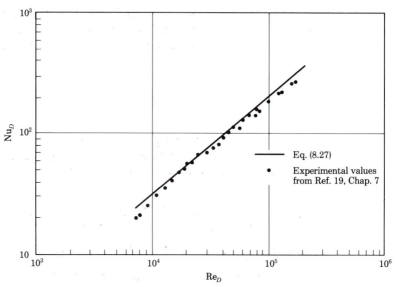

FIGURE 8.1. Turbulent heat transfer in pipes. Comparison of the Dittus-Beolter equation with experimental data.

$$\mathrm{Nu}_D = 0.027 \mathrm{Re}_D^{0.8} \mathrm{Pr}^{1/3} \left(\frac{\mu}{\mu_s}\right)^{0.14},$$

$$\left[\begin{array}{l} 0.7 < \mathrm{Pr} < 160 \\ 10^4 < \mathrm{Re}_D < 10^6 \\ \text{properties, except } \mu_s, \text{ at } t_b \\ \mu_s \text{ at } t_s \end{array}\right]. \qquad (8.28)\bigstar$$

The foregoing two equations offer certain simplicity for computations, particularly Eq. (8.27), since knowledge of t_s is not required. However, errors of as much as $\pm 20\%$ may be encountered. Petukhov (Ref. 14) has recently produced a much more accurate correlation which agrees well with the data of many workers:

$$\mathrm{Nu}_D = \frac{(f/8)\mathrm{Re}_D \mathrm{Pr}}{1.07 + 12.7 \sqrt{f/8}(\mathrm{Pr}^{2/3} - 1)} \left(\frac{\mu}{\mu_s}\right)^n,$$

$$\left[\begin{array}{l} f = (1.82 \log_{10} \mathrm{Re}_D - 1.64)^{-2} \\ n = 0.11 \text{ for liquids, } t_s > t_b \\ n = 0.25 \text{ for liquids, } t_s < t_b \\ n = 0 \text{ for gases} \\ 0.5 < \mathrm{Pr} < 200 \text{ for 6\% accuracy} \\ 200 < \mathrm{Pr} < 2000 \text{ for 10\% accuracy} \\ 10^4 < \mathrm{Re}_D < 5 \times 10^6 \\ 0 < \mu/\mu_s < 40 \\ \text{properties, except } \mu_s, \text{ at } t_b \\ \mu_s \text{ at } t_s. \end{array}\right. \qquad (8.29)\bigstar$$

The equation recommended for f in Eq. (8.29) is for smooth pipes. For rough pipes f may be found from the Moody diagram in Fig. 7.14.

Equations (8.27) through (8.29) presume the existence of fully developed flow and that the effect of the starting length is negligible. They should be applied only when $L/D \gtrsim 60$. As written, then, these equations give *local* values of h since they are based on the local bulk temperature of the fluid. The bulk temperature changes with distance as the fluid flows along the pipe and the determination of this change is described later; however, for an average h over a given pipe length, it is generally suitable to use the above correlations based on the average bulk temperature between inlet and outlet. This is shown later in Example 8.4.

For short tubes it may be necessary to account for starting length effects. As noted in Chapter 7, not much information is available on turbulent starting lengths other than the general statement that they are usually much shorter than in laminar flow. Nusselt (Ref. 15) found the following expression to be applicable for shorter tubes:

$$\mathrm{Nu}_D = 0.036 \mathrm{Re}_D^{0.8} \mathrm{Pr}^{1/3} \left(\frac{D}{L}\right)^{1/18},$$

(8.30)

$$\begin{bmatrix} 10 < L/D < 400 \\ \text{properties at } t_b \end{bmatrix}.$$

EXAMPLE 8.3

Water flows with a bulk temperature of 30°C at a velocity of 1 m/s in a tube with an inside diameter of 2.5 cm. The tube surface is maintained at 50°C. Find the heat transfer coefficient using (a) Eq. (8.29); (b) Eq. (8.28); (c) Eq. (8.27).

Solution. At $t_b = 30°C$ and $t_s = 50°C$, Table A.3 gives

$$\mu = 0.7978 \times 10^{-3} \text{ kg/m-s}, \qquad \nu = 0.8012 \times 10^{-6} \text{ m}^2/\text{s},$$

$$k = 0.6150 \text{ W/m-°C}, \qquad \mathrm{Pr} = 5.42,$$

$$\mu_s = 0.5471 \times 10^{-3} \text{ kg/m-s}.$$

Thus,

$$\mathrm{Re}_D = \frac{1 \times 0.025}{0.8012 \times 10^{-6}} = 3.12 \times 10^4, \text{ turbulent.}$$

(a) Equation (8.29) yields

$$f = (1.82 \log_{10} \mathrm{Re}_D - 1.64)^{-2} = 0.02338,$$

$$\mathrm{Nu}_D = \frac{(f/8)\mathrm{Re}_D \, \mathrm{Pr}}{1.07 + 12.7 \sqrt{f/8}(\mathrm{Pr}^{2/3} - 1)}\left(\frac{\mu}{\mu_s}\right)^{0.11} = 206.0,$$

$$h = \frac{\mathrm{Nu}k}{D} = 206.0 \times \frac{0.6150}{0.025} = 5066 \text{ W/m}^2\text{-°C}.$$

(b) Equation (8.28) yields

$$Nu_D = 0.027 Re_D^{0.8} Pr^{1/3} \left(\frac{\mu}{\mu_s} \right)^{0.14} = 196.9,$$

$$h = 4844 \ \text{W/m}^2\text{-°C}.$$

This value is within 4% of that in part (a).

(c) Equation (8.27) yields

$$Nu_D = 0.023 Re_D^{0.8} Pr^{0.4} = 178.1,$$

$$h = 4381 \ \text{W/m}^2\text{-°C}.$$

This value differs by 14% from that predicted by the more accurate Eq. (8.29); however, Eq. (8.27) has been applied beyond its specified limits since $t_s - t_b = 20°C > 6°C$. ∎

EXAMPLE 8.4

Steam enters a 3-in schedule 40 pipe at 600 psia, 800°F. It flows through the pipe with an average velocity of 20 ft/s, leaving at a temperature of 600°F. The pipe wall is maintained uniformly at 500°F. Assuming that the pressure drop along the pipe is negligible (see Example 8.5), find the heat transfer coefficient (a) at the entrance conditions; (b) at the exit conditions; (c) at the mean of the entrance and exit conditions. Compare the results in part (c) with the mean of the results in parts (a) and (b).

Solution. Table 8.1 gives $D = 3.068$ in. Then, since $t_b - t_s > 60°C$, Eq. (8.27) is not applicable, so use Eq. (8.28) to find h. (Equation (8.29) may as well be used with slightly different results.)

(a) At 600 psia, $t_b = 800°F$, Table A.5 gives

$$\rho = 0.8409 \ \text{lb}_m/\text{ft}^3 \qquad \mu = 6.176 \times 10^{-2} \ \text{lb}_m/\text{ft-h}, \qquad Pr = 0.96,$$

$$k = 3.55 \times 10^{-2} \ \text{Btu/h-ft-°F}, \qquad \mu_s = 4.367 \times 10^{-2} \ \text{lb}_m/\text{ft-h}.$$

Thus,

$$Re_D = \frac{20 \times 3600 \times (3.068/12) \times 0.8409}{0.06176} = 2.506 \times 10^5,$$

$$Nu_D = 0.027 Re_D^{0.8} Pr^{1/3} \left(\frac{\mu}{\mu_s} \right)^{0.14} = 583.1,$$

$$h = 583.1 \times \frac{0.0355}{3.068/12} = 81.0 \ \text{Btu/h-ft}^2\text{-°F}.$$

(b) At 600 psia, $t_b = 600°F$,

$$\rho = 1.0575 \ \text{lb}_m/\text{ft}^3, \qquad \mu = 4.991 \times 10^{-2} \ \text{lb}_m/\text{ft-h},$$

$$Pr = 1.11, \qquad k = 2.96 \times 10^{-2} \ \text{Btu/h-ft-°F},$$

$$\text{Re}_D = 3.900 \times 10^5,$$

$$\text{Nu}_D = 846.2, \qquad h = 98.0 \text{ Btu/h-ft}^2\text{-}°\text{F}.$$

(c) At 600 psia, $t_b = 700°\text{F}$,

$$\rho = 0.9323 \text{ lb}_m/\text{ft}^3, \qquad \mu = 5.592 \times 10^{-2} \text{ lb}_m/\text{ft-h},$$

$$\text{Pr} = 1.01, \qquad k = 3.212 \times 10^{-2} \text{ Btu/h-ft-}°\text{F},$$

$$\text{Re}_D = 3.069 \times 10^5,$$

$$\text{Nu}_D = 687.8, \qquad h = 86.4 \text{ Btu/h-ft}^2\text{-}°\text{F}.$$

The mean of parts (a) and (b) is $h = 89.5$ Btu/h-ft²-°F, which is within 4% of that found at the mean conditions in part (c). Hence, values of h are frequently calculated at mean conditions between entrance and exit. ∎

Heat Transfer to Liquid Metals in Pipes or Tubes

Other than the correlation given in Eq. (8.2) for laminar flow past a flat plate, the correlations given so far in this chapter have been limited to fluids with Prandtl numbers greater than about 0.6. This limitation includes all gases and most liquids. However, liquid metals are a notable exception. Owing to their very high thermal conductivity, liquid metals exhibit Prandtl numbers several orders of magnitude lower than other liquids, usually of the order of 0.01, as may be observed in Table A.9. As noted earlier, the principal influence of the low value of Pr is that the thermal boundary layer is much thicker than the velocity layer. Thus, the fluid velocity is nearly uniform at its undisturbed value through much of the thermal boundary layer, altering considerably the solution of the energy equation.

The Martinelli analogy (Ref. 16) presents a detailed analytical expression for fully developed turbulent flow in tubes which applies to a wide range of Pr, including the liquid metals. The resulting expression, however, is difficult to use and is not always reliable. For this reason, most current engineering calculations for liquid metal heat transfer in tubes are carried out using any of a number of empirical or semiempirical relations.

For fluids of low Pr, the significant heat transfer parameter is the Peclet number defined as

$$\text{Peclet number} = \text{Pe} = \text{PrRe} = \frac{Ul}{\alpha}. \tag{8.31}$$

Seban and Shimazaki (Ref. 17) applied the Martinelli analogy to the case of fixed surface temperature, and determined that for very low Prandtl numbers, the results are well approximated by the formula:

$$\text{Nu}_D = 5.0 + 0.025\text{Pe}_D^{0.8},$$

$$\begin{bmatrix} \text{Pe}_D > 100 \\ L/D > 60 \\ \text{properties at } t_b \end{bmatrix}. \tag{8.32}★$$

Other correlations for constant surface temperature include that due to Azer and Chao (Ref. 18):

$$\text{Nu}_D = 5.0 + 0.05\text{Pe}_D^{0.77}\text{Pr}^{0.25},$$

$$\begin{bmatrix} \text{Pe}_D > 15{,}000 \\ \text{Pr} < 0.1 \\ \text{properties at } t_b \end{bmatrix}, \tag{8.33}\bigstar$$

and that due to Sliecher et al. (Ref. 19):

$$\text{Nu}_D = 4.8 + 0.0156\text{Pe}_D^{0.85}\text{Pr}^{0.08},$$

$$\begin{bmatrix} 0.004 < \text{Pr} < 0.1 \\ \text{Re}_D < 5 \times 10^5 \\ \text{properties at } t_b \end{bmatrix}. \tag{8.34}\bigstar$$

Heat Transfer in Tubes of Finite Length

The discussions so far in this section has been limited to methods of predicting surface heat transfer coefficients for flows inside circular tubes. Some of the recommended relations (such as in laminar flow) give average values of h to be used over a given tube length, while others (such as in the fully developed turbulent cases) give local values of h which may be used to find applicable averages as in Example 8.4. All these relations depend on knowing the bulk fluid temperature, or more usually, the average of the bulk temperature as the fluid moves along the pipe. In order to apply these relations meaningfully and in order to determine the total heat transferred over a given pipe length (or, conversely, to determine the required length needed to transfer a given amount of heat) one needs to know how the bulk temperature of the fluid depends on the axial location in the pipe.

The above question is answered relatively easily by considering the case of fixed pipe wall temperature depicted in Fig. 8.2. Here a fluid, flowing at the mass rate $\dot{m}$, enters the tube with a bulk temperature t_{b_i}. The tube has a constant wall temperature t_s, taken as greater than t_{b_i} for purposes of discussion. As the fluid flows along the pipe its bulk temperature increases due to heat transfer to the fluid from the wall. At the end of the pipe, i.e., at L, the temperature has become t_{b_o}.

The way in which t_b varies with x, the distance along the pipe, can be deduced by writing a heat balance on an element of the tube dx long, as noted in Fig. 8.2. If h is the surface heat transfer coefficient and C is the circumference of the pipe, a heat balance on dx yields

$$dq = h(C\, dx)(t_s - t_b) = \dot{m}c_p\, dt_b. \tag{8.35}$$

where dt_b is the increase of the fluid bulk temperature over the distance dx. The above may be rewritten in the following form since t_s is constant:

$$\frac{d(t_b - t_s)}{t_b - t_s} = -\frac{hC}{\dot{m}c_p}\, dx. \tag{8.36}$$

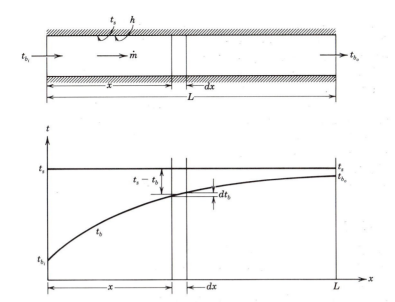

FIGURE 8.2. Temperature variation in a pipe with constant wall temperature.

When integrated from $x = 0$, $t_b = t_{b_i}$ to $x = x$, $t_b = t_b$, Eq. (8.36) gives, for h taken constant,

$$\ln\left(\frac{t_b - t_s}{t_{b_i} - t_s}\right) = -\frac{hC}{\dot{m}c_p}x,$$

$$\frac{t_b - t_s}{t_{b_i} - t_s} = \exp\left(-\frac{hC}{\dot{m}c_p}x\right). \qquad (8.37)$$

Equation (8.37) shows that the bulk temperature varies exponentially with distance along the pipe.

To find the total heat transferred over a pipe length L, Eq. (8.35) may be integrated to yield

$$q = hCL(t_s - t_b)_m = \dot{m}c_p(t_{b_o} - t_{b_i}), \qquad (8.38)$$

where $(t_s - t_b)_m$ is the mean of the temperature difference $(t_s - t_b)$ over the length L. This mean temperature difference is found by evaluating Eq. (8.37) for $x = L$, $t_b = t_{b_o}$:

$$\ln\left(\frac{t_{b_o} - t_s}{t_{b_i} - t_s}\right) = -\frac{hCL}{\dot{m}c_p}.$$

The factor $hCL/\dot{m}c_p$ is found from Eq. (8.38) to be $(t_{b_o} - t_{b_i})/(t_s - t_b)_m$, so that the above expression becomes

$$(t_s - t_b)_m = \frac{(t_s - t_{b_i}) - (t_s - t_{b_o})}{\ln\left(\dfrac{t_s - t_{b_i}}{t_s - t_{b_o}}\right)}. \qquad (8.39)$$

Thus, the appropriate mean value to use for $(t_s - t_b)$ as it varies from $(t_s - t_{b_i})$ to $(t_s - t_{b_o})$ is the *log mean* as given in Eq. (8.39). Then the total heat transferred over the pipe length L is given by Eq. (8.38).

The log mean appearing here will be encountered again later in Chapter 12 when dealing with heat exchangers.

In the discussion above, it should be remembered that it has been assumed that the surface heat transfer coefficient h is a constant over the pipe length. Thus some suitable average must have been computed previously. The formulation given in Eqs. (8.38) and (8.39) may be used to find the required length of pipe to raise the bulk temperature to a desired value or to find the outlet bulk temperature for a given pipe length as discussed in Example 8.5.

EXAMPLE 8.5

(a) Find the required length of pipe for the data of Example 8.4.
(b) Verify the statement given in Example 8.4 that the pressure drop in the pipe is negligible.
(c) Outline the calculation procedure to be used if the pipe length had been specified rather than the outlet bulk temperature, it being desired to find t_{b_o}.

Solution.

(a) For the heat transfer coefficient, use the h found for the average conditions in part (c) of Example 8.4, $h = 86.4$ Btu/h-ft^2-°F. For $t_s = 500$°F, $t_{b_i} = 800$°F, $t_{b_o} = 600$°F, Eq. (8.39) gives

$$(t_s - t_b)_m = \frac{(500 - 800) - (500 - 600)}{\ln\left(\dfrac{500 - 800}{500 - 600}\right)} = -182.0°F.$$

(Note that the arithmetic mean would be -150°F, considerably different.) The mass flow rate at the average conditions is

$$\dot{m} = \frac{\pi}{4} D^2 \rho U_m = \frac{\pi}{4}\left(\frac{3.068}{12}\right)^2 \times 0.9323 \times 20 \times 3600 = 3446 \text{ lb}_m/\text{h}.$$

Thus, Eq. (8.38) gives, with c_p evaluated at the average conditions,

$$q = 86.4 \times \pi\left(\frac{3.068}{12}\right) \times L \times (-182.0)$$

$$= 3446 \times 0.5821(600 - 800)$$

$$q = 4.01 \times 10^5 \text{ Btu/h},$$

$$L = 31.8 \text{ ft}.$$

(b) For the Reynolds number at the average conditions of $\text{Re}_D = 3.069 \times 10^5$, Fig. 7.14 for $\epsilon/D = 0$, gives $f = 0.014$. [Equation (7.114) gives $f = 0.0147$.] Thus, by Eq. (7.108) the pressure drop is

$$\Delta P = \frac{1}{2} \rho U^2 \frac{L}{D} f = \frac{1}{2}(0.9323) \frac{20^2}{32.2}\left(\frac{31.8}{3.068/12}\right)\frac{0.014}{144}$$

$$= 0.07 \text{ psia.}$$

Thus, the pressure drop is justifiably neglected.

(c) If the pipe length had been specified rather than the outlet temperature, t_{b_o}, then the same basic procedure is applied. However, an iterative calculation must be performed. One first assumes a value for t_{b_o}; then the average h is calculated, as in Example 8.4, for conditions at $(t_{b_i} + t_{b_o})/2$. The mean temperature difference $(t_s - t_b)_m$ is calculated, using the assumed t_{b_o} in Eq. (8.39). Then Eq. (8.38) is used with the given L to calculate t_{b_o}. This calculated value of t_{b_o} is then compared with the assumed value and the calculations repeated until a satisfactory agreement is obtained. ∎

8.5

FORCED CONVECTION IN NONCIRCULAR SECTIONS

The correlations of Sec. 8.4 were all limited to flows in circular pipes and tubes. Many engineering applications occur in which noncircular cross sections are involved. Both analytical and empirical information on forced convection in noncircular sections are sparse. In some instances the correlations for circular passages may be used if the Nusselt and Reynolds numbers are based on the *hydraulic*, or *effective*, diameter defined as four times the cross-sectional area divided by the wetted perimeter:

$$D_h = \frac{4A}{P}. \tag{8.40}$$

Laminar Flow

For the case of laminar flow in noncircular sections, even the use of the hydraulic diameter does not yield very accurate data, particularly if the section involves sharp corners. Thus, little information is available. However, the limiting case of fully developed flow in noncircular sections (i.e., corresponding to $\text{Nu}_D = 3.66$ for the circular section) has been solved numerically for a number of cases by several workers. Kays (Ref. 1) summarizes these findings for rectangular ducts with constant surface temperature as given in Table 8.1. In Table 8.1 the limiting, constant Nusselt number is that based on the hydraulic diameter which, for a rectangular duct of dimension $a \times b$, is

$$\text{Nu}_{D_h} = \frac{hD_h}{k},$$

$$D_h = \frac{2ab}{a + b},$$

in accord with Eq. (8.40).

TABLE 8.1   Nusselt Numbers of Fully Developed Laminar Flow in Rectangular Ducts with Constant Wall Temperature

b/a	Nu_{D_h}
1.0	3.61
1.43	3.73
2.0	4.12
3.0	4.79
4.0	5.33
8.0	6.49
∞	8.24

For other cross sections and for the starting length solution in a few cases, the reader is referred to Refs. 1 and 5.

Turbulent Flow

For turbulent flow, which may be taken to occur for $\mathrm{Re}_{D_h} \gtrsim 2300$, the correlations quoted in Sec. 8.4 for circular tubes may be applied for reasonable results if Nu_D and Re_D are replaced with Nu_{D_h} and Re_{D_h}, respectively. In view of the uncertainties involved, there is probably no reason to employ relations any more sophisticated than the Dittus–Boelter formula given in Eq. (8.27). In particular, this latter expression has been found to give useful predictions for the physically significant case of fully developed turbulent flow in the annular space between two pipes. In this instance the hydraulic diameter is easily shown to be

$$D_h = D_o - D_i$$

where D_o and D_i are the outside and inside diameters, respectively. The h thus found from Eq. (8.27) is applicable at either surface of the annulus.

8.6

FORCED CONVECTION FOR EXTERNAL FLOW NORMAL TO TUBES, TUBE BANKS, AND NONCIRCULAR SECTIONS

The last four sections have dealt with forced convective heat transfer for either external flow past flat surfaces or internal flow in pipes and ducts. Of equal engineering importance are the cases resulting from the flow of a fluid external to a nonflat surface. The flow of fluids normal to single cylinders, other noncircular shapes, and over banks of tubes are cases occurring frequently in engineering practice. The flow of fluids around such bluff bodies introduces special complications.

Flow Normal to a Single Circular Cylinder

Figure 8.3(a) depicts a circular cylinder placed normal to a fluid stream, the properties of which at points well upstream of the cylinder are denoted by U_∞, t_∞, and ρ_∞. Classical ideal, inviscid flow theory predicts a pressure distribution around the surface of the cylinder as suggested in Fig. 8.3(b). The pressure at the forward stagnation point equals the sum of the free stream static pressure and the free stream dynamic pressure. Then, the pressure first drops as the fluid speeds up around the forward part of the cylinder. Since the ideal fluid is not retarded by surface friction, it slows around the rear half of the cylinder exactly in reverse of the acceleration around the forward half. Consequently, the pressure rises around the rear half of the cylinder.

The behavior of a boundary layer in a region of increasing pressure, such as on the rear of a cylinder in cross flow, is quite different than those discussed previously for flat surfaces. As a viscous boundary layer moves into a region of rising pressure, its velocity profile becomes successively more and more flattened. The retarded layer, deficient in momentum, has increasing difficulty overcoming the rising pressure. Such a process is depicted in Fig. 8.4. Eventually a point is reached at which

$$\left(\frac{\partial v_x}{\partial y}\right)_{y=0} = 0.$$

Downstream of this point reverse flow sets in, and the boundary layer pulls away, or "separates," from the surface. This separation causes a profound change in the pressure distribution along the surface.

Figure 8.3(b) indicates typical pressure distributions in real fluid flows around a circular cylinder compared with the ideal fluid distribution. Significant differences are seen to exist between the laminar and turbulent cases. A laminar layer will separate at a polar coordinate of about 80° from the forward stagnation point; a turbulent layer, having, on the average, more kinetic energy, will remain

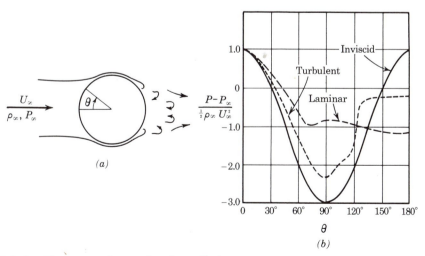

FIGURE 8.3. Flow normal to a circular cylinder.

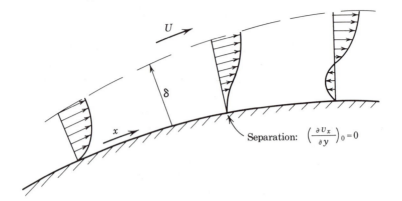

FIGURE 8.4. Boundary layer flow in an adverse pressure gradient.

attached up to about 110°. The separation phenomenon is accompanied by the creation and shedding of eddies in the wake of the cylinder, as suggested in Fig. 8.3(a).

The peculiar flow processes just described must have significant influence on the local heat transfer for a cylinder whose surface is maintained at a temperature different from that of the fluid stream. This problem was investigated extensively by Giedt (Ref. 20) for air flowing past heated cylinders. Figure 8.5 shows typical results. At the lower Reynolds numbers the boundary layer is laminar and a minimum value of the local Nusselt number is seen to occur at the separation point. Subsequent to separation, a rise in Nu is seen to occur as transverse mixing is enhanced by the creation of the eddies mentioned above. At higher Reynolds numbers a first minimum in Nu is observed when the laminar–turbulent transition takes place and a second when separation occurs.

The above processes are sufficiently complex that an analytical description is virtually impossible. Also, it is usually the average heat transfer coefficient, rather than the local, which is desired for engineering applications. Hence, empirical correlations are normally used.

Zhukauskas (Ref. 21) has performed the most recent and most comprehensive study of flow normal to single tubes and banks of tubes. For single tubes the following correlation yields rather accurate results when used in conjunction with Table 8.2 and when the mean film temperature is defined as $t_m = (t_\infty + t_s)/2$:

$$\mathrm{Nu}_D = C\mathrm{Re}_D^m \mathrm{Pr}^n \left(\frac{\mathrm{Pr}_\infty}{\mathrm{Pr}_s}\right)^{1/4},$$

$$\begin{bmatrix} 0.7 < \mathrm{Pr} < 500 \\ 1 < \mathrm{Re}_D < 10^6 \\ C \text{ and } m \text{ from Table 8.2} \\ n = 0.37, \ \mathrm{Pr} \leq 10 \\ n = 0.36, \ \mathrm{Pr} > 10 \\ \text{properties at } t_m, \text{ except } \mathrm{Pr}_\infty, \mathrm{Pr}_s \\ \mathrm{Pr}_\infty \text{ at } t_\infty, \mathrm{Pr}_s \text{ at } t_s \end{bmatrix}. \qquad (8.41)\bigstar$$

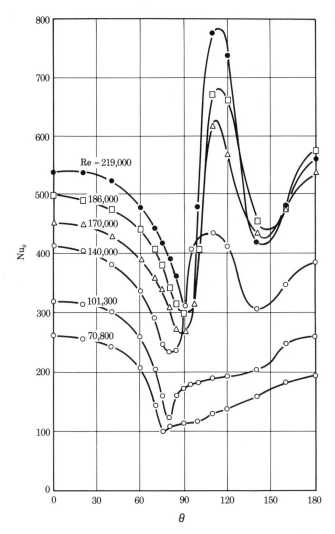

FIGURE 8.5. Circumferential variation of the local heat transfer coefficient for cross flow of air over a cylinder. (From W. H. Geidt, "Variation of Point Unit Heat Transfer Coefficient Around a Cylinder Normal to an Air Stream," *Trans. ASME,* Vol. 71, 1949, p. 375. Used by permission.)

TABLE 8.2 Values of *C* and *m* for Zhukauskas' Equation for Flow Normal to Single Circular Cylinders*

Range of Re_D	C	m
1–40	0.75	0.4
40–1000	0.51	0.5
10^3–2×10^5	0.26	0.6
2×10^5–10^6	0.076	0.7

*From Ref. 21.

8.6 FORCED CONVECTION NORMAL TO TUBES AND BANKS

Churchill and Bernstein (Ref. 22) have proposed, based on an analysis of many data covering wide ranges of Pr and Re_D, the following correlation:

$$Nu_D = 0.3 + \frac{0.62Re_D^{1/2}Pr^{1/3}}{[1 + (0.4/Pr)^{2/3}]^{1/4}}\left[1 + \left(\frac{Re_D}{2.82 \times 10^5}\right)^{5/8}\right]^{4/5},$$

$$\begin{bmatrix} \text{for all } Re_D Pr > 0.2 \\ \text{properties at } t_m \end{bmatrix}.$$

$$(8.42)\bigstar$$

Flow Normal to Noncircular Cylinders

The results for flow normal to noncircular cylinders is less complete than that above for circular cylinders. For gases, Jakob (Ref. 2) recommends the use of the following correlation for the shapes shown in Table 8.3:

$$Nu_D = CRe_D^m Pr^{1/3},$$

$$\begin{bmatrix} C \text{ and } m \text{ from Table 8.3} \\ \text{properties at } t_m \end{bmatrix}.$$

$$(8.43)\bigstar$$

TABLE 8.3 Values of C and m for Forced Convection of Gases to Noncircular Cylinders*

Geometry	Re_D	C	m
	$5 \times 10^3 – 10^5$	0.102	0.675
	$5 \times 10^3 – 10^5$	0.246	0.588
	$5 \times 10^3 – 1.95 \times 10^4$ $1.95 \times 10^4 – 10^5$	0.160 0.0385	0.638 0.782
	$5 \times 10^3 – 10^5$	0.153	0.638
	$4 \times 10^3 – 1.5 \times 10^4$	0.228	0.731

*From Ref. 2.

Flow over Banks of Tubes

Many heat exchangers, air conditioning cooling or heating coils, etc., involve a bank (or bundle) of circular tubes over which a fluid may flow. Typical geometric arrangements are shown in Fig. 8.6. Figure 8.6(a) depicts the *aligned* arrangement of a tube bank in which successive rows of tubes are aligned with one another, as viewed with respect to the velocity of the incident flow. Also employed is the *staggered* arrangement shown in Fig. 8.6(b) wherein successive rows are displaced transversely with respect to the oncoming flow. In either arrangement, the significant bank dimensions are the tube diameter, D, the longitudinal spacing, S_L, and the transverse tube spacing, S_T.

The flow pattern through a tube bank is generally quite complex. The heat transfer to or from a particular tube is not only influenced by the boundary layer separation phenomenon discussed for a single tube but also influenced by effects such as the "shading" of one tube by another, the impingement of the wake of one tube on another, etc. The heat transfer for any particular tube is then not only determined by the incident fluid conditions, U_∞ and t_∞, but also by D, S_L,

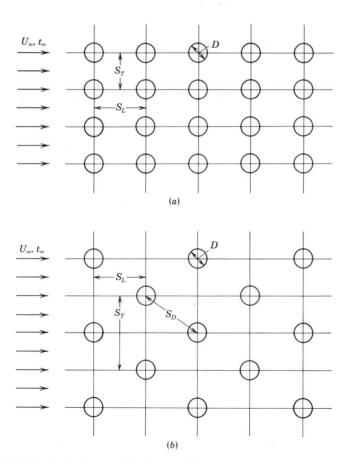

(a)

(b)

FIGURE 8.6. Tube banks: (a) aligned; (b) staggered.

8.6 FORCED CONVECTION NORMAL TO TUBES AND BANKS

S_T and the tube position in the bank. The heat transfer coefficient for the first row of tubes is much like that for a single cylinder in cross flow, while that for tubes in the inner rows is generally larger because of the wake generated by the previous tubes.

For heat transfer correlations, in tube banks, the Reynolds number used is defined as

$$Re_{D_m} = \frac{U_m D}{\nu} \tag{8.44}$$

where U_m is the maximum average fluid velocity occurring at the minimum free area of the bank. For the aligned tube arrangement, U_m is clearly

$$U_m = U_\infty \frac{S_T}{S_T - D}, \tag{8.45}$$

while for the staggered arrangement, U_m is gven by Eq. (8.45) when the diagonal spacing, $S_D = [S_L^2 + (S_T/2)^2]^{1/2}$ is such that $2(S_D - D) > S_T - D$. For $2(S_D - D) < S_T - D$, U_m is then

$$U_m = U_\infty \frac{S_T}{2(S_D - D)}. \tag{8.46}$$

Generally, one wishes to know an *average* heat transfer coefficient for the entire tube bank. For many years the correlation of Grimson (Ref. 23) has been used and may still be expected to give useful results. However, the more recent and extensive work of Zhukauskas summarized in Ref. 21 is recommended here. For *many* rows of tubes, say 20 or more, this correlation for the average heat transfer coefficient is

TABLE 8.4 Values of C and n for Zhukauskas' Equation for Flow over Banks of Tubes*

Re_{D_m} Range	Aligned		Staggered	
	C	n	C	n
10–100	0.8	0.4	0.9	0.4
100–1000	see note 1		see note 1	
1000–2 × 10⁵	$S_T/S_L < 0.7$: see note 2		$S_T/S_L < 2$: $0.35(S_T/S_L)^{1/5}$	0.6
	$S_T/S_L > 0.7$: 0.27	0.63	$S_T/S_L > 2$: 0.40	0.6
2 × 10⁵–10⁶	0.021	0.84	0.022	0.84

Notes:
1. In this range of Re_D use the correlation for a single tube, Eq. (8.4).
2. In this range of Re_D for aligned patterns, the heat transfer is poor, and designs of this geometry are not recommended.
*From Ref. 21.

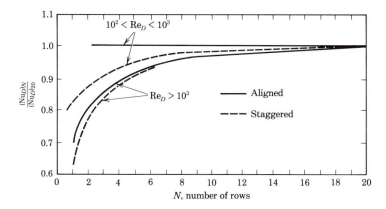

FIGURE 8.7. Correction factor for one number of rows in tube banks. (From A. Zhukauskas, "Heat Transfer for Tubes in Crossflow," *Advances in Heat Transfer,* **New York, Academic Press, 1972, p. 93. Used by permission.)**

$$\mathrm{Nu}_D = C\mathrm{Re}_{D_m}^n \mathrm{Pr}^{0.36}\left(\frac{\mathrm{Pr}_\infty}{\mathrm{Pr}_s}\right)^{1/4}, \quad \begin{bmatrix} 0.7 < \mathrm{Pr} < 500 \\ 10 < \mathrm{Re}_{D_m} < 10^6 \\ C \text{ and } n \text{ from Table 8.4} \\ \text{properties at } t_m \text{ except } \mathrm{Pr}_\infty, \mathrm{Pr}_s \\ \mathrm{Pr}_\infty \text{ at } t_\infty, \mathrm{Pr}_s \text{ at } t_s \end{bmatrix}. \quad (8.47)\bigstar$$

As noted earlier, the above correlation is for the inner rows of a bank, or for banks of many rows. Using 20 rows as a normalizing case, Zhukauskas recommends the use of Fig. 8.7 wherein the correction factor

$$\frac{(\mathrm{Nu}_D)_{N \text{ rows}}}{(\mathrm{Nu}_D)_{20 \text{ rows}}}$$

is plotted. Thus, Eq. (8.47) is used to determine Nu_D for 20 rows and Fig. 8.7 to correct the result for cases with less than 20 rows.

REFERENCES

1. KAYS, W. M., and M. E. CRAWFORD, *Convective Heat and Mass Transfer,* 2nd ed., New York, McGraw-Hill, 1980.
2. JAKOB, M., *Heat Transfer,* Vol. 1 (1949) and Vol. 2 (1957), New York, Wiley.
3. KNUDSEN, J. G., and D. L. KATZ, *Fluid Dynamics and Heat Transfer,* New York, McGraw-Hill, 1958.
4. ECKERT, E. R. G., and R, M. DRAKE, *Analysis of Heat and Mass Transfer,* New York, McGraw-Hill, 1972.
5. ROHSENOW, W. M., and J. P. HARTNET, eds., *Handbook of Heat Transfer,* New York, McGraw-Hill, 1973.
6. CHURCHILL, S. W., and H. OZOE, "Correlations for Laminar Forced Convection in Flow over an Isothermal Flat Plate and in Developing and Fully Developed Flows in an Isothermal Tube," *J. Heat Transfer, Trans. ASME,* Vol. 95, 1973, p. 416.
7. ECKERT, E. R. G., "Engineering Relations for Friction and Heat Transfer to Surfaces in High Velocity Flow," *J. Aerosp. Sci.,* Vol. 22, 1955, p. 585.

8. Wimbrow, W. R., "Experimental Investigations of Temperature Recovery Factors on Bodies of Revolution at Supersonic Speeds," *NACA Tech. Note 1975,* Washington, D.C., 1949.

9. Eckert, E. R. G., "Survey on Heat Transfer at High Speeds," *Aeronaut. Res. Lab. Rep. 189,* Office of Aerospace Research, Wright-Patterson Air Force Base, Ohio, 1961.

10. Lin, C. C., ed., *Turbulent Flows and Heat Transfer, High Speed Aerodynamics and Jet Propulsion,* Vol. V, Princeton, N.J., Princeton U.P., 1959.

11. Hausen, H., "Darstellung des Wärmeüberganges in Rohren durch vergallgemeinerte Potenzbeziehungen," *Z.A.V.D.I. Beihefte Verfahrenstech.,* No. 4, 1943, p. 91.

12. Sieder, E. N., and E. G. Tate, "Heat Transfer and Pressure Drop of Liquids in Tubes," *Ind. Eng. Chem.,* Vol. 28, 1936, p. 1429.

13. Dittus, F. W., and L. M. K. Boelter, *Univ. Calif., Berkley, Publ. Eng.,* Vol. 2., 1930, p. 443.

14. Petukhov, B. S., "Heat Transfer and Friction in Turbulent Pipe Flow with Variable Physical Properties," *Advances in Heat Transfer,* Vol. 6, New York, Academic Press, 1970, 504.

15. Nusselt, W., "Der Wärmeaustausch zwischen Wand und Wasser im Roher," *Forsch. Geb. Ingenieurwes.,* Vol. 2, 1931, p. 309.

16. Martinelli, R. C., "Heat Transfer to Molten Metals," *Trans. ASME,* Vol. 69, No. 8, 1947, p. 447.

17. Seban, R. A., and T. T. Shimazaki, "Heat Transfer to a Fluid Flowing Turbulently in a Smooth Pipe with Walls at Constant Temperature," *Trans. ASME,* Vol. 73, 1951, p. 803.

18. Azer, N. Z., and B. T. Chao, "Turbulent Heat Transfer in Liquid Metals—Fully Developed Pipe Flow with Constant Wall Temperature," *Int. J. Heat Mass Transfer,* Vol. 3, 1961, p. 77.

19. Sleicher, C. A., A. S. Awad, and R. H. Notter, "Temperature and Eddy Diffusivity Properties in N_aK," *Int. J. Heat Mass Transfer,* Vol. 16, 1973, p. 1565.

20. Giedt, W. H., "Investigation of the Variation of Point Unit Heat Transfer Coefficient Around a Cylinder Normal to an Air Stream," *Trans. ASME,* Vol. 71, 1949, p. 375.

21. Zhukauskas, A., "Heat Transfer from Tubes in Crossflow," *Advances in Heat Transfer,* Vol. 8, New York, Academic Press, 1972, p. 93.

22. Churchill, S. W., and M. Bernstein, "A Correlating Equation for Forced Convection from Gases and Liquids to a Circular Cylinder in Crossflow," *J. Heat Transfer, Trans. ASME,* Vol. 94, 1977, p. 300.

23. Grimson, E. D., "Correlation and Utilization of New Data on Flow Resistance and Heat Transfer for Cross Flow of Gases over Tube Banks," *Trans. ASME,* Vol. 59, 1937, p. 583.

PROBLEMS

8.1 Glycerine at 40°C flows at 3 m/s past a flat surface maintained at 10°C. If the plate is 0.3 m long and 0.3 m wide, find the heat transferred from one side of the plate using both Eqs. (8.1) and (8.3). What is the drag force for one side? Neglect the effects of viscous dissipation.

8.2 Water at 90°C flows with a velocity of 1 m/s past a flat surface maintained at 50°C. Calculate the local heat transfer coefficient at 0.01, 0.05, 0.1, and

0.15 m from the leading edge. Find the average heat transfer coefficient for a total plate length of 0.15 m. Neglect the effects of viscous dissipation.

8.3 Atmospheric air at 20°C flows with a velocity of 25 m/s past a flat surface maintained at 80°C. The surface is 15 cm long. Using both Eqs. (8.1) and (8.3), find the average heat transfer coefficient for the surface. Neglect the effects of viscous dissipation.

8.4 Air at 15°C flows with a velocity of 25 m/s past a flat plate, 50 cm long, maintained at 75°C. Find (a) the local heat transfer coefficient at locations 12.5, 25, 37.5, and 50 cm from the leading edge; and (b) the average heat transfer coefficient for the plate. Neglect the effects of viscous dissipation.

8.5 A plate has the dimensions of 3 in. × 18 in. It is maintained at 190°F and is placed in an airstream at 1 atm, 50°F, flowing at 90 ft/s. Find the total heat transferred from one side of the plate if (a) its 18-in. edge is the leading edge, and (b) its 3-in. edge is the leading edge. Neglect the effects of viscous dissipation.

8.6 Nitrogen at atmospheric pressure and 50°C flows with a velocity of 20 m/s past a flat plate held at 0°C. The plate is 0.8 m long and 0.2 m wide. Find (a) the total heat transferred and the total drag force on one side of the plate, and (b) the fraction of the heat transferred and drag force in the laminar portion of the boundary layer. Neglect the effects of viscous dissipation.

8.7 Justify the neglection of the effects of viscous dissipation in Probs. 8.1 through 8.6.

8.8 The lubricating oil listed in Table A.4 flows at 60°C with a velocity of 35 m/s past a flat surface maintained at 20°C. Find the heat transfer coefficient and the heat flux (in magnitude and direction) at a point 10 cm from the leading edge if (a) viscous dissipation is neglected, and (b) if viscous dissipation is not neglected.

8.9 Compute the recovery temperature, the local heat transfer coefficient, and the local heat flux that may be expected to occur at a point 0.3 m behind the leading edge of the fin of a supersonic plane traveling at a speed of 660 m/s at (a) sea level (1 atm, 15°C), and (b) 20,000-ft altitude (0.46 atm, −25°C). The fin surface is to be maintained at 100°C in both cases.

8.10 Air at an altitude of 1500 m (84 kN/m², 5°C) flows past a flat surface maintained at 10°C with a velocity of 200 m/s. Find the local heat flux at a point 2.5 cm from the leading edge if (a) viscous dissipation is neglected, and (b) if viscous dissipation is not neglected.

8.11 A flat plate 2 ft long and 1 ft wide is placed in an airstream at 0.50 psia, −40°F, flowing with a velocity of 3020 ft/s. How much cooling must be provided to maintain the plate at 100°F if viscous dissipation is accounted for?

8.12   Air at a pressure of 35 kN/m² and a temperature of $-40°C$ flows past a flat plate 10 cm long and 30 cm wide with a free stream Mach number of 3.0. The plate is maintained at 150°C. Estimate the total heat transferred to or from the plate.

8.13   Verify the algebra showing that the formulation in Eq. (8.21) is equivalent to that in Eq. (8.17).

8.14   The lubricating oil listed in Table A.4 flows with an average velocity of 0.3 m/s and an average bulk temperature of 160°C in a tube 2.3 cm in diameter, the surface of which is maintained at 140°C. Find the average surface heat transfer coefficient if the tube is (a) 2 m long, and (b) 6 m long.

8.15   Atmospheric pressure air flows with a mean bulk temperature of 50°C through a tube 0.6 cm in diameter at a mass rate of 0.5 kg/h. If the tube surface is maintained at 90°C, and it is 0.3 m long, find the average surface heat transfer coefficient.

8.16   Air at atmospheric pressure flows at the volume rate of 1 m³/s with an average bulk temperature of 15°C through an air conditioning duct 20 cm in diameter with a surface temperature of 30°C. Determine the average surface heat transfer coefficient using Eqs. (8.27) through (8.29).

8.17   Water at 80°F enters a 2-in. schedule 40 pipe and leaves at 120°F. It flows at the rate of 50,000 lb_m/h. The pipe surface is maintained at 140°F. Find the heat transfer coefficient at the pipe surface for (a) the entrance conditions, (b) the exit conditions, and (c) the average of the entrance and exit conditions.

8.18   Superheated steam at 4000 kN/m² pressure and 500°C flows at the rate of 1.25 kg/s through an 8-in., schedule 80 pipe with a surface temperature of 450°C. Find the heat transfer coefficient at the pipe surface.

8.19   Water in a circulating chilled water air conditioning system enters the chilling unit at 15°C and leaves at 5°C. The tubes in the chilling unit are 3.8 cm in diameter, the surface temperature is 2°C, and the average water velocity is 0.6 m/s. Estimate the heat transfer coefficient at the tube surface.

8.20   Air at 40 atm. pressure flows at a velocity of 6 m/s through a tube 2.5 cm in diameter at a mean bulk temperature of 250°C. The tube surface is a 80°C. Calculate the expected heat transfer coefficient at the tube surface.

8.21   For fully developed flow in a circular tube with a surface temperature of 150°C, find the surface heat transfer coefficient in each case noted below if the mean flow velocity is 6 m/s in each case:
(a) Air, 90°C, 1 atm pressure, tube diameter = 2.5 cm.
(b) Air, 90°C, 1 at pressure, tube diameter = 1.25 cm.
(c) Air, 90°C, 10 atm pressure, tube diameter = 2.5 cm.

(d) Water, 90°C, 10 atm pressure, tube diameter = 2.5 cm.

(e) Lubricating oil, 90°C, tube diameter = 2.5 cm.

8.22 Liquid sodium at 538°C flows at the rate of 6 m/s through a pipe 2.5 cm in diameter. Estimate the convective heat transfer coefficient using Eqs. (8.32) through (8.34).

8.23 Repeat Prob. 8.22 for mercury at 316°C.

8.24 A 44% potassium, 56% sodium, liquid metal mixture is used as the heat transfer medium in a nuclear reactor. What heat transfer coefficient can be expected at the inner surface of a pipe of 2.5 cm I.D. if the medium is at 371°C and flows with a velocity of 3 m/s?

8.25 Natural gas at a pressure of 100 psia (approximated as methane) flows at 90°F through a $\frac{3}{4}$-in.-diameter pipe at the rate of 10 ft/s. If the pipe wall is at 80°F, estimate the surface heat transfer coefficient.

8.26 Water flowing at the rate of 2.6 kg/s is heated from 7°C to 20°C as it flows through a smooth 2-in. schedule 40 pipe. The inside pipe wall temperature is 90°C. What is the required length of pipe?

8.27 Water flowing at the mass rate of 0.75 kg/s is to be cooled from 50°C to 20°C. Which of the following pipes offers the least pressure loss?
(a) The water flows through a 1.25-cm-I.D. pipe maintained at 10°C.
(b) The water flows through a 2.5-cm-I.D. pipe maintained at 15°C.

8.28 Superheated steam at 6000 kN/m² pressure, 400°C, enters a pipe 2.5 cm in diameter with a velocity of 12 m/s. The pipe wall is maintained at 500°C. How long must the pipe be to heat the steam to 430°C?

8.29 Water at 10°C and flowing at the mass rate of 2.5 kg/s enters a pipe 5 cm in diameter. The pipe wall is maintained at 80°C. If the pipe is 5 m long, what is the outlet temperature of the water?

8.30 Superheated steam flows at a mass rate of 2 kg/s through a pipe 8 cm in diameter. The steam enters at 4000 kN/m² pressure and 450°C. The pipe is 10 m long and has its surface maintained at 350°C. Find the outlet temperature of the steam.

8.31 Air at atmospheric pressure and 30°C flows with a velocity of 5 m/s through a square air conditioning duct, 20 cm × 20 cm, the surface of which is at 20°C. Estimate the surface heat transfer coefficient.

8.32 Superheated steam flows through the annulus formed between a 4-in. and a 6-in. schedule 40 steel pipe. The steam is at 200 psia, 400°F, and its velocity is 5 ft/s. Water at 120°F is flowing at 2 ft/s through the inner pipe. Estimate the heat transferred, per foot of pipe length, to the water.

8.33 A velocity-measuring device consists of a wire 0.025 cm in diameter. By measuring the electrical input to the wire to maintain it at a given temperature, the velocity of air flowing normal to it may be estimated. If the wire is 1.25 cm long and has its surface maintained at 100°C, how much heat does it give up to atmospheric air at 15°C flowing normal to it with a velocity of 10 m/s?

8.34 Hot combustion gases, having properties close to those of nitrogen, flow at a pressure of 175 kN/m² and a temperature of 60°C normal to a 5-cm-O.D. cylinder at a velocity of 15 m/s. What is the surface heat transfer coefficient if the pipe is held at 120°C?

8.35 Repeat Prob. 8.34 for the flow of the same gases at the same conditions past a square pipe, 5 cm on the side, placed with one side normal to the flow. The pipe is, again, at 120°C.

8.36 A fluid at 40°C flows with a velocity of 10 m/s normal to a cylinder heated to 120°C and having a diameter of 5 mm. Find the heat transfer coefficient at the cylinder surface if the fluid is (a) helium at 1 atm, (b) air at 1 atm, and (c) water.

8.37 A tube bank uses an aligned arrangement with $S_T = S_L = 2.0$ cm and tubes 1.0 cm in diameter. There are eight rows of tubes. The tube surfaces are at 80°C, and atmospheric air at 20°C approaches the bank, normal to the tubes, with an undisturbed velocity of 5 m/s. Estimate the average surface heat transfer coefficient for the bank of tubes. How much heat, per square meter of frontal area of the bank normal to the incident flow, is transferred to the air?

8.38 Repeat Prob. 8.37 if a staggered arrangement is used, maintaining the same S_T and S_L.

8.39 A tube bank uses an aligned arrangement with $S_T = 2.5$ cm, $S_L = 2.0$ cm, and tubes 1.5 cm in diameter. There are four rows of tubes. The tube surfaces are at 5°C, and atmospheric air at 25°C approaches the bank, normal to the tubes, with an undisturbed velocity of 6 m/s. Estimate the average heat transfer coefficient for the bank.

8.40 Repeat Prob. 8.39 if a staggered arrangement is used, maintaining the same S_T and S_L.

8.41 Water at 95°C flows in a 1-in. 16-gage copper tube with a velocity of 0.6 m/s. The outside surface is exposed to air at 20°C through a surface heat transfer coefficient of 8.5 W/m²-°C. Determine the heat loss from a 1-m length of the tube.

8.42 Superheated steam at 400 psia, 600°F, flows with a velocity of 20 ft/s through a 6-in. steel schedule 40 pipe. The outer surface is exposed to a stream of air (1 atm, 70°C) flowing normal to the pipe at 40 ft/s. Estimate the rate of heat loss, per foot of length, from the pipe.

CHAPTER 9

Heat Transfer by Free Convection

INTRODUCTORY REMARKS

The examples of convective heat transfer considered up to now have all been examples of *forced* convection between a fluid and a solid body wherein the fluid motion relative to the solid surfaces was caused by an external input of work (i.e., by means of a pump, fan, propulsive motion of an aircraft, etc.). This chapter will be concerned with *free* or *natural* convection in which the fluid velocity at points remote from the body will be essentially zero. Near the body there will be some fluid motion if the body is at a temperature different from that of the free fluid. If such is the case, there will be a density difference between the fluid near the surface and that far removed from the surface. This density difference will produce a positive or negative buoyant force (depending on whether the surface is hotter or colder than the fluid) in the fluid near the surface. The buoyant force results in a fluid motion, substantially in the vertical direction, past the surface—with consequent convective heat transfer taking place. The force of gravity is, then, the driving force which produces the fluid motion and maintains the convective process.

   The heating of rooms and buildings by the use of "radiators" is a familiar example of heat transfer by free convection. Heat losses from hot pipes, ovens, etc., surrounded by cooler air are due to free convection, at least in part.

THE GOVERNING EQUATIONS OF FREE CONVECTION

For purposes of discussion let the heated surface, at temperature t_s, be a vertical plane. The fluid far removed from the surface is at zero velocity, and its temperature is t_f. If one assumes that $t_s > t_f$, the distribution of the fluid velocity and fluid temperature in a direction normal to the surface will be something like that shown in Fig. 9.1. The temperature decreases (from the value imposed at the surface) asymptotically to the value of the fluid far away. The fluid velocity,

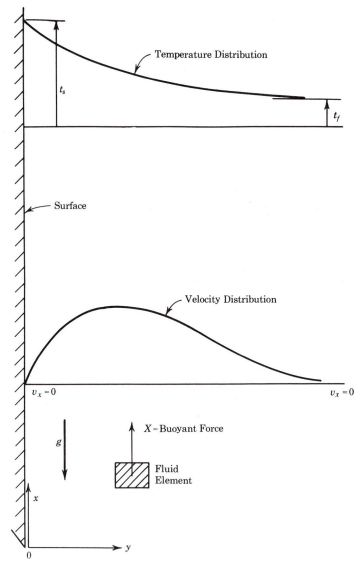

FIGURE 9.1. **Temperature and velocity distributions near a heated vertical surface in free convection flow.**

on the other hand, must somehow increase from zero at the wall to a maximum value in the immediate neighborhood of the wall and then decrease, asymptotically, to a zero value far away. This distribution of velocity is as shown in Fig. 9.1.

The motion of the fluid is restricted to a region close to the surface. For fluids of low viscosity this region is relatively thin, and it has been found that the simplifications leading to the boundary layer equations of motion and energy [Eqs. (6.30) and (6.31)] are applicable. However, the equation of motion must be modified. Reference to Sec. 6.9 will show that in the development of the equation or motion, Eq. (6.30), the body forces X and Y (defined in Sec. 6.5 in the derivation of the Navier–Stokes equations) were neglected. This was perfectly satisfactory for application to problems of forced convection, wherein the body force due to gravity would be expected to be small in comparison to the inertia forces of the motion. In free convection, however, it is the body force which sustains the fluid motion, and, hence, it certainly must be included. Upon selecting the x-coordinate axis as parallel to the surface, as shown in Fig. 9.1, and the y-coordinate axis as normal to it, the body forces per unit mass will be

$$\text{In the } y \text{ direction, } Y = 0, \tag{9.1}$$

$$\text{In the } x \text{ direction, } X = \frac{1}{\rho} g(\rho_f - \rho)$$

as long as one adopts the convention that the pressure term in the equation of motion represents the difference between the local fluid pressure and the pressure in the fluid, at the same x, far removed from the heated surface. In Eq. (9.1) ρ_f denotes the density of the undisturbed fluid (at temperature t_f) far removed from the surface. The coefficient of volume expansion, β, defined in Sec. 2.6, may be used to replace ρ by temperature:

$$\begin{aligned} X &= \frac{1}{\rho} g(\rho_f - \rho) \\ &= g\beta(t - t_f). \end{aligned} \tag{9.2}$$

The boundary layer equation for motion parallel to the surface is then obtained by adding Eq. (9.2) to Eq. (6.30):

$$v_x \frac{\partial v_x}{\partial x} + v_y \frac{\partial v_x}{\partial y} = -\frac{1}{\rho} \frac{dP}{dx} + \nu \frac{\partial^2 v_x}{\partial y^2} + g\beta(t - t_f). \tag{9.3}$$

The pressure gradient, dP/dx, may, in some cases, be neglected.

The boundary layer equation of energy, Eq. (6.31) needs no alteration other than the readily justified neglection of the viscous dissipation term $(\mu/\rho c_p)(\partial v_x/\partial y)^2$. Thus, one must also satisfy

$$v_x \frac{\partial t}{\partial x} + v_y \frac{\partial t}{\partial y} = \alpha \frac{\partial^2 t}{\partial y^2}. \tag{9.4}$$

For the determination of the velocity and temperature fields near the heated surface, and subsequently the heat transfer coefficient h, the last two equations would have to be solved simultaneously to satisfy the conditions

$$\text{At } y = 0: \quad t = t_s, \, v_x = 0.$$

(9.5)

$$\text{At } y = \infty: \quad t = t_f, \, v_x = 0.$$

The problem as set forth by Eqs. (9.3) through (9.5) proves to be quite complex, even for the simple case of a vertical plane surface. The additional complexity involved here over those encountered in the case of forced convection past a plane plate is the additional term just included in the equation of motion. In the case of forced convection this term was missing, and, as was pointed out in Sec. 7.1, this fact made it possible to solve for the velocity distribution in the hydrodynamic boundary layer without consideration of the thermal boundary layer because of the absence of any term in the equation of motion which contained, or was a function of, the fluid temperature. In the problem of free convection this is not the case, and the velocity and temperature distributions must be found, simultaneously.

In like fashion, integral expressions may be written for the free convection boundary layer. The development of the integral momentum equation in Sec. 6.10 neglected the presence of the buoyancy term. Inclusion of this effect in the summation of forces in the development leading to Eq. (6.36) yields the additional term

$$\int_0^\delta g\beta(t - t_f) \, dy$$

on the right side of that equation. If the pressure gradient is neglected and if the fact that $U(x) = 0$ is used, Eq. (6.36) becomes

$$\frac{d}{dx} \int_0^\delta v_x^2 \, dy = \int_0^\delta g\beta(t - t_f) \, dy - \nu\left(\frac{\partial v_x}{\partial y}\right)_0.$$

(9.6)

The integral energy equation, Eq. (6.38), remains unchanged:

$$\frac{d}{dx} \int_0^{\delta_t} (t - t_f)v_x \, dy = -\alpha\left(\frac{\partial t}{\partial y}\right)_0.$$

(9.7)

The same boundary conditions, Eq. (9.5), must be satisfied. The solution of the integral equations, Eqs. (9.6) and (9.7), is likewise complicated by the fact that the equations are "coupled." That is, since temperature appears in both the energy and momentum expressions, a simultaneous solution must be accomplished.

The similarity analysis of Sec. 6.7 showed that the presence of buoyancy forces were accounted for by the Grashof number, Gr.

$$\text{Gr} = \frac{l^3 g\beta \, \Delta t}{\nu^2},$$

(9.8)

wherein l is a characteristic length and Δt is the surface–fluid temperature difference $(t_s - t_f)$. The general prediction that convective heat transfer is of the form

$$Nu = f(Re, Gr, Ec, Pr)$$

reduces to the following for "pure" free convection (i.e., no external stream and no dissipation):

$$Nu = f(Gr, Pr).$$

The analyses given in the following sections will yield relations of this form.

One would expect, then, that the Grashof number will perform for free convection much the same function that the Reynolds number does for forced convection. This is, indeed, the case; however another parameter, the Rayleigh number is also used for this purpose. The Rayleigh number, Ra, is defined as

$$Ra = GrPr, \tag{9.9}$$

$$= \frac{l^3 g \beta \, \Delta t \, Pr}{\nu^2},$$

simply the product of the previously defined Grashof and Prandtl numbers. In terms of the Rayleigh number and correlations for free convection would be expected to take the form

$$Nu = f(Ra, Pr). \tag{9.10}$$

All free convection flow is not limited to laminar flow. In fact, most free convective flows arise from instabilities created by heated fluid rising past cooler fluid which must descend. Such disturbances may be amplified in the resultant boundary layers and, eventually, turbulence may be established. The laminar–turbulent transition depends on the relative magnitude of the buoyant forces and the resisting viscous forces—as expressed through the Grashof or Rayleigh numbers. A detailed analysis of free convection stability is rather complex (Ref. 1); however, as a general rule one may expect that transition will occur for a critical Rayleigh number of

$$Ra_{x,c} \cong 10^9. \tag{9.11}$$

9.3

ANALYTICAL SOLUTIONS OF FREE CONVECTION PAST VERTICAL PLANE SURFACES

Extensive work has been carried out in the analytical investigation of free convection problems. Two such solutions for laminar flows will be presented here to illustrate the application of the differential equations [Eqs. (9.3) and (9.4)] and the application of the integral equations [Eqs. (9.6) and (9.7)]. An application of the integral method to the turbulent case will also be shown. Although more complex, the basic ideas in application of these equations are the same as in the forced convection cases. Hence, many details will be omitted in the discussions to follow.

Solution of the Laminar Case by Differential Equations

Schmidt and Beckmann (Ref. 2) analyzed the problem of laminar free convection past a vertical flat surface, as depicted in Fig. 9.1, by application of Eqs. (9.3) through (9.5):

$$v_x \frac{\partial v_x}{\partial x} + v_y \frac{\partial v_x}{\partial y} = \nu \frac{\partial^2 v_x}{\partial y^2} + g\beta(t - t_f). \tag{9.3}$$

$$v_x \frac{\partial t}{\partial x} + v_y \frac{\partial t}{\partial y} = \alpha \frac{\partial^2 t}{\partial y^2}. \tag{9.4}$$

$$\text{At } y = 0: \quad t = t_s, \ v_x = 0. \tag{9.5}$$

$$\text{At } y = \infty: \quad t = t_f, \ v_x = 0.$$

The x coordinate has been taken as positive measured along the plate in the direction of the buoyant force. The pressure gradient term has been neglected.

The above equations may be made ordinary equations by the introduction of a similarity parameter:

$$\eta = C \frac{y}{x^{1/4}},$$

$$C = \left[\frac{g\beta(t_s - t_f)}{4\nu^2} \right]^{1/4}.$$

The reasoning behind the selection of this form for the similarity parameter is rather involved but not unlike that leading to the definition of the Blasius parameter in Sec. 7.2. The parameter η may also be written as

$$\eta = \left(\frac{g\beta \, \Delta t}{4\nu^2} \right)^{1/4} \frac{y}{x^{1/4}}$$

$$= \left(\frac{1}{4} \right)^{1/4} \left(\frac{g\beta \, \Delta t \, x^3}{\nu^2} \right)^{1/4} \frac{y}{x}$$

$$= \frac{1}{\sqrt{2}} \, \text{Gr}_x^{1/4} \frac{y}{x},$$

where Gr_x denotes the *local* Grashof number. The notation

$$\Delta t = t_s - t_f$$

will be used throughout the remaining analysis.

The velocity component parallel to the plate surface is next assumed to be of the form

$$v_x = 4\nu C^2 x^{1/2} F'(\eta), \tag{9.12}$$

in which $F(\eta)$ is an as-yet-undetermined function of only η. Continuity considerations then require that v_y satisfy

$$v_y = \nu C x^{-1/4} [\eta F'(\eta) - 3F(\eta)]. \tag{9.13}$$

Finally, one defines a dimensionless temperature

$$\varphi = \frac{t - t_f}{t_s - t_f} = \frac{t - t_f}{\Delta t} \tag{9.14}$$

and assumes that it is a function of η *only*. When Eqs. (9.12) through (9.14) are substituted into Eqs. (9.3) through (9.5), considerable algebra leads finally to

$$F''' + 3FF'' - 2(F')^2 + \varphi = 0,$$
$$\varphi'' + 3\mathrm{Pr}F\varphi' = 0, \tag{9.15}$$
$$F'(0) = 0, \quad \varphi(0) = 1,$$
$$F'(\infty) = 0, \quad \varphi(\infty) = 0.$$

The fact that the system of equations has been converted into a set of ordinary equations involving η only justifies the assumptions made above. However, the equations are still coupled, and they must be solved simultaneously. Presume for the moment that this solution has been accomplished; then one part of the solution would be the dimensionless temperature gradient at the wall, $\varphi'(0)$. This quantity is necessarily a function of only the Prandtl number, since that is the only parameter in Eq. (9.15). Make the definition

$$f(\mathrm{Pr}) = -\varphi'(0). \tag{9.16}$$

Then the heat transfer coefficient is

$$h_x = \frac{-k(\partial t/\partial y)_0}{t_s - t_f},$$

$$= -k\varphi'(0)\frac{d\eta}{dy} = -k\varphi'(0)\left(\frac{1}{\sqrt{2}}\right)\mathrm{Gr}_x^{1/4}\frac{1}{x}, \tag{9.17}$$

$$= \frac{1}{\sqrt{2}}\frac{k}{x}\mathrm{Gr}_x^{1/4}f(\mathrm{Pr}).$$

Integration over a finite plate length L yields the average h:

$$h = \frac{4}{3}\left(\frac{1}{\sqrt{2}}\right)\frac{k}{L}\mathrm{Gr}_L^{1/4}f(\mathrm{Pr}), \tag{9.18}$$

in which the plate length Grashof number has been introduced:

$$\mathrm{Gr}_L = \frac{g\beta\,\Delta t\,L^3}{\nu^2}. \tag{9.19}$$

In dimensionless forms the above expressions for the local and average heat transfer coefficients are

$$\mathrm{Nu}_x = \frac{h_x x}{k} = \frac{1}{\sqrt{2}}\mathrm{Gr}_x^{1/4}f(\mathrm{Pr}), \tag{9.20}$$

$$\mathrm{Nu}_L = \frac{hL}{k} = \frac{4}{3}\left(\frac{1}{\sqrt{2}}\right)\mathrm{Gr}_L^{1/4}f(\mathrm{Pr}). \tag{9.21}$$

The foregoing equations are still incomplete in that the function $f(\text{Pr}) = -\varphi'(0)$ is undetermined. Equations (9.15) have been solved by Ostrach (Ref. 3), and the resulting velocity and temperature profiles are shown in Figs. 9.2 and 9.3. The function $f(\text{Pr}) = -\varphi'(0)$ is implied in the slopes of the curves in Fig. 9.3 and may also be found in Ref. 4. An empirical fit to these results is

$$f(\text{Pr}) = 0.676\text{Pr}^{1/2}(0.861 + \text{Pr})^{-1/4},$$

so that the local and average Nusselt numbers are

$$\text{Nu}_x = 0.478\text{Gr}_x^{1/4}\text{Pr}^{1/2}(0.861 + \text{Pr})^{-1/4},$$ (9.22)

$$\text{Nu}_L = 0.637\text{Gr}_L^{1/4}\text{Pr}^{1/2}(0.861 + \text{Pr})^{-1/4}.$$

In terms of the Rayleigh number:

$$\text{Nu}_x = 0.478\text{Ra}_x^{1/4}\left(1 + \frac{0.861}{\text{Pr}}\right)^{-1/4},$$ (9.23)

$$\text{Nu}_L = 0.637\text{Ra}_L^{1/4}\left(1 + \frac{0.861}{\text{Pr}}\right)^{-1/4}.$$

Equations (9.22) and (9.23) are found to give reliable results for laminar free convection past vertical plates if the fluid properties are all evaluated at the mean film temperature, $t_m = (t_s + t_f)/2$, except β. For liquids β should be evaluated at t_m, but for gases it should be found at t_f. However, certain empirical relations

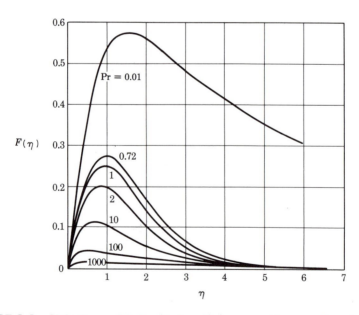

FIGURE 9.2. Velocity profile for laminar free convection on a flat plate. (From S. Ostrach, "An Analysis of Laminar Free Convection Flow and Heat Transfer About a Flat Plate Parallel to the Direction of the Generating Body Force," *NACA Tech. Note 2635*, Washington, D.C., 1952.)

HEAT TRANSFER BY FREE CONVECTION

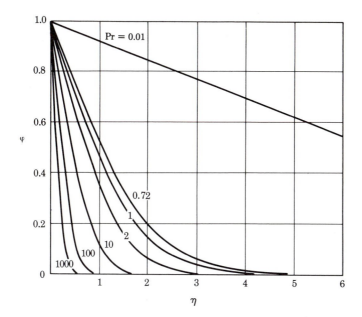

FIGURE 9.3. Temperature profile for laminar free convection on a flat plate. (From S. Ostrach, "An Analysis of Laminar Free Convection Flow and Heat Transfer About a Flat Plate Parallel to the Direction of the Generating Body Force," *NACA Tech. Note 2635*, Washington, D.C., 1952.)

are actually recommended for practical calculations and will be given in a later section.

Solution of the Laminar Case by Integral Equations

The problem just discussed has been solved also by Eckert (Ref. 5) by use of the integral equations, Eqs. (9.6) and (9.7):

$$\frac{d}{dx} \int_0^\delta v_x^2 \, dy = \int_0^\delta g\beta(t - t_f) \, dy - v\left(\frac{\partial v_x}{\partial y}\right)_0, \tag{9.6}$$

$$\frac{d}{dx} \int_0^{\delta_t} (t - t_f)v_x \, dy = -\alpha\left(\frac{\partial t}{\partial y}\right)_0. \tag{9.7}$$

The same basic procedure is used as was used for the integral analysis of the forced convection problem. First, one assumes logical forms of the temperature and velocity profiles. For simplicity in analysis it will be assumed that the thermal and velocity layers are virtually the same (i.e., $\delta \approx \delta_t$). The form of the temperature profile sketched in Fig. 9.1 suggests a dimensionless profile of the form

$$\varphi = \frac{t - t_f}{t_s - t_f} = \left(1 - \frac{y}{\delta}\right)^2. \tag{9.24}$$

The selection of a form for the velocity profile is more difficult since an obvious characteristic velocity for nondimensionalizing does not exist. The form of the velocity profile of Fig. 9.1 suggests an expression of the form

$$v_x = v_x^* \frac{y}{\delta}\left(1 - \frac{y}{\delta}\right)^2,$$

(9.25)

where v_x^* is an unknown quantity having the dimensions of a velocity. Differentiation of Eq. (9.25) will show that v_x has a maximum value, call it $v_{x_{\max}}$, at $y/\delta = \frac{1}{3}$, and this maximum is

$$v_{x_{\max}} = \frac{4}{27} v_x^*.$$

Thus, the assumed profile in Eq. (9.25) may be written

$$v_x = \frac{27}{4} v_{x_{\max}} \frac{y}{\delta}\left(1 - \frac{y}{\delta}\right)^2.$$

(9.26)

The quantity $v_{x_{\max}}$ is still unknown and would be expected to be a function of the distance from the leading edge, x.

Equations (9.24) and (9.26) may be put in more convenient forms by introducing the parameter

$$\eta = \frac{y}{\delta},$$

(9.27)

so that

$$\varphi = \frac{t - t_f}{t_s - t_f} = (1 - \eta)^2,$$

(9.28)

$$\frac{v_x}{v_{x_{\max}}} = \frac{27}{4}\eta(1 - \eta)^2.$$

(9.29)

In terms of these parameters, Eqs. (9.6) and (9.7) become

$$\frac{d}{dx}\int_0^1 v_{x_{\max}}^2 \left(\frac{v_x}{v_{x_{\max}}}\right)^2 \delta\, d\eta = \int_0^1 g\beta\, \Delta t\, \varphi\delta\, d\eta - v\frac{d}{d\eta}\left(\frac{v_x}{v_{x_{\max}}}\right)_0 \frac{v_{x_{\max}}}{\delta}$$

and

$$\frac{d}{dx}\int_0^1 v_{x_{\max}}\Delta t\, \varphi\, \frac{v_x}{v_{x_{\max}}}\delta\, d\eta = -\alpha\, \Delta t\left(\frac{d\varphi}{d\eta}\right)_0 \frac{1}{\delta}.$$

Substitution of the assumed forms for φ and $v_x/v_{x_{\max}}$, Eqs. (9.28) and (9.29) yield, for the momentum equation,

$$\frac{(27/4)^2}{105}\frac{d}{dx}(v_{x_{\max}}^2\delta) = \frac{g\beta\, \Delta t\, \delta}{3} - \frac{27}{4}\frac{vv_{x_{\max}}}{\delta},$$

(9.30)

and for the energy equation,

$$\frac{27/4}{30} \frac{d}{dx}(v_{x_{max}}\delta) = \frac{2\alpha}{\delta}. \tag{9.31}$$

Equations (9.30) and (9.31) are a coupled set of equations in the unknown functions of x, δ, and $v_{x_{max}}$. A solution is obtained by assuming a power law of the form

$$v_{x_{max}} = C_1 x^m, \tag{9.32}$$

$$\delta = C_2 x^n,$$

in which C_1, C_2, m, and n are yet to be determined. When these expressions are introduced into Eqs. (9.30) and (9.31), there results

$$\frac{(27/4)^2}{105} C_1^2 C_2 (2m + n) x^{2m+n-1} = \frac{g\beta \Delta t C_2}{3} x^n - \frac{27}{4} \nu \frac{C_1}{C_2} x^{m-n} \tag{9.33}$$

and

$$\frac{27}{4} \frac{C_1 C_2 (m + n)}{30} x^{m+n-1} = \frac{2\alpha}{C_2} x^{-n}. \tag{9.34}$$

Since Eqs. (9.33) and (9.34) must be satisfied for all values of x, the exponents of the individual terms in each must be equal:

$$2m + n - 1 = n = m - n,$$

$$m + n - 1 = -n.$$

These are satisfied by

$$m = \tfrac{1}{2},$$

$$n = \tfrac{1}{4},$$

so Eqs. (9.33) and (9.34) may be solved simultaneously for C_1 and C_2:

$$C_1 = \frac{80}{27 \sqrt{15}} \nu \left(\frac{20}{21} + \mathrm{Pr}\right)^{-1/2} \left(\frac{g\beta \Delta t}{\nu^2}\right)^{1/2},$$

$$C_2 = (240)^{1/4} \left(\frac{20}{21} + \mathrm{Pr}\right)^{1/4} \left(\frac{g\beta \Delta t}{\nu^2}\right)^{-1/4} \mathrm{Pr}^{-1/2}.$$

Since $\delta = C_2 x^n$, the following relation for the layer growth is obtained:

$$\frac{\delta}{x} = (240)^{1/4} \mathrm{Gr}_x^{-1/4} \mathrm{Pr}^{-1/2} \left(\frac{20}{21} + \mathrm{Pr}\right)^{1/4}. \tag{9.35}$$

Now, the local heat transfer coefficient is

$$h_x = \frac{-k(\partial t/\partial y)_0}{\Delta t} = -k\varphi'(0)\frac{1}{\delta}.$$

Equation (9.28) gives $\varphi'(0) = -2$, so

$$h_x = 2\frac{k}{\delta} \qquad \text{or} \qquad \mathrm{Nu}_x = \frac{h_x x}{k} = 2\frac{x}{\delta}.$$

Thus, Eq. (9.35) gives

$$Nu_x = \frac{2}{(240)^{1/4}} Gr_x^{1/4} Pr^{1/2}\left(\frac{20}{21} + Pr\right)^{-1/4},$$

$$Nu_x = 0.508 Gr_x^{1/4} Pr^{1/2}(0.952 + Pr)^{-1/4}, \qquad (9.36)$$

$$Nu_x = 0.508 Ra_x^{1/4}\left(1 + \frac{0.952}{Pr}\right)^{-1/4}.$$

The average Nusselt number for a plate length L is readily found to be

$$Nu_L = 0.678 Gr_L^{1/4} Pr^{1/2}(0.952 + Pr)^{-1/4}, \qquad (9.37)$$

$$Nu_L = 0.678 Ra_x^{1/4}\left(1 + \frac{0.952}{Pr}\right)^{-1/4}.$$

Equations (9.36) and (9.37) compare favorably with the results of the differential analysis given in Eqs. (9.22) and (9.23).

Turbulent Free Convection past Plane Surfaces

For Rayleigh numbers in excess of about 10^9, free convection boundary layers are observed to be in a turbulent state of flow. Eckert and Jackson (Ref. 6) investigated this case, again using the integral analysis as just described for laminar flow. In fact, the analysis is so similar that only the barest details will be given.

The integral momentum equation must be altered in that the term $-\nu(\partial v_x/\partial y)_0$ must be replaced by $-\tau_0/\rho$. Also, in the energy equation $-\alpha(\partial t/\partial y)_0$ must be replaced by $(q/A)_0$. Then the following assumptions are made:

1. The thermal and velocity layer thicknesses are the same.
2. The plate is entirely in turbulent flow.
3. The dimensionless temperature and velocity profiles follow laws analogous to those used for laminar flow with the exception that the one-seventh power law is introduced:

$$v_x = v_x^*\left(\frac{y}{\delta}\right)^{1/7}\left(1 - \frac{y}{\delta}\right)^4,$$

$$\varphi = \frac{t - t_f}{t_s - t_f} = 1 - \left(\frac{y}{\delta}\right)^{1/7}.$$

4. The skin friction at the surface follows the Blasius law for forced flow, Eq. (7.62):

$$\frac{\tau_0}{\frac{1}{2}\rho v_x^{*2}} = 0.045\left(\frac{\nu}{v_x^*\delta}\right)^{1/4}.$$

5. Reynolds' analogy may be applied to relate τ_0 to $(q/A)_0$.

With these assumptions, an identical sequence of analytical steps was applied to the integral equations as was just carried out for the laminar case. The details are omitted here, but the final results for the local and average heat transfer were

$$\mathrm{Nu}_x = 0.0295\mathrm{Gr}_x^{2/5}\mathrm{Pr}^{7/15}(1 + 0.494\mathrm{Pr}^{2/3})^{-2/5},$$ (9.38)

$$\mathrm{Nu}_L = 0.0246\mathrm{Gr}_L^{2/5}\mathrm{Pr}^{7/15}(1 + 0.494\mathrm{Pr}^{2/3})^{-2/5}.$$

These results agree well with experimental data for air and water, but they have not been checked extensively for other fluids.

9.4

WORKING CORRELATIONS FOR FREE CONVECTION

The analyses just shown illustrate the techniques employed to predict heat transfer coefficients in free convection. Similar analyses have been applied to numerous other geometric configurations, and many experimental investigations have been carried out for a vast number of situations. This section will be devoted to a presentation, with a minimum of discussion, of the results of some of these studies that may be used for practical engineering calculations. As in the case of forced convection, the applicable range of the parameters involved will be noted as will the recommendation for the evaluation of the fluid properties. All correlations will be given in the forms suggested in Sec. 9.1:

$$\mathrm{Nu} = f(\mathrm{Gr}, \mathrm{Pr}),$$

$$\mathrm{Nu} = f(\mathrm{Ra}, \mathrm{Pr}).$$

Free Convection past Vertical Plane Surface

From an analysis of the data of many experimenters, McAdams (Ref. 7) proposed a correlation for the plate length Nusselt number on vertical plane surfaces of the form

$$\mathrm{Nu}_L = C\mathrm{Ra}_L^m,$$ (9.39)

in which the coefficient C and the exponent m are functions of Ra_L, the Rayleigh number based on the total plate length L. Values of these quantities are not reported here since a more accurate correlation will be quoted shortly. However, it is interesting to note that in the laminar range ($10^4 < \mathrm{Ra}_L < 10^9$) McAdams recommended that $m = \frac{1}{4}$ in accord with the forms of Eqs. (9.22), (9.23), and (9.37). Also, for the turbulent range ($10^9 < \mathrm{Ra}_L$) McAdams suggested $m = \frac{1}{3}$. This latter fact is interesting in that since the Rayleigh number involves L^3 and the Nusselt number simply L, then Eq. (9.39) shows that h is independent of L in this range. Such behavior is not uncommon in other free convection correlations.

Recently Churchill and Chu (Ref. 8) have performed an extensive correlation of a great number of workers and recommend the following relation:

$$Nu_L = 0.68 + 0.670Ra_L^{1/4}\left[1 + \left(\frac{0.492}{Pr}\right)^{9/16}\right]^{-4/9}, \quad 0 < Ra_L < 10^9,$$

$$Nu_L = \left\{0.825 + 0.387Ra_L^{1/6}\left[1 + \left(\frac{0.492}{Pr}\right)^{9/16}\right]^{-8/27}\right\}^2, \quad 10^9 < Ra_L,$$

$$(9.40)\star$$

$$\begin{bmatrix} 0 < Pr < \infty \\ \text{properties, except } \beta, \text{ at } t_m \\ \beta \text{ at } t_m \text{ for liquids, at } t_f \text{ for gases} \end{bmatrix}.$$

Actually, the second of the above two equations may be applied over the entire range of Ra; however, better accuracy is obtained from the use of the first equation. In the application of the above correlations it is useful to remember that for gases, the coefficient of volume expansion, β, is equal to the reciprocal of the absolute temperature, T_f, if the ideal gas approximation is used.

The form of Eqs. (9.40) is not unlike that of the analytical predictions in Sec. 9.3 or the form suggested by McAdams in Eq. (9.39); however, the additive constant 0.68 is not present in these other forms. The presence of this constant results from the need at very low Ra, where the boundary layer approximations do not hold, to approach a nonzero value of Nu as the heat transfer becomes predominately that of conduction in the ambient fluid.

Free Convection past Vertical Cylinders

If the surface in question is a vertical cylinder, one would expect very little difference between the heat transfer coefficient in this instance and that for the vertical plane just considered as long as the circumferential curvature of the cylinder is not great. Indeed, this is the case, and Gebhart (Ref. 1) shows that the correlation for a vertical plate may be applied to the case of the vertical cylinder as long as

$$\frac{D}{L} > \frac{35}{Gr_L^{1/4}}.$$

EXAMPLE 9.1

A vertical hot oven door, 0.5 m high, is at 200°C and is exposed to atmospheric pressure air at 20°C. Estimate the average heat transfer coefficient at the surface of the door.

Solution. For $t_m = (200 + 20)/2 = 110°C$, Table A.6 gives $\nu = 24.10 \times 10^{-6}$ m²/s, $k = 0.03194$ W/m-°C, $Pr = 0.704$. At $t_f = 20°C$, $\beta = 1/293.15°K$. The characteristic length and temperature difference are $L = 0.5$ m and $\Delta t = 180°C$. Thus, with $g = 9.8$ m/s²,

$$Gr_L = \frac{L^3 g \beta \, \Delta t}{\nu^2} = \frac{(0.5)^3 \times 9.8 \times (1/293.15) \times 180}{(24.10 \times 10^{-6})^2} = 1.30 \times 10^9,$$

$$Ra_L = Gr_L Pr = 1.30 \times 10^9 \times 0.704 = 9.12 \times 10^8.$$

Equation (9.40) gives

$$Nu_L = 0.68 + 0.670(9.12 \times 10^8)^{1/4} \left[1 + \left(\frac{0.492}{0.704} \right)^{9/16} \right]^{-4/9},$$

$$= 89.96 = \frac{hL}{k},$$

$$h = 5.75 \text{ W/m}^2\text{-°C } (1.01 \text{ Btu/h-ft}^2\text{-°F}).$$

Had Eq. (9.23) been applied there would result

$$Nu_L = 90.07, \qquad h = 5.79 \text{ W/m}^2\text{-°C } (1.02 \text{ Btu/h-ft}^2\text{-°F}),$$

and Eq. (9.37) gives

$$Nu_L = 95.13, \qquad h = 6.07 \text{ W/m}^2\text{-°C } (1.07 \text{ Btu/h-ft}^2\text{-°F}).$$

Thus, the analytical predictions agree well with those from the empirical correlation. ∎

Free Convection Around Horizontal Plates

The cases considered so far had the principal body dimension in line with the gravity vector, i.e., vertical. The flow produced by the free convection was, then, parallel to the surface regardless of whether the surface was hotter or colder than the ambient fluid. However, in the case of free convection about heated, or cooled, horizontal plates, the principal body dimension is normal to the direction of the bouyant forces and considerable differences occur in the flow patterns. Consider the situations depicted in Fig. 9.4, that of a heated surface facing down or facing up.

In the case of a heated plate facing down, as depicted in Fig. 9.4(a), the less dense, heated, fluid forms at the bottom surface of the plate and, seeking a way to rise, flows laterally to the edge and then up. A flow pattern, somewhat as sketched, is then established. However, if the heated surface faces up, as in Fig. 9.4(b), the less dense fluid forms on the top surface, and its tendency to

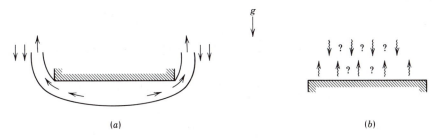

FIGURE 9.4. Heated plates facing (a) down; (b) up.

rise is inhibited by the denser, cooler, fluid above it that tends to move down. Thus, in this case an unstable situation is created, and the exact nature of the flow pattern is strongly influenced by other effects—such as possible small inclinations of the plate surface, the influence of fluid motion on the other side of the plate, etc. This same instability is observed in other free convective situations in which there is a heated surface with a large dimension placed at a significant angle to the gravity vector (90° in this case, but perhaps inclined at large angles in other cases). Of particular interest are the rectangular enclosures considered later.

Because of the above effects, reliable correlations for horizontal plates are difficult to find. Based on the work of McAdams (Ref. 7) and Goldstein et al. (Ref. 9), the following recommendations are made:

Characteristic length:

$$L_c = \frac{\text{plate area}}{\text{plate perimeter}}.$$

Heated plate facing up, cooled plate facing down:

$$\text{Nu}_{L_c} = 0.54\text{Ra}_{L_c}^{1/4}, \qquad 2.6 \times 10^4 < \text{Ra}_{L_c} < 10^7,$$

$$\text{Nu}_{L_c} = 0.15\text{Ra}_{L_c}^{1/3}, \qquad 10^7 < \text{Ra}_{L_c} < 3 \times 10^{10}. \tag{9.41}★$$

Heated plate facing down, cooled plate facing up:

$$\text{Nu}_{L_c} = 0.27\text{Ra}_{L_c}^{1/4}, \qquad 3 \times 10^5 < \text{Ra}_{L_c} < 3 \times 10^{10},$$

$$\begin{bmatrix} \text{properties, except } \beta, \text{ at } t_m \\ \beta \text{ at } t_m \text{ for liquids, } t_f \text{ for gases} \end{bmatrix}.$$

Free Convection past Inclined Plates

The availability of correlations for inclined plates is quite limited. For surfaces inclined at angles of 60°, *or less,* with the vertical, Churchill and Chu (Ref. 8) recommend that Eqs. (9.40) be used with g replaced with $g \cos\theta$ (θ = angle with vertical) for $\text{Ra}_L < 10^9$. For $\text{Ra}_L > 10^9$, this modification is not used.

For inclinations with the vertical exceeding about 60°, the situation becomes quite complex and the reader is referred to the works reported in Refs. 11 and 12.

Free Convection Around Long Horizontal Cylinders

For horizontal cylinders sufficiently long that end effects may be neglected, McAdams (Ref. 7) again suggested a correlation of the form

$$\text{Nu}_D = C\text{Ra}_D^m,$$

where the Nusselt and Rayleigh numbers are based on the cylinder diameter.

However, Churchill and Chu (Ref. 13), in another extensive analysis, recommend the following relation which is valid for a wide range of Ra_D:

$$Nu_D = \left\{0.60 + 0.387Ra_D^{1/6}\left[1 + \left(\frac{0.559}{Pr}\right)^{9/16}\right]^{-8/27}\right\}^2,$$

$$\begin{bmatrix} 0 < Pr < \infty \\ 10^{-5} < Ra_D < 10^{12} \\ \text{properties, except } \beta, \text{ at } t_m \\ \beta \text{ at } t_m \text{ for liquids, } t_f \text{ for gases} \end{bmatrix}. \qquad (9.42)\bigstar$$

Free Convection Around Spheres

For free convection about a sphere, Yuge (Ref. 14) determined the correlation:

$$Nu_D = 2 + 0.43Ra_D^{1/4},$$

$$\begin{bmatrix} Pr \approx 1 \\ 1 < Ra_D < 10^5 \\ \text{properties, except } \beta, \text{ at } t_m \\ \beta \text{ at } t_m \text{ for liquids, } t_f \text{ for gases} \end{bmatrix}. \qquad (9.43)\bigstar$$

The lead constant of 2 in Eq. (9.43) results from the fact that for no buoyant forces ($Ra_D \to 0$), all heat loss from the sphere is by conduction in the ambient fluid.

EXAMPLE 9.2

A horizontal 4-in schedule 40 pipe has a surface temperature of 200°F. It is placed in a quiescent fluid at 40°F. Find the free convective surface heat transfer coefficient if the fluid is (a) atmospheric air; (b) water.

Solution. Table B.1 gives $D = 4.500$ in.

(a) at $t_m = 120°F$, Table A.6 gives $\nu = 0.6902$ ft²/h, $k = 0.01602$ Btu/h-ft-°F, and $Pr = 0.709$. At $t_f = 40°F$, $\beta = 1/(40 + 460) = 1/(500°R)$. Thus,

$$Gr_D = \frac{(4.5/12)^3 \times 32.2 \times (3600)^2 \times (1/500)(200 - 40)}{(0.6902)^2}$$

$$= 1.48 \times 10^7,$$

$$Ra_D = 1.05 \times 10^7,$$

so that Eq. (9.42) yields

$$Nu_D = \frac{hD}{k} = 28.7,$$

$$h = 1.22 \text{ Btu/h-ft}^2\text{-°F} (6.93 \text{ W/m}^2\text{-°C}).$$

(b) For water at $t_m = 120°F$. Table A.3 gives $\nu = 0.02185$ ft²/h, $k = 0.3693$ Btu/h-ft-°F, Pr = 3.65, $\beta = 0.253 \times 10^{-3}$. Similar calculations as above give

$$Gr_D = 1.87 \times 10^9,$$

$$Ra_D = 6.81 \times 10^9,$$

$$Nu_D = 25.7,$$

$$h = 252 \text{ Btu/h-ft}^2\text{-°F} (1431 \text{ W/m}^2\text{-°C}).$$

Note that the free convective heat transfer coefficient in water is more than 200 times as great as that in air under similar conditions. ∎

Free Convection in Inclined Rectangular Enclosures

The configuration formed by a rectangular enclosure in which two of the opposite surfaces are maintained at different temperatures and the other two are adiabatic is one of considerable interest for heat exchange by free convection between the two surfaces of given temperature. The air space between the walls of a building or in the space between the layers of glass in double glazed windows are cases of this geometry with the heated surfaces in a vertical position and are of significant engineering interest. The space enclosed between the absorbing surface of a solar collecter and its glass, or plastic, cover plate is an example of this geometry in which the isothermal surfaces are inclined. Current wide interest in solar energy collection has led to recent extensive studies of free convective exchange in such enclosures.

Figure 9.5 illustrates the basic geometry and nomenclature to be considered. The hotter of the two surfaces is denoted by t_h and the cooler by t_c. These two surfaces are parallel and spaced a distance L apart. Their dimension normal to

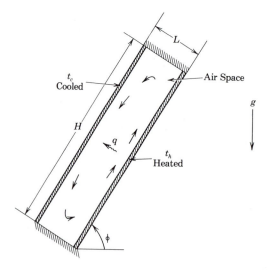

FIGURE 9.5. Free convection in an inclined rectangular enclosure.

L is given by H and the dimension normal to the plane of the paper is presumed to be large when compared with the gap spacing L. The other two surfaces forming the rectangular enclosure are taken to be adiabatic. An important geometric parameter of the enclosure is the *aspect ratio,* Ar:

$$Ar = \frac{H}{L}. \tag{9.44}$$

The gravity vector is taken as vertical and the inclination of the *heated surface* with the horizontal is denoted by the angle φ. Thus, for $\varphi = 90°$, the air gap is vertical; for $\varphi = 0°$ it is horizontal with the heated surface on the bottom—an unstable situation, as noted earlier. For $\varphi = 180°$ the layer is also horizontal, but is in the stable configuration with the cooler surface on the bottom.

The effective heat transfer coefficient for the enclosure is defined as the h for which the heat flow, per unit of area of the isothermal surfaces, q/A, is given by

$$\frac{q}{A} = h(t_h - t_c). \tag{9.45}$$

The Rayleigh and Nusselt numbers for this configuration are defined as

$$Ra_L = \frac{L^3 g \beta (t_h - t_c) Pr}{\nu^2}, \tag{9.46}$$

$$Nu_L = \frac{hL}{k}. \tag{9.47}$$

For given values of the geometric parameters and the surface temperatures, the value of h, or Nu, is strongly dependent on the inclination angle φ, which determines the form of the free convective flow pattern set up between the two surfaces. Based mainly on the studies of Hollands and his coworkers reported in Refs. 15 and 16 and on the work of Arnold et al., in Ref. 17, the following recommendations are made for various ranges of the inclination φ. In all instances the fluid properties are evaluated at the average temperature:

$$t_{av} = \frac{t_h + t_c}{2}; \tag{9.48}$$

$\varphi = 90°$: $\quad Nu_{90} = $ largest of: Nu_1, Nu_2, Nu_3

$$Nu_1 = 0.0605 Ra_L^{1/3},$$

$$Nu_2 = \left\{ 1 + \left[\frac{0.104 Ra_L^{0.293}}{1 + (6310/Ra_L)^{1.36}} \right]^3 \right\}^{1/3},$$

$$Nu_3 = 0.242 \left(\frac{Ra_L}{Ar} \right)^{0.272},$$

$$\left[\begin{matrix} 5 < Ar < 110 \\ 10^2 < Ra_L < 2 \times 10^7 \end{matrix} \right].$$

$$(9.49)\bigstar$$

$\varphi = 60°$: $Nu_{60} =$ largest of: Nu_1, Nu_2

$$Nu_1 = \left[1 + \left(\frac{0.0936 Ra_L^{0.314}}{1 + G} \right)^7 \right]^{1/7},$$

$$G = \frac{0.5}{[1 + (Ra_L/3160)^{20.6}]^{0.1}},$$

$$Nu_2 = \left(\frac{0.104 + 0.175}{Ar} \right) Ra_L^{0.283}, \qquad (9.50)\bigstar$$

$$\begin{bmatrix} 5 < Ar < 110 \\ 10^2 < Ra_L < 2 \times 10^7 \end{bmatrix}.$$

$90° > \varphi > 60°$:   Linear interpolation between Nu_{90} and Nu_{60}. $(9.51)\bigstar$

$60° > \varphi \geq 0°$: $Nu = 1 + 1.44[1 - G]^{\bullet}[1 - G(\sin 1.8\varphi)^{1/6}]$

$$+ [0.664 G^{-1/3} - 1]^{\bullet},$$

$$G = \frac{1708}{Ra_L} \cos \varphi, \qquad\qquad (9.52)\bigstar$$

Notation $[\ ]^{\bullet}$ means that the bracketed term should be made zero if negative,

$$\begin{bmatrix} Ar > 12 \\ 0 < Ra_L < 10^5 \end{bmatrix}.$$

$90° < \varphi < 180°$: $Nu = 1 + [Nu_{90} - 1] \sin \varphi,$

$$\begin{bmatrix} \text{all } Ar \\ 10^3 < Ra_L < 10^6 \end{bmatrix}. \qquad (9.53)\bigstar$$

9.5

MIXED FREE AND FORCED CONVECTION

The analyses and correlations of Chapters 7 and 8 were restricted to "pure forced" convection by neglecting buoyant forces in the governing equations. The analyses and correlations presented thus far in this chapter have been limited to "pure free" convection by specifying a zero fluid velocity at points far removed from the surface in question. There are, however, a number of practical applications in which neither of these conditions are fully met and one must account for mixed flow state in which both buoyant and inertia forces are significant. For example, forced flow in a very large diameter horizontal gas pipelines may simultaneously experience vertical free convection effects if significant vertical temperature gradients exist.

Referring to the similarity analysis of Sec. 6.7, the fundamental solution for the dimensionless temperature distribution given in Eq. (6.18) accounted for the effects of buoyancy with the term

$$\frac{lg\beta \, \Delta t}{U^2}.$$

This term originated [refer to the equations preceding Eq. (6.18)] by dividing the coefficient of the buoyancy term in the equation of motion by the coefficient of the inertia term. Thus, the above dimensionless group represents a measure of the ratio of buoyant forces to inertia forces. This grouping was then subsequently recast in terms of the Grashof and Reynolds numbers:

$$\frac{\text{buoyancy}}{\text{inertia}} \approx \frac{lg\beta \, \Delta t}{U^2} = \frac{l^3 g\beta \, \Delta t}{\nu^2}\left(\frac{\nu}{Ul}\right)^2$$

$$\approx \frac{\text{Gr}}{\text{Re}^2}.$$

Hence, the magnitude of the ratio of buoyant forces to the inertia forces is measured by Gr/Re^2. This parameter is used to delineate the free, forced, and mixed convection regimes:

$$\frac{\text{Gr}}{\text{Re}^2} \ll 1: \quad \text{pure forced convection.}$$

$$\frac{\text{Gr}}{\text{Re}^2} \approx 1: \quad \text{mixed convection.}$$

$$\frac{\text{Gr}}{\text{Re}^2} \gg 1: \quad \text{pure free convection.}$$

As might be expected, analytical or empirical correlations for the mixed regime when $\text{Gr}/\text{Re}^2 \approx 1$ are sparse.

Mixed Convection past Vertical Plane Surfaces

Convective heat transfer in the mixed regime for forced flow past a flat surface for which buoyant forces act in the same direction as the free stream velocity, as suggested in Fig. 9.6, has been examined in a numerical analyis by Lloyd and Sparrow (Ref. 18). Their results are compared in Fig. 9.6, for a wide range of Pr, with the solutions for pure free and forced convection. As would be expected, the mixed flow results approach those for forced convection as $\text{Gr}_x/\text{Re}_x^2$ becomes small and those for free convection as $\text{Gr}_x/\text{Re}_x^2$ becomes large. The effect of buoyancy on the deviation of mixed convection from pure forced convection is more pronounced for fluids of low Prandtl numbers. The tabulation of Table 9.1 shows the value of $\text{Gr}_x/\text{Re}_x^2$ below which the use of the pure forced convection result of Eq. (8.1) may be used with less than 5% deviation. One may use Fig. 9.6 to similarly estimate the value of $\text{Gr}_x/\text{Re}_x^2$ above which pure free convection relations may be used.

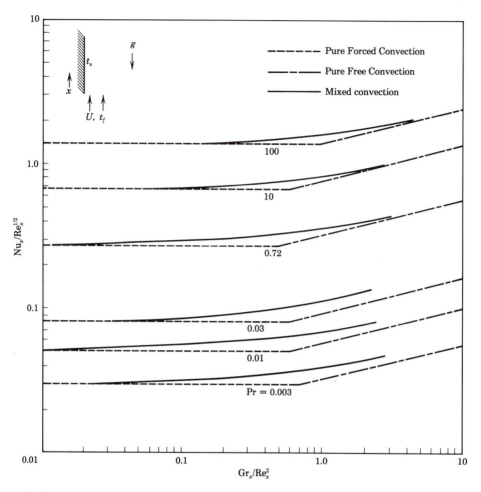

FIGURE 9.6. Combined free and forced convection past a vertical flat surface. (From J. R. Lloyd and E. M. Sparrow, *Int. J. Heat Mass Transfer*, Vol. 13, 1970, p. 434. Used by permission.)

Mixed Convection in Horizontal Pipes

Forced flow in horizontal pipes of large diameter may experience buoyancy effects when $Gr_D/Re_x^2 \approx 1$. Metais and Eckert (Ref. 19) examined this case and delineated the regions in which free convection contributions are significant. Their study indicated that laminar flow may be expected if $Re_D < 2000$ when

TABLE 9.1 Maximum Values of Gr_x/Re_x^2 for Neglection of Buoyancy on Forced Convection past a Vertical Plate with Less Than 5% Error

Pr	100	10	0.72	0.03–0.003
Gr_x/Re_x^2	0.24	0.13	0.08	0.056–0.05

$\text{Ra}_D(D/L) < 2 \times 10^4$ or if $\dot{\text{Re}}_D < 800$ when $\text{Ra}_D(D/L) > 2 \times 10^4$. With these criteria for the presence of laminar or turbulent flow, the following recommendations are made for mixed convection in horizontal pipes (Refs. 19 and 20):

Laminar:

$$\text{Nu}_D = 1.75 \left\{ \left[\left(\frac{D}{L}\right)\text{Re}_D\text{Pr} \right] + 0.012 \left[\left(\frac{D}{L}\right)\text{Re}_D\text{Pr}\text{Gr}_D^{1/3} \right]^{4/3} \right\}^{1/3} \left(\frac{\mu}{\mu_s}\right)^{0.14},$$

$$(9.54)\bigstar$$

$$\begin{bmatrix} \text{Re}_D < 2000 \text{ when } \text{Ra}_D(D/L) < 2 \times 10^4 \\ \text{Re}_D < 800 \text{ when } \text{Ra}_D(D/L) > 2 \times 10^4 \\ \text{properties, except } \mu_s, \text{ at mean } t_b \\ \mu_s \text{ at } t_s \end{bmatrix}.$$

Turbulent:

$$\text{Nu}_D = 4.69\text{Re}_D^{0.27}\text{Pr}^{0.21}\text{Gr}_D^{0.07}\left(\frac{D}{L}\right)^{0.36},$$

$$(9.55)\bigstar$$

$$\begin{bmatrix} \text{Re}_D > 2000 \text{ when } \text{Ra}_D(D/L) < 2 \times 10^4 \\ \text{Re}_D > 800 \text{ when } \text{Ra}_D(D/L) > 2 \times 10^4 \\ \text{properties at mean } t_b \end{bmatrix}.$$

EXAMPLE 9.3

If the oven door described in Example 9.1 is subjected to an upward forced flow of air, find the minimum free stream air velocity for which free convection effects may be neglected.

Solution.   The calculations of Example 9.1 showed

$$\text{Gr}_L = 1.30 \times 10^9.$$

For the air temperature in question, $\text{Pr} = 0.704$. At this value of Pr, Table 9.1 shows that less than 5% error will result by treating the flow as pure forced convection if

$$\frac{\text{Gr}_L}{\text{Re}_L^2} < 0.08,$$

or

$$\text{Re}_L > 1.27 \times 10^5.$$

Thus,

$$\frac{U \times 0.5}{24.1 \times 10^{-6}} > 1.27 \times 10^5$$

$$U > 6.1 \text{ m/s.} \qquad \blacksquare$$

REFERENCES

1. GEBHART, B., *Heat Transfer,* 2nd ed., New York, McGraw-Hill, 1971.
2. SCHMIDT, E., and W. BECKMANN, "Das Temperatur und Geschwindigkeitsfeld vor einer Wärmer abgebenden senkrechten Platte bei näturicher Konvection," *Tech. Mech. Thermodynamik,* Vol. 1, 1930, p. 341.
3. OSTRACH, S., "An Analysis of Laminar Free Convection Flow and Heat Transfer About a Flat Plate Parallel to the Generating Body Force," *NACA Rep. 1111,* Washington, D.C., 1953.
4. EDE, A. J., "Advances in Free Convection," *Advances in Heat Transfer,* Vol. 4, New York, Academic Press, 1967.
5. ECKERT, E. R. G., *Introduction to Heat and Mass Transfer,* New York, McGraw-Hill, 1963.
6. ECKERT, E. R. G., and T. W. JACKSON, "An Analysis of Turbulent Free Convection Boundary Layer on a Flat Plate," *NACA Tech. Note 2207,* Washington, D.C., 1950.
7. MCADAMS, W. H., *Heat Transmission,* 3rd ed., New York, McGraw-Hill, 1954.
8. CHURCHILL, S. W., and H. H. S. CHU, "Correlating Equations for Laminar and Turbulent Free Convection from a Vertical Plate," *Int. J. Heat Mass Transfer,* Vol. 18, 1975, p. 1323.
9. GOLDSTEIN, R. J., E. M. SPARROW, and D. C. JONES, "Natural Convection Mass Transfer Adjacent to the Horizontal Plates," *Int. J. Heat Mass Transfer,* Vol. 16, 1973, p. 1025.
10. LLOYD, J. R., and W. R. MORAN, "Natural Convection Adjacent to Horizontal Surfaces of Various Planforms." *ASME Paper 74-WA/HT-66,* 1974.
11. VLIET, G. C., "Natural Convection Local Heat Transfer on Constant-Heat-Flux Inclined Surfaces." *J. Heat Transfer, Trans. ASME,* Vol. 91, 1969. p. 511.
12. FUJII, T., and H. IMURA, "Natural Convection Heat Transfer from a Plate with Arbitrary Inclination," *Int. J. Heat Mass Transfer,* Vol. 15, 1972, p. 755.
13. CHURCHILL, S. W., and H. H. S. CHU, "Correlating Equations for Laminar and Turbulent Free Convection from a Horizontal Cylinder," *Int. J. Heat Mass Transfer,* Vol. 18, 1975, p. 1049.
14. YUGE, T., "Experiments on Heat Transfer from Spheres Including Combined Natural and Forced Convection," *J. Heat Transfer, Trans. ASME,* Vol. 82, 1960, p. 214.
15. ELSHERBING, S. M., G. D. RAITHBY, and K. G. T. HOLLANDS, "Heat Transfer by Natural Convection Across Vertical and Inclined Air Layers," *J. Heat Transfer, Trans, ASME,* Vol. 104, 1982, p. 96.
16. HOLLANDS, K. G. T., T. E. UNNY, G. D. RAITHBY, and L. KONICEK, "Free Convection Heat Transfer Across Inclined Air Layers," *J Heat Transfer, Trans. ASME,* Vol. 98, 1976, p. 189.
17. ARNOLD, J. N., I. CATTON, and D. K. EDWARDS, "Experimental Investigation of Natural Convection in Inclined Rectangular Regions of Differing Aspect Ratios," *J. Heat Transfer, Trans. ASME,* Vol. 98, 1976, p. 67.
18. LLOYD, J. R., and E. M. SPARROW, "Combined Forced and Free Convection Flow on Vertical Surfaces," *Int. J. Heat Mass Transfer,* Vol. 13, 1970, p. 434.
19. METAIS, B., and E. R. G. ECKERT, "Forced, Mixed and Free Convection Regimes," *J. Heat Transfer, Trans. ASME,* Vol. 86, 1964, p. 295.
20. BROWN, C. K., and W. H. GAUVIN, "Combined Free and Forced Convection, Parts I and II," *Can. J. Chem. Eng.,* Vol. 43, 1965, pp. 306, 313.

PROBLEMS

9.1 Verify the albegra leading from Eqs. (9.3) and (9.4) to Eq. (9.15).

9.2 Verify the algebra leading from Eqs. (9.6) and (9.7) to Eqs. (9.30) and (9.31).

9.3 Verify the algebra leading from Eqs. (9.30) and (9.31) to Eqs. (9.36) and (9.37).

9.4 A vertical plate 20 cm long and 10 cm wide is placed in still atmospheric air at 40°C. The plate is maintained at 90°C. Find the average heat transfer coefficient at the plate surface by use of (a) Eq. (9.22) or (9.23), (b) Eq. (9.37), and (c) Eq. (9.40).

9.5 Repeat Prob. 9.4 if the fluid is helium.

9.6 If the plate of Prob. 9.4 is placed vertically in water at 40°C, find the average heat transfer coefficient by use of (a) Eq. (9.38), and (b) Eq. (9.40).

9.7 For the plate of Prob. 9.4, find the local heat transfer coefficient, the boundary layer thickness, the maximim velocity, and the distance away from the surface at which the maximum velocity occurs, all at a location (a) 10 cm from the leading edge, and (b) 20 cm from the leading edge.

9.8 A plate 0.3 m high is maintained at 150°C and is placed vertically in air at 40°C. Find the average free convection heat transfer coefficient on the plate surface if the air pressure is (a) 0.1 atm, (b) 1.0 atm, (c) 5.0 atm, and (d) 10.0 atm.

9.9 Estimate the coefficient of free convection between a vertical plate 0.6 m high maintained at 120°C in nitrogen at −15°C, 70 atm.

9.10 A cylinder 7.5 cm in diameter and 2.0 m long is placed in atmospheric air at 40°C. The surface of the cylinder is maintained at 90°C. Find the total heat loss from the cylinder by free convection if the cylinder is (a) horizontal, and (b) vertical.

9.11 A circular cylinder 6 ft long and 6 in. in diameter is maintained with a surface temperature of 450°F. It is placed in atmospheric air at 75°F. Estimate the total heat loss from the cylinder if it is (a) horizontal, and (b) vertical.

9.12 A circular horizontal electrical heater plate is 10 cm in diameter. Its upper and lower surfaces are maintained at 200°C in a tank containing lubricating oil at 20°C. Calculate the total required heat input to the plate.

9.13 Repeat Prob. 9.12 if the plate is placed in atmospheric air at 20°C.

9.14 A plate 25 cm long and 12 cm wide is maintained at 90°C in atmospheric air at 40°C. Find the total heat transferred from both sides of the plate if it is placed so that the 25-cm dimension (a) is vertical or (b) makes an angle of 25° with the vertical.

9.15   A plate 25 cm long and 12 cm wide is maintained at 90°C in water at 40°C. Find the total heat transferred from both sides of the plate if it is placed so that the 25-cm dimension (a) is vertical or (b) makes an angle of 25° with the vertical.

9.16   Find the average coefficient of free convective heat transfer between a horizontal 14-cm diameter cylinder with a surface temperature of 90°C and atmospheric air at 30°C.

9.17   Find the average coefficient of free convective heat transfer between a vertical plate, the length of which equals the half-circumference of the cylinder in Prob. 9.16. The surface and air temperature are the same as in Prob. 9.16. Discuss possible reasons for the closeness between the answers of this problem and Prob. 9.16.

9.18   A horizontal cylinder with a surface temperature of 200°C is placed in atmospheric air at 40°C. Find the free convective heat transfer coefficient for diameters of (a) 0.025 cm, (b) 0.25 cm, (c) 2.5 cm, and (d) 25 cm.

9.19   A horizontal 12-in. schedule 40 steam pipe passes through a room in which the air is at 100°F. The pipe surface is at 600°F. What is the heat loss by free convection per foot of pipe length?

9.20   Find the coefficient of free convective heat transfer between a horizontal 10-cm pipe with a surface temperature of 40°C and (a) methane at $-20°C$, 350 kN/m², and (b) water at 90°C.

9.21   An electrical heating element consists of a wire 0.125 cm in diameter. It is immersed horizontally in a bath of lubricating oil at 40°C and the wire surface is maintained at 150°C. Find the free convection heat transfer coefficient.

9.22   A horizontal electrical wire 0.475 cm in diameter dissipates 13 W of heat per meter of length to surrounding atmospheric air at 25°C. Estimate the surface temperature of the wire.

9.23   A horizontal cylinder of unknown diameter is maintained with a surface temperature of 45°C in atmospheric air at 30°C. The heat loss, per unit surface of area, is found to be 158 W/m². What will be the heat loss if the surface temperature is raised to 100°C?

9.24   The heat loss by free convection from a horizontal, electrically heated wire to ambient atmospheric air at 50°F is known to be 350 Btu/h-ft² when the wire temperature is 150°F. What is the rate of heat loss, per unit surface area, when the wire temperature is raised to 300°F?

9.25 A straight horizontal fin consists of a circular bar 0.6 cm in diameter and 10 cm long. The fin base is at 150°C, and the fin extends into atmospheric air at 50°C. Applying the results of this chapter and Sec. 3.12, estimate the heat loss from the fin to the air if the bar is made of 0.5% carbon steel.

9.26 A cast iron rod 0.75 in. in diameter and 10 in. long extends horizontally into atmospheric air at 90°F from a heat source at 300°F. Estimate the rate of heat exchange between the rod and the air.

9.27 A sphere of 2.5 cm diameter has an internal heat source which maintains its surface at 90°C. If the sphere exchanges heat by free convection with a fluid at 20°C, find the rate of heat flow, in watts, if the fluid is (a) atmospheric air, (b) water, and (c) ethylene glycol. Assume that Eq. (9.43) holds.

9.28 A double glazed window is 1.25 m high and 0.8 m wide. The two layers of glass are 6 cm apart. If the two glass layers are at 20°C and -10°C, what is the heat loss by free convection through the air space?

9.29 Repeat Prob. 9.28 if the air space thickness is changed to (a) 3 cm, and (b) 8 cm.

9.30 A solar collector panel, 2 m × 1 m, consists of a lower absorber plate at 70°C with a glass cover plate at 30°C. The air space between the two surfaces is 5 cm thick. If the air in this space is at atmospheric pressure, find the heat loss between the two surfaces by free convection if the 2-m dimension of the hotter surface is inclined at (a) 60° or (b) 45° with the horizontal.

9.31 Repeat Prob. 9.30 if the air space between the two surfaces is evacuated to a pressure of 0.25 atm.

9.32 A rectangular space of the geometry shown in Fig. 9.5 contains air at one atm pressure. The following conditions exist: $H = 1$ m, $L = 5$ cm, $t_h = 80$°C, $t_c = 40$°C. Calculate the heat transfer coefficient due to free convection for values of φ ranging from 0 to 180°, by increments of 20°.

9.33 Atmospheric air at 20°C flows upward along a flat surface 0.4 m high maintained at 50°C. What is the minimum flow velocity for which the effect of buoyancy will be less than 5%?

CHAPTER 10

Heat Transfer in Condensing and Boiling

INTRODUCTORY REMARKS

The convective processes discussed in Chapters 6 through 9 have all been limited to cases in which the fluid medium remained in a single phase—liquid or gas. A great number of practical engineering applications exists in which a phase change occurs simultaneously with the heat transfer process. The importance of the processes of condensing and boiling to the production of power from vapor cycles, the production of refrigeration, the design of petrochemical processes, etc., is obvious.

While the solid–gas phase change process is of current importance in applications such as the thermal protection of spacecraft and missiles, and the solid–liquid phase change process is important for thermal storage considerations, these applications will not be treated here and only the vapor–liquid phase change processes of condensing and boiling will be discussed. Since fluid motion is still involved, condensing and boiling are classified as convective mechanisms; however, there are profound differences between these mechanisms and single phase convective heat transfer. There are significant differences between the various fluid properties in the two phases (such as density, viscosity, conductivity, specific heat) that influence the fluid mechanical processes markedly. At the same time there is a release or consumption of latent heat that also has profound influence on the observed heat transfer rates during phase change, so that heat transfer coefficients that are orders of magnitude greater than those in single phase convection are commonly observed.

First, the process of the condensation of a vapor will be discussed. Subsequently, the process of boiling will be considered. Because of the great complexity of these processes, much of these considerations will be descriptive in nature.

10.2

GENERAL COMMENTS REGARDING CONDENSATION

The process of condensation of a vapor is usually accomplished by allowing it to come in contact with a surface the temperature of which is maintained at a value lower than the saturation temperature of the vapor for the pressure at which it exists. The removal of thermal energy from the vapor causes it to release its latent heat of vaporization and, hence, to condense onto the surface.

The appearance of the liquid phase on the cooling surface, either in the form of individual drops or in the form of a continuous film, offers a greater thermal resistance to the further removal of heat from the remaining vapor. In most cases of condensation, the condensate is removed by the action of gravity. Naturally, then, the rate of removal of condensate, and with it the rate of heat removal from the vapor, is greater for vertically placed surfaces than for horizontal surfaces. In order to avoid the accumulation of large amounts of liquid at the lower end of a condensing surface it should be short; hence, horizontal cylinders are particularly applicable for such purposes. Thus, most condensing equipment consists of assemblies of horizontal tubes around which the vapor to be condensed is allowed to flow. The cool temperature of the outer tube surface is maintained by circulating a colder medium, usually water, through the inside of the tube.

Two general types of condensation have been observed to occur in practice. The first, called *film condensation,* usually occurs when a vapor, relatively free of impurities, is allowed to condense on a clean surface. Under these conditions, it is found that the condensate will appear as a continuous film all over the surface and it will flow off the surface under the action of gravity.

The second type of condensation, less frequently observed, is known as *dropwise condensation,* and has been observed to occur either on highly polished surfaces or on surfaces contaminated with certain fatty acids. In this case the condensate is found to appear in the form of individual drops. These drops increase in size and combine with one another until their size is great enough that their weight causes them to run off the surface, leaving the surface exposed for the formation of a new drop.

Since there will be much less of a liquid barrier between the condensing surface and the vapor, it should not be surprising to know that heat transfer rates are five to ten times greater for dropwise condensation than for film condensation. Hence, it would appear that dropwise condensation would be preferred for commercial applications. Some of the details of dropwise condensation are discussed in a later section, but it is sufficient to state here that it is a difficult mode to maintain in practice so that conservative design calculations are usually made on the basis of presuming that film condensation will exist. This presumption has the advantage that film condensation is much easier to subject to analysis, as described next.

ANALYSIS OF LAMINAR FILM CONDENSATION ON VERTICAL SURFACES

Figure 10.1 depicts the physical situation for film condensation on a vertical flat surface. The surface is maintained at some temperature, t_s, which is lower than the saturation temperature of the neighboring vapor. Heat removal from the vapor causes condensation on the surface and a film of condensate forms and flows down the surface under the influence of gravity. The film is presumed to originate at the top of the surface and increases in thickness as it flows down due to the accumulation of condensate. The film thickness, $\delta(x)$, is thus a function of the distance along the surface, x. (The dimension y is measured normal to the surface.) If one can determine the shape of the velocity profile and how δ varies with x, then the mass flow rate of condensate may be found as a function

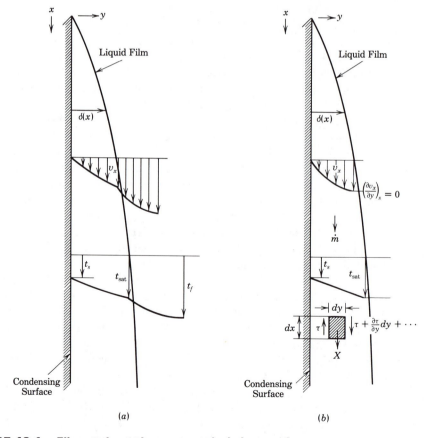

FIGURE 10.1. Film condensation on a vertical plane surface.

of position. This flow rate is, in turn, related to the amount of vapor condensed, so that one may eventually find the heat transfer coefficient as a function of x.

Figure 10.1(a) depicts the temperature and velocity profile in the film for a general situation in which the ambient vapor may be superheated. Thus a temperature gradient may exist in the vapor near the liquid–vapor interface. The interface is at the saturation temperature, t_{sat}, and then subcooling of the liquid film occurs to the surface value t_s. Likewise, if the ambient vapor is moving relative to the liquid at the interface, some interfacial shear may exist as suggested by the velocity gradient shown at $y = \delta$.

Nusselt (Ref. 1) simplified the problem by invoking a number of assumptions. These assumptions were: (1) the film flow is laminar; (2) the vapor is not superheated; (3) the film grows in thickness so slowly that the transfer of heat through it is purely by conduction, giving a linear temperature distribution; (4) there is no interfacial shear, so that $\partial v_x/\partial y = 0$ at $y = \delta$; and (5) the film moves slowly enough to neglect inertia forces within it.

Figure 10.1(b) illustrates some of the assumptions. The neglection of inertia forces means that a force balance on an element of the film, dx by dy, gives (for a unit depth into the plane of the paper)

$$-\left(\frac{\partial \tau}{\partial y}\, dy\right) dx = X(dx\, dy).$$

In the above, τ is the shear stress shown acting on the dx sides of the element and X is the body force, per unit volume, acting on the element. If the flow in the film is laminar then $\tau = \mu(\partial v_x/\partial y)$, v_x being the flow velocity parallel to the surface. The body force is that of gravity acting on the element, less the buoyancy resulting from the displacement of the vapor. Thus,

$$X = g(\rho_l - \rho_v),$$

so that the above force balance becomes

$$\frac{\partial^2 v_x}{\partial y^2} = -\frac{(\rho_l - \rho_v)g}{\mu_l}. \tag{10.1}$$

In Eq. (10.1), and all subsequent equations in this chapter, the subscript l refers to properties of the liquid component and the subscript v to those of the vapor component.

Integration of Eq. (10.1) twice and application of the conditions that $v_x = 0$ at $y = 0$ and $\partial v_x/\partial y = 0$ at $y = \delta$ yield

$$v_x = \frac{g(\rho_l - \rho_v)\delta^2}{\mu_l}\left[\frac{y}{\delta} - \frac{1}{2}\left(\frac{y}{\delta}\right)^2\right]. \tag{10.2}$$

The mass flow rate in the film $\dot{m}$, at any section x, is then readily found (again assuming unit depth):

$$\dot{m}(x) = \int_0^\delta \rho_l v_x\, dy,$$

$$= \frac{g\rho_l(\rho_l - \rho_v)\delta^3}{3\mu_l}. \tag{10.3}$$

Equation (10.3) interrelates the film thickness, δ, and the mass flow rate, $\dot{m}$, but both are still unknown functions of x. These may be related by noting that the heat flux into the wall may be written in terms of each. By use of Fourier's law and the assumption of a linear temperature distribution in the film, one has for the heat flux:

$$\frac{q}{A} = -\frac{k_l(t_s - t_{sat})}{\delta},$$

$$= k_l \frac{\Delta t}{\delta},$$

$$\tag{10.4}$$

where $\Delta t = t_{sat} - t_s$ has been introduced. Also, noting that the heat flux into the wall over a distance dx must result in the condensation of an amount of vapor equal to $d\dot{m}$, one may also state, for unit depth, that

$$\frac{q}{A} = h_{fg} \frac{d\dot{m}}{dx} \tag{10.5}$$

in which h_{fg} represents the latent heat of vaporization in accord with usual thermodynamics notation.

Equating Eqs. (10.4) and (10.5) and using Eq. (10.3) gives

$$k_l \frac{\Delta t}{\delta} = h_{fg} \frac{d\dot{m}}{dx} = h_{fg} \frac{g\rho_l(\rho_l - \rho_v)\delta^2}{\mu_l} \frac{d\delta}{dx},$$

$$\delta^3 \frac{d\delta}{dx} = \frac{\mu_l k_l \, \Delta t}{g\rho_l(\rho_l - \rho_v)h_{fg}}.$$

Integration of this latter equation with the condition that $\delta = 0$ at $x = 0$, gives the following for the dependence of the film thickness on the distance along the surface:

$$\delta = \left[\frac{4\mu_l k_l \, \Delta t \, x}{g\rho_l(\rho_l - \rho_v)h_{fg}} \right]^{1/4}. \tag{10.6}$$

This result may be substituted into Eq. (10.4) to find the heat flux, q/A; however, it is customary to define the local heat transfer coefficient so that

$$\frac{q}{A} = h_x(t_{sat} - t_s). \tag{10.7}$$

Thus, Eqs. (10.4), (10.6), and (10.7) give

$$h_x = \left[\frac{g\rho_l(\rho_l - \rho_v)h_{fg}k_l^3}{4\mu_l x \, \Delta t} \right]^{1/4} \tag{10.8}$$

In terms of the usual local Nusselt number, the above is

$$Nu_x = \frac{h_x x}{k_l} = \frac{1}{\sqrt{2}} \left[\frac{g\rho_l(\rho_l - \rho_v)h_{fg}x^3}{\mu_l k_l \, \Delta t} \right]^{1/4} \tag{10.9}$$

Integrating for an average h and an average Nusselt number on a plate of total length L gives

$$h = \frac{4}{3}\left[\frac{g\rho_l(\rho_l - \rho_v)h_{fg}k_l^3}{4\mu_l L \, \Delta t}\right]^{1/4},$$

$$\mathrm{Nu}_L = \frac{hL}{k_l} = 0.943\left[\frac{g\rho_l(\rho_l - \rho_v)h_{fg}L^3}{\mu_l k_l \, \Delta t}\right]^{1/4}. \tag{10.10}$$

In spite of the several simplifying assumptions made, Nusselt's analysis resulting in Eqs. (10.9) and (10.10) yields rather good results when compared with experiments. More refined analyses of laminar film condensation are summarized in Ref. 2, wherein the effects of some of these assumptions are investigated. The most significant refinements take into account the energy removed in subcooling the film temperature profiles. Sparrow and Gregg (Ref. 3), in a detailed analysis of laminar film condensation on a vertical surface, show that Nusselt's analysis is quite accurate for fluids with Prandtl numbers greater than those for liquid metals as long as the Jakob number is small. The Jakob number, a measure of the relative magnitude of the film subcooling, is defined as

$$\text{Jakob number} = \mathrm{Ja} = \frac{c_{pl}\,\Delta t}{h_{fg}}. \tag{10.11}$$

Thus, for modest amounts of subcooling, Nusselt's equations are satisfactory for a majority of practical applications.

Adjustments to Nusselt's equations to account for the effects of values of Ja other than zero, and for other effects not discussed here, will be summarized in the next section under the heading of "working correlations." That section will also present recommended relations for other geometries, many of which are the results of analyses and some the results of experiment.

10.4

WORKING CORRELATIONS FOR FILM CONDENSATION

This section presents a summary of analytical and empirical relations recommended for engineering calculations. In these correlations, the mean film temperature is defined as $t_m = (t_{sat} + t_s)/2$, and usually all liquid properties are evaluated at this temperature except h_{fg}, which should be found at t_{sat}.

The Effect of Turbulence

The analysis presented in Sec. 10.3 presumed the existence of laminar flow in the liquid film. This assumption is generally true for short surfaces for which an excessively thick layer of liquid does not build up. However, in many practical applications, the surface may be long enough that the film becomes sufficiently thick to cause a transition to turbulent flow. A Reynolds number must be defined for the film in order to characterize the transition to turbulence.

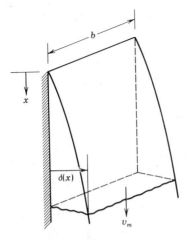

FIGURE 10.2

Referring to Fig. 10.2, a local Reynolds number in the condensate film may be defined as

$$\mathrm{Re} = \frac{v_m D_h \rho_l}{\mu_l},$$

in which v_m is the mean velocity in the film at some location x, and D_h is the hydraulic diameter defined in Eq. (8.40). For a plate of width b, as in Fig. 10.2, the flow area at the location in question is $b\delta$, δ being the local film thickness, and the wetted perimeter is b. Thus the hydraulic diameter is $4b\delta/b = 4\delta$, and the local Reynolds number is

$$\mathrm{Re} = \frac{4v_m \rho_l \delta}{\mu_l}.$$

The product $v_m \rho_l \delta$ is recognized as the mass flow rate in the film, per unit of film width, and is given the symbol $\Gamma = v_m \rho_l \delta$, so that the Reynolds number customarily used in film condensation work is

$$\mathrm{Re}_\Gamma = \frac{4\Gamma}{\mu_l}. \tag{10.12}$$

For a flat plate, Γ is identical to the quantity $\dot{m}$ discussed in Sec. 10.3; for condensation on a vertical cylinder, Γ is the mass flow condensed divided by the circumference πD; etc. The quantity Γ, and hence Re_Γ, is obviously a function of position along the surface in the direction of the flow of the film; hence Γ and Re_Γ have their greatest values at the lower end of the condensing surface. Since all the mass condensed must pass the lower end, one may also write, for a total plate length L with an average heat transfer coefficient h,

$$\mathrm{Re}_{\Gamma_{max}} = \frac{4\Gamma_{max}}{\mu_l},$$

$$= \frac{4hL\,\Delta t}{h_{fg}\mu_l}. \tag{10.13}$$

Experiment indicates that the critical value of Re_Γ for film condensation is about 1800, laminar films occurring below this value and turbulent ones above it.

Film Condensation on Vertical or Inclined Plates and on Vertical Cylinders

For laminar film condensation on vertical plates, McAdams (Ref. 4) recommends that Nusselt's equations, Eq. (10.10), be employed with the constant 0.943 empirically adjusted to 1.13. Further, Rohsenow (Ref. 5), suggests that subcooling of the film be accounted for by replacing the latent heat of vaporization with

$$h'_{fg} = h_{fg}(1 + 0.68Ja),$$

where Ja is the Jakob number defined in Eq. (10.11). Finally, it is found that the Nusselt relation holds for film condensation on the upper side of a plate inclined at an angle φ with the horizontal if g is replaced with $g \sin \varphi$. Thus, the recommended relation for laminar film condensation on vertical or inclined plates is

$$Nu_L = 1.13 \left[\frac{g'\rho_l(\rho_l - \rho_v)h'_{fg}L^3}{\mu_l k_l \,\Delta t} \right]^{1/4},$$

$$\begin{bmatrix} Pr > 0.5 \\ Re_{\Gamma_{max}} < 1800 \\ Ja < 1.0 \\ g' = g \sin \varphi \\ h'_{fg} = h_{fg}(1 + 0.68Ja) \\ \text{liquid properties at } t_m \\ h_{fg} \text{ at } t_{sat} \end{bmatrix}.$$

(10.14)★

When turbulence occurs in the film, McAdams further suggests the use of the following empirical relation:

$$Nu_L = 0.0077 \left[\frac{g'\rho_l(\rho_l - \rho_v)L^3}{\mu_l^2} \right]^{1/3} Re_{\Gamma_{max}}^{0.4},$$

$$\begin{bmatrix} Pr > 0.5 \\ Re_{\Gamma_{max}} > 1800 \\ g' = g \sin \varphi \\ \text{liquid properties at } t_m \end{bmatrix}.$$

(10.15)★

For most cases of practical interest, the relative magnitudes of the liquid and vapor densities enables the use of the approximation

$$\rho_l(\rho_l - \rho_v) \approx \rho_l^2$$

in Eqs. (10.14) and (10.15) without much error.

Equations (10.14) and (10.15) must be used with caution for plates of small inclination since an obviously incorrect result occurs for $\varphi \to 0$. Horizontal plates are treated later.

Equations (10.14) and (10.15) may be used for *vertical* cylinders as long as the film thickness is quite small compared with the cylinder radius, $\delta \ll R$. These relations *may not* be used for inclined cylinders. Horizontal cylinders are treated later.

EXAMPLE 10.1

Saturated steam at 70 kN/m² (10 psia) condenses on a vertical surface 0.2 m wide and maintained at 40°C. Find the mass rate at which steam is condensed on one side of the surface if it is (a) 1 m high; (b) 3 m high.

Solution. At 70 kN/m², the Steam Tables give $t_{\text{sat}} = 89.95°C$, $h_{\text{fg}} = 2283.3$ kJ/kg. Thus, from the given data, $\Delta t = 49.95°C$, $t_m = 65°C$. At t_m, Table A.3 gives

$$\rho_l = 980.5 \text{ kg/m}^3, \qquad \mu_l = 0.4338 \times 10^{-3} \text{ kg/m-s},$$

$$k_l = 0.6553 \text{ W/m-°C}, \qquad c_{pl} = 4.187 \text{ kJ/kg-°C}.$$

(a) For $L = 1$ m, assume that the flow is laminar and verify this assumption later. Thus, using $\rho_l (\rho_l - \rho_v) \approx \rho_l^2$, Eq. (10.14) gives

$$\text{Ja} = \frac{4.187 \times 49.95}{2283.3} = 0.0916,$$

$$h'_{\text{fg}} = 2283.3(1 + 0.68 \times 0.0916) = 2425.5 \text{ kJ/kg},$$

$$\text{Nu}_L = 1.13 \left[\frac{9.8 \times (980.5)^2 \times 2425.5 \times 1000 \times 1^3}{0.4338 \times 10^{-3} \times 0.6553 \times 49.95} \right]^{1/4} = 7157.2,$$

$$h = 7157.2 \frac{0.6553}{1} = 4690 \text{ W/m}^2\text{-°C}.$$

Thus, the condensed steam is

$$\dot{m} = \frac{4690 \times (0.2 \times 1) \times 49.95}{2283.3 \times 1000} = 0.0205 \text{ kg/s } (162 \text{ lb}_m/\text{h}).$$

To verify that the flow is laminar, find $\Gamma_{\text{max}} = 0.0205/0.2 = 0.103$ kg/m-s, so that

$$\text{Re}_{\Gamma_{\text{max}}} = \frac{4 \times 0.103}{0.4338 \times 10^{-3}} = 949.$$

Thus, the flow is laminar as assumed.

(b) For $L = 3$ m, the flow may well be turbulent so that Eq. (10.15) must be used. Since this equation involves the unknown $\text{Re}_{\Gamma_{\text{max}}}$, an iterative calculation may appear necessary. However, the second definition of Re_Γ in Eq. (10.13) may be substituted into Eq. (10.15) to avoid iteration:

$$\frac{hL}{k} = 0.0077 \left(\frac{g\rho_l^2 L^3}{\mu_l^2} \right)^{1/3} \left(\frac{4hL \, \Delta t}{h_{\text{fg}}\mu_l} \right)^{0.4}.$$

Substitution of the known values of all but h yields

$$h = 4335 \text{ W/m}^2\text{-}°\text{C}.$$

Then identical calculations as above for $\dot{m}$, Γ_{max}, and $\text{Re}_{\Gamma_{max}}$ give

$$\dot{m} = 0.0569 \text{ kg/s} (452 \text{ lb}_m/\text{h}),$$

$$\Gamma_{max} = 0.2485 \text{ kg/s-m},$$

$$\text{Re}_{\Gamma_{max}} = 2623,$$

showing the film to be turbulent as assumed. ∎

Laminar Film Condensation on the Top of a Horizontal Plate

As noted above, when a plate is in a horizontal position, the analysis of Nusselt fails since the driving force $g \sin \varphi \to 0$ as $\varphi \to 0$. In such an instance one must recast the analysis to account for the fact that the only driving force is that due to hydrostatic pressure difference created by the film being thicker in the center of the plate than at the edge. This problem was analyzed by Clifton and Chapman (Ref. 6), with the following result:

$$\text{Nu}_L = 2.43 \left(\frac{g \rho_l^2 h_{fg} L^3}{\mu_l k_l \, \Delta t} \right) F \left(\frac{\text{Ja}}{\text{Pr}} \right),$$

$$\begin{bmatrix} \text{liquid properties at } t_m \\ h_{fg} \text{ at } t_{sat} \\ F(\text{Ja/Pr}) \text{ from Table 10.1} \end{bmatrix}.$$

(10.16)★

The function $F(\text{Ja/Pr}) = F(k_l \, \Delta t / \mu_l h_{fg})$ is a complex function tabulated in Table 10.1.

Film Condensation on the Outside of Horizontal Cylinders

Analyses similar to that of Nusselt for vertical surfaces may be carried out for other shapes of interest. One very important case is that of condensation on the outside of a horizontal cylinder as illustrated in Fig. 10.3(a). The condensate

TABLE 10.1 Function F for Use in Eq. (10.16)

$\dfrac{\text{Ja}}{\text{Pr}} = \dfrac{k_l \, \Delta t}{\mu_l h_{fg}}$	F
0.176	0.329
0.381	0.322
0.698	0.320
2.27	0.285
4.08	0.264

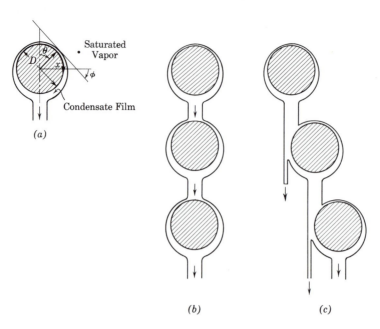

FIGURE 10.3. Film condensation on horizontal cylinders.

film forms at the top of the cylinder and flows around it, thickening as it goes. Nusselt analyzed this problem by representing the film at any polar location θ as that along an inclined flat surface making an angle φ with the horizontal. Thus, Nusselt used Eq. (10.8) with g replaced with $g \sin \varphi$. Since $\varphi = \pi/2 - \theta$ and $\theta = x/(D/2)$, Eq. (10.8) may be written as a function of x and then integrated around the cylinder to obtain the average h and Nu. Performing this integration numerically, Nusselt obtained the result noted below. This result agrees well when compared with experiment. The question of transition to turbulence rarely arises in practical applications involving horizontal cylinders.

$$\text{Nu}_D = \frac{hD}{k} = 0.728 \left[\frac{g\rho_l(\rho_l - \rho_v)h'_{fg}D^3}{\mu_l k_l \Delta t} \right]^{1/4}, \tag{10.17}\bigstar$$

$$\begin{bmatrix} \text{Pr} > 0.5 \\ \text{Ja} < 1.0 \\ h'_{fg} = h_{fg}(1 + 0.68\text{Ja}) \\ \text{liquid properties at } t_m \\ h_{fg} \text{ at } t_{sat} \end{bmatrix}.$$

If one uses the vertical plate relation for a vertical cylinder, one may note that the ratio of the heat transfer coefficient for a horizontal cylinder to what it would have in a vertical position is

$$\frac{h_{horiz}}{h_{vert}} \frac{D}{L} = \frac{0.728}{1.13} \left(\frac{D^3}{L^3} \right)^{1/4}, \tag{10.18}$$

$$\frac{h_{horiz}}{h_{vert}} = 0.644 \left(\frac{L}{D} \right)^{1/4}.$$

In commercial condensers the ratio L/D may be in the range of 50 to 100, or even more. Thus as much as twice the condensing capacity may be expected for the horizontal case than for the vertical case—a result of the fact that a much thinner film forms in the horizontal position.

The correlation of Eq. (10.17) applies only to a single cylinder. Most commercial condensers involve banks of tubes. If the tubes are arranged in a vertical pattern as suggested in Fig. 10.3(b), the condensate from one tube drips onto that of the next, and so on. Thus, the film thickness is greater for the lower tubes, increasing the resistance to heat transfer. For a vertical stack of N tubes, Nusselt showed that the average coefficient is reduced according to

$$\frac{(Nu_D)_N}{(Nu_D)_1} = \left(\frac{1}{N}\right)^{1/4}, \tag{10.19}$$

where the subscripts N and 1 refer to N tubes and a single tube. Thus, improved performance in a condenser can be achieved by using a staggered arrangement as suggested in Fig. 10.3(c). For the analyses of complex tube bank arrangements the reader is referred to Ref. 7.

EXAMPLE 10.2

Saturated steam at 1 psia condenses on a horizontal tube, 1 in. in diameter, maintained with a surface temperature of 95°F. Find the amount of steam condensed, per foot of tube length.

Solution. At 1 psia, the Steam Tables give $t_{sat} = 101.7°F$, $h_{fg} = 1036.0$ Btu/lb$_m$. Thus, $\Delta t = 6.7°F$ and $t_m = 98°F$. Table A.3 gives

$$\rho_l = 62.02 \text{ lb}_m/\text{ft}^3, \qquad \mu_l = 1.687 \text{ lb}_m/\text{ft-hr},$$
$$k_l = 0.3607 \text{ Btu/h-ft-°F}, \qquad c_{pl} = 0.998 \text{ Btu/lb}_m\text{-°F}.$$

Thus, Eq. (10.17) yields

$$Ja = \frac{0.998 \times 6.7}{1036.0} = 0.0065,$$

$$h'_{fg} = 1036.0(1 + 0.68 \times 0.0065) = 1040.5 \text{ Btu/lb}_m,$$

$$Nu_D = 0.728 \left[\frac{32.2 \times (3600)^2 \times (62.02)^2 \times 1036.0 \times (1/12)^3}{1.687 \times 0.3607 \times 6.7}\right]^{1/4} = 507,$$

$$h = 507 \times \frac{0.3607}{1/12} = 2196 \text{ Btu/h-ft}^2\text{-°F}.$$

The steam condensed, per unit length, is then

$$\frac{\dot{m}}{L} = \frac{q}{L}\frac{1}{h_{fg}} = \pi Dh \frac{\Delta t}{h_{fg}}$$

$$= \frac{\pi \times 2196 \times 6.7}{12 \times 1036} = 3.72 \text{ lb}_m/\text{h-ft} (1.54 \times 10^{-3} \text{ kg/s-m}).$$

■

EXAMPLE 10.3

Saturated Freon-12 at 50°C condenses on a horizontal 3-cm-diameter tube. Find the condensing heat transfer coefficient.

Solution. From the given data, $t_{sat} = 50°C$, $t_m = 45°C$, $\Delta t = 10°C$. At 50°C, Ref. 8 gives $h_{fg} = 121.43$ kJ/kg. At $t_m = 45°C$, Table A.4 gives

$$\rho_l = 1236.55 \text{ kg/m}^3, \quad c_{pl} = 1.01175 \text{ kJ/kg-°C}, \quad k_l = 0.068 \text{ W/m-°C},$$

$$\nu_l = 0.191 \times 10^{-6} \text{ m}^2/\text{s}, \quad \mu_l = \rho_l \nu_l = 2.362 \times 10^{-4} \text{ kg/m-s}.$$

Thus, Eq. (10.17) gives

$$\text{Ja} = \frac{1.01175 \times 10}{121.43} = 0.0833,$$

$$h'_{fg} = h_{fg}(1 + 0.68\text{Ja}) = 128.31 \text{ kJ/kg},$$

$$\text{Nu}_D = 0.728 \left[\frac{9.8 \times (1236.55)^2 \times 128.31 \times 1000 \times (0.03)^3}{2.362 \times 10^{-4} \times 0.068 \times 10} \right]^{1/4}$$

$$= 548.9,$$

$$h = 548.9 \times \frac{0.068}{0.03} = 1244 \text{ W/m}^2\text{-°C} \ (96.8 \text{ Btu/h-ft}^2\text{-°F}). \qquad \blacksquare$$

Condensation Inside Horizontal Tubes

The process of the condensation of a vapor flowing inside a horizontal cylindrical tube is one of importance in refrigeration cycles and chemical and petrochemical processes. In a comprehensive review of the work in this field, and after additional experimental investigations, Akers et al. (Ref. 9) report that for condensing vapors in either horizontal or vertical tubes, the following relations correlate the data within about 20%:

$$\text{Nu}_D = 5.03\text{Re}_G^{1/3}\text{Pr}^{1/3}, \qquad \text{Re}_G < 5 \times 10^4, \qquad (10.20)\bigstar$$

$$\text{Nu}_D = 0.0265\text{Re}_G^{0.8}\text{Pr}^{1/3}, \qquad \text{Re}_G > 5 \times 10^4.$$

These expressions appear very similar to those for forced convection without phase change inside of tubes, as discussed in Chapter 7. However, the Reynolds number must be given a new interpretation to account for the existence of the two-phase flow and to express the relative amounts of each phase.

The customary diameter Reynolds number is defined as

$$\text{Re}_D = \frac{U_m \rho D}{\mu}.$$

Instead of the mean velocity, one introduces the *mass velocity*, G, which is the mass rate of flow per unit cross-sectional area:

$$G = U_m \rho.$$

Then in terms of the mass velocity, the Reynolds number is

$$\text{Re}_G = \frac{DG}{\mu}. \tag{10.21}$$

The correlation of Akers et al., given in the equations above, is written in terms of a Reynolds number, Re_G, based on an *equivalent mass velocity, G_E,* defined as

$$G_E = G_l + G_v\left(\frac{\rho_l}{\rho_v}\right)^{1/2}. \tag{10.22}$$

In this expression G_l and G_v are the mass velocities, based on the full pipe area, of the liquid and vapor phases, respectively. The symbols ρ_l and ρ_v are the densities of the liquid and vapor phases, respectively. Equation (10.20) results from choosing a model for the flow inside the tube which replaces the vapor core by an analogous liquid which exerts the same shearing stress at the liquid–vapor interface.

10.5

EFFECTS OF NONCONDENSABLE GASES AND VAPOR VELOCITY ON CONDENSATION

The analyses and correlations presented in Secs. 10.3 and 10.4 presume that the vapor to be condensed was the only component present. Thus, the liquid–vapor interfacial temperature was taken to be t_{sat} at the pressure of the bulk vapor. When a noncondensable gas, such as air, is also present, the saturation pressure in the bulk gas–vapor medium is that corresponding to the partial pressure of the vapor in this mixture. In addition, the vapor, in moving toward the condensing surface, must diffuse through the noncondensable component. Thus, a decreasing gradient in the partial pressure of the vapor component must exist as the interface is approached. Consequently, the saturation temperature at the interface will be lower than that in the bulk mixture. The diffusion process and the lowered saturation temperature reduces the condensation rate and, hence, the observed heat transfer coefficient. Rather complex theories exist to predict these effects (Ref. 2) but are beyond the scope of this text; however, the effect of noncondensables may be significant. Typical calculations show that the condensing heat transfer coefficient for steam may be reduced by as much as 50% when air in amounts as low as 1%, by mass, is present. Consequently, it is customary practice in steam condenser design to provide means of removing as much of the noncondensable gases as possible through venting or the use of ejectors.

The simplified Nusselt analysis presumed no shear stress at the liquid–vapor interface. In some applications, particularly for flows inside tubes, the ambient vapor may be flowing with a significant velocity relative to the liquid film. This vapor velocity may be accompanied by significant shear stresses at the interface. Obviously, profoundly different results may occur between the cases when the vapor flows in the same direction as the liquid or in the opposite direction. Again, these effects have been examined by a number of workers, and the reader is referred to Ref. 2 for a summary of their results.

10.6

DROPWISE CONDENSATION

As noted at the outset of Sec. 10.2, two basic types of condensation are observed to occur, film and dropwise. All the subsequent discussion was limited to film condensation. It has been observed that when traces of oil, or other similar substances, are present on the condensing surface so that the liquid does not readily wet it, the condensate breaks into droplets—forming what is known as "dropwise condensation." The drops thus formed grow as condensation progresses and either coalesce with neighboring drops or run off the surface after their weight overcomes the restraining forces of surface tension. As the drops run off, a greater fraction of the condensing surface is left free of liquid, thus enhancing the heat transfer process.

Since the entire surface is not covered with a continuous liquid film, heat transfer in the dropwise condensation mode is observed to be five to ten times as great as in the film mode. Thus, if dropwise condensation could be maintained in industrial condensing equipment, significant savings in the size of such devices could be realized. Consequently, a great deal of time and effort has been devoted to the study of this mechanism and in seeking ways to promote its occurrence. The subject is too complex and the results too tentative to be treated adequately here, but the reader may wish to consult Chapter 12 of Ref. 2 for a summary of the subject. The results of studies to date indicate that dropwise condensation is promoted by the use of highly polished surfaces or by the injection of various fatty acids such as oleic or stearic acids. However, the effectiveness of additives lasts for only a few hundred hours, due to the cleansing effects of steam. Consequently, the use of dropwise promoters may require continual injection which would become expensive or incur undesired effects in the rest of the plant, or cycle, using the condenser.

As a result of all the above, most industrial condensing equipment is still designed on the presumption that film condensation will be present.

10.7

HEAT TRANSFER DURING THE BOILING OF A LIQUID

The process of the purposeful conversion of a liquid into a vapor is one of obvious engineering importance. The production of large quantities of steam for electrical power generation through the use of vapor cycles is an example that most readily comes to mind. Many processes involved in the refining of petroleum and the manufacture of chemicals include the vaporization of a liquid.

Careful experiments have shown that if a liquid, such as water, is distilled and completely degasified under vacuum (i.e., if it contains *no* impurities, not even dissolved gases) it will undergo the liquid–vapor phase change without the appearance of bubbles when it is heated in a clean, smooth vessel. Under normally encountered engineering conditions, however, the presence of impurities, dissolved gases, and surface irregularities causes the appearance of vapor bubbles on the heating surface when the rate of heat input is great enough. The creation of these bubbles, their subsequent growth and detachment, etc., profoundly influence the heat transfer process taking place.

Boiling may occur under various conditions. When the heating surface is submerged in an otherwise quiescent pool of liquid, and heat is transferred to the liquid by free convection and bubble agitation, the process is termed *pool boiling*. This term is used in contrast to boiling which may occur simultaneously with fluid motion induced by externally imposed pressure differences and which is referred to as *forced convection boiling*. Most of the discussion to follow concentrates on pool boiling.

As noted in the foregoing, most boiling encountered in normal engineering applications eventually results in the formation, growth, collapse, etc., of vapor bubbles. In order to understand some of the peculiarities of pool boiling it is advantageous to first note some facts concerning the mechanics and thermodynamics of bubbles. Consider the existence of a spherical bubble of radius R in a liquid as depicted in Fig. 10.4. The pressure of the vapor inside the bubble, P_v, must exceed that in the surrounding liquid, P_l, because of the surface tension acting on the liquid–vapor interface. If σ represents the surface tension (force per unit length) then a force balance on an equatorial plane gives

$$\pi R^2(P_v - P_l) = 2\pi R\sigma,$$

or   (10.23)

$$P_v - P_l = \frac{2\sigma}{R}.$$

Several aspects may be noted from Eq. (10.23). To create a bubble of small radius, it would be necessary to develop very large pressures in the vapor. Hence, bubbles are usually found to form at pits existing in surface irregularities where a bubble of finite initial radius may form, or, more likely, at points on the surface where gases dissolved in the surface of the liquid come out of solution. Also, consider the situation when the bubble is in thermal equilibrium—i.e., when the liquid and vapor have the same temperature and the bubble is not growing. The temperature of the vapor in the bubble must be the saturation

FIGURE 10.4. Spherical vapor bubble in a liquid.

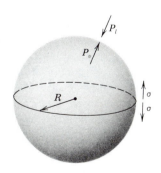

temperature at the pressure P_v. Since the liquid is at the same temperature it is then at a temperature in excess of the saturation temperature at P_l since $P_l < P_v$. Thus, the liquid surrounding the bubble is in a metastable state. That is, the liquid is *superheated* at a temperature in excess of t_{sat} for the liquid, typically taken as the boiling point at the given ambient pressure. Thus, near the heating surface where bubbles are formed, liquid superheat temperatures are found to exist. To cause a bubble to grow, the liquid superheat will exceed that predicted by Eq. (10.23), and liquid superheats of the order of 0.5 to 8°C above the ambient saturation temperature are commonly encountered in the boiling of water. If the bubble detaches from the surface and moves up into cooler liquid, it may recondense and collapse if the superheat is insufficient to maintain it. The foregoing discussion assumed the existence of spherical bubbles. In real cases, the bubbles created on a boiling surface will be nonspherical, depending on local surface irregularities and the contact angle the bubble makes with the surface—a quantity dependent on the way in which the liquid "wets" the surface.

Pool Boiling

With the above comments in mind it is now possible to describe the various processes that may occur during the pool boiling of a liquid. For a given ambient pressure over a pool of liquid, heated from below by a horizontal surface, the boiling is termed *subcooled pool boiling* if the temperature of the bulk of the liquid is below that of saturation at the given pressure. In this instance, bubbles formed at the surface rise and are eventually recondensed in the liquid. All vapor produced is simply that by evaporation at the free surface. Continued heating will raise the temperature of the bulk of the pool, until one reaches the condition in which the bulk liquid is at, or slightly exceeds, the saturation temperature. This condition is known as *saturated pool boiling*.

Saturated pool boiling has been studied extensively. The reader may wish to consult Refs. 10 and 11 for summaries of these works. In general, the heat flux through the heating surface during saturated pool boiling is a complex function of the temperature excess, $t_s - t_{sat}$, by which the surface temperature exceeds the ambient pressure saturation temperature. The necessity for the existence of a surface temperature greater than t_{sat} has already been discussed. Figure 10.5 shows the typical dependence of the heat flux, q/A, on $\Delta t = t_s - t_{sat}$. The data of Fig. 10.5 are those typically observed for water at atmospheric pressure and were obtained using an electrically heated wire which serves the dual purpose of a heating surface and a resistance thermometer for the measurement of t_s. This curve is similar to those observed for other liquids and other pressures.

Figure 10.5 displays several regimes of pool boiling in which the physical mechanisms taking place are quite different. If one defines the boiling heat transfer coefficient as based on the surface-saturation temperature excess so that the heat flux is

$$\frac{q}{A} = h(t_s - t_{sat}) = h\,\Delta t, \tag{10.24}$$

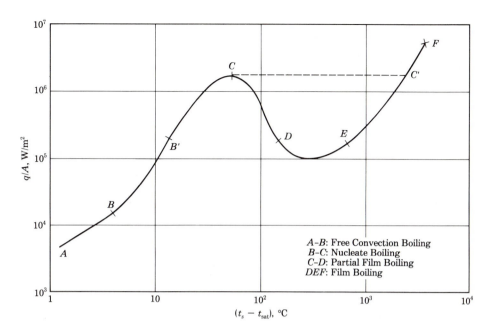

FIGURE 10.5.   Typical boiling curve for water.

then q/A depends not only on Δt directly, but also in accord with how h depends on Δt. In the range of low Δt (approximately 5°C), region A-B in Fig. 10.5, no bubbles are observed even though the liquid is slightly superheated. Heat is transported from the heating surface to the bulk of the liquid mainly by free convection effects. All vapor produced is by evaporation at the liquid surface. In this regime, called *free convection boiling*, h is proportional to $(\Delta t)^{1/4}$ in accord with the results of Chapter 9, so that q/A varies as $(\Delta t)^{5/4}$. Note, however, that the magnitude of the heat flux shown in Fig. 10.5 is much larger than those normally encountered for free convection without phase change.

Continued increase in Δt causes the process to enter the regime of *nucleate boiling* denoted by B-C in Fig. 10.5. In this region (up to Δt of the order of 50°C) bubbles are found to appear on the heating surface. The bubbles first appear at certain favorable sites, or nuclei. As Δt increases, more and more sites are activated. In the lower end of the region B-C, the bubbles leave the surface and may recondense in the bulk of the liquid. As Δt is increased more, the bubbles rise all the way to the liquid surface to release vapor. This bubble activity causes the liquid near the heating surface to become highly agitated and a marked increase in the heat flux occurs, increasing to Δt raised to the third or fourth power. Since the liquid phase is such a better conductor than the vapor, most of the heat transfer in nucleate boiling is from the surface into the liquid and then from the liquid into the vapor of the bubble rather than from the surface directly into the vapor. Eventually, as Δt increases further, so many bubble nucleation sites become activated that they begin to interfere with one another, causing parts of the surface to become covered with vapor and inhibiting liquid motion near the surface. Thus, the increase of q/A is slowed, as suggested by point B' in Fig. 10.5, until a maximum point is reached. This upper limit of the nucleate boiling regime, point C, is referred to as the *peak heat flux* (also called the *critical heat flux* or the *boiling crisis*).

Additional increase of Δt beyond that at the peak heat flux causes the process to enter the regime of *transition boiling,* or *partial film boiling,* denoted by *C-D* in Fig. 10.5. At any location on the surface, an unstable condition exists in which the process oscillates between nucleate boiling and *film boiling* in which the surface is blanketed with a film of vapor. Eventually, the regime of *stable film boiling, D-E-F* in Fig. 10.5, is reached in which the surface is completely covered with a blanket of vapor. Heat transfer takes place by conduction through the vapor and then into the liquid. Since the vapor is a poorer conductor than the liquid, the heat flux in the film boiling region, *D-E,* is measurably less than that at the peak heat flux. If Δt is further increased, above, say, 150°C, q/A again increases as heat exchange by radiation comes into play. Eventually, a point is reached at which the surface melts—point F in Fig. 10.5.

The preceding discussion was based on the presumption that the temperature of the heating surface could be maintained at a particular value—that is, that Δt was the independent variable. With Δt thus under control, the boiling curve in Fig. 10.5 could be traversed as described, and it would appear that the desired operating point for a commercial vapor-producing system would be at the peak heat flux of point C. However, in many applications it is the heat flux that is the independent variable (as for an electrically heated wire or a boiler with a given rate of heat production from fuel burning) and the surface temperature assumes a value compatible with q/A. However, the boiling curve is multivalued when viewed with q/A as the independent variable, so that more than one possible surface temperature may correspond to a given heat flux. Also, the partial film boiling region is unstable, so that for a given q/A, the only equilibrium points possible are along *A-B-C* or *D-E-F* in Fig. 10.5. If one were to operate too close to the peak heat flux at point C, a slight increase in q/A may make the system choose to operate in the film boiling regime—say at point C'. This action would result in a dramatic increase in the surface temperature, producing important mechanical design problems. It is even possible that the point C' may be so close to the surface melting point, that failure of the system could result.

10.8

WORKING CORRELATIONS FOR POOL BOILING

All the discussion in Sec. 10.7 should indicate that boiling heat transfer is a very complex subject. Both the analytical and empirical investigations of boiling are quite involved, and much remains to be done to thoroughly understand it. While this matter is still the object of much current research, this section presents the most reliable information available at this time.

Nucleate Pool Boiling

In the nucleate pool boiling regime, the most widely accepted correlation is that due to Rohsenow (Ref. 12):

$$\frac{q}{A} = \mu_l h_{fg} \left[\frac{g(\rho_l - \rho_v)}{g_c \sigma} \right]^{1/2} \left(\frac{c_{pl} \, \Delta t}{C_{sf} h_{fg} Pr_l^s} \right)^3,$$

$$\begin{bmatrix} s = 1.0 \text{ for water} \\ s = 1.7 \text{ for other liquids} \\ \text{properties at } t_{sat} \end{bmatrix}.$$

(10.25)★

In Eq. (10.25) the subscript l refers to the physical properties of the *saturated liquid* and v to those of the *saturated vapor*. The latent heat of vaporization is given by h_{fg}, σ is the surface tension at the liquid–vapor interface, g is the acceleration of gravity, and g_c is the dimensional constant discussed in Sec. 1.7 and is included in the event an inconsistent system of units is employed. The constant C_{sf} is an empirically determined quantity that is dependent on the composition and roughness of the heating surface and the way in which the liquid wets the surface. Some values of C_{sf} are given in Table 10.2 for water as compiled by Holman (Ref. 13) from a number of sources. Reference 13 may be consulted for values of C_{sf} for other surface-liquid combinations.

For water, the surface tension in Eq. (10.25) may be calculated from (Ref. 14):

$$\sigma, \text{N/m} = 0.2358 \left(1 - \frac{T}{647.15} \right)^{1.256}$$

$$\times \left[1 - 0.625 \left(1 - \frac{T}{647.15} \right) \right],$$

(10.26)★

in which T is the absolute temperature in degrees Kelvin. For other substances the reader may seek values of σ from various handbooks.

The peak heat flux described in Sec. 10.7 is an important quantity, and it may be predicted from the correlation of Zuber (Ref. 15):

$$\left(\frac{q}{A} \right)_{max} = \frac{\pi}{24} h_{fg} \rho_v \left[\frac{\sigma g \, g_c (\rho_l - \rho_v)}{\rho_v^2} \right]^{1/4} \left(1 + \frac{\rho_v}{\rho_l} \right)^{1/2}$$

(10.27)★

The notation in Eq. (10.27) is the same as that for Eq. (10.25).

TABLE 10.2 Values of C_{sf} for Water and Various Surfaces

Surface	C_{sf}
Brass	0.0060
Copper-polished	0.0130
Copper-scored	0.0068
Platinum	0.0130
Stainless steel	
Chemically etched	0.0133
Ground and polished	0.0080
Mechanically polished	0.0132
Teflon pitted	0.0058

EXAMPLE 10.4

Nucleate boiling of water at atmospheric pressure occurs on the bottom of a polished copper pan maintained at 118°C. Find (a) the boiling heat flux; (b) the maximum heat flux.

Solution. At t = 100°C, the Steam Tables give ρ_l = 958.31 kg/m³, ρ_v = 0.5978 kg/m³, h_{fg} = 2257.0 kJ/kg, and Table A.3 gives

$$c_{pl} = 4.215 \text{ kJ/kg-°C}, \qquad \mu_l = 0.2822 \times 10^{-3} \text{ kg/m-s}, \qquad Pr_l = 1.76.$$

When Eq. (10.26) is evaluated at T = 373.15°K, one obtains σ = 0.05892 N/m.

(a) The boiling heat flux is given by Eq. (10.25 with s = 1.0, C_{sf} = 0.013, and Δt = 18°C:

$$\frac{q}{A} = 0.2822 \times 10^{-3} \times 2257.0 \left[\frac{9.8 \times (958.31 - 0.5978)}{0.05892} \right]^{1/2}$$

$$\times \left[\frac{4.215 \times 18}{0.013 \times 2257.0 \times 1.76} \right]^3,$$

$$= 806.2 \text{ kW/m}^2 \ (2.556 \times 10^5 \text{ Btu/h-ft}^2).$$

(b) The peak heat flux is given by Eq. (10.27):

$$\left(\frac{q}{A} \right)_{max} = \frac{\pi}{24} \times 2257.0 \times 0.5978$$

$$\times \left[\frac{0.05892 \times 9.8 \times (958.31 - 0.5978)}{(0.5978)^2} \right]^{1/4}$$

$$\times \left[1 + \frac{0.5978}{958.31} \right]^{1/2},$$

$$= 1108.1 \text{ kW/m}^2 \ (3.513 \times 10^5 \text{ Btu/h-ft}^2). \qquad \blacksquare$$

Film Pool Boiling

When the state of full film boiling is established, the heating surface is completely covered by a blanket of vapor. The temperature of the surface becomes rather large since all heat transfer must be done through the poorly conducting vapor layer.

For film boiling on a submerged horizontal cylinder, the convective picture is that of a layer of vapor flowing upward past a cylinder and bounded by a large liquid domain. Bromley (Ref. 16) likened this situation to that of condensation on a horizontal cylinder—a liquid layer flowing downward past a cylinder and bounded by a large vapor domain. Following the same analysis as Nusselt did for the condensation problem, Bromley obtained the following for film boiling on a horizontal cylinder:

$$\text{Nu}_D = \frac{h_b D}{k_v} = 0.62 \left[\frac{g\rho_v(\rho_l - \rho_v)(h_{fg} + 0.4c_{pv}\,\Delta t)D^3}{\mu_v k_v\,\Delta t} \right]^{1/4},$$

$$\left[\begin{array}{l} h_{fg} \text{ and } \rho_l \text{ at } t_{sat} \\ \text{other properties at } (t_s + t_{sat})/2 \end{array} \right]. \qquad (10.28)\bigstar$$

In Eq. (10.28), the symbol h_b is used to denote the *boiling* heat transfer coefficient in order to distinguish it from the radiation coefficient to be defined next.

When the surface temperature is high enough to produce film boiling, it is usually hot.enough to cause an appreciable heat transfer by radiation. While radiative heat transfer is not discussed until the next chapter, it is possible to account for its effect here by quoting some relations to be developed in Chapter 11. The radiative heat transfer in this case may be expressed as

$$\left(\frac{q}{A}\right)_r = h_r(t_s - t_{sat}),$$

where h_r is a radiation coefficient defined as

$$h_r = \frac{\sigma\epsilon(T_s^4 - T_{sat}^4)}{T_s - T_{sat}}. \qquad (10.29)$$

In Eq. (10.29) the capital T's are absolute temperature, ϵ is the surface emissivity, and σ is the Stefan–Boltzmann radiation constant (not surface tension!):

$$\sigma = 5.67 \times 10^{-8}\ \text{W/m}^2\text{-}^\circ\text{K}^4. \qquad (10.30)$$

Bromley suggested that the combined heat transfer by boiling and radiation in the case of film boiling on a horizontal surface could be calculated by defining a total heat transfer coefficient given by the empirical relation

$$h = h_r + h_b\left(\frac{h_b}{h}\right)^{1/3}. \qquad (10.31)$$

In Eq. (10.31), h_r and h_b are found from Eqs. (10.28) and (10.29). Then the total heat flux from the cylinder is

$$\frac{q}{A} = h(t_s - t_{sat}). \qquad (10.32)$$

Quite obviously, the use of Eq. (10.31) requires an iterative calculation.

EXAMPLE 10.5
Film boiling occurs on the surface of a horizontal 1-cm-diameter cylinder with a surface temperature of 300°C when it is submerged in a water bath at 1 atm, 100°C. If the surface emissivity is 0.8, estimate the surface heat flux.

Solution. At the liquid saturation temperature of 100°C, the Steam Tables give $\rho_l = 598.3\dot{1}$ kg/m^3, $h_{fg} = 2257.0$ kJ/kg. The mean film temperature is $t_m = (100 + 300)/2 = 200$°C, and the vapor properties are to be evaluated at this temperature and a pressure of 1 atm (101.35 kN/m^2). The entries in Table A.5 for a pressure of 100 kN/m^2, 200°C, are sufficiently accurate for these purposes:

$$\rho_v = 0.4603 \text{ kg/m}^3, \qquad c_{pv} = 1.979 \text{ kJ/kg-°C},$$

$$\mu_v = 16.18 \times 10^{-6} \text{ kg/m-s}, \qquad k_v = 33.37 \times 10^{-3} \text{ W/m-°C}.$$

Equation (10.28) gives the boiling heat transfer coefficient:

$$\frac{h_b D}{k_v} = 0.62 \left[\frac{\begin{array}{c} 9.8 \times 0.4603 \times (598.31 - 0.4603) \\ \times [2257.0 + 0.4 \times 1.979 \times 200](0.01)^3 \times 1000 \end{array}}{16.18 \times 10^{-6} \times 33.37 \times 10^{-3} \times 200} \right]^{1/4}$$

$$= 27.69,$$

$$h_b = 92.4 \text{ W/m}^2\text{-°C}.$$

With $\epsilon = 0.8$ and $\sigma = 5.67 \times 10^{-8}$ W/m²-°K⁴, Eq. (10.29) gives the radiation coefficient as

$$h_r = \frac{5.67 \times 10^{-8} \times 0.8 \times (573.15^4 - 373.15^4)}{573.15 - 373.15} = 20.1 \text{ W/m}^2\text{-°C}.$$

The combined heat transfer coefficient is given by Eq. (10.31):

$$h = h_r + h_b \left(\frac{h_b}{h} \right)^{1/3}$$

$$= 20.1 + 92.4 \left(\frac{92.4}{h} \right)^{1/3}.$$

Iterative solution of the above gives

$$h = 108 \text{ W/m}^2\text{-°C},$$

so that the heat flux is

$$\frac{q}{A} = h(t_s - t_{\text{sat}})$$

$$= 108(300 - 100)$$

$$= 21.600 \text{ W/m}^2 \ (6.85 \times 10^3 \text{ Btu/h-ft}^2). \qquad \blacksquare$$

10.9

FORCED CONVECTION BOILING

Forced convection boiling is most often encountered in the case of forced flow of a liquid through a tube with boiling occurring at the inner surface. Bubble growth and the influence of the bubble dynamics on the heat transfer mechanism are strongly influenced by the forced velocity of the flow. If a subcooled liquid enters a tube maintained at a surface temperature in excess of t_{sat}, the heat transfer is initially accomplished by forced convection in accord with the relations of Chapter 8. However, boiling is soon initiated, and bubbles begin to appear on the surface as the fluid moves down the tube. The bubbles grow, break away from the surface and are taken into the mainstream of the liquid.

This is known as the *bubbly-flow* regime and is associated with a sharply increased heat transfer coefficient. As the flow continues down the tube, the fraction of vapor becomes so large that slugs of vapor are formed, separated by slugs of liquid. This *slug-flow* regime is followed by the *annular-flow* regime, in which so much vapor exists that it flows in the central core of the pipe with the liquid in an annular film at the surface. The heat transfer rate continues to rise through the bubbly-, slug-, and annular-flow regimes. However, dry spots begin to occur on the surface, and the heat transfer coefficient drops as the flow passes into the *mist-flow* regime. In this latter regime, the flow is mainly vapor with the liquid phase suspended in it in a mist of liquid droplets. Eventually all the mist evaporates, and the vapor begins to superheat.

A complete summary of the state of knowledge of forced convection boiling and two-phase flow is given in Ref. 2. Needless to say, the above description of the forced boiling process is sufficiently complicated to preclude the presentation of the many correlations which have been developed for the various flow regimes. As a general simple rule, Rohsenow and Griffith (Ref. 17) suggest that forced convection boiling be treated by superimposing the two effects of forced convection and nucleate boiling, at least up to the slug-flow regime. That is, the total heat transfer is to be found from

$$\left(\frac{q}{A}\right)_{\text{total}} = \left(\frac{q}{A}\right)_{\text{boiling}} + \left(\frac{q}{A}\right)_{\text{conv.}} \tag{10.33}$$

In Eq. (10.33), the boiling contribution is found from Eq. (10.25) and the convective contribution is found by use of the Dittus–Boelter equation in Eq. (8.27) with the constant 0.023 changed to 0.019.

REFERENCES

1. NUSSELT, W., "Die Oberflächenkondensation des Wasserdampfes," *Z.V.D.I.*, Vol. 60, 1916, p. 569.

2. ROHSENOW, W. M., and J. P. HARTNETT, eds., *Handbook of Heat Transfer*, New York, McGraw-Hill, 1973.

3. SPARROW, E. M., and J. L. GREGG, "A Boundary-Layer Treatment of Laminar Film Condensation, *J. Heat Transfer, Trans. ASME*, Vol. 81, 1959, p. 13.

4. McADAMS, W. H., *Heat Transmission*, 3rd ed., New York, McGraw-Hill, 1954.

5. ROHSENOW, W. M., "Heat Transfer and Temperature Distribution in Laminar Film Condensation," *Trans. ASME*, Vol. 78, 1956, p. 1645.

6. CLIFTON, J. V., and A. J. CHAPMAN, "Condensation of a Pure Vapor on a Finite Size Horizontal Plate," *ASME Paper 67-WA/HT-18*, New York, 1967.

7. DEVORE, A., "How to Design Multitube Condensers," *Petroleum Refiner*, Vol. 38, No. 6, June 1959, p. 205.

8. VAN WYLEN G. J., and R. E. SONNTAG, *Fundamentals of Classical Thermodynamics*, 2nd ed., New York, Wiley, 1978.

9. AKERS, W. W., H. A. DEANS, and O. K. CROSSER, "Condensing Heat Transfer Within Horizontal Tubes," *Chem. Eng. Prog. Symp. Ser.*, Vol. 55, No. 29, 1958, p. 171.

10. LEPPERT, G., and C. C. PITTS, "Boiling," *Advances in Heat Transfer*, Vol. 1, New York, Academic Press, 1964.

11. ROHSENOW, W. M., ed., *Developments in Heat Transfer,* Cambridge, Mass. MIT Press, 1964.

12. ROHSENOW, W. M., "A Method of Correlating Heat Transfer Data for Surface Boiling Liquids," *Trans. ASME,* Vol. 74, 1952, p. 969.

13. HOLMAN, J. P., *Heat Transfer,* 5th ed., New York, McGraw-Hill, 1981.

14. *ASME Steam Tables,* 3rd ed., New York, Am. Soc. Mech. Engrs., 1977.

15. ZUBER, N., "On the stability of Boiling Heat Transfer," *Trans. ASME,* Vol. 80, 1958, p. 711.

16. BROMLEY, L. A., "Heat Transfer in Stable Film Boiling," *Chem. Eng. Prog.,* Vol. 46, 1950, p. 221.

17. ROHSENOW, W. M., and P. GRIFFITH, "Correlation of Maximum Heat Flux Data for Boiling of Saturated Liquids," *AIChE–ASME Heat Transfer Symp.,* Louisville, Ky., 1955.

PROBLEMS

10.1   A vertical plate 20 cm wide and 1 m high is maintained at 65°C and is exposed to saturated steam at 1 atm pressure. Find the total heat transferred and the amount of steam condensed, per hour, from both sides of the plate.

10.2   A vertical $\frac{3}{4}$-in. tube is 15 ft long and has its surface maintained at 85°F. Steam at 90°F is condensing on its surface. How much steam is condensed in 1 h?

10.3   For the data of Prob. 10.1, find the thickness of the condensate film and the maximum film velocity at the bottom edge of the plate and at a location halfway down the plate.

10.4   A vertical plate (1 m high, 0.5 m wide) is maintained at 80°C and exposed to saturated atmospheric pressure steam. Estimate the local condensing heat transfer coefficient at the middle of the plate and at the bottom edge of the plate.

10.5   A vertical plate is maintained at 54°C while condensing saturated steam at 1 atm pressure. Determine the average condensing heat transfer coefficient if the plate is (a) 1 m high, and (b) 2.5 m high.

10.6   Saturated ammonia at 658 kN/m² pressure is condensing on a vertical surface 0.6 m high maintined at 8°C. Find the average condensing heat transfer coefficient.

10.7   Saturated Freon-12 at 25°C condenses on the outer surface of a vertical 10-cm-diameter pipe of 1 m length. The pipe has a uniform surface temperature of 15°C. Determine the total mass rate of condensation.

10.8   A circular tube 1.6 cm in diameter and 1.8 m long has a surface temperature of 45°C. Saturated steam at 55°C is condensing on its surface. Find the mass of steam condensed, per hour, if the tube is (a) horizontal, and (b) vertical.

10.9   Saturated steam condenses on the outer surface of a horizontal 1-in.-diameter tube with a surface temperature of 60°F. Find the condensing heat transfer coefficient if the steam pressure is (a) 14.7 psia, (b) 50 psia, and (c) 100 psia.

10.10 Saturated Freon-12 at 40°C condenses on a horizontal 2.5-cm tube with a surface temperature of 32°C. Find the rate at which the Freon is being condensed, per meter of tube length.

10.11 Saturated ammonia at 90°F condenses on the outer surface of a horizontal 1-in.-O.D. tube maintined at 80°F. Find the condensing heat transfer coefficient.

10.12 Saturated steam at a pressure of 15 kN/m² condenses on the outer surface of a horizontal $\frac{5}{8}$-in. tube with a surface temperature of 45°C. Estimate the rate at which steam condenses, per meter of tube length.

10.13 Saturated steam at atmospheric pressure condenses on a 10×10 square array of horizontal tubes, 0.8 cm in diameter. If the tube surfaces are at 27°C, find the mass rate of condensation, per unit length of the tubes.

10.14 A horizontal pipe carrying cold water has a surface temperature of 3°C as it passes through a room where the air is at 35°C with a relative humidity of 75%. The pipe has a diameter of 6 cm and is 10 m long. Assuming that the moisture in the air condenses as a saturated vapor at its partial pressure in the air, estimate the mass rate of condensation.

10.15 Steam at 1 psia condenses on a horizontal bank of $\frac{5}{8}$-in. tubes. The bank is eight tubes high and the tubes are not staggered. The tube surfaces are at 96°F. Find the average heat transfer coefficient for the bank of tubes. What percentage of the steam condensed is condensed by the top row of tubes?

10.16 What is the peak heat flux for water boiling at (a) 1 atm pressure, and (b) 10 atm pressure?

10.17 Calculate the heat flux and the heat transfer coefficient for nucleate pool boiling of water at atmospheric pressure on a mechanically polished stainless steel surface when the surface temperature is (a) 110°C, and (b) 120°C.

10.18 Repeat Prob. 10.17 if the surface is brass.

10.19 The bottom of a polished copper pan is maintained at 117°C and is used to boil water at 1 atm pressure. If the pan is 0.2 m in diameter, what is the power required?

10.20 It is desired to boil 2 kg/h of water at atmospheric pressure in a polished copper pan 20 cm in diameter. What must the surface temperature be?

10.21 Calculate the heat flux and the heat transfer coefficient for water boiling at 14.7 psia pressure if the surface is ground and polished stainless steel maintained at (a) 220°F, and (b) 230°F.

10.22 A platinum wire is submerged in water at a pressure of 500 kN/m². If the wire surface is 10°C hotter than the water saturation temperature, find the heat flux from the wire.

10.23 An electrical heating element consists of a horizontal rod 0.6 cm in diameter. It is maintained at a surface temperature of 260°C in a pool of atmospheric pressure water. Assuming that the element has a surface emissivity of 0.9 and that stable film boiling exists, estimate the power required to operate the heater, per meter of length.

10.24 A steel bar 2.5 cm in diameter has a surface emissivity of 1.0. It is heated to a temperature of 450°C and then suddenly thrust into a water bath at 1 atm pressure. Estimate the initial rate of heat transfer from the bar, per unit length.

10.25 Water at atmospheric pressure flows with a bulk temperature of 95°C and an average velocity of 1.5 m/s through a brass tube 1.5 cm in diameter. If the tube surface is at 110°C, find the combined rate of heat transfer due to forced convection and boiling, per unit of length.

CHAPTER 11

Heat Transfer by Radiation

INTRODUCTORY REMARKS

The fundamental physical phenomenon which forms the basis of all heat transfer studies is the observed fact that the temperature of a body, or a portion of a body, which is hotter than its surroundings tends to decrease with time. This temperature decrease indicates a flow of energy from the body. The entire discussion of the foregoing ten chapters has been limited to cases in which some physical medium was necessary for the transport of the energy from the high temperature source to the low temperature sink, leading to the mechanisms of conduction and convection. Generally speaking, the rate of flow of the thermal energy in these instances was proportional to the difference in temperature between the source and the sink.

If, however, a heated body is physically isolated from its cooler surroundings (i.e., by a vacuum), its temperature is still observed to decrease in time, again showing a loss of energy. In this case an entirely different energy transfer mechanism is taking place, and it is called *thermal radiation*.

A body need not be heated to exhibit the loss of energy by radiation. The "thermal" radiation just referred to is one aspect of a more general phenomenon which might be termed *radiant energy*. The emission of other forms of radiant energy may be caused when a body is excited by such means as an oscillating electrical current, electronic or neutronic bombardment, chemical reaction, etc. Also, when radiant energy strikes a body and is absorbed, it may manifest itself in the form of thermal internal energy, a chemical reaction, an electromotive force, etc., depending on the nature of the incident radiation and the substance of which the body is composed. In either case, emission or absorption, this book will be concerned only with thermal radiation—i.e., radiation produced by or which produces thermal excitation of a body.

Several theories have been proposed to explain the transport of energy by radiation. One theory holds that the body emits discrete *packets,* or *quanta,* of energy and has been successful in explaining the experimental facts observed in such cases as the photoelectric emission of electrons, thermal radiation emission, etc. Another theory asserts that radiation may be represented as an electromagnetic wave motion and has been useful in explaining such phenomena as interference of light, polarization of light, etc. At the present time a dual theory is generally accepted, giving radiant energy the characteristics of a wave motion as well as discontinuous emission.

Whichever theory is used, it is convenient to classify all electromagnetic radiant energy emissions in terms of the wavelength when considered as wave motions propagating at the velocity of light—3×10^8 m/s. In this manner, while thermal radiation is found to consist of electromagnetic radiation covering the entire spectrum of wavelengths, most of it is concentrated in the band of wavelengths between 1×10^{-1} and 1×10^2 μm.* Other well-known electromagnetic waves are listed below with their approximate wavelength bands indicated. The subdivision of the electromagnetic spectrum into such bands is somewhat arbitrary.

Cosmic rays:	up to 4×10^{-7} μm
Gamma rays:	4×10^{-7} to 1.4×10^{-4} μm
X-rays:	1×10^{-5} to 2×10^{-2} μm
Ultraviolet rays:	5×10^{-3} to 3.9×10^{-1} μm
Visible light:	3.9×10^{-1} to 7.8×10^{-1} μm
Solar radiation:	1×10^{-1} to 3.0 μm
Infrared radiation:	7.8×10^{-1} to 1×10^3 μm
Thermal radiation:	1×10^{-1} to 1×10^2 μm
Hertzian waves:	1×10^2 to 5×10^{10} μm
Radio waves:	1×10^7 to 5×10^{10} μm

Before proceeding further, it will be necessary to define a number of terms and properties which are used to characterize radiant energy. The following discussion applies mainly to thermal radiation, although some of the concepts are applicable to other forms of radiant emission.

*Most electromagnetic radiation wavelengths are expressed in terms of the micrometer, μm, or 1×10^{-6} m. This unit is also called the *micron*.

BASIC DEFINITIONS

General definitions will be made first. Further discussion will show a need for refinement of these definitions to account for a multitude of special considerations, such as the dependence of a quantity on direction or on wavelength.

As noted in the foregoing section, thermal radiation is defined as that electromagnetic radiation between wavelengths of about 1×10^{-1} and 1×10^2 μm. If the thermal radiation emitted by a surface were to be decomposed into its spectum over the wavelength band, it would be found that the radiation is not equally distributed over all wavelengths. Likewise, radiation incident on a surface, reflected by a surface, absorbed by a surface, etc., may be wavelength dependent. This wavelength dependency will be generally different from case to case, surface to surface, etc. The wavelength dependency of any radiative quantity or surface property will be referred to as a *spectral* dependency. The adjective *monochromatic* will be used to qualify a radiative quantity as one applicable at a single wavelength (e.g., monochromatic emission) while the adjective *total* will indicate that the phenomenon concered has been evaluated over the entire thermal radiation spectrum (e.g., total emission).

Similarly, radiative quantities and surface properties may exhibit *directional* dependencies. For example, a given surface may, because of roughness, emit radiation preferentially in certain directions. In radiation terminology, the adjective *directional* is used to qualify a quantity applicable to a single direction, and the adjective *hemispherical* applies to a quantity when the effect in question has been summed over *all* directions above the surface involved. Except for the quantity *intensity,* to be defined shortly, the presentation in this text will be limited to hemispherical phenomena only and directional effects will not be considered. Generally, the adjective hemispherical will not be applied, but will be understood.

Emissive Power, Radiosity, and Irradiation

Emissive Power.   The term *emissive power* is used to denote the *emitted* thermal radiation leaving a surface, per unit time, per unit area of surface. The *total hemispherical* emissive power of a surface is all the emitted energy, summed over all directions and all wavelengths, and is usually denoted by the symbol E. The total emissive power is found to be dependent upon the temperature of the emitting surface, the substance of which the surface is composed, and the nature of the surface structure (i.e., roughness, etc.).

In general, the emissive power of a given surface element may be both spectrally and directionally dependent. As noted earlier, this text will not consider directional effects, and the emissive power will always be understood to be hemispherical—summed over all directions in the hemisphere above the surface. However, it will be necessary to consider the way in which the emission from a surface is distributed among the wavelengths in the thermal band. The *monochromatic* emissive power, symbolized by E_λ, is then defined as the rate, per unit of area, at which a surface emits thermal radiation at a particular wavelength, λ. Thus, the total and monochromatic hemispherical emissive powers are related by

$$E = \int_0^\infty E_\lambda \, d\lambda \qquad (11.1)$$

and the functional dependence of E_λ on λ must be known to evaluate E.

It should be noted that the emissive power, total or monochromatic, consists only of *original* emission leaving a surface. It does not include any energy leaving a surface that is the result of the reflection of any incident radiation. The quantity *radiosity* is used to account for such reflections, as noted next.

Radiosity. *Radiosity* is the term used to indicate all the radiation *leaving* a surface, per unit time and unit area. The symbols J_λ and J are used to denote the monochromatic and total radiosities, the monochromatic radiosity being that at a single wavelength and the total being that summed over all wavelengths:

$$J = \int_0^\infty J_\lambda \, d\lambda. \qquad (11.2)$$

The radiosity differs from the emissive power in that it includes reflected energy as well as original emission. Again, the work presented here presumes that the radiosity has been summed over all directions over a surface.

Irradiation. *Irradiation* is the term used to denote the rate, per unit area, at which thermal radiation is incident upon a surface (from all directions). The irradiation incident upon a surface is the result of emissions and reflections from other surfaces, and may thus be spectrally dependent. The symbols G_λ and G are used to denote the monochromatic and total irradiation, respectively, and

$$G = \int_0^\infty G_\lambda \, d\lambda. \qquad (11.3)$$

Clearly, the emissive power, radiosity, and irradiation of a surface are interrelated by the reflective, absorptive, and transmissive properties of a surface, as discussed next.

Absorptivity, Reflectivity, and Transmissivity

When radiation is incident on a surface, part of it may be reflected away from the surface, part of it may be absorbed by the surface, and part may be transmitted through the surface. These fractions of reflected, absorbed, and transmitted energy are interpreted as surface properties called *reflectivity, absorptivity,* and *transmissivity,* respectively. Depending on whether or not one is concerned with the surface behavior with respect to incident energy at a specific wavelength or summed over all wavelengths, both *monochromatic* and *total* values of these properties may be defined. The following symbols are used:

ρ_λ = monochromatic reflectivity = fraction reflected at wavelength λ,

ρ = total reflectivity = fraction reflected at all wavelengths,

α_λ = monochromatic absorptivity = fraction absorbed at wavelength λ,

α = total absorptivity = fraction absorbed at all wavelengths,

τ_λ = monochromatic transmissivity = fraction transmitted at wavelength λ,

τ = total transmissivity = fraction transmitted at all wavelengths.

Interrelations between the monochromatic and total properties will be deduced in a later section.

Using the above definitions, energy conservation gives

$$\rho_\lambda + \alpha_\lambda + \tau_\lambda = 1, \tag{11.4}$$

$$\rho + \alpha + \tau = 1.$$

In general both the monochromatic and total surface properties are dependent on the surface composition, its roughness, etc., and on its temperature. The monochromatic properties are dependent on the wavelength of the incident radiation, and the total properties are dependent on the spectral distribution of the incident energy.

In the case of gases, these properties are also dependent on the geometrical size and shape of the gas bulk through which the radiation passes. Most gases have high values of τ and low values of α and ρ. For instance, air at atmospheric pressure is virtually transparent to thermal radiation, so that one may take $\alpha \approx \rho \approx 0$ and $\tau \approx 1$. Other gases, notably water vapor and carbon dioxide, may be highly absorptive to thermal radiation—at least at certain wavelengths.

Most solids, except for glass, encountered in engineering practice are opaque to thermal radiation, so that $\tau \approx 0$. The initial parts of this chapter will be devoted to the study of the radiant exchange between opaque solid surfaces separated by thermally transparent media. The questions which arise when the surfaces are separated by absorbing gases will be considered at the end of the chapter.

For thermally opaque solid surfaces, one has

$$\rho + \alpha = 1. \tag{11.5}$$

From the definitions of emissive power, radiosity, and irradiation, one has the useful relations

$$J = E + \rho G, \tag{11.6}$$

$$= E + (1 - \alpha)G.$$

Monochromatic versions of Eqs. (11.5) and (11.6) may also be written.

The particular behavior of real surfaces with respect to ρ, α, and τ will be discussed following further definitions and the discussion of ideal blackbody radiation.

Intensity of Radiation

In the consideration of the exchange of radiant energy between surfaces it is necessary to define a quantity, known as the *intensity* of radiation. The intensity describes the directional distribution of radiant energy leaving a surface (either by emission or reflection) or the directional distribution of the energy incident upon a surface. A rigorous definition of radiation intensity may be made in terms of the radiation passing a point in space in a particular direction. However, for ease in understanding, a definition of intensity will be made in terms of the energy leaving a "point source" on a surface and how it distributes directionally in the hemisphere above the surface. Whether the energy originates from the surface by original emission or by reflection will be left unspecified for the moment. Also, the definition given in the following may be made for a given wavelength, resulting in a monochromatic intensity, or as a quantity summed over all wavelengths, resulting in a total intensity.

Consider, as illustrated in Fig. 11.1, a small emitting surface of area ΔA, the center of which is denoted as the point Q. Let Δq represent the *rate* at which radiant energy leaves ΔA. It will be useful to think in terms of a radiant energy *flux*. The average flux leaving ΔA is defined as

$$f_{av} = \frac{\Delta q}{\Delta A}.$$

The radiant flux from the *point source* at Q is defined as

$$f_Q = \lim_{\Delta A \to 0} \frac{\Delta q}{\Delta A}. \tag{11.7}$$

It is important to remember that the flux just defined is based on the energy *leaving* a surface (reflected or emitted) and is calculated per unit area of the surface from which it leaves.

Now, the radiant flux from ΔA, or Q, fills the half-space above ΔA. The

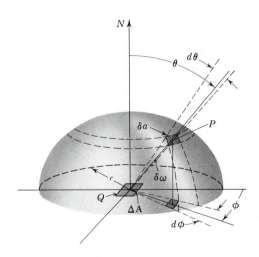

FIGURE 11.1

intensity of the radiation at a point in space due to the emission from this point source is defined as the radiant energy passing the spatial point per unit time, per unit solid angle subtended at Q, per unit area of radiating surface projected normal to the direction which the point in space makes with the emitting point. In terms of the drawing in Fig. 11.1, the radiation intensity at some point in space, say P, due to the radiation leaving ΔA may be defined in terms of the radiation falling on an element of the spherical surface (center at Q, radius r) which passes through P. If δa represents an element of the spherical surface surrounding the point P, only a fraction of Δq strikes δa. Let $\delta(\Delta q)$ represent this fraction of the energy leaving ΔA which falls on δa. In reference to Fig. 11.1, N is the normal to the plane containing ΔA, the radiating surface, θ is the angle between N and the line connecting P and Q, while $\delta\omega = \delta a / r^2$ is the solid angle subtended by δa at the point Q.

By definition, the intensity at the element δa due to the radiation from ΔA, call it $I_{\delta a / \Delta A}$, is

$$I_{\delta a / \Delta A} = \frac{\delta(\Delta q)}{\delta\omega(\Delta A \cos \theta)}.$$

The intensity at δa due to the radiation coming from the *point source Q* is, by use of Eq. (11.7),

$$I_{\delta a / Q} = \frac{\delta f}{\delta\omega} \frac{1}{\cos \theta},$$

$$= \frac{\delta f}{\delta a} \frac{r^2}{\cos \theta}. \tag{11.8}$$

In Eq. (11.8), the flux f is that flux measured at the *radiating* source, Q, not at the receiving area δa.

The intensity at the *point P*, in the direction θ, due to the radiation originating from Q is then defined as the limit of Eq. (11.8) as $\delta a \to 0$:

$$I = \frac{df}{d\omega} \frac{1}{\cos \theta},$$

$$= \frac{df}{da} \frac{r^2}{\cos \theta}. \tag{11.9}$$

The quantity $df/d\omega$ represents the radiated flux, per unit spatial solid angle, while df/da is the flux at P of the flux leaving Q—that is, df/da is the areal density at P of the flux emanating from Q.

Equations (11.9) may be rewritten in the following forms for calculation of the fraction of the flux leaving the surface that is contained in the solid angle $d\omega$ or intercepted by the receiving element da:

$$df = I \cos \theta \, d\omega, \tag{11.10}$$

$$= I \cos \theta \, \frac{da}{r^2}.$$

Since all of the flux leaving the surface at Q must pass into the hemisphere above it, as depicted in Fig. 11.1, then

$$f = \int_h I \cos \theta \, d\omega, \tag{11.11}$$

in which the notation $\int_h$ is used to denote an integration taken over all the solid angle of the hemisphere. In terms of the polar angle θ and the azimuthal angle φ (i.e., longitude) shown in Fig. 11.1,

$$d\omega = \frac{da}{r^2} = \frac{r \, d\theta(r \sin \theta \, d\varphi)}{r^2},$$

$$= \sin \theta \, d\varphi \, d\theta.$$

Thus, Eq. (11.11) is also

$$f = \int_0^{2\pi} \int_0^{\pi/2} I \sin \theta \cos \theta \, d\theta \, d\varphi. \tag{11.12}$$

Intensity, it is seen, is an inherently directional quantity and in general I may depend on the two angles θ and φ. Thus knowledge of $I = I(\theta, \varphi)$ is necessary in order to evaluate f from Eq. (11.11) or (11.12). Radiation is termed *diffuse* if I is uniform—i.e., independent of θ and φ. Thus, for diffuse radiation

$$f = I \int_0^{2\pi} \int_0^{\pi/2} \sin \theta \cos \theta \, d\theta \, d\varphi,$$

$$= I\pi, \tag{11.13}$$

and the surface flux and the intensity are related in a very simple way. Also, Eq. (11.9) shows that if I is constant, the flux intercepted by an area element varies as $\cos \theta$ and inversely as r^2—the familiar *Lambert's law* of diffuse radiation.

The preceding definition and discussion of intensity were based on the assumption that the radiant flux, f, was leaving the surface and flowing *outward* in the direction given by θ and φ and contained in the solid angle $d\omega$. In the event one is interested only in the original emission leaving a surface, the flux is the hemispherical emissive power, E, and the associated intensity will be symbolized by $I_e(\theta, \varphi)$. Thus,

$$E = \int_h I_e(\theta, \varphi) \cos \theta \, d\omega = \int_0^{2\pi} \int_0^{\pi/2} I_e(\theta, \varphi) \sin \theta \cos \theta \, d\theta \, d\varphi. \tag{11.14}$$

A surface is termed a *diffuse emitter* if $I_e(\theta, \varphi)$ is uniform. Then,

$$E = \pi I_e. \tag{11.15}$$

If one is interested in all the energy leaving a surface, the flux is the hemispherical radiosity, J, and the associated intensity is denoted by $I_{e+r}(\theta, \varphi)$ to indicate that both reflected energy and original emission are involved. Then,

$$J = \int_h I_{e+r}(\theta, \varphi) \cos \theta \, d\omega = \int_0^{2\pi} \int_0^{\pi/2} I_{e+r}(\theta, \varphi) \sin \theta \cos \theta \, d\theta \, d\varphi. \tag{11.16}$$

If a surface is a diffuse emitter *and* a *diffuse reflector,* then

$$J = \pi I_{e+r}. \tag{11.17}$$

The geometric variables associated with these concepts are illustrated in Fig. 11.2(a).

In addition to the radiant energy flowing outward from a surface as just discussed, one may also associate with any direction (θ, φ) an inwardly directed intensity that would result from radiation originating (by emission or reflection) from other surfaces. If $I_i(\theta, \varphi)$ represents the rate at which energy is incident upon a surface, per unit area of intercepting surface normal to the direction (θ, φ), per unit solid angle, then the same relations given in Eqs. (11.11) and (11.12) apply with I_i replacing I. The flux intercepted by the surface is the irradiation defined earlier. Thus, as suggested in Fig. 11.2(b),

$$G = \int_h I_i(\theta, \varphi) \cos \theta \, d\omega = \int_0^{2\pi} \int_0^{\pi/2} I_i(\theta, \varphi) \sin \theta \cos \theta \, d\theta \, d\varphi. \tag{11.18}$$

If the incident energy is diffuse so that I_i is uniform, then

$$G = \pi I_i. \tag{11.19}$$

As mentioned earlier, all the foregoing definitions for intensity, its relation to emissive power, radiosity, irradiation, etc., may be made in the monochromatic sense as well as the total sense. That is, all the energy fluxes are defined for a given wavelength, and the associated intensities are monochromatic ones (i.e., $I_\lambda, I_{\lambda,e}, I_{\lambda,e+r}, I_{\lambda,i}$). The integrations given in Eqs. (11.14) through (11.19) then yield the monochromatic emissive power, radiosity, and irradiation. Equations (11.1) through (11.3) relate these latter quantities to their total counterparts.

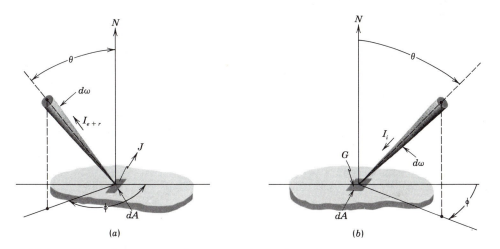

FIGURE 11.2. Angular relations for radiation intensity.

BLACKBODY RADIATION

In order to describe the radiation characteristics and properties of real surfaces, it is useful to define an ideal surface for purposes of comparison. Such an ideal surface is the *perfect blackbody*. A perfect blackbody is defined as one which absorbs *all* incident radiation regardless of the spectral distribution or directional character of the incident radiation—i.e., $\alpha_\lambda = \alpha = 1$ or $\rho_\lambda = \rho = 0$. The term "black" is used since dark surfaces normally show high values of absorptivity. Since a blackbody absorbs all incident radiation, the only radiation leaving a blackbody surface is original emission. The emissive power of a blackbody, to be denoted by E_b, will be seen to depend on the surface temperature only. This dependence on temperature and the spatial distribution of blackbody radiation as given by the intensity function $I_b(\theta, \varphi)$ will be developed in the following discussion. Also the spectral distribution of the emitted energy as given by the dependence of the monochromatic emissive power, $E_{b\lambda}$, on wavelength, is of great importance and will be discussed in some detail.

While a perfect blackbody is an idealization, it is possible to produce radiation in the laboratory that is very nearly the same as that which would originate from such a surface. Imagine a hollow cavity, such as depicted in Fig. 11.3(a), the walls of which are maintained at a uniform temperature. If a small hole (small compared to the size of the cavity) is provided in the wall, then any ray of incident radiant energy entering the hole will undergo a large number of internal reflections—with a portion of the radiation being absorbed at each reflection. Very little of the incident beam ever finds its way out of the small hole, and thus the plane of the hole appears to be a perfect absorber. Thus, the radiation found to be emanating from the hole will appear to be that coming from a blackbody.

This concept of blackbody radiation may also be explained by consideration of the situation depicted in Fig. 11.3(b). Here, one imagines a cavity, the walls of which have been heated to a *uniform* temperature. The surface of the cavity

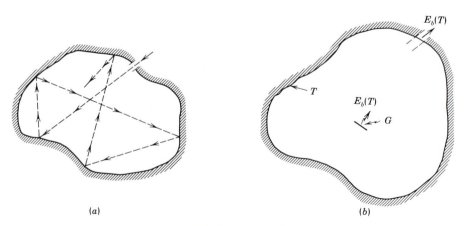

(a)   (b)

FIGURE 11.3

is not necessarily black, and thus the space is filled with a radiation field which is the result of original emission and reflections from the walls. Since the walls are maintained at a uniform temperature, the resultant radiant field is in thermal equilibrium with the walls. Now, imagine a very small wafer of material to be placed inside the cavity—small enough that its presence only negligibly alters the existing radiation field. Further, presume that the material of which the wafer is composed is a perfect blackbody and that it is allowed to come to thermal equilibrium with the walls—that is, the wafer temperature is the same as that of the walls. If $E_b(T)$ is the emissive power of a blackbody at temperature T, then $E_b(T)$ is the flux of energy away from the surface of the wafer. Since thermal equilibrium exists, the energy flux incident on the wafer, i.e., the irradiation G, must be the same since it is all absorbed. Changing the orientation or position of the wafer does not change this result, and one may say that $G = E_b(T)$ throughout the cavity. Thus one concludes that the irradiation field produced in a uniformly heated cavity, regardless of the cavity surface properties, is uniform and equal to the emissive power of an ideal blackbody at the temperature of the walls. If a very small hole were made in the wall of the cavity, the radiation escaping from it would be that which would be emitted by a blackbody.

The Stefan–Boltzmann Law for Blackbody Emissive Power

In the preceding discussion showing that the radiation field in an isothermal enclosure is the emissive power of a blackbody at the same temperature, the only surface property involved was the temperature. Thus, one concludes that the emissive power of an ideal blackbody is a function of temperature only. Based on experiments, Josef Stefan in 1879 suggested that the total emissive power of a blackbody is proportional to the fourth power of the absolute temperature. Later, Ludwig Boltzmann applied the principles of classical thermodynamics and analytically derived the same fact. Thus the following dependence of E_b on T is called the *Stefan–Boltzmann law:*

$$E_b = \sigma T^4, \tag{11.20}$$

in which σ is the Stefan–Boltzmann constant with the value

$$\sigma = 5.670 \times 10^{-8} \text{ W/m}^2\text{-}^\circ\text{K}^4,$$

$$= 0.1714 \times 10^{-8} \text{ Btu/h-ft}^2\text{-}^\circ\text{R}^4.$$

The temperature of the black surface is the *absolute* temperature. For computational purposes it is convenient to note that

$$E_b = \sigma T^4 = 5.670 \times \left(\frac{T}{100}\right)^4 \text{ W/m}^2 (T \text{ in } ^\circ\text{K}),$$

$$= 0.1714 \left(\frac{T}{100}\right)^4 \text{ Btu/h-ft}^2 (T \text{ in } ^\circ\text{R}).$$

The Stefan–Boltzmann law for the *total* emissive power gives the total energy emitted by a blackbody, summed over all wavelengths. It is also derivable from Planck's law for the monochromatic emissive power as discussed later.

Spatial Characteristics of Blackbody Radiation

The earlier discussion not only showed that the irradiation field in an isothermal cavity is equal to E_b but also that the irradiation is the same for *all* planes of *any* orientation within the cavity. Thus, the intergral expression of Eq. (11.18) must yield the same result for all locations and orientations. It may then be shown that the intensity of blackbody radiation, I_b, is uniform. Thus, blackbody radiation is diffuse and

$$E_b = \pi I_b. \tag{11.21}$$

Spectral Distribution of Blackbody Radiation

The Stefan–Boltzmann law gives the *total* emissive power of an ideal black-body. The distribution of the emission from a blackbody over the complete spectrum of wavelengths as given by the monochromatic emissive power $E_{b\lambda} = E_{b\lambda}(\lambda, T)$ is also of great interest. A typical distribution of $E_{b\lambda}$ is shown in Fig. 11.4. The monochromatic emission (emitted energy per unit time and area at *a* wavelength) is seen to rise from very small values for short wave-lengths, reach a peak value, and then fall again to small values as the wavelength becomes large. Most of the emission at temperatures encountered in engineering applications lies in the wavelength band between 1×10^{-1} and 1×10^{2} μm, as noted earlier. The area under the $E_{b\lambda}$ versus λ curve is the total emissive power given in Eqs. (11.1) and (11.20):

$$E_b = \int_0^\infty E_{b\lambda}(\lambda, T)\, d\lambda = \sigma T^4. \tag{11.22}$$

Since E_b is a function of only the surface temperature, it follows that the monochromatic blackbody emissive power depends only on temperature *and* wavelength. The analytic prediction of the observed dependence of $E_{b\lambda}$ on T and λ was one of the central problems of interest to physicists of the late nineteenth century. Its eventual discovery by Max Planck in 1901 marked the first achievement of his quantum theory of radiation. Planck's law is given as

$$E_{b\lambda}(\lambda, T) = \frac{C_1}{\lambda^5 (e^{C_2/\lambda T} - 1)}, \tag{11.23}$$

where the accepted values of the two constants are

$$C_1 = 3.7413 \times 10^8 \text{ W-}\mu\text{m}^4/\text{m}^2,$$
$$C_2 = 1.4388 \times 10^4 \ \mu\text{m-}°\text{K}.$$

The units of C_1 and C_2 indicate that λ in Eq. (11.23) is to be expressed in μm. Integration of Eq. (11.23) according to Eq. (11.22) will show that the Ste-fan–Boltzmann constant is related to C_1 and C_2:

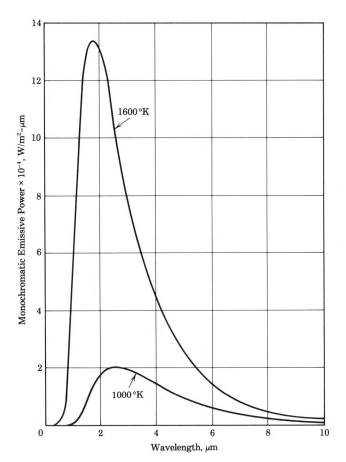

FIGURE 11.4. Monochromatic emissive power of an ideal blackbody at different temperatures.

$$\sigma = \left(\frac{\pi}{C_2}\right)^4 \frac{C_1}{15}. \tag{11.24}$$

Figure 11.4 shows a plot of Planck's equation for two different temperatures. The area under these curves is proportional to T^4 in accord with the Stefan–Boltzmann law. As may be deduced from Eq. (11.23), the monochromatic blackbody emissive power exhibits a peak value that increases with T and which occurs at shorter wavelengths as T increases. *Wien's law* may be deduced from Planck's, and states that the maximum value of $E_{b\lambda}$ occurs at the wavelength given by

$$\lambda_{max}T = C_3 \tag{11.25}$$

$$= 2897.8 \ \mu\text{m-}°\text{K}.$$

For use in performing certain calculations to be described later it is useful to have tabulated values of the monochromatic blackbody emissive power. This is most efficiently done by noting that the ratio $E_{b\lambda}/\sigma T^5$ is a function of the product λT only:

$$\frac{E_{b\lambda}}{\sigma T^5} = \frac{C_1}{\sigma} \frac{1}{(\lambda T)^5 (e^{C_2/\lambda T} - 1)}. \tag{11.26}$$

Equation (11.26) is displayed graphically in Fig. 11.5 and is tabulated in Table A.12. For given λ and T, $E_{b\lambda}$ may be readily evaluated.

Also of great subsequent use is the fraction of the total emission (i.e., the area under the curves in Fig. 11.4) that occurs between specified values of the wavelength. This fraction is readily determined from the partial integral of $E_{b\lambda}$ between $\lambda = 0$ and any value λ. This partial integral, when divided by the total emission is defined as the quantity $F_{0-\lambda}(T)$:

$$F_{0-\lambda}(T) = \frac{\int_0^\lambda E_{b\lambda}\, d\lambda}{\int_0^\infty E_{b\lambda}\, d\lambda} = \frac{\int_0^\lambda E_{b\lambda}\, d\lambda}{\sigma T^4},$$

$$= \int_0^{\lambda T} \frac{C_1 d(\lambda T)}{\sigma(\lambda T)^5 (e^{C_2/\lambda T} - 1)}. \tag{11.27}$$

This latter function is also seen to be a function of only the product λT and is also displayed in Fig. 11.5 and tabulated in Table A.12. The fraction of energy emitted between two wavelengths, λ_1 and λ_2, by a blackbody at temperature T is thus

$$F_{\lambda_1 - \lambda_2}(T) = F_{0-\lambda_2}(T) - F_{0-\lambda_1}(T).$$

EXAMPLE 11.1

A blackbody has a surface temperature of 600°K. Find the total emissive power, the wavelength of the maximum monochromatic emissive power, the magnitude of this maximum, and the fraction of the total emission that occurs between the wavelengths of 2.0 and 6.0 μm.

Solution. At T = 600°K, Eq. (11.20) gives the total emissive power to be

$$E_b = 5.670\left(\frac{600}{100}\right)^4 = 7348 \text{ W/m}^2.$$

Wien's law in Eq. (11.25) gives the wavelength of the maximum monochromatic emissive power to be

$$\lambda_{max} = \frac{2897.8}{600} = 4.83 \ \mu\text{m}.$$

For $\lambda T = \lambda_{max}T = 2897.8 \ \mu\text{m-°K}$, Table A.12 gives

$$\left(\frac{E_{b\lambda}}{\sigma T^5}\right)_{max} = 22.688 \times 10^{-5} \ (\mu\text{m-°K})^{-1},$$

$$(E_{b\lambda})_{max} = 5.670 \times 10^{-8} \times (600)^5 \times 22.692 \times 10^{-5}$$

$$= 1.00 \times 10^3 \text{ W/m}^2\text{-}\mu\text{m}.$$

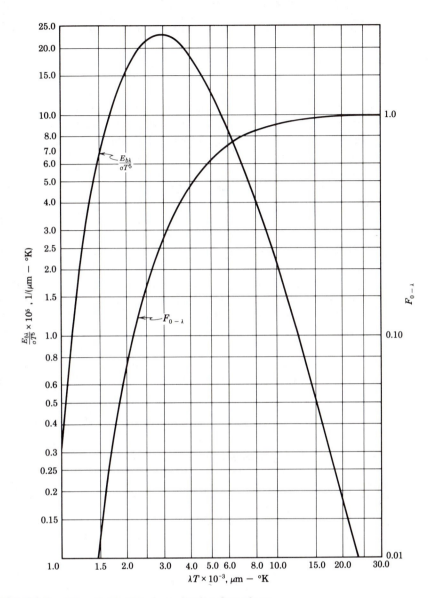

FIGURE 11.5. Planck blackbody radiation functions.

At the wavelength of 2.0 μm, $\lambda_1 T = 2 \times 600 = 1200$ μm-$°$K and at $\lambda_2 = 6.0$ μm, $\lambda_2 T = 3600$ μm-$°$K. Table A.12 gives

$$F_{0-\lambda_1} = 0.00213$$

$$F_{0-\lambda_2} = 0.40360,$$

so the fraction of the total emission that occurs between 2.0 and 6.0 μm is

$$F_{6.0-2.0} = 0.40360 - 0.00213 = 0.40147. \qquad \blacksquare$$

Calculations similar to those just illustrated in Example 11.1 reveal the dramatic effect of the surface temperature on the spectral distribution of blackbody radiation as shown in Table 11.1. The temperature of 300°K was chosen as representative of typically encountered ambient temperatures, and 5800°K will be shown later to approximate the emission of the sun. The effect of temperature on the total emissive power is profound in keeping with the fourth-power requirement of the Stefan–Boltzmann law. However, the effect of temperature on the spectal distribution is equally dramatic. Emission at typical ambient temperatures is seen to be predominately at the larger wavelengths and almost entirely in the infrared portion on the spectrum. As temperatures corresponding to the sun's emission are approached, it may be noted that significant portions occur in the ultraviolet and visible portions of the spectrum. It is worth noting that the wavelength of the maximum emission from the sun occurs almost exactly in the middle of the visible spectrum as sensed by the human eye. The last two rows of data, the fractions of the emission that occurs above and below a wavelength of about 4 μm, are particularly interesting as far as analyses associated with solar phenomena are concerned. Surfaces at typically ambient temperatures of about 300°K emit almost all their energy at wavelengths in excess of about 4 μm, while the sun emits almost all its energy at wavelengths less than 4 μm. Thus, a cool surface subjected to solar irradiation will emit energy of a markedly different spectral character than that incident upon it. This "uncoupling" of the radiation may produce a significant difference between the emissive and solar-absorptive properties of real surfaces as will be illustrated later.

TABLE 11.1 Effect of Temperature on Blackbody Emission

Surface temperature, °K	300	800	1600	5800
Total emissive power, W/m^2	459.2	23,220	3.71×10^5	64.16×10^6
Wavelength of maximum emission, μm	9.66	3.62	1.81	0.500
Fraction of emission in the band				
Ultraviolet (5×10^{-3}–3.9×10^{-1} μm)	0.000	0.000	0.000	0.112
Visible Light (3.9×10^{-1}–7.8×10^{-1} μm)	0.000	0.000	0.003	0.456
Infared (7.8×10^{-1}–1×10^3)	1.000	1.000	0.997	0.432
Fraction of emission				
Below $\lambda = 4$ μm	0.002	0.318	0.769	0.990
Above $\lambda = 4$ μm	0.998	0.682	0.231	0.010

RADIATION CHARACTERISTICS OF NONBLACK SURFACES

The monochromatic and total emissive powers of blackbody radiation as given by the Planck and Stefan–Boltzmann laws, together with the diffuse nature of blackbody radiation intensity, may be used as the ideal basis to which the radiative characteristics of real surfaces may be compared. The radiant characteristics of nonblack, real, surfaces differ from the ideal blackbody in several important ways. The monochromatic emissive power of a real surface at a given temperature differs from that of a blackbody at the same temperature both in amount and in spectral distribution. A nonblack surface exhibits an absorptivity less than unity, and its value may be dependent on the wavelength of the incident radiation. Thus, a nonblack surface reflects, perhaps with spectral preference, some incident radiation, so that its surface radiosity consists of more than original emission.

Nonblack surfaces may also exhibit nondiffuse behavior, so that the intensity of emission or of reflected incident energy may not be constant as is the case for blackbody radiation. A proper, rigorous approach to the description of real surfaces would be that which first defines *monochromatic directional* properties and then proceeds to the *total* and *hemispherical* properties by appropriate integration over the spectrum and all directions. However, the presentation here will ignore possible directional dependencies by either defining simply the *hemispherical* properties or by assuming diffuseness when appropriate.

Emissivity, Monochromatic and Total

If $E_\lambda(\lambda, T)$ represents the hemispherical monochromatic emissive power of a nonblack surface maintained at a temperature T and measured at a particular wavelength λ, then the *hemispherical monochromatic emissivity* is defined as the ratio of $E_\lambda(\lambda, T)$ to the hemispherical monochromatic emissive power of a blackbody at the same T and λ:

$$\epsilon_\lambda(\lambda, T) = \frac{E_\lambda(\lambda, T)}{E_{b\lambda}(\lambda, T)}. \qquad (11.28)$$

Note that no statement has been made as to whether the monochromatic emissivity, ϵ_λ, is greater or less than 1. That is, in this definition of ϵ_λ no assumption has been made as to whether or not a blackbody is a better or poorer emitter than a nonblack one. The fact that ϵ_λ must be less than 1 will be demonstrated later by use of Kirchhoff's law. It should be emphasized that ϵ_λ is a *hemispherical* emissivity, the emissive powers E_λ and $E_{b\lambda}$ being the sum of the energy emitted in all directions in the hemisphere above the surface.

Most real surfaces exhibit, to some degree, the characteristics of a "selective emitter" in that ϵ_λ is different for different wavelengths of the emitted energy. Also, ϵ_λ may be dependent on the surface temperature both in magnitude and spectral dependence, hence the notation $\epsilon_\lambda(\lambda, T)$; however, this temperature dependence may be small and is often ignored. Figures 11.6 and 11.7 show typical values of ϵ_λ, as functions of wavelength, for a number of surfaces. It may be noted that some surfaces exhibit dramatic variations of ϵ_λ with λ, while others show fairly constant values of ϵ_λ. Similar data are available in the literature for other surfaces and Refs. 1 through 4 may be consulted in this regard.

The hemispherical *total* emissivity is defined as the ratio of the total emissive power of a nonblack surface to that for a blackbody at the same temperature:

$$\epsilon(T) = \frac{E(T)}{E_b(T)},$$ (11.29)

so that the emissive power of a nonblack surface is readily calculated when ϵ is known:

$$E(T) = \epsilon\sigma T^4.$$ (11.30)

The total emissivity may be expressed in terms of the monochromatic as follows:

$$\epsilon(T) = \frac{E(T)}{E_b(T)} = \frac{\int_0^\infty E_\lambda(T)\, d\lambda}{\int_0^\infty E_{b\lambda}(\lambda, T)\, d\lambda},$$

$$= \frac{\int_0^\infty \epsilon_\lambda(\lambda, T) E_{b\lambda}(\lambda, T)\, d\lambda}{\int_0^\infty E_{b\lambda}(\lambda, T)\, d\lambda},$$ (11.31)

$$= \frac{\int_0^\infty \epsilon_\lambda(\lambda, T) E_{b\lambda}(\lambda, T)\, d\lambda}{\sigma T^4}.$$

Since the monochromatic emissivity is a function of λ and T, Eq. (11.31) shows the total emissivity to depend on the surface temperature only. The total emissivity may be measured directly on the basis given in Eq. (11.29). Table A.10 lists values of the total emissivity for various surfaces, including the temperature dependency, and Fig. 11.8 illustrates graphically some typical data.

The total emissivity may be evaluated from the monochromatic emissivity by application of Eq. (11.31). If the $\epsilon_\lambda(\lambda, T)$ function is known, then Eq. (11.31) may be evaluated, probably numerically, using the Planck function in Eq. (11.23) or (11.26). If the ϵ_λ versus λ data are complex, as in Figs. 11.6 and 11.7, or known in numerical form, the integral of Eq. (11.31) may be expressed in terms of the function $F_{0-\lambda}$ given in Eq. (11.27) in a finite difference form. The spectrum of ϵ_λ versus λ may be divided into intervals $(\lambda_1, \lambda_2, \ldots, \lambda_i, \ldots, \lambda_n)$ in which the monochromatic emissivity may be taken as approximately constant. If $\bar{\epsilon}_{\lambda_{i,i+1}}$ represents the average value of ϵ_λ between the wavelengths λ_i and λ_{i+1}, then Eq. (11.31) may be written in the form

$$\epsilon_\lambda(\lambda, T) = \sum \bar{\epsilon}_{\lambda_{i,i+1}}[F_{0-\lambda_{i+1}}(T) - F_{0-\lambda_i}(T)].$$ (11.32)

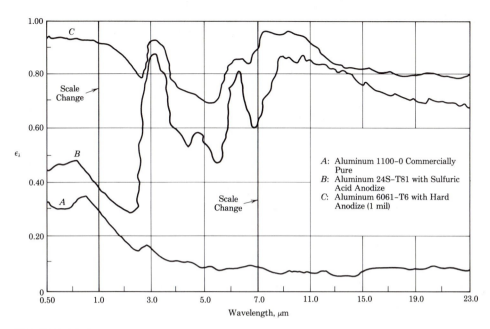

FIGURE 11.6. Monochromatic emissivity of various surfaces. (From D. K. Edwards, K. E. Nelson, R. D. Roddick, and J. T. Gier, "Basic Studies on the Use and Control of Solar Energy," *Univ. Calif. Dept. Eng. Rep. 60–93,* 1960. Used by permission.)

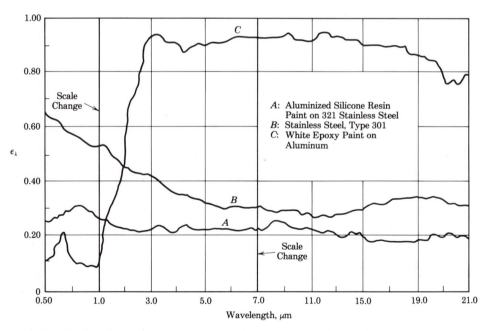

FIGURE 11.7. Monochromatic emissivity of various surfaces. (From D. K. Edwards, K. E. Nelson, R. D. Roddick, and J. T. Gier, "Basic Studies on the Use and Control of Solar Energy," *Univ. Calif. Dept. Eng. Rep. 60–93,* 1960. Used by permission.)

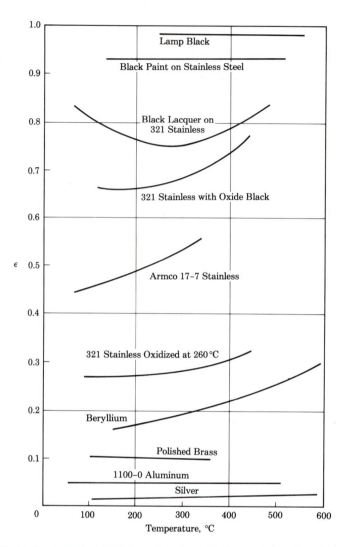

FIGURE 11.8. Total emissivity of various surfaces. (Based on data in Table A.10.)

Use of this relation is shown in Example 11.2.

A special type of nonblack surface, called the *ideal gray body*, is defined as one for which the monochromatic emissivity is independent of wavelength; i.e., the ratio of E_λ to $E_{b\lambda}$ is the same for all wavelengths of emitted energy at the given temperature. Thus, for the ideal gray surface, one readily obtains from Eq. (11.31):

$$\epsilon(T) = \epsilon_\lambda(T). \qquad (11.33)$$

There are other rather profound simplifications that result from a surface exhibiting gray behavior, but these must be reserved until after the discussion of the property known as absorptivity. The assumption of a gray surface, ϵ_λ independent of λ, means that the curves of the monochromatic emissive power for a blackbody and a gray surface at the same temperature are affine to one another, there being no shift in the peak of the curve as illustrated in Fig. 11.9. Also shown in Fig. 11.9 is the monochromatic emissive power of a typical nonblack, nongray surface.

EXAMPLE 11.2

A nonblack surface exhibits a monochromatic emissivity as shown in Fig. 11.10. If the surface temperature is 1500°K, find the total emissivity.

Solution. Equation (11.32) gives

$$\epsilon = \bar{\epsilon}_{0,\lambda_1}[F_{0-\lambda_1}(T) - 0] + \bar{\epsilon}_{\lambda_1,\lambda_2}[F_{0-\lambda_2}(T) - F_{0-\lambda_1}(T)]$$
$$+ \bar{\epsilon}_{\lambda_2,\infty}[1 - F_{0-\lambda_2}(T)],$$

Where the facts that the function $F_{0-\lambda} = 0$ for $\lambda = 0$ and $F_{0-\lambda} = 1$ as $\lambda \to \infty$ are used. The data of Fig. (11.10) give $\bar{\epsilon}_{0,\lambda_1} = 0.2$, $\bar{\epsilon}_{\lambda_1,\lambda_2} = 0.8$, $\bar{\epsilon}_{\lambda_2,\infty} = 0.4$, $\lambda_1 = 2$ μm, $\lambda_2 = 5$ μm. Thus, Table A.12 gives

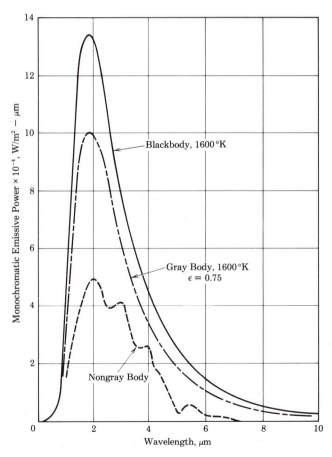

FIGURE 11.9. Monochromatic emissive power of a blackbody, a gray body, and a nongray body.

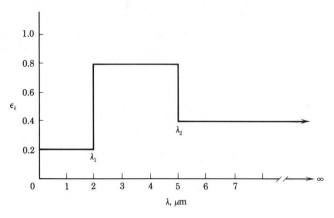

FIGURE 11.10.   Example 11.2.

At $\lambda_1 T = 2 \times 1500 = 3000 \ \mu m$, $F_{0-\lambda_1} = 0.273$,

At $\lambda_2 T = 5 \times 1500 = 7500 \ \mu m$, $F_{0-\lambda_2} = 0.834$.

Thus,

$\epsilon = 0.2 \times 0.273 + 0.8(0.834 - 0.273) + 0.4(1 - 0.834) = 0.570.$ ■

Absorptivity, Monochromatic and Total

One of the most significant ways in which real surfaces differ from the ideal blackbody is that real surfaces do not, generally, absorb all incident energy. Like emissivity, the absorptive character of a surface may exhibit both directional and spectral selectivity. Once again ignoring directional preferences by considering only the *hemispherical* irradiation incident upon a surface, G_λ, the *hemispherical monochromatic absorptivity* is defined as the fraction of this incident radiation that is absorbed:

$$\alpha_\lambda(\lambda, T) = \frac{[G_\lambda(\lambda)]_{abs}}{G_\lambda(\lambda)}. \tag{11.34}$$

In this definition G_λ is the irradiation, from all directions, incident on the surface at *a* particular wavelength and can be expressed in terms of the incident intensity by use of Eq. (11.18):

$$G_\lambda(\lambda) = \int_h I_{\lambda_i}(\lambda, \theta, \varphi) \cos \theta \ d\omega. \tag{11.35}$$

In this latter expression $I_{\lambda_i}(\lambda, \theta, \varphi)$ represents the incident *monochromatic* intensity, which, in general, is a function of direction as well as λ. Equation (11.34) presumes this function is known and the integral has been evaluated.

The notation $\alpha_\lambda(\lambda, T)$ in Eq. (11.34) is used to emphasize the fact that in addition to the obvious dependence of absorptivity on wavelength, experiment shows that it is also dependent on the surface temperature. It would appear that on the basis of the above definition, the monochromatic absorptivity α_λ is an additional surface property that would require experimental determination as a function of λ and T. However, as later discussion will show, Kirchhoff's law relates α_λ to ϵ_λ, defined previously. For the time being, however, α_λ will continue to be written as a separate entity.

The hemispherical *total* absorptivity is defined as the sum of the absorbed irradiation over all wavelengths divided by the total incidence. With this definition, the total absorptivity may be written in the following ways, using Eqs. (11.34) and (11.35):

$$\alpha(T, \text{ source}) = \frac{\int_0^\infty [G_\lambda(\lambda)]_{abs} \, d\lambda}{\int_0^\infty G_\lambda(\lambda) \, d\lambda}$$

$$= \frac{\int_0^\infty \alpha_\lambda(\lambda, T) G_\lambda(\lambda) \, d\lambda}{\int_0^\infty G_\lambda(\lambda) \, d\lambda} \qquad (11.36)$$

$$= \frac{\int_0^\infty \alpha_\lambda(\lambda, T) \int_h I_{\lambda_i}(\lambda, \theta, \varphi) \cos \theta \, d\omega \, d\lambda}{\int_0^\infty \int_h I_{\lambda_i}(\lambda, \theta, \varphi) \cos \theta \, d\omega \, d\lambda}.$$

The notation $\alpha(T, \text{ source})$ is used to emphasize the fact that the total absorptivity depends not only on the absorbing surface temperature as implied in the function $\alpha_\lambda(\lambda, T)$, but also on the spectral and spatial characteristics of the incident radiation. In this sense the total absorptivity of a surface differs significantly from the total emissivity already defined. The total emissivity was seen to be a function of only the surface and its temperature; the total absorptivity is *also* a function of the source of the radiation incident on the surface. Thus, the total absorptivity is not a simple surface property that can be tabulated as can emissivity—a separate tabulation would have to be made for each source. This is normally not done except for a unique source, the sun, as noted in a later section.

It should be noted that a fundamental fact associated with the definition of the monochromatic and total absorptivities is that they are quantities necessarily less than or equal to 1.

One could continue in the line of reasoning presented for ϵ and α to define monochromatic and total values for the reflectivity and transmissivity of a surface. Reference 1 presents a comprehensive treatment of such formulations. However, most applications will be made to opaque surfaces so that $\tau = 0$, and the reflectivity may be deduced from

$$\rho_\lambda = 1 - \alpha_\lambda,$$

$$\rho = 1 - \alpha.$$

Kirchhoff's Law

The properties emissivity and absorptivity discussed in the foregoing may be related by what is known as Kirchhoff's law. It has already been established at the beginning of Sec. 11.3 that the irradiation field in an isothermal enclosure is equal to the emissive power of a blackbody at the temperature of the enclosure, i.e.,

$$G = E_b(T).$$

If a small nonblack surface is placed in the enclosure (small enough that it does not alter the irradiation field) with total emissivity ϵ and total absorptivity α, and allowed to come to the same temperature as the enclosure, then thermal equilibrium requires that the absorbed and emitted energies be equal:

$$\alpha G = E(T) = \epsilon E_b(T).$$

Since $G = E_b(T)$, as noted above, one has

$$\alpha E_b(T) = \epsilon E_b(T), \tag{11.37}$$

$$\alpha = \epsilon.$$

Equation (11.37) is one form of the statement of Kirchhoff's law. Since $\alpha \leq 1$, then $\epsilon \leq 1$; that is, a blackbody, a perfect absorber, is also a perfect emitter.

It should be noted that the form of Kirchhoff's law just quoted in Eq. (11.37) is subject to the restrictions of *thermal equilibrium* in an *isothermal enclosure*. This also includes the condition of diffuseness since it was shown in Sec. 11.3 that the radiant field in an isothermal enclosure is diffuse. Extensions of Eq. (11.37) to other situations (as attractive as they may be) must be made with caution. Reference 1 presents an excellent and detailed exposition of Kirchhoff's law, including directional and spectral effects. In the case of *monochromatic* radiation, a derivation similar to that just shown yields the fact that

$$\alpha_\lambda(\lambda, T) = \epsilon_\lambda(\lambda, T) \tag{11.38}$$

as long as *either* the incident radiation is diffuse *or* the surface is such that α_λ and ϵ_λ have no directional dependencies. As long as these conditions are met, then known data for $\epsilon_\lambda(\lambda, T)$ such as are shown in Figs. 11.6 and 11.7 may be used to evaluate the total absorptivity from Eq. (11.36). Thus, additional physical data are not necessary; however, one still needs to know the spectral nature of the incident source.

The monochromatic form of Kirchhoff's law in Eq. (11.38) may be used with Eqs. (11.31) and (11.36) to deduce under what conditions that the total form of this law, Eq. (11.37), will be valid. Repeating Eq. (11.31) and rewriting Eq. (11.36) with the use of the monochromatic form of Kirchhoff's law:

$$\epsilon(T) = \frac{\displaystyle\int_0^\infty \epsilon_\lambda(\lambda, T) E_{b\lambda}(\lambda, T)\, d\lambda}{\displaystyle\int_0^\infty E_{b\lambda}(\lambda, T)\, d\lambda}, \tag{11.31}$$

$$\alpha(T, \text{ source}) = \frac{\int_0^\infty \epsilon_\lambda(\lambda, T) G_\lambda(\lambda) \, d\lambda}{\int_0^\infty G_\lambda(\lambda) \, d\lambda}. \tag{11.39}$$

Comparison of the above two equations shows that Kirchhoff's law in the total sense,

$$\alpha(T) = \epsilon(T), \tag{11.40}$$

occurs when *either* of the two following conditions is met:

1. If the incident irradiation on the receiving surface has a spectral distribution the same as that of the emission from a blackbody at the *same temperature*, i.e., $G_\lambda(\lambda) = E_{b\lambda}(\lambda)$.
2. If the receiving surface is an ideal *gray surface*, i.e., ϵ_λ is *not* a function of λ so that it may be moved from under the integral signs in Eqs. (11.31) and (11.39).

The first of the above two conditions is a restriction on the source of the irradiation, which must be that of a blackbody at the same temperature as the receiver. This was the case of the isothermal enclosure assumption leading to Eq. (11.37), but it is of little interest since thermal equilibrium seldom exists in heat transfer applications. The second condition, that of grayness, is a condition on the nature of the surface rather than the incident radiation. While there are numerous engineering applications in which the assumption of a gray surface may not be made (such as in the solar applications to be discussed next) there are many instances in which this idealization may be made. In such cases heat transfer calculations can be carried out using the total form of Kirchhoff's law, Eq. (11.40), even though thermal equilibrium does not exist. Such analyses are carried out extensively in Secs. 11.6 through 11.10.

11.5

SOLAR RADIATION AND SOLAR ABSORPTIVITY

Many applications of current engineering interest involving thermal radiation are ones in which radiation from the sun plays an important role. Many of these applications involve the use of surfaces for which the simplifying assumption of grayness cannot be made. Hence, some characterization needs to be made of the solar radiation received at the earth.

Figure 11.11 illustrates the physical situation at hand. The sun, located at a distance from the earth equal to the earth's orbit radius r_e, radiates from its surface in all directions. The radiation received at a surface element on the earth's surface appears to be virtually unidirectional, striking the surface at a solar angle of θ_s with the surface normal. As viewed from the surface, the sun subtends a solid angle $\Delta\omega$ which is

$$\Delta\omega = \frac{\pi r_s^2}{r_e^2},$$

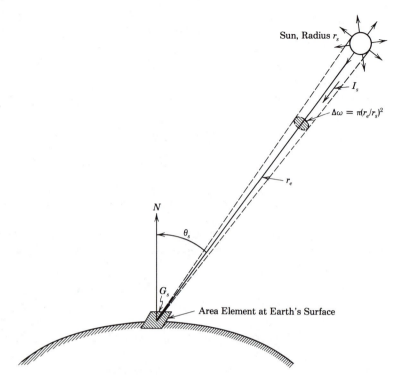

FIGURE 11.11. Geometric relations for solar irradiation.

where r_s is the sun's radius. Let the incident intensity from the sun, along the direction θ_s, be denoted by I_s (total, summed over all wavelengths) and let it be desired to find the total irradiation, G_s, falling on the surface. From the definition given in Eq. (11.18), the irradiation is

$$G_s = \int_h I_s(\theta, \varphi) \cos \theta \, d\omega,$$

where the integration is to be made over the hemisphere above the surface element. Because of the fact that all the energy incident on the element is contained in the solid angle $\Delta\omega$, the above integral is zero except in that solid angle. When the integration is taken in $\Delta\omega$, the polar angle θ varies slightly but may be taken as virtually constant and equal to θ_s, the polar angle of the center of the sun. Likewise, the solar intensity varies with θ, but because of the great distances involved, it may also be taken as constant in $\Delta\omega$. Thus, the irradiation received at the surface is

$$G_s \cong I_s \cos \theta_s \, \Delta\omega$$

$$= I_s \cos \theta_s \left(\frac{\pi r_s^2}{r_e^2} \right).$$

The product $I_s(\pi r_s^2 / r_e^2)$ is the irradiation the surface would recieve were it placed normal to the sun's rays. Thus,

$$G_s = G_{sn} \cos \theta_s, \tag{11.41}$$

$$G_{sn} = I_s \left(\frac{\pi r_s^2}{r_e^2} \right).$$

Spectral examination of the sun's radiation outside the earth's atmosphere indicates that its monochromatic intensity distributes among the wavelengths much as blackbody emission. There are some differences of course, and the earth's atmosphere further alters the spectral distribution by absorption and scattering. However for present purposes, it is adequate to treat the sun as a blackbody emitter. Thus, one may replace I_s in Eq. (11.41) with the blackbody intensity, $I_b(T_s)$, which by use of the diffuseness of blackbody radiation given in Eq. (11.21), may be expressed in terms of the blackbody emissive power at the sun's apparent surface temperature, T_s:

$$I_s \cong I_b(T_s),$$

$$= \frac{E_b(T_s)}{\pi}.$$

Thus, the sun's normal irradiation at the earth's surface is

$$G_{sn} = E_b(T_s) \frac{r_s^2}{r_e^2},$$

$$= \sigma T_s^4 \frac{r_s^2}{r_e^2}. \tag{11.42}$$

The normal intensity received at the earth, just outside the atmosphere, varies with the time of year since the earth's orbit is not circular. For purposes here, it is sufficient to use a yearly average value of

$$G_{sn} = 1396.9 \text{ W/m}^2, \tag{11.43}$$

$$= 442.8 \text{ Btu/h-ft}^2.$$

Using this value, an average value of the earth's orbit radius of 149×10^6 km, and the sun's radius of 0.70×10^6 km, Eq. (11.42) yields an equivalent solar temperature of

$$T_s = \left[\frac{1396.9}{5.67 \times 10^{-8}} \left(\frac{149}{0.7} \right)^2 \right]^{1/4},$$

$$= 5780°K \cong 5800°K.$$

Thus, for many engineering applications it is adequate to represent the sun's radiation received at the earth's surface, outside the atmosphere, as

blackbody radiation at $T_s = 5800°K$ (10,400°R),

$$G_{sn} = 1396.9 \text{ W/m}^2 \ (442.8 \text{ Btu/h-ft}^2), \tag{11.44}$$

$$G_s = G_{sn} \cos \theta_s, \ \theta_s = \text{angle of incidence.}$$

Absorption by the earth's atmosphere depletes the normal component just calculated. The amount of depletion is dependent upon the location on the earth's surface, the time of year and day, as well as local atmospheric conditions. Also, scattering of the solar beam occurs in the atmosphere, so that a "diffuse" component is often significant in addition to the depleted direct beam. These details will not be treated here, and the reader may wish to consult Refs. 5 and 6 for more detailed information.

Solar Absorptivity

The discussion of Sec. 11.4 emphasized the fact that the total absorptivity of a surface depends on the spectral character of the incident irradiation as well as the nature of the receiving surface, while total emissivity is dependent on the surface alone. Since the sun represents a unique radiation source, values of the absorptivity of a surface exposed to the sun, the *solar absorptivity*, are unique surface properties that may be evaluated and tabulated. These values are necessary for use in the analysis of radiant heat transfer involving solar phenomena. Unless it is gray, the solar absorptivity of a surface at a modest, earthbound temperature may differ significantly from its emissivity since the spectral distribution of the energy from the high temperature sun is vastly different from that of the emission at the lower temperature of the surface. If the surface is an ideal gray surface, then its emissivity and solar absorptivity will be the same.

Table A.11 quotes measured values of the solar absorptivity, α_s, for a number of surfaces, along with the emissivity. The ratio α_s/ϵ, a quantity whose significance will be seen shortly, is also tabulated, and a wide range of variation of α_s and α_s/ϵ is observed. Note, in particular, that α_s/ϵ may be much greater or much less than the gray case for which $\alpha_s/\epsilon = 1.0$.

The difference between α_s and ϵ, or the magnitude of α_s/ϵ, depends on the spectral dependence of the monochromatic emissivity $\epsilon_\lambda(\lambda, T)$. This dependence is best seen by returning to the definitions of the total emissivity and total absorptivity. Equation (11.31) for ϵ and its evaluation in terms of the partial integrals, $F_{0-\lambda}$, in Eq. (11.32) (as used in Example 11.1) are

$$
\epsilon(T) = \frac{\int_0^\infty \epsilon_\lambda(\lambda, T)E_{b\lambda}(\lambda, T)\, d\lambda}{\sigma T^4},
$$

$$
= \Sigma\, \bar{\epsilon}_{\lambda i, i+1}[F_{0-\lambda_{i+1}}(T) - F_{0-\lambda_i}(T)].
$$

(11.45)

The solar absorptivity is given by Eq. (11.39) if the irradiation is taken to be that of blackbody at the solar temperature T_s. The resulting expression may also be written in terms of the function $F_{0-\lambda}$ over intervals in which ϵ_λ is taken constant:

$$
\alpha_s(T) = \frac{\int_0^\infty \epsilon_\lambda(\lambda, T)E_{b\lambda}(\lambda, T_s)\, d\lambda}{\sigma T_s^4},
$$

$$
= \Sigma\, \bar{\epsilon}_{\lambda i, i+1}[F_{0-\lambda_{i+1}}(T_s) - F_{0-\lambda_i}(T_s)].
$$

(11.46)

Even though the function $\epsilon_\lambda(\lambda, T)$ is the same in both Eqs. (11.45) and (11.46), the values of the integrals will generally be different since the function $E_{b\lambda}$, or $F_{0-\lambda}$, will be different—being evaluated at the surface temperature T for ϵ or at the solar temperature T_s for α_s. The difference is illustrated in Example 11.3.

EXAMPLE 11.3

Find the solar absorptivity of the surface described in Example 11.2. Compare the result with the emissivity found in that example.

Solution. Using Eq. (11.46), the solar absorptivity, in terms of the notation of Example 11.2, is

$$\alpha_s = \bar{\epsilon}_{0,\lambda_1}[F_{0-\lambda_1}(T_s) - 0] + \bar{\epsilon}_{\lambda_1,\lambda_2}[F_{0-\lambda_2}(T_s) - F_{0-\lambda_1}(T_s)]$$

$$+ \bar{\epsilon}_{\lambda_2,\infty}[1 - F_{0-\lambda_2}(T_s)].$$

Using $T_s = 5800°K$, Table A.12 gives

$$\text{At } \lambda_1 T_s = 2 \times 5800 = 11,600 \ \mu\text{m-}°\text{K}, \ F_{0-\lambda_1} = 0.940,$$

$$\text{At } \lambda_2 T_s = 5 \times 5800 = 29,000 \ \mu\text{m-}°\text{K}, \ F_{0-\lambda_2} = 0.995.$$

Thus,

$$\alpha_s = 0.2 \times 0.940 + 0.8[0.995 - 0.940] + 0.4[1 - 0.995] = 0.234.$$

The solar absorptivity is seen to be much smaller than the emissivity at 1500°K, due primarily to the fact that so much more of the solar spectrum is concentrated in the spectral region where ϵ_λ is small. ∎

The ratio α_s/ϵ, mentioned earlier, gives an indication of a surface's solar absorbing capability compared with its emitting capability. A large value of α_s/ϵ would be desirable for a solar collector surface since the surface should be able to absorb more energy than it would lose by reradiation. The reverse might be true for the surface of a spacecraft from which it might be desired to reject a net amount of heat in the presence of solar irradiation. These facts are sometimes expressed in terms of the equilibrium temperature that a surface would achieve if placed, adiabatically isolated, in the solar environment. In such a case the absorbed solar radiation would equal the emission, if any convectiuon losses are ignored. Thus, if T^* represents the equilibrium temperature of the surface:

$$\alpha_s G_s = E(T^*),$$

$$\alpha_s G_{sn} \cos \theta_s = \epsilon E_b(T^*) = \epsilon \sigma T^*,$$

or

$$T^* = \left(\frac{\alpha_s}{\epsilon} \frac{G_{sn} \cos \theta_s}{\sigma} \right)^{1/4}, \tag{11.47}$$

where θ_s is the angle between the surface normal and the sun's rays. Since G_{sn} and σ are fixed, the equilibrium temperature, T^*, depends on θ_s and the ratio α_s/ϵ. Large values of α_s/ϵ result in large T^*, while low values give low T^*.

In many applications the surface temperature may be fixed at a value different from T^*—as, for example, in a solar collector where it is desired to collect energy at a given temperature. In such a case, the *net* flux of energy *into* a surface maintained at temperature T is

$$\frac{q_{net}}{A} = \alpha_s G_{sn} \cos \theta_s - \epsilon T^4,$$

$$= \alpha_s G_{sn} \cos \theta_s \left[1 - \left(\frac{T}{T^*} \right)^4 \right].$$

(11.48)

For a specified collection temperature, T, the net flux absorbed would be positive only when a surface is chosen with α_s/ϵ large enough that $T^* > T$. Collection efficiency is improved the greater is T^*. Exactly the reverse is true if one desires, as perhaps in a spacecraft, to reject a net amount of heat in the presence of the sun.

The above analysis is made more complicated if simultaneous heat losses or gains to an ambient convecting fluid are included. Such problems are deferred to Chapter 13, where combined heat transfer modes are discussed.

EXAMPLE 11.4

A surface is placed normal to the sun's rays outside the atmosphere. Find the equilibrium temperature the surface would achieve if thermally isolated and the net flux absorbed if the surface is held at 130°C when the surface is (a) silicon-coated aluminum foil; (b) anodized aluminum; (c) white epoxy paint on aluminum.

Solution. Table A.11 gives

(a) Silicon on Al: $\alpha_s = 0.522$, $\epsilon = 0.12$,
(b) Anodized Al: $\alpha_s = 0.923$, $\epsilon = 0.841$,
(c) White epoxy on Al: $\alpha_s = 0.246$, $\epsilon = 0.882$.

With $G_{sn} = 1396.9$ W/m² from Eq. (11.44), $\theta_s = 0$, and $T = 130 + 273 = 403°$K, the equilibrium temperature and heat flux are calculated from Eqs. (11.47) and (11.48). For case (a):

$$T^* = \left(\frac{0.522}{0.12} \frac{1396.9}{5.67 \times 10^{-8}} \right)^{1/4} = 572°K = 299°C \ (570°F)$$

$$\frac{q_{net}}{A} = 0.522 \times 1396.9 \left[1 - \left(\frac{403}{572} \right)^4 \right]$$

$$= 550 \text{ W/m}^2 \ (174 \text{ Btu/h-ft}^2).$$

Similar calculations for the other cases give:

(b) $T^* = 406°K = 133°C \ (271°F)$,

$$\frac{q_{net}}{A} = 37.7 \text{ W/m}^2 \ (11.9 \text{ Btu/h-ft}^2).$$

(c) $T^* = 289°K = 15°C$ (60°F),

$$\frac{q_{\text{net}}}{A} = -964 \text{ W/m}^2 \ (-305 \text{ Btu/h-ft}^2).$$

These calculations illustrate the effect of α_s/ϵ on the performance of a surface in the solar environment. In case (a) with a high value of $\alpha_s/\epsilon = 0.522/0.12 = 4.35$ the surface absorbs a net amount of heat, while in case (c) with a low value of $\alpha_s/\epsilon = 0.246/0.882 = 0.28$, a surface at the same temperature rejects a net amount of heat. ■

EXAMPLE 11.5

Find the equilibrium temperature of a thermally isolated surface, placed normal to the sun's irradiation, (a) if the monochromatic emissivity of the surface is the step function shown in Fig. 11.12(a); (b) repeat the problem if the distribution is reversed, as in Fig. 11.12(b).

Solution. For the notation in Fig. 11.12, Eqs. (11.45) and (11.46) give the total emissivity and solar absorptivity to be the following when T_s is the solar temperature and T^* is the unknown equilibrium surface temperature:

$$\alpha_s = \epsilon_{\lambda_1} F_{0-\lambda_c}(T_s) + \epsilon_{\lambda_2}[1 - F_{0-\lambda_c}(T_s)],$$

$$\epsilon = \epsilon_{\lambda_1} F_{0-\lambda_c}(T^*) + \epsilon_{\lambda_2}[1 - F_{0-\lambda_c}(T^*)].$$

The equilibrium temperature is, by Eq. (11.47),

$$T^* = \left(\frac{\alpha_s}{\epsilon} \frac{G_{sn}}{\sigma}\right)^{1/4}.$$

The solar absorptivity α_s depends only on the known solar temperature $T_s = 5800°K$; however, ϵ depends on the unknown equilibrium temperature T^*. Thus, α_s may be determined from the first of the above equations and then T^* found by iterative calculation using the second and third equations.

(a) For $\epsilon_{\lambda_1} = 0.1$, $\epsilon_{\lambda_2} = 0.9$, $\lambda_c = 4 \ \mu\text{m}$, one has

$$\lambda_c T_s = 4 \times 5800 = 23{,}200,$$

$$F_{0-\lambda_c} = 0.990,$$

$$\alpha_s = 0.1 \times 0.990 + 0.9 \times (1 - 0.990) = 0.108.$$

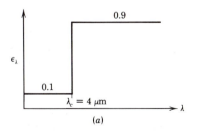

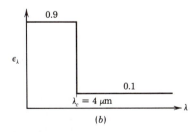

FIGURE 11.12. Example 11.5.

Thus,

$$\epsilon = 0.1 F_{0-4}(T^*) + 0.9[1 - F_{0-4}(T^*)],$$

$$T^* = \left(\frac{0.108}{\epsilon} \frac{1396.9}{5.67 \times 10^{-8}}\right)^{1/4},$$

and iterative calculations yield

$$T^* = 233°K = -40°C \ (-40°F),$$

$$F_{0-\lambda_c} = 0.0002,$$

$$\epsilon = 0.900.$$

(b) Identical calculations for $\epsilon_{\lambda_1} = 0.9$, $\epsilon_{\lambda_2} = 0.1$, $\lambda_c = 4$ μm give

$$\alpha_s = 0.892,$$

$$T^* = 577°K = 304°C \ (579°F),$$

$$F_{0-\lambda_c} = 0.1217,$$

$$\epsilon = 0.197.$$

In case (a), the surface had a very low value of ϵ_λ for wavelengths less than 4 μm, where most of the sun's emission is concentrated (as shown in Table 11.1) and a high ϵ_λ in the infrared region where it does most of its emitting. Thus, it is a poor solar absorber but a good low temperature emitter. In case (b), exactly the reverse is true. Such selective surfaces are very useful in many solar energy and spacecraft applications. ∎

11.6

DIFFUSE RADIANT EXCHANGE BETWEEN GRAY INFINITE PARALLEL PLANES

A general problem of radiation heat transfer that will eventually be treated is that of radiant exchange between a collection of surfaces of arbitrary location and arbitrary size and shape. One of the fundamental aspects of such a problem is the geometric question of how much of the radiation leaving one surface is actually incident on another. However, in order to concentrate initially on some of the physical processes taking place, the simplified case of the radiant exchange between the opposing faces of two parallel infinite planes will be treated first. In this instance one is assured of the fact that all the radiation leaving one surface is incident on the other. The assumption of infinite planes implies the neglection of edge effects in the case of finite surfaces and is often a justifiable simplification if the space between the planes is small compared to the area of the planes.

Consider, then, two parallel infinite planes, as suggested in Fig. 11.13, each of a known temperature on the opposing surfaces. Even though the surfaces may not be black, knowing their temperatures is equivalent to knowing the

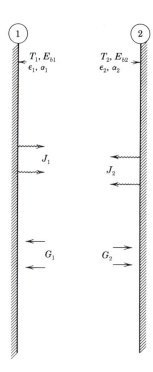

FIGURE 11.13. Infinite parallel gray planes.

associated blackbody emissive powers: $E_{b1} = \sigma T_1^4$, $E_{b2} = \sigma T_2^4$. Then the actual surface emissive powers are

$$E_1 = \epsilon_1 E_{b1} = \epsilon_1 \sigma T_1^4$$

$$E_2 = \epsilon_2 E_{b2} = \epsilon_2 \sigma T_2^4.$$

This practice of using specified equivalent blackbody emissive power in the place of specified temperature, even when the surfaces are nonblack, will be used throughout the remainder of this chapter and has the advantage that the resultant heat exchange is expressible as a linear function of the E_b's rather than a nonlinear function of the T's.

Return now to the case depicted in Fig. 11.13. If the opposing faces of the two planes have specified temperatures, (i.e., E_b's) and known values of the surface properties ϵ_1, α_1, ϵ_2, α_2, let it be desired to find the *net* rate of heat flow *away* from each surface—i.e., the heat that must be supplied, or removed, to maintain the surfaces at the specified temperatures. One could determine this flow of energy by deducing how much of the emission from a surface is reflected back and absorbed through a multiplicity of interreflections (and partial absorptions) with the other surface. However, this energy flow may be deduced more simply by algebraic means. An elementary heat balance on, say, surface 1 for the *net* radiant energy leaving it, q_1, is

$$q_1 = A_1(J_1 - G_1),$$

where J_1 and G_1 represent the radiosity and irradiation, respectively, of that surface. Since the surfaces are infinite in extent, it only makes sense to work in terms of the heat flux, q_1/A_1. Also, since all the energy flowing away from surface 1 must end up at surface 2, no energy being lost out the ends, one has

$$\frac{q_1}{A_1} = -\frac{q_2}{A_2} = J_1 - G_1, \tag{11.49}$$

where q_2 represents the net radiant energy leaving A_2. Since the two planes are infinite in extent and parallel, the irradiation on one must be the radiosity of the other:

$$G_1 = J_2, \tag{11.50}$$
$$G_2 = J_1.$$

In addition, the definition of radiosity, the energy leaving a surface gives, as in Eq. (11.6),

$$J = E + \rho G.$$

For an opaque surface, the reflectivity ρ is $\rho = 1 - \alpha$; however, if the surface is diffuse and *gray,* then Kirchhoff's law holds for the total properties so that $\alpha = \epsilon$, or

$$J = E + (1 - \epsilon)G,$$
$$= \epsilon E_b + (1 - \epsilon)G.$$

When applied to the two surfaces in question,

$$J_1 = \epsilon_1 E_{b1} + (1 - \epsilon_1)G_1, \tag{11.51}$$
$$J_2 = \epsilon_2 E_{b2} + (1 - \epsilon_2)G_2.$$

Equations (11.50) and (11.51) are four equations in J_1, J_2, G_1, and G_2 and may be solved for J_1 and G_1 for use in Eq. (11.49):

$$J_1 = \frac{\epsilon_1 E_{b1} + (1 - \epsilon_1)\epsilon_2 E_{b2}}{1 - (1 - \epsilon_1)(1 - \epsilon_2)},$$

$$G_1 = \frac{\epsilon_2 E_{b2} + (1 - \epsilon_2)\epsilon_1 E_{b1}}{1 - (1 - \epsilon_1)(1 - \epsilon_2)}. \tag{11.52}$$

Equation (11.49) then yields the desired heat flows in terms of the specified properties and E_b's (i.e., temperature):

$$\frac{q_1}{A_1} = -\frac{q_2}{A_2} = \frac{\epsilon_1\epsilon_2 E_{b1} - \epsilon_1\epsilon_2 E_{b2}}{1 - (1 - \epsilon_1)(1 - \epsilon_2)},$$

$$= \frac{E_{b1} - E_{b2}}{(1/\epsilon_1) + (1/\epsilon_2) - 1}, \tag{11.53}$$

$$= \frac{\sigma(T_1^4 - T_2^4)}{(1/\epsilon_1) + (1/\epsilon_2) - 1}.$$

EXAMPLE 11.6

Two opposed, parallel, infinite planes are maintained at 300°F and 400°F, respectively. (a) What is the net heat flux between the two planes if one has an emissivity of 0.7 and the other an emissivity of 0.8? Does it matter which plane has which emissivity? (b) What is the net heat flux between the planes if they are black? (c) Repeat part (a) if the temperature difference is doubled by raising the 400°F plane to 500°F.

Solution.

(a) The two surface temperatures are $T_1 = 300 + 460 = 760°R$, $T_2 = 400 + 460 = 860°R$. In English units $\sigma = 0.1714 \times 10^{-8}$ Btu/h-ft²-°R⁴; thus,

$$E_{b1} = 0.1714 \left(\frac{760}{100}\right)^4 = 571.8 \text{ Btu/h-ft}^2,$$

$$E_{b2} = 0.1714 \left(\frac{860}{100}\right)^4 = 937.6 \text{ Btu/h-ft}^2.$$

With $\epsilon_1 = 0.7$ and $\epsilon_2 = 0.8$, Eq. (11.53) gives

$$\frac{q_1}{A_1} = -\frac{q_2}{A_2} = \frac{571.8 - 937.6}{(1/0.7) + (1/0.8) - 1}$$

$$= -\frac{365.8}{1.679} = -217.9 \text{ Btu/h-ft}^2 \ (687.4 \text{ W/m}^2).$$

In this example it is immaterial which plane has which emissivity since the emissivities are independent of temperature.

(b) If the planes are black,

$$\frac{q_1}{A_1} = -\frac{q_2}{A_2} = -365.8 \text{ Btu/h-ft}^2 \ (1154 \text{ W/m}^2).$$

(c) If $T_2 = 500 + 460 = 960°R$, $E_{b2} = 1455.8$ Btu/h-ft², and

$$\frac{q_1}{A_1} = -\frac{q_2}{A_2} = \frac{571.8 - 1455.8}{1.679}$$

$$= -526.5 \text{ Btu/h-ft}^2 \ (1661 \text{ W/m}^2).$$

Doubling the temperature difference increases the heat flow by a factor of 2.4. ∎

Plane Radiation Shields

Parts (a) and (b) of Example 11.6 show the great influence that a surface emissivity different from unity has on the heat exchange between a pair of planes. This fact is often used to reduce the loss, or gain, of radiant energy from a surface by use of what is known as a radiation shield.

The radiant exchange for two planes, A_1 at T_1 and A_2 at T_2, has already been shown to be

$$\left(\frac{q_1}{A_1}\right)_0 = \mathcal{F}_{1-2}(E_{b1} - E_{b2}),$$

where

$$\frac{1}{\mathcal{F}_{1-2}} = \frac{1}{\epsilon_1} + \frac{1}{\epsilon_2} - 1,$$

and the notation $(q_1/A_1)_0$ is used to signify the heat flux when nothing is placed between the planes.

Now let a third plane, a radiation shield (call it A_s with emissivity ϵ_s on each side), be placed between planes A_1 and A_2. The shield A_s comes to an equilibrium temperature T_s (or equivalent blackbody emissive power $E_{bs} = \sigma T_s^4$) such that the heat it exchanges with plane A_1 equals the negative of that it exchanges with A_2. Application of Eq. (11.53) for the heat exchanged by A_s with A_1 gives

$$-\frac{q_{s-1}}{A_s} = \mathcal{F}_{1-s}(E_{b1} - E_{bs}),$$

$$\frac{1}{\mathcal{F}_{1-s}} = \frac{1}{\epsilon_1} + \frac{1}{\epsilon_s} - 1.$$

and that exchanged by A_s with A_2 is

$$\frac{q_{s-2}}{A_s} = \mathcal{F}_{s-2}(E_{bs} - E_{b2}),$$

$$\frac{1}{\mathcal{F}_{s-2}} = \frac{1}{\epsilon_s} + \frac{1}{\epsilon_2} - 1.$$

Equating the last two expressions for the shield heat flux gives the equilibrium blackbody emissive power of the shield, or its temperature, to be

$$E_{bs} = \frac{\mathcal{F}_{1-s}E_{b1} + \mathcal{F}_{s-2}E_{b2}}{\mathcal{F}_{1-s} + \mathcal{F}_{s-2}},$$

$$T_s^4 = \frac{\mathcal{F}_{1-s}T_1^4 + \mathcal{F}_{s-2}T_2^4}{\mathcal{F}_{1-s} + \mathcal{F}_{s-2}}.$$

The factors $\mathcal{F}_{1-s}$ and $\mathcal{F}_{s-2}$ are functions of only the emissivities of the surfaces and the shield as noted above. Since the heat flow away from surface A_1 with the shield in place is

$$\left(\frac{q_1}{A_1}\right)_s = -\frac{q_{s-1}}{A_s} = \mathcal{F}_{1-s}(E_{b1} - E_{bs}),$$

then Eq. (11.55) gives, after some algebra,

$$\left(\frac{q_1}{A_1}\right)_s = \frac{E_{b1} - E_{b2}}{\dfrac{1}{\mathscr{F}_{1-s}} + \dfrac{1}{\mathscr{F}_{s-2}}},$$

$$(11.56)$$

$$= \frac{E_{b1} - E_{b2}}{\dfrac{1}{\epsilon_1} + \dfrac{2}{\epsilon_s} + \dfrac{1}{\epsilon_2} - 2}.$$

Comparison of Eqs. (11.54) and (11.56) shows that since $\epsilon_s < 1$, the heat flow with the shield in place, $(q_1/A_1)_s$, is always less than that with no shield, $(q_1/A_1)_0$. The exact amount of the reduction of the heat flow depends on the relative orders of magnitude of the emissivities ϵ_1, ϵ_2, and ϵ_s.

The amount of reduction in heat flow is best expressed as the ratio

$$\frac{(q_1/A_1)_s}{(q_1/A_1)_0} = \frac{1/\mathscr{F}_{1-2}}{\dfrac{1}{\mathscr{F}_{1-s}} + \dfrac{1}{\mathscr{F}_{s-2}}},$$

$$(11.57)$$

$$= \frac{\dfrac{1}{\epsilon_1} + \dfrac{1}{\epsilon_2} - 1}{\dfrac{1}{\epsilon_1} + \dfrac{2}{\epsilon_s} + \dfrac{1}{\epsilon_2} - 2}.$$

A feeling for the effectiveness of a shield is obtained by looking at the special case when all emissivities are the same. In this instance $\epsilon_1 = \epsilon_2 = \epsilon_s$, the above becomes

$$\frac{(q_1/A_1)_s}{(q_1/A_1)_0} = \frac{1}{2};$$

that is, a single shield of the same emissivity as the surfaces to be shielded reduces the heat flow by one-half. It may be shown that N shields of equal emissivities will reduce the heat flow according to

$$\frac{(q_1/A_1)_N}{(q_1/A_1)_0} = \frac{1}{N + 1}.$$

$$(11.58)$$

11.7

THE SHAPE FACTOR FOR DIFFUSE RADIANT EXCHANGE BETWEEN FINITE SURFACES

The simplicity of the problems just discussed in Sec. 11.6 was due primarily to the fact that all the radiation leaving one plane was incident on the other. Thus the radiosity of one surface was known to be the irradiation upon the other. In many applications, such as the radiant exchange between the several surfaces of a room, a furnace, a spacecraft, etc., this simplification does not hold. Thus, it is now necessary to consider the general case of a collection of finite, non-parallel, nonplanar surfaces that are exchanging radiant energy.

The Shape Factor

Of basic importance is the *direct* exchange of energy between two surfaces—that fraction of the energy leaving one surface which strikes the second directly, none being considered which is transferred by reflection or reradiation from other surfaces that may be present. This direct flux of energy between one surface and another is expressed in terms of the *shape factor,* which is found by determining the exchange between differential area elements in each surface and then performing simultaneous integrations over both surfaces. When Lambert's law is presumed to hold, the result is dependent solely upon the geometry of the problem.

The geometry of the situation is illustrated in Fig. 11.14. The two surfaces in question are denoted by A_1 and A_2. These surfaces are located arbitrarily in space, are of arbitrary shape, and are not necessarily plane. Elements of each surface are denoted by dA_1 and dA_2 with corresponding normals N_1 and N_2. The line connecting dA_1 to dA_2 has length r and makes polar angles θ_1 and θ_2 with the two normals N_1 and N_2, respectively. According to Eq. (11.10), the fraction of the flux leaving dA_1 which is intercepted by dA_2 is

$$df = I_1 \cos \theta_1 \, d\omega_{2-1},$$

in which I_1 represents the intensity of the radiation leaving dA_1 and $d\omega_{2-1}$ is the solid angle subtended by dA_2 at dA_1. Projecting dA_2 normal to the radial line r gives $d\omega_{2-1} = (dA_2 \cos \theta_2)/r^2$, so that

$$df = I_1 \frac{\cos \theta_1 \cos \theta_2}{r^2} \, dA_2.$$

Since df is the portion of the *flux* from dA_1 intercepted by dA_2, it may be written as

$$df = \frac{dq_{1 \to 2}}{dA_1} = I_1 \frac{\cos \theta_1 \cos \theta_2}{r^2} \, dA_2,$$

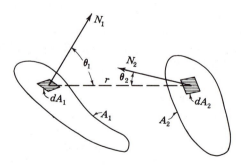

FIGURE 11.14

where $q_{1\rightarrow 2}$ represents the energy leaving all of A_1 that strikes all of A_2. Thus, $q_{1\rightarrow 2}$ is found by integrating the above expression over both the areas:

$$q_{1\rightarrow 2} = \int_{A_1}\int_{A_2} I_1 \frac{\cos\theta_1\cos\theta_2}{r^2}\, dA_2\, dA_1.$$

If the radiosity of the originating surface, J_1, is uniform over that surface, the total energy leaving it is $A_1 J_1$, so the *fraction* of the energy leaving A_1 which strikes A_2 is

$$F_{1-2} = \frac{q_{1\rightarrow 2}}{A_1 J_1}. \tag{11.59}$$

If, in addition, the radiant energy leaving A_1 is *diffuse* (both the emissive and reflected portions), then I_1 is a constant and Eq. (11.17) states that $J_1 = \pi I_1$. Thus, the two equations above give

$$F_{1-2} = \frac{1}{A_1}\int_{A_1}\int_{A_2}\frac{\cos\theta_1\cos\theta_2}{\pi r^2}\, dA_2\, dA_1. \tag{11.60}$$

The factor F_{1-2} is called the diffuse *configuration factor* or *shape factor* of A_1 with respect to A_2—the fraction of the energy leaving A_1 which *directly* strikes A_2. As may be seen from its representation in Eq. (11.60), the shape factor for diffuse radiation is a purely geometric property of the two surfaces involved. The order of the subscripts on F_{1-2} is quite important since F_{2-1} will be used to represent the fraction of energy leaving A_2 which directly strikes A_1. In general, $F_{1-2}\neq F_{2-1}$, since reasoning similar to that given above yields

$$F_{2-1} = \frac{1}{A_2}\int_{A_2}\int_{A_1}\frac{\cos\theta_2\cos\theta_1}{\pi r^2}\, dA_1\, dA_2. \tag{11.61}$$

Some Properties of the Shape Factor

Certain useful properties of the shape factor may be deduced from the definition given by Eqs. (11.59) and (11.60). These properties will be useful for the determination of shape factors for specific geometries and for the analysis of heat exchange between surfaces.

The Reciprocal Property. The most important property of the shape factor is readily deducible from Eq. (11.60) and (11.61). For two surfaces, A_1 and A_2, in general $F_{1-2}\neq F_{2-1}$; however, the following reciprocal relation exists:

$$A_1 F_{1-2} = A_2 F_{2-1}. \tag{11.62}$$

The Additive Property. If one of the two surfaces, say A_i, is divided into subareas $A_{i_1}, A_{i_2}, \ldots, A_{i_n}$ (so that $\Sigma\, A_{i_n} = A_i$), the application of Eqs. (11.59) and (11.60) will lead to the following relation:

$$A_i F_{i-j} = \sum_n A_{i_n} F_{i_n-j}. \tag{11.63}$$

The reciprocal relation given above then yields

$$F_{j-i} = \sum_n F_{j-i_n}.$$ (11.64)

Thus, if the transmitting surface is subdivided, the shape factor for that surface with respect to a receiving surface is not simply the sum of the individual shape factors, although the AF product is expressed by such a sum. The shape factor from a surface to a subdivided receiving surface *is* simply the sum of the individual shape factors.

If surface A_i is subdivided into n parts $(A_{i_1}, A_{i_2}, \ldots, A_{i_n})$ and surface A_j into m parts $(A_{j_1}, A_{j_2}, \ldots, A_{j_m})$, the above reasoning leads to

$$A_i F_{i-j} = \sum_n \sum_m A_{i_n} F_{i_n - j_m}.$$ (11.65)

These additive properties are useful in finding the shape factor for complex shapes by subdividing the surfaces into subsections for which the shape factor is either known or more simply evaluated. This is discussed in detail in Appendix E.

The Enclosure Property. If the interior surface of a completely enclosed space, such as a room, is subdivided into n parts—each part having a finite area $A_1, A_2, \ldots, A_n$, then

$$\sum_{j=1}^n F_{i-j} = 1, \qquad i = 1, 2, \ldots, n.$$ (11.66)

The above representation admits the shape factors $F_{1-1}, F_{2-2}, \ldots, F_{n-n}$ since some of the surfaces may "see" themselves if they are convex. That is, for example, the definition of the shape factor in Eq. (11.60) does not preclude the possibility that the two incremental areas, dA_1 and dA_2, may be on the same surface.

If a surface A_1, is completely enclosed by a second surface, A_2, and if A_1 does not see itself ($F_{1-1} = 0$), then $F_{1-2} = 1$. Then the reciprocal relation of Eq. (11.62) gives

$$F_{2-1} = \frac{A_1}{A_2}.$$ (11.67)

The Calculation of Shape Factors

The definition of the shape factor is given by

$$F_{1-2} = \frac{1}{A_1} \int_{A_1} \int_{A_2} \frac{\cos\theta_1 \cos\theta_2}{\pi r^2} \, dA_2 \, dA_1.$$

The two integrals in this equation are double integrals—each taken over a surface. Calculation of F_{1-2} for a specific geometrical setup will require that a coordinate system be established for each surface. Then the location of dA_1 and dA_2 on A_1 and A_2, respectively, may be expressed in terms of these coordinates.

Likewise, the angles θ_1 and θ_2 and the connecting distance r may also be expressed in terms of the coordinate locations of dA_1 and dA_2. Finally, the area elements themselves must be expressed in terms of differentials of the coordinates—resulting in a twofold integral for each of the double integrals. Thus, in terms of geometrical coordinates, the expression for F_{1-2} becomes, usually, a fourfold integral.

For even the simplest geometrical shapes the evaluation of the fourfold integral becomes quite involved. Since such integrations are principally mathematical in nature, the details of shape factor evaluation, for a selection of physically significant cases, are given in Appendix E. The resultant expressions for these cases are presented for convenience in Figs. 11.15 through 11.19. Figures 11.15 and 11.16 present the shape factors for directly opposed rectangles and perpendicular rectangles with a common edge; Fig. 11.17 gives F_{1-2} for parallel disks, and Fig. 11.18 presents F_{2-1} and F_{2-2} for coaxial cylinders of finite length. Two cases of importance in the present-day analysis of radiant exchange between a planetary body and elements of a spacecraft or satellite are given in Fig. 11.19—the shape factors between a large sphere and either a small plane element or a small sphere.

More complex shapes often may be reduced to simpler cases by the application of "shape factor algebra"—i.e., the use of some of the shape factor properties discussed in the preceding section, particularly the additive property. Appendix E illustrates the manner in which this reduction may be carried out. The material

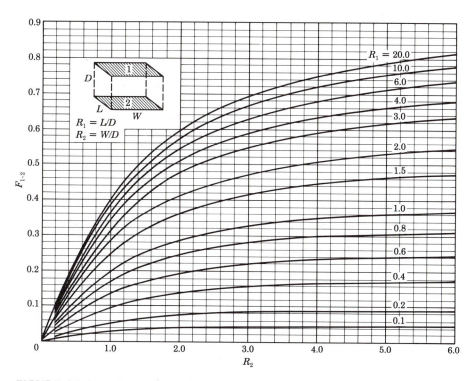

FIGURE 11.15. Shape factor for parallel, directly opposed, rectangles.

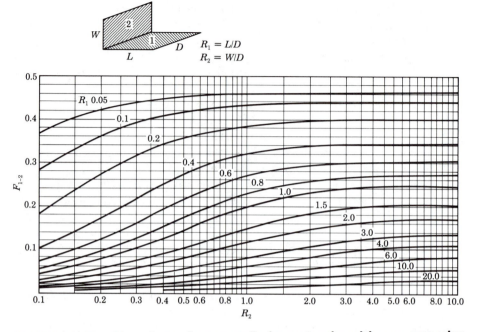

FIGURE 11.16. Shape factor for perpendicular rectangles with a common edge.

which follows in this chapter presumes that the reader is familiar with the methods of Appendix E and has mastered the methods of determining the shape factor for a given geometry.

Such mathematical methods as Stokes' theorem may be applied in some cases to reduce the double area integrals to contour integrals (Ref. 7). For many geometries, numerical methods utilizing digital computers may be advantageous (Ref. 8).

EXAMPLE 11.7

Find the shape factor F_{1-2} for the geometry shown in Fig. 11.20.

Solution. Define the additional areas A_3, A_4, A_5, and A_6 as shown in the figure. Application of the principles of shape factor algebra, as illustrated in Appendix E, gives

$$A_1 F_{1-2} = A_5 F_{5-2} + 2A_6 F_{6-2},$$

$$= A_5 F_{5-2} + 2 \times \tfrac{1}{2}[A_{(5,6)} F_{(5,6)-(2,3)} - A_5 F_{5-2} - A_6 F_{6-3}],$$

$$= A_{(5,6)} F_{(5,6)-(2,3)} - A_6 F_{6-3}.$$

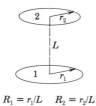

$$R_1 = r_1/L \qquad R_2 = r_2/L$$

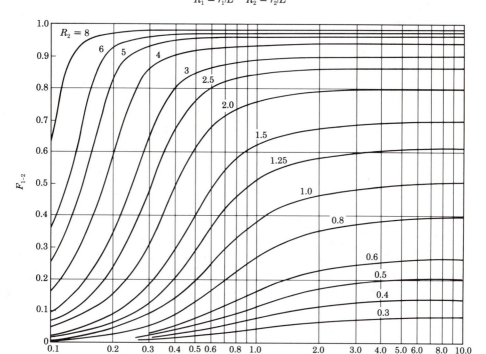

FIGURE 11.17. Shape factor for parallel concentric disks.

For $F_{(5,6)-(2,3)}$, Fig. 11.16 or Eq. (E.3) gives

$$R_1 = \tfrac{12}{8}, \qquad R_2 = \tfrac{3}{8}, \qquad F_{(5,6)-(2,3)} = 0.085,$$

and for F_{6-3},

$$R_1 = \tfrac{12}{2}, \qquad R_2 = \tfrac{3}{2}, \qquad F_{6-3} = 0.051.$$

$$F_{1-2} = \frac{1}{10 \times 12} (8 \times 12 \times 0.85 - 2 \times 12 \times 0.051)$$

$$= 0.058. \qquad \blacksquare$$

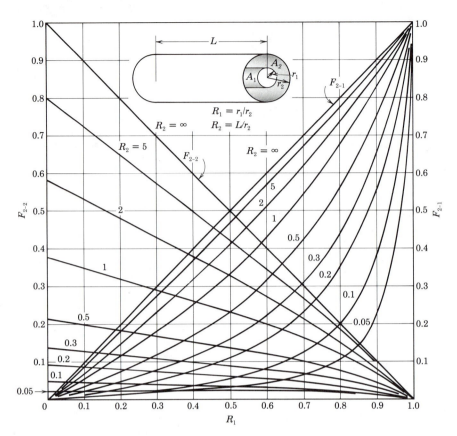

FIGURE 11.18. Shape factors for concentric cylinders of finite length.

11.8

RADIANT EXCHANGE AMONG BLACK SURFACES AND IN BLACK ENCLOSURES

The general problem of radiative exchange between a collection of surfaces, perhaps gray, of either specified temperature or specified heat flux will be treated in some detail in the next section. A useful introduction to that problem is the special case in which the surfaces are black and of known temperature. In this instance one knows that the only energy leaving a surface is original emission, there being no reflection.

Consider, then, a system of two or more black surfaces, A_1, A_2, . . . , A_n. Each of the surfaces has a known specified temperature, T_1, T_2, . . . , T_n or equivalently, known emissive power E_{b1}, E_{b2}, . . . , E_{bn}. The surfaces may, or may not, form a complete enclosure. If they do not, imagine them to be situated in an irradiation-free environment. That is, any radiation leaving a surface that does not strike another in the collection may be presumed lost and no irradiation is incident upon a surface from any source that is not one of the black surfaces

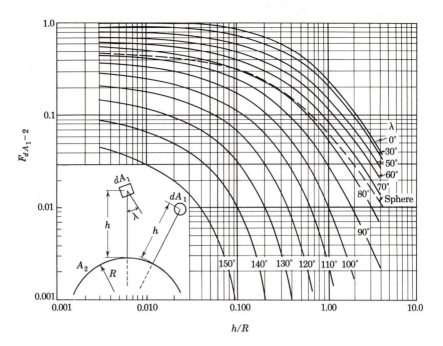

FIGURE 11.19. Shape factor for infinitesimal planes and spheres in the presence of a large sphere.

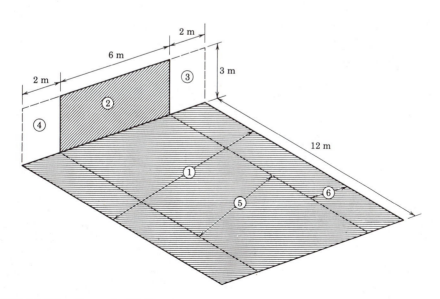

FIGURE 11.20. Example 11.7.

in the collection. Let it be desired to find the net rate of energy flow away from any one of the surfaces.

Since all the surfaces involved are black, the net energy flow away from a surface is the difference between its emissive power and the irradiation upon it. For the ith surface of the collection

$$q_i = A_i E_{bi} - A_i G_i.$$

The energy falling on the surface (and absorbed), $A_i G_i$, comes from the other surfaces of the collection. Thus, from the definition of the shape factor,

$$A_i G_i = \sum_{j=1}^{n} (A_j J_j) F_{j-i},$$

$$= \sum_{j=1}^{n} A_j F_{j-i} E_{bj}.$$

In the above J_j represents the radiosity of one of the surfaces in the collection and the fact that $J_j = E_{bj}$ for a black surface has been used. The summation is to be taken over *all* the surfaces in the collection, including the ith surface in the event it sees itself and $F_{i-i} \neq 0$. Thus, the net heat flow away from A_i is

$$q_i = A_i E_{bi} - \sum_j A_j F_{j-i} E_{bj},$$

$$= A_i E_{bi} - \sum_j A_i F_{i-j} E_{bj}, \qquad (11.68)$$

$$= A_i \left(E_{bi} - \sum_j F_{i-j} E_{bj} \right).$$

In arriving at Eq. (11.68) the general reciprocity of the shape factor, $A_i F_{i-j} = A_j F_{j-i}$, has been used. Since the geometry of the collection of surfaces is known, all the shape factors may be determined; thus, Eq. (11.68) enables one to calculate the heat flow away from each surface, given all the E_b's.

If the system of surfaces forms a complete enclosure, one has from Eq. (11.66)

$$\sum_j F_{i-j} = 1$$

if $j = i$ is included in the summation. Then, Eq. (11.68) takes the form

$$q_i = A_i \left(\sum_j F_{i-j} E_{bi} - \sum_j F_{i-j} E_{bj} \right)$$

$$= \sum_j A_i F_{i-j} (E_{bi} - E_{bj}). \qquad (11.69)$$

In the above equations, and in those to follow, the abbreviated notation $\sum_j$ has been introduced to indicate a summation to be taken over all surfaces in the enclosure, i.e., $j = 1, 2, \ldots, n$.

In a complete enclosure the net heat given up by a surface, say q_i, must be absorbed by the other surfaces of the enclosure. Thus, the term $A_i F_{i-j}$

$(E_{bi} - E_{bj})$ is often termed the "exchange" of energy between A_i and A_j, denoted by q_{ij}:

$$q_{ij} = A_i F_{i-j}(E_{bi} - E_{bj}), \tag{11.70}$$

$$q_i = \sum_j q_{ij}.$$

Clearly, for a complete enclosure energy conservation requires

$$\sum_i q_i = 0.$$

If a system of black surfaces does not form a complete enclosure, as described in Eq. (11.68), it may be treated as one, and Eq. (11.69) applied, by enclosing the system with a fictitious surface (call it A_s for "space") with $E_{bs} = 0$, $\epsilon_s = 1$.

EXAMPLE 11.8

If for the geometry of Example 11.7, the surface A_1 is maintained at 40°C and A_2 at 5°C, find the energy exchange between A_1 and A_2 and the net energy given up by A_1 if the surfaces are located in a radiation-free environment.

Solution. For $T_1 = 40 + 273 = 313°K$ and $T_2 = 5 + 273 = 278°K$, $E_{b1} = 544.2 \ W/m^2$ and $E_{b2} = 338.7 \ W/m^2$. Thus, Eq. (11.70) gives the exchange between A_1 and A_2 to be

$$q_{12} = A_1 F_{1-2}(E_{b1} - E_{b2})$$

$$= (10 \times 12)(0.058)(544.2 - 338.7)$$

$$= 1430.3 \ W \ (4880 \ Btu/h).$$

By Eq. (11.68), the net energy given up by A_1 is

$$q_1 = A_1(E_{b1} - F_{1-2}E_{b2})$$

$$= (10 \times 12)(544.2 - 0.058 \times 338.7)$$

$$= 62,950 \ W \ (214,800 \ Btu/h). \qquad \blacksquare$$

EXAMPLE 11.9

An outlet shoe store with a display window in the front is shown, with dimensions, in Fig. 11.21. The store is to be heated by making the floor a black radiant heating panel at 45°C. The glass window acts as a black plane at 10°C and the other walls and ceiling act as black planes at 25°C. Find the net heat given up by the floor. What difference results if the ceiling height is raised to 4.5 m, the other dimensions remaining the same?

Solution. If the floor is designated as A_1 and the window as A_2, then the geometric arrangement of these surfaces is identical to that considered in Example 11.8. Thus, from that example

$$F_{1-2} = 0.058.$$

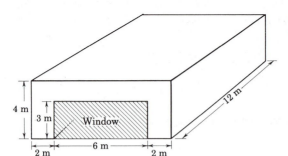

FIGURE 11.21. Geometry for Example 11.9.

Since $F_{1-1} = 0$, the shape factor between the floor and all the walls (and the ceiling) except the window is

$$F_{1-w} = 1 - 0.058 = 0.942.$$

With $T_1 = 45 + 273 = 318°K$, $T_2 = 10 + 273 = 283°K$, $T_w = 25 + 273 = 298°K$, one has $E_{b1} = 579.8$, $E_{b2} = 363.7$, $E_{bw} = 447.1$ W/m². Thus, since there is a complete enclosure, Eq. (11.69) gives

$$q_1 = A_1[F_{1-2}(E_{b1} - E_{b2}) + F_{1-w}(E_{b1} - E_{bw})]$$

$$= (10 \times 12)[0.058(579.8 - 363.7) + 0.942(579.8 - 447.1)]$$

$$= 16,473 \text{ W } (56,210 \text{ Btu/h}).$$

Equation (11.68) will give identical results.

Changing the ceiling height to any value greater than the window height does not alter the results since the factor F_{1-w} remains equal to $1 - F_{1-2}$. ∎

Equations (11.69) and (11.70) suggest that for an enclosure of black surfaces, the radiant heat exchange may be represented by an electrical network analogy. In such an analogy, each surface is represented as a nodal point with its potential maintained at the blackbody emissive power, E_{bi}. Each node is connected to all other nodes through resistances which are $\mathcal{R}_{ij} = 1/A_iF_{i-j} = 1/A_jF_{j-i}$. Equation (11.70) notes that the current flow in each resistor is $(E_{bi} - E_{bj})/\mathcal{R}_{ij}$ and Eq. (11.69) expresses the fact that the current flow at any node equals the sum of the currents in the resistors connected to it. Figure 11.22 shows the equivalent network for Example 11.9.

11.9

RADIANT EXCHANGE IN ENCLOSURES OF DIFFUSE, GRAY SURFACES

The problem of the radiant exchange between parallel infinite planes considered in Sec. 11.6 was relatively simple since the shape factors between the surfaces were all unity, and the analysis could concentrate on the effects of multiple reflections due to grayness of the surfaces. The analysis just presented in Sec. 11.8, on the other hand, concentrated on the geometrical effects associated with

FIGURE 11.22. Electrical net-work analogy for Example 11.9.

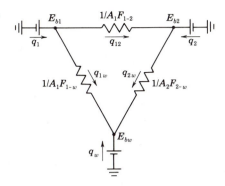

finite-sized surfaces but avoided the complication of multiple interreflections by considering only black surfaces. Thus, the energy flow from one surface to another was known to be due to emissive power only, and all energy incident on a surface was absorbed there.

This section will consider (following the methods of Refs. 9 and 10) the general case in which the surfaces involved are of finite size so that $F_{i-j} \neq 1$. Also, the possibility that the surfaces, or some of them, may be gray will be accounted for. The analysis will presume that the collection of surfaces forms a complete enclosure. The radiant exchange between the various surfaces and windows of a room, between the surfaces in a furnace, etc., are examples of a complete enclosure. In some cases one may be interested in the exchange be-tween a collection of surfaces that do not form an enclosure, but radiate at one another in an otherwise radiation-free environment—as, for example, between portions of a spacecraft in deep space. In such instances, the system may be made an enclosure by adding an enclosing surface with $E_b = 0$ and $\alpha = \epsilon = 1$. Thus, the assumption of a complete enclosure will not be unduly restrictive.

All surfaces will be taken as diffuse emitters and reflectors that are gray so that the diffuse shape factor may be used and Kirchhoff's law in the total sense, $\alpha = \epsilon$, may be used. Even if a surface has a uniform temperature so that its emissive power is uniform, the irradiation upon it (and hence its radiosity if gray) may not be uniform, depending on how the various surfaces in the col-lection "see" one another. The analysis presented here will assume that E_b, G, and J of each surface is uniform. If for a given enclosure this is not the case, it will be presumed that the surfaces may be subdivided into smaller portions (thus increasing the number of surfaces involved) until the assumption can be made.

The entire analysis of this section presumes the existence of the steady state so that all quantities are time independent.

Basic Equations and the General Problem

For a thermally opaque surface the definition of radiosity, emissive power, and irradiation given in Sec. 11.2 led to Eq. (11.6)

$$J_i = E_i + \rho_i G_i,$$

$$= E_i + (1 - \alpha_i)G_i,$$

$$= \epsilon_i E_{bi} + (1 - \alpha_i)G_i,$$

and when Kirchhoff's law is applied,

$$J_i = \epsilon_i E_{bi} + (1 - \epsilon_i)G_i. \tag{11.71}$$

In Eq. (11.71) the subscript i denotes any one of the surfaces of the enclosure. A heat balance made on surface A_i shows that the heat flow *away* from that surface q_i, is

$$q_i = A_i(J_i - G_i). \tag{11.72}$$

If Eq. (11.71) is used to eliminate G_i, the irradiation, one has

$$q_i = \frac{A_i \epsilon_i}{1 - \epsilon_i}(E_{bi} - J_i), \tag{11.73a}$$

$$E_{bi} = \frac{1 - \epsilon_i}{A_i \epsilon_i} q_i + J_i. \tag{11.73b}$$

The second form given in Eq. (11.73b) will be of use later; however, it is sufficient here to note that Eq. (11.73a) expresses the heat flow away from A_i in terms of its equivalent blackbody emissive power (i.e., temperature, since $E_{bi} = \sigma T_i^4$) and the radiosity leaving A_i.

A second form of the heat flow away from a surface in an enclosure may be obtained by considering its interaction with the other surfaces of the enclosure. If A_j represents another surface, and if n is the total number of surfaces in the enclosure, then the irradiation falling on A_i must come from these other surfaces:

$$A_i G_i = \sum_{j=1}^{n} F_{j-i} A_j J_j.$$

In the above, the diffuse shape factor has been used, consistent with the assumption of diffuseness made earlier. The general reciprocity property gives

$$A_i G_i = \sum_{j=1}^{n} A_i F_{i-j} J_j. \tag{11.74}$$

Introduced into Eq. (11.72), Eq. (11.74) yields

$$q_i = A_i J_i - \sum_j A_i F_{i-j} J_j, \tag{11.75}$$

where the abbreviated notation $\sum_j$ means that the summation is to be made over *all* the surfaces in the enclosure—including A_i in the event A_i sees itself and $F_{i-i} \neq 0$. Equation (11.75) may be written in another form noting that for the enclosure $\sum_j F_{i-j} = 1$. Thus,

$$q_i = A_i J_i \sum_j F_{i-j} - \sum_j A_i F_{i-j} J_j,$$

$$q_i = \sum_j A_i F_{i-j}(J_i - J_j). \tag{11.76}$$

Equations (11.75) and (11.76) give an alternative expression for q_i from that in Eq. (11.73), the heat flow being expressed in terms of the radiosity of a surface, J_i, and the radiosities of all the other surfaces of the enclosure. Equations (11.73a) and (11.76) may be equated to yield another important relation:

$$\frac{A_i \epsilon_i}{1 - \epsilon_i} (E_{bi} - J_i) = \sum_j A_i F_{i-j} (J_i - J_j). \tag{11.77}$$

Equations (11.73), (11.76) and (11.77) express the radiant exchange among surfaces of an enclosure in terms of the emissive powers and radiosities of the participating surfaces. These equations form the basis of the *radiosity method* of enclosure analysis. The general problem of an enclosure analysis may be stated as follows: Given an enclosure of surfaces of known geometry and known surface properties, and given the emissive power (i.e., temperature) of some of the surfaces and given the heat flux away from the others, let it be desired to find the heat flow away from the surfaces on which E_{bi} is specified and the temperature (i.e., E_{bi}) of the surfaces for which q_i is specified. In a general sense, this problem may be solved in the following way: For each of the surfaces on which E_{bi} is known, write an equation of the form of Eq. (11.77); on each of the surfaces on which q_i is known, write an equation of the form of Eq. (11.76). There results, then, n equations in the n unknown radiosities which can be solved for the J_i's. Then Eq. (11.73b) can be used to find E_{bi} on each surface for which q_i is known and Eq. (11.76) can be used to find q_i for each surface for which E_{bi} is known.

The actual carrying out of the general procedure just mentioned may involve complications such as some of the equations becoming indeterminate when a surface is black, etc.; however these matters will be treated later when detailed procedures are discussed. Also, the methods of actually implementing the procedure may differ depending on whether one is dealing with just a few surfaces or with a large number of surfaces or on whether one is seeking a closed-form analytic solution or a numerical answer for a particular situation. These different aspects will be treated in the material which follows.

As indicated in the foregoing discussion, conditions commonly imposed on surfaces in an enclosure include either the specification of a fixed surface temperature (or E_{bi}) or the specification of a known heat flow at the surface. In the first case one wishes to find the heat flow necessary to maintain the given emissive power and in the second case one seeks the resultant equilibrium temperature of the surface. While there are applications in radiant exchange analyses in which a surface heat flow of some known value is specified, the most commonly encountered case is that in which the flux is known to be zero—giving rise to the "adiabatic, reradiating" surface defined next. Most of the examples considered in what follows will be limited to cases in which the only heat flux specification applied is that of zero.

The Adiabatic, Reradiating Surface

In an enclosure, an *adiabatic, reradiating surface* is defined as one which is thermally isolated so that the net heat flow away from the surface is zero. Such a surface interacts radiatively with the other surfaces of the enclosure, absorbing and reflecting incident irradiation and reemitting the absorbed energy. Through this interaction it is allowed to come to an equilibrium emissive power (or temperature) compatible with its radiative environment so that its net heat flow is zero. Thus, for an adiabatic surface $J_i = G_i$. The radiation shield considered at the end of Sec. 11.6 was such an adiabatic, reradiating surface. The refractory walls in a furnace which serve to reflect or absorb and reradiate energy from the fire are also examples of adiabatic surfaces.

The enclosure problem to be treated in the remainder of this chapter will consist of collections of surfaces, some of which are of known temperature and some of which are adiabatic surfaces. It will be desired to find the equilibrium temperature of these adiabatic surfaces. Equation (11.73b) shows that for an adiabatic surface, the equilibrium temperature is readily found from its radiosity:

$$\sigma T_{ad}^4 = E_{bad} = J_{ad}, \tag{11.78}$$

where the subscript ad is used to denote an *adiabatic* surface. Equation (11.78) shows that the equilibrium temperature, or equivalent blackbody emissive power, is independent of the surface properties. Finding the radiosity of an adiabatic surface is tantamount to finding its temperature; hence, the basic enclosure problem remains the same: Find the radiosities of all the surfaces in the enclosure.

The Network Analogy

The carrying out of the analysis of the radiation exchange in an enclosure may be facilitated by noting a generalization of the network analogy developed in Sec. 11.8 for black surfaces. Equations (11.73a) and (11.76) showed the net heat flow from a surface in an enclosure to be

$$q_i = \frac{A_i \epsilon_i}{1 - \epsilon_i}(E_{bi} - J_i) = \frac{E_{bi} - J_i}{(1 - \epsilon_i)/A_i \epsilon_i},$$

$$= \sum_j A_i F_{i-j}(J_i - J_j) = \sum_j \frac{J_i - J_j}{1/A_i F_{i-j}}.$$

As noted by Oppenheim in Ref. 11, the first of these equations suggest that the heat flow at a surface may be represented as the current flow in a resistor of magnitude $(1 - \epsilon_i)/A_i \epsilon_i$ connected between nodes, one of which is maintained at a potential equal to the blackbody emissive power of the surface, E_{bi}, and the other at a potential equal to the surface radiosity. The second equation suggests that the exchange between a surface and others in the enclosure may be represented as the current flows in resistors connecting the radiosity node to all other radiosity nodes through resistors of magnitude $1/A_i F_{i-j}$. Figure 11.23 depicts the representation of a surface maintained at a given E_{bi}.

If a surface is black, one knows that $E_{bi} = J_i$, so that the two nodes become one and no resistor connects them. If a surface is an adiabatic reradiator, the node is not maintained at a given E_{bi}, but is allowed to "float" and reach some equilibrium radiosity, in keeping with the concept discussed in the preceding

FIGURE 11.23. Electrical
network analogy for a gray
surface.

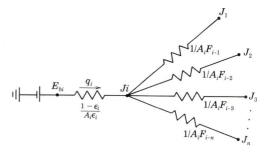

section. Figure 11.24 illustrates the network for an enclosure with a black surface of known temperature, A_1, two gray surfaces of known temperature, A_2 and A_3, and a reradiating surface, A_4.

The utility of the network analogy will be illustrated in the next section, in which enclosure analysis is discussed in some detail.

Direct Solution of Enclosures of Black, Gray, and Adiabatic Surfaces

The general concepts discussed in the foregoing will now be applied to the solution of specified enclosure problems. The details of implementation will differ depending on the magnitude of the particular problem at hand. If a rather large number of surfaces are involved so that the computations may require the use of a digital computer, matrix methods may be applied as described in the next section. However, for a modest number of surfaces in which one seeks the answer for a specific case or desires a closed-form solution, a more direct approach may be used. Such a direct solution is described here.

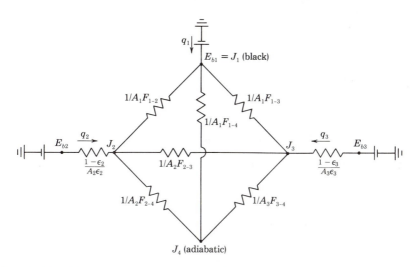

FIGURE 11.24. Network analogy for an enclosure of two gray surfaces, one black surface, and one adiabatic surface.

Consider, then, the case in which one is given an enclosure of n surfaces of known geometry. Some of the surfaces have specified temperature, or E_{bi}, and it is desired to find the heat flows at these surfaces. Call these surfaces "active surfaces" and number them $i = 1, 2, \ldots, n_1$, where $n_1 \le n$. The active surfaces may be black or perhaps gray with known ϵ_i. Let the remaining surfaces, if any, be adiabatic surfaces with $q_i = 0$, and number them $i = n_i + 1$, $n_1 + 2, \ldots, n$. It is desired to find the equilibrium temperatures, or E_{bi}, of these adiabatic surfaces. It is presumed that all the necessary shape factors, F_{i-j}, between all surfaces of the enclosure have been found according to the methods of Sec. 11.7. To find the desired heat flows at the n_1 active surfaces and the equilibrium temperatures at the $n - n_1$ adiabatic surfaces, it is necessary to find the n radiosities of all the surfaces. One proceeds as follows:

1. For each active surface write Eq. (11.77), the E_{bi}'s being known. If a particular active surface is black, Eq. (11.77) becomes indeterminate, but it is known that $J_i = E_{bi}$, effectively reducing the number of unknown J_i's. Thus, one writes:

   For $i = 1, 2, \ldots, n_1; j = 1, 2, \ldots, n_1, \ldots, n$

   If gray: $$\frac{A_i \epsilon_i}{1 - \epsilon_i}(E_{bi} - J_i) = \sum_j A_i F_{i-j}(J_i - J_j), \qquad (11.79)$$

   If black: $J_i = E_{bi}$.

2. For each adiabatic surface write Eq. (11.76) knowing $q_i = 0$:

   For $i = n_1 + 1, \ldots, n; j = 1, 2, \ldots, n_1, \ldots, n$ $\qquad (11.80)$

   $$\sum_j A_i F_{i-j}(J_i - J_j) = 0.$$

3. If one prefers, the analogous electrical network for the enclosure may be drawn, and steps 1 and 2 are equivalent to writing current balances at each of the nodes with unknown J_i. No balance is necessary at a black active node since $J_i = E_{bi}$ there.
4. Steps 1 and 2, or 3, yield n equations in the n unknown J_i's (or fewer when black active surfaces are present). Solve these equations for the unknown J_i's.
5. For each active surface, apply Eq. (11.76) [or Eq. (11.73a) if gray] to find the desired q_i. This is equivalent to finding the current flow at each active E_{bi} node if the network analogy is used. Thus,

   For $i = 1, 2, \ldots, n_1; j = 1, 2, \ldots, n_1, \ldots, n$

   If gray: $$q_i = \frac{A_i \epsilon_i}{1 - \epsilon_i}(E_{bi} - J_i), \qquad (11.81)$$

   If gray or black: $q_i = \sum_j A_i F_{i-j}(J_i - J_j).$

6. For each adiabatic surface find the equilibrium temperature according to Eq. (11.78):

$$\text{For } i = n_1 + 1, \ldots, n: \quad T_i = \left(\frac{E_{bi}}{\sigma}\right)^{1/4} = \left(\frac{J_i}{\sigma}\right)^{1/4} \quad (11.82)$$

In general, the above procedure involves solving a set of simultaneous equations equal to the number of gray active surfaces plus the number of adiabatic surfaces. If this number is small the procedure can be carried out by hand calculation, or even algebraically in some cases. If a large number of surfaces are present matrix solution, as described in the next section, may be desirable. Clearly, if there are no adiabatic surfaces present, one needs only to find the radiosities of the active gray surfaces present. If the actives are all black, one need find only the radiosities of the adiabatic surfaces. If the enclosure consists of only black active surfaces, the method degenerates to the simple case considered in Sec. 11.8.

EXAMPLE 11.10

For the shoe store considered in Example 11.9 and shown in Fig. 11.21, presume that the floor (call it A_1) acts as a black plane at 45°C, the window (call it A_2) acts as a black plane at 10°C, and the rest of the walls and ceiling (call them A_3) act as a single adiabatic reradiating surface. Find the net heat flow at the floor and window, and the equilibrium temperature of the walls.

Solution. From the geometry of Example 11.9, one has $A_1 = 120$ m², $A_2 = 18$ m². With $T_1 = 45°C = 318°K$ and $T_2 = 10°C = 283°K$, one has $E_{b1} = 579.8$, $E_{b2} = 363.7$ W/m². Example 11.8 showed $F_{1-2} = 0.058$; thus $F_{2-1} = A_1 F_{1-2}/A_2 = 0.387$. Then since $F_{1-1} = F_{2-2} = 0$, one has

$$F_{1-3} = 1.0 - 0.058 = 0.942,$$

$$F_{2-3} = 1.0 - 0.387 = 0.613.$$

So

$$A_1 F_{1-2} = A_2 F_{2-1} = 6.96 \text{ m}^2,$$

$$A_1 F_{1-3} = A_3 F_{3-1} = 113.0 \text{ m}^2,$$

$$A_2 F_{2-3} = A_3 F_{3-2} = 11.0 \text{ m}^2.$$

Application of Eq. (11.79) at the two active surfaces is unnecessary since they are black and it is known that

$$J_1 = E_{b1} = 579.8 \text{ W/m}^2, \quad J_2 = E_{b2} = 363.7 \text{ W/m}^2.$$

For the adiabatic surface, Eq. (11.80) gives

$$A_3 F_{3-1}(J_3 - J_1) + A_3 F_{3-2}(J_3 - J_2) = 0,$$

$$113.0(J_3 - 579.8) + 11.0(J_3 - 363.7) = 0,$$

so

$$J_3 = 560.63 \text{ W/m}^2.$$

Thus, the equilibrium temperature of the walls, A_3, is

$$T_3 = 100 \times \left(\frac{560.63}{5.67}\right)^{1/4} = 315°K = 42°C \ (108°F).$$

The heat flows at A_1 and A_2 are, from Eq. (11.81),

$$q_1 = A_1F_{1-2}(J_1 - J_2) + A_1F_{1-3}(J_1 - J_3) = 3670 \text{ W (12,520 Btu/h)},$$

$$q_2 = A_2F_{2-1}(J_2 - J_1) + A_2F_{2-3}(J_2 - J_3) = -3670 \text{ W } (-12,520 \text{ Btu/h}).$$

The heat flows must satisfy $q_1 = -q_2$ since A_3 is adiabatic.

In this elementary example, the network analogy, shown in Fig. 11.25, can be used to show that by adding resistances in parallel and series

$$q_1 = \frac{E_{b1} - E_{b2}}{\{A_1F_{1-2} + [(1/A_1F_{1-3}) + (1/A_2F_{2-3})]^{-1}\}^{-1}} = 3670 \text{ W.} \qquad \blacksquare$$

EXAMPLE 11.11

A rectangle 5 ft × 10 ft is maintained at $T_1 = 500°F$ and has an emissivity of $\epsilon_1 = 0.7$. It is parallel and directly opposed to a second rectangle of the same size, 5 ft away. The second rectangle has a temperature of $T_2 = 900°F$ and $\epsilon_2 = 0.9$. Find the net heat flow at each of the two surfaces (considering only the opposed faces) if (a) they are located in a radiation-free environment; (b) they are connected by a single adiabatic surface.

Solution. The electrical network analogy for this example is shown in Fig. 11.26. For the two active surfaces one has (using $\sigma = 0.1714 \times 10^{-8}$ Btu/h-ft^2-°R^4)

$$A_1 = 50 \text{ ft}^2, \qquad T_1 = 500°F = 960°R, \qquad E_{b1} = 1456 \text{ Btu/h-ft}^2,$$

$$A_2 = 50 \text{ ft}^2, \qquad T_2 = 900°F = 1360°R, \qquad E_{b2} = 5864 \text{ Btu/h-ft}^2.$$

Figure 11.15, or Eq. (E.2), yields $F_{1-2} = F_{2-1} = 0.285$. If A_3 is used to denote the radiation-free space in part (a) or the adiabatic surface in part (b), one has $F_{1-3} = F_{2-3} = 1.0 - 0.285 = 0.715$ since $F_{1-1} = F_{2-2} = 0$. Thus,

$$\frac{A_1\epsilon_1}{1 - \epsilon_1} = 116.67 \text{ ft}^2, \qquad \frac{A_2\epsilon_2}{1 - \epsilon_2} = 450.0 \text{ ft}^2,$$

$$A_1F_{1-2} = A_2F_{2-1} = 14.25 \text{ ft}^2,$$

$$A_1F_{1-3} = A_2F_{2-3} = A_3F_{3-1} = A_3F_{3-2} = 35.75 \text{ ft}^2.$$

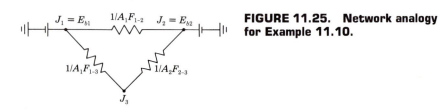

FIGURE 11.25. Network analogy for Example 11.10.

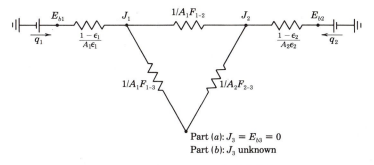

FIGURE 11.26. Network analogy for Example 11.11.

(a) In the case in which the surfaces are located in a radiation-free space, complete the enclosure by representing the space as a third active black surface at $T_3 = E_{b3} = 0$. No adiabatic surfaces are present, so only heat balances need to be made at the three active ones. Thus, Eq. (11.79), or use of the network, gives

$$\frac{A_1\epsilon_1}{1 - \epsilon_1}(E_{b1} - J_1) = A_1F_{1-2}(J_1 - J_2) + A_1F_{1-3}(J_1 - J_3),$$

$$\frac{A_2\epsilon_2}{1 - \epsilon_2}(E_{b2} - J_2) = A_2F_{2-1}(J_2 - J_1) + A_2F_{2-3}(J_2 - J_3),$$

$$J_3 = E_{b3} = 0,$$

or

$$116.67(1456 - J_1) = 14.25(J_1 - J_2) + 35.75J_1,$$

$$450.0(5864 - J_2) = 14.25(J_2 - J_1) + 35.75J_2,$$

from which

$$J_1 = 1474 \text{ Btu/h-ft}^2 \qquad J_2 = 5319 \text{ Btu/h-ft}^2.$$

The heat flows at each surface are then found from Eq. (11.81):

$$q_1 = \frac{A_1\epsilon_1}{1 - \epsilon_1}(E_{b1} - J_1) = 116.67(1456 - 1474)$$

$$= -2103 \text{ Btu/h (616 W)},$$

$$q_2 = \frac{A_2\epsilon_2}{1 - \epsilon_2}(E_{b2} - J_2) = 450.0(5854 - 5319)$$

$$= 244{,}980 \text{ Btu/h (71,800 W)}.$$

The heat flows q_1 and q_2 are not equal in magnitude since some energy is lost to the radiation-free space.

(b) In the case in which the surfaces are enclosed by a single adiabatic surface, Eq. (11.79) applied to A_1 and A_2 and Eq. (11.80) applied to A_3 gives

$$\frac{A_1\epsilon_1}{1 - \epsilon_1}(E_{b1} - J_1) = A_1F_{1-2}(J_1 - J_2) + A_1F_{1-3}(J_1 - J_3),$$

$$\frac{A_2\epsilon_2}{1 - \epsilon_2}(E_{b2} - J_2) = A_2F_{2-1}(J_2 - J_1) + A_2F_{2-3}(J_2 - J_3),$$

$$A_3F_{3-1}(J_3 - J_1) + A_3F_{3-2}(J_3 - J_2) = 0,$$

or

$$116.67(1456 - J_1) = 14.25(J_1 - J_2) + 35.75(J_1 - J_3),$$

$$450.0(5864 - J_2) = 14.25(J_2 - J_1) + 35.75(J_2 - J_3),$$

$$35.75(J_3 - J_1) + 35.75(J_3 - J_2) = 0.$$

This time a system of three equations in three unknown radiosities must be solved. The results are

$$J_1 = 2357 \text{ Btu/h-ft}^2, \qquad J_2 = 5630 \text{ Btu/h-ft}^2, \qquad J_3 = 3993 \text{ Btu/h-ft}^2.$$

For the active surfaces, Eq. (11.81) gives

$$q_1 = \frac{A_1\epsilon_1}{1 - \epsilon_1}(E_{b1} - J_1) = -105,100 \text{ Btu/h } (-30,800 \text{ W}),$$

$$q_2 = \frac{A_2\epsilon_2}{1 - \epsilon_2}(E_{b2} - J_2) = 105,100 \text{ Btu/h } (30,800 \text{ W}).$$

The heat flows, q_1 and q_2, are equal in magnitude since there are just two active surfaces and no energy is lost to the surroundings.

Equation (11.82) gives the equilibrium temperature of A_3, the adiabatic surface, to be

$$T_3 = \left(\frac{J_3}{\sigma}\right)^{1/4} = 100 \times \left(\frac{3993}{0.1714}\right)^{1/4} = 1235°\text{R} = 775°\text{F } (413°\text{C}). \blacksquare$$

EXAMPLE 11.12

Figure 11.27 depicts a furnace which constitutes an enclosure. The furnace is 5 m high, 10 m wide, and 20 m long. The floor of the furnace, A_1, acts as a black plane at $T_1 = 200°\text{C} = 473°\text{K}$; the back wall (5 m × 20 m) acts as a gray plane, A_2, with $T_2 = 400°\text{C} = 673°\text{K}$ and $\epsilon_2 = 0.4$. The two 5 m × 10 m ends act as a single adiabatic surface, A_3, and the remaining 10 m × 20 m ceiling and 5 m × 20 m front wall act as a second adiabatic surface, A_4. Find the heat flow at the two active surfaces (A_1 and A_2) and the equilibrium temperature of the adiabatic surfaces (A_3 and A_4).

Solution. The analogous electrical network is also shown in Fig. 11.27. Application of the methods of Sec. 11.7 and Appendix E yields the following shape factors: $F_{1-1} = F_{2-2} = 0$, $F_{1-2} = 0.167$, $F_{1-3} = 0.158$, $F_{1-4} = 0.675$, $F_{2-3} = 0.168$, $F_{2-4} = 0.498$, $F_{3-4} = 0.484$. Thus,

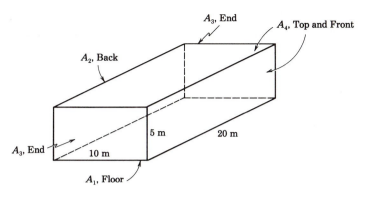

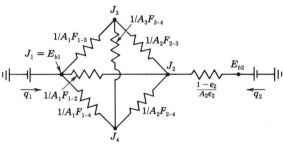

FIGURE 11.27. Geometry and network analogy for Example 11.12.

$$A_1 = 200 \text{ m}^2, \quad A_2 = 100 \text{ m}^2, \quad A_3 = 100 \text{ m}^2, \quad A_4 = 300 \text{ m}^2,$$

$$\frac{\epsilon_2 A_2}{1 - \epsilon_2} = 66.67 \text{ m}^2,$$

$$A_1 F_{1-2} = A_2 F_{2-1} = 33.4 \text{ m}^2; \quad A_1 F_{1-3} = A_3 F_{3-1} = 31.6 \text{ m}^2;$$

$$A_1 F_{1-4} = A_4 F_{4-1} = 135.0 \text{ m}^2; \quad A_2 F_{2-3} = A_3 F_{3-2} = 16.8 \text{ m}^2;$$

$$A_2 F_{2-4} = A_4 F_{4-2} = 49.8 \text{ m}^2; \quad A_3 F_{3-4} = A_4 F_{4-3} = 48.4 \text{ m}^2.$$

Equation (11.79), or current balances at the two active nodes, gives

$$J_1 = E_{b1},$$

$$\frac{A_2 \epsilon_2}{1 - \epsilon_2}(E_{b2} - J_2) = A_2 F_{2-1}(J_2 - J_1) + A_2 F_{2-3}(J_2 - J_3)$$

$$+ A_2 F_{2-4}(J_2 - J_4),$$

while Eq. (11.80), or current balances at the adiabatic nodes, A_3 and A_4, gives

$$A_3 F_{3-1}(J_3 - J_1) + A_3 F_{3-2}(J_3 - J_2) + A_3 F_{3-4}(J_3 - J_4) = 0,$$

$$A_4 F_{4-1}(J_4 - J_1) + A_4 F_{4-2}(J_4 - J_2) + A_4 F_{4-3}(J_4 - J_3) = 0.$$

With $T_1 = 473°K$ and $T_2 = 673°K$, one has $E_{b1} = 2838$ W/m^2, $E_{b2} = 11,632$ W/m^2. Thus, the above equations give the following three equations in J_2, J_3, and J_4:

$$66.67(11,632 - J_1) = 33.4(J_2 - 2838) + 16.8(J_2 - J_3) + 49.8(J_2 - J_4),$$

$$31.6(J_3 - 2838) + 16.8(J_3 - J_2) + 48.4(J_3 - J_4) = 0,$$

$$135.0(J_4 - 2838) + 49.8(J_4 - J_3) + 48.4(J_4 - J_3) = 0.$$

The solution to these equations is

$$J_2 = 6811, \qquad J_3 = 4081, \qquad J_4 = 3944 \text{ W/m}^2.$$

Equation (11.81) gives, for active surfaces

$$q_1 = A_1F_{1-2}(J_1 - J_2) + A_1F_{1-3}(J_1 - J_3) + A_1F_{1-4}(J_1 - J_4)$$

$$= -321,350 \text{ W } (-1,096,500 \text{ Btu/h-ft}^2),$$

$$q_2 = \frac{\epsilon_2 A_2}{1 - \epsilon_2}(E_{b2} - J_2) = 321,350 \text{ W } (1,096,500 \text{ Btu/h-ft}^2).$$

Equation (11.82) gives the equilibrium temperatures of the adiabatic surfaces to be

$$T_3 = 100 \times \left(\frac{4081}{5.67}\right)^{1/4} = 517.9°K = 244.9°C \text{ (473°F)},$$

$$T_4 = 100 \times \left(\frac{3944}{5.67}\right)^{1/4} = 513.6°K = 240.4°C \text{ (465°F)}.$$

The closeness of the values of J_3 and J_4 (or T_3 and T_4) suggest that the two adiabatic surfaces might have been treated as a single surface without much error and with a considerable saving in algebra. ■

Matrix Solution of Enclosures of Black, Gray, and Adiabatic Surfaces

The preceding examples illustrate the method that may be followed when the enclosure consists of only a few surfaces of unknown radiosity. If the number of gray active and adiabatic surfaces exceed about three, the solution for the unknown radiosities may become cumbersome unless matrix techniques are applied. In order to solve for the J's by matrix inversion, it is useful to recast the governing equations into forms useful for digital computer application and which do not become indeterminate when an active surface is black.

Equation (11.79) resulted from equating Eqs. (11.73a) and (11.76). However, for present purposes, it is desirable to use Eq. (11.75) in place of (11.76):

$$q_i = A_i J_i - \sum_j A_i F_{i-j} J_j.$$

If the Kronecker delta function δ_{ij} is introduced ($\delta_{ij} = 1$ when $i = j$, $\delta_{ij} = 0$ when $i \neq j$), the above is

$$\frac{q_i}{A_i} = \sum_j (\delta_{ij} - F_{i-j}) J_j. \tag{11.83}$$

When equated with Eq. (11.73a) some algebra gives

$$\frac{A_i \epsilon_i}{1 - \epsilon_i} (E_{bi} - J_i) = A_i \sum_j (\delta_{ij} - F_{i-j}) J_j \tag{11.84}$$

$$E_{bi} = \sum_j \frac{\delta_{ij} - (1 - \epsilon_i) F_{i-j}}{\epsilon_i} J_j.$$

Equations of the form of Eq. (11.84) may be written for each surface on which E_{bi} (i.e., T_i) is known, and equations of the form of Eq. (11.83) may be written for the adiabatic surfaces with $q_i/A = 0$. This set of equations may be solved for the unknown surface radiosities by matrix inversion routines normally found in most digital computing installations.

For the case of an enclosure of n surfaces, the first n_1 of which have specified E_{bi} and the remaining $n - n_1$ of which are adiabatic, the resulting set of n equations in the n radiosities may be represented by the matrix equation

$$[A][J] = [C]. \tag{11.85}$$

In Eq. (11.85), $[J]$ is the matrix of the radiosities,

$$[J] = \begin{bmatrix} J_1 \\ J_2 \\ \vdots \\ J_i \\ \vdots \\ J_n \end{bmatrix},$$

and $[C]$ is the matrix of the constant terms,

$$[C] = \begin{bmatrix} C_1 \\ C_2 \\ \vdots \\ C_i \\ \vdots \\ C_n \end{bmatrix}.$$

In the matrix $[C]$, Eqs. (11.84) and (11.83) show that

$$C_i = E_{bi}, \quad i = 1, 2, \ldots, n_1,$$
$$C_i = 0, \quad i = n_1 + 1, \ldots, n. \tag{11.86}$$

The coefficient matrix $[A]$ is

$$[A] = \begin{bmatrix} a_{11} & a_{12} & \cdots & a_{1j} & \cdots & a_{1n} \\ a_{21} & a_{22} & \cdots & a_{2j} & \cdots & a_{2n} \\ & & & & & \\ \cdot & & & & & \\ \cdot & & & & & \\ a_{i1} & a_{i2} & \cdots & a_{ij} & \cdots & a_{in} \\ \cdot & & & & & \\ \cdot & & & & & \\ a_{n1} & a_{n2} & \cdots & a_{nj} & \cdots & a_{nn} \end{bmatrix},$$

In which the elements a_{ij} are formed, from Eqs. (11.83) and (11.84), according to the rule:

For $i = 1, 2, \ldots, n_1;$ $\quad j = 1, 2, \ldots, n_1, \ldots, n$

$$a_{ij} = \frac{\delta_{ij} - (1 - \epsilon_i)F_{i-j}}{\epsilon_i} \tag{11.87}$$

For $i = n_1 + 1, \ldots, n;$ $\quad j = 1, 2, \ldots, n_1, \ldots, n$

$$a_{ij} = \delta_{ij} - F_{i-j}.$$

By inverting the matrix $[A]$, to obtain its inverse $[A]^{-1}$, the radiosities are found from

$$[J] = [A]^{-1}[C]. \tag{11.88}$$

Once the radiosities are known, the desired heat fluxes at the active surfaces are found from application of Eq. (11.83):

For $i = 1, 2, \ldots, n_1;$ $\quad j = 1, 2, \ldots, n_1, \ldots, n$ $\qquad$ (11.89)

$$\frac{q_i}{A_i} = \sum_j (\delta_{ij} - F_{i-j})J_j$$

and the equilibrium temperatures of the adiabatic surfaces from:

For $i = n_1 + 1, \ldots, n$ $\qquad$ (11.90)

$$T_i = \left(\frac{J_i}{\sigma}\right)^{1/4}.$$

Nothing in the above procedure fails if one of the active surfaces is black, one obtaining $E_{bi} = J_i$ automatically.

The method just outlined may be extended to cases in which the heat fluxes at the "nonactive" surfaces are given values other than zero; however, this generalization is not presented here for brevity and to avoid complications which might result when the specified fluxes violate the energy conservation principle. Reference 9 may be consulted in regard to this general problem.

EXAMPLE 11.13

Repeat Example 11.12 using the matrix formulation.

Solution. For the enclosure defined in Example 11.12, $A_1 = 200$ m^2 and $A_2 = 100$ m^2 are active surfaces with $E_{b1} = 2838$ W/m^2, $E_{b2} = 11,632$ W/m^2, $\epsilon_1 = 1.0$, $\epsilon_2 = 0.4$. The adiabatic surfaces are $A_3 = 100$ m^2 and $A_4 = 300$ m^2. Example 11.12 gave the following shape factors: $F_{1-1} = F_{2-2} = 0$, $F_{1-2} = 0.167$, $F_{1-3} = 0.158$, $F_{1-4} = 0.675$, $F_{2-3} = 0.168$, $F_{2-4} = 0.498$, $F_{3-4} = 0.484$. Application of shape factor reciprocity gives $F_{2-1} = 0.334$, $F_{3-1} = 0.316$, $F_{3-2} = 0.168$, $F_{4-1} = 0.450$, $F_{4-2} = 0.166$, $F_{4-3} = 0.161$. Then $F_{3-3} = 1 - 0.316 - 0.168 - 0.484 = 0.032$ and $F_{4-4} = 1 - 0.450 - 0.166 - 0.161 = 0.223$.

According to the rule of Eq. (11.87), the coefficient matrix is

$$[A] = \begin{bmatrix} \dfrac{1 - (1 - \epsilon_1)F_{1-1}}{\epsilon_1} & \dfrac{-(1 - \epsilon_1)F_{1-2}}{\epsilon_1} & \dfrac{-(1 - \epsilon_1)F_{1-3}}{\epsilon_1} & \dfrac{-(1 - \epsilon_1)F_{1-4}}{\epsilon_1} \\[2mm] \dfrac{-(1 - \epsilon_2)F_{2-1}}{\epsilon_2} & \dfrac{1 - (1 - \epsilon_2)F_{2-2}}{\epsilon_2} & \dfrac{-(1 - \epsilon_2)F_{2-3}}{\epsilon_2} & \dfrac{-(1 - \epsilon_2)F_{2-4}}{\epsilon_2} \\[2mm] -F_{3-1} & -F_{3-2} & 1 - F_{3-3} & -F_{3-4} \\ -F_{4-1} & -F_{4-2} & -F_{4-3} & 1 - F_{4-4} \end{bmatrix}$$

$$= \begin{bmatrix} 1 & 0 & 0 & 0 \\ -0.5010 & 2.500 & -0.2520 & -0.7470 \\ -0.3160 & -0.1680 & 0.9680 & -0.4840 \\ -0.4500 & -0.1660 & -0.1610 & 0.7770 \end{bmatrix}.$$

The inverse of this matrix is

$$[A]^{-1} = \begin{bmatrix} 1 & 0 & 0 & 0 \\ 0.5482 & 0.4518 & 0.2118 & 0.5663 \\ 0.8587 & 0.1413 & 1.2187 & 0.8950 \\ 0.8742 & 0.1258 & 0.2978 & 1.5934 \end{bmatrix}.$$

The constant matrix, from eq. (11.86), is

$$[C] = \begin{bmatrix} 2838 \\ 11,632 \\ 0 \\ 0 \end{bmatrix},$$

so that the radiosities are

$$\begin{bmatrix} J_1 \\ J_2 \\ J_3 \\ J_4 \end{bmatrix} = [A]^{-1}[C] = \begin{bmatrix} 2838 \\ 6811 \\ 4081 \\ 3944 \end{bmatrix}.$$

These radiosities are the same as those found in Example 11.12.

The above coefficient matrix and its inverse were rather simple in the first row since A_1 was a black surface. However, had A_1 been gray the matrix solution would be no more complex, but the solution by the method of Example 11.12 would become quite involved. ∎

11.10

SOME CLOSED-FORM SOLUTIONS FOR ENCLOSURES OF DIFFUSE, GRAY SURFACES

The methods outlined in Sec. 11.9 for the radiant exchange in enclosures of gray (or black) diffuse surfaces may be carried out algebraically when the number of unknown surface radiosities does not exceed two or three. In many instances the results may be written directly from examination of the corresponding electrical network analogy. The results of some of these analyses are summarized in the following and cover cases of some practical interest.

Two Gray Surfaces and a Single Adiabatic Surface

A case of some practical interest is that in which two active gray surfaces, of known E_b or temperature, are enclosed by a single adiabatic surface. The heat source and sink in a furnace enclosed by refractory walls is a typical example of such an enclosure. Figure 11.28 shows the electrical network for this case. The two active surfaces are denoted by A_1 and A_2 and the single adiabatic surface by A_r (the subscript r denoting the refractory surface). Summing the resistances in series and parallel between the nodes of specified E_{b1} and E_{b2} gives the heat flow from the two surfaces:

$$q_1 = -q_2$$

$$= \frac{E_{b1} - E_{b2}}{\dfrac{1 - \epsilon_1}{A_1\epsilon_1} + \left\{ A_1 F_{1-2} + \left[\left(\dfrac{1}{A_1 F_{1-r}}\right) + \left(\dfrac{1}{A_2 F_{2-r}}\right) \right]^{-1} \right\}^{-1} + \dfrac{1 - \epsilon_2}{A_2\epsilon_2}}.$$

$$(11.91)$$

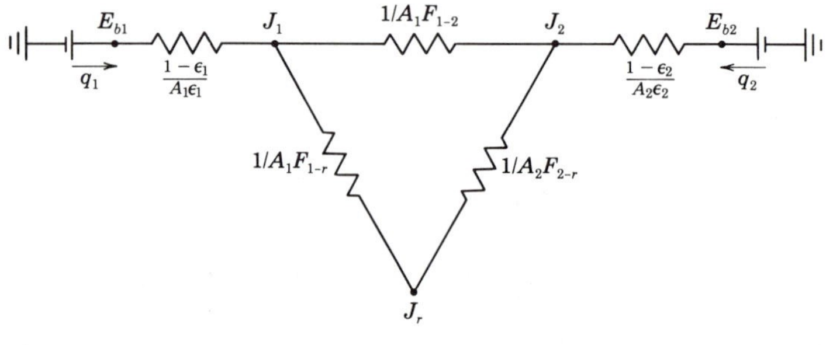

FIGURE 11.28.   Network analogy for two gray surfaces and one adiabatic surface.

The equilibrium radiosity, and hence temperature, of the single adiabatic surface is found by applying Eq. (11.80) or summing currents at J_r:

$$A_1 F_{1-r}(J_r - J_1) + A_2 F_{2-r}(J_r - J_2) = 0,$$ (11.92)

$$\sigma T_r^4 = J_r = \frac{A_1 F_{1-r} J_1 + A_2 F_{2-r} J_2}{A_1 F_{1-r} + A_2 F_{2-r}}.$$

The J_1 and J_2 in Eq. (11.92) are found from Eq. (11.81) or the current flows between E_{b1} and J_1, and E_{b2} and J_2:

$$q_1 = \frac{A_1 \epsilon_1}{1 - \epsilon_1}(E_{b1} - J_1); \qquad J_1 = E_{b1} - \frac{1 - \epsilon_1}{A_1 \epsilon_1} q_1,$$ (11.93)

$$q_2 = -q_1 = \frac{A_2 \epsilon_2}{1 - \epsilon_2}(E_{b2} - J_2); \qquad J_2 = E_{b2} + \frac{1 - \epsilon_2}{A_2 \epsilon_2} q_1.$$

With q_1 known from Eq. (11.91), Eqs. (11.92) and (11.93) permit calculation of J_r or T_r.

If the active surfaces are black, Eq. (11.91) is correspondingly simplified, and the effect of the surface grayness is seen to be that of reducing the net heat flow. When A_1 and A_2 are black, the determination of J_r from Eq. (11.92) is simplified since J_1 and J_2 are simply replaced by E_{b1} and E_{b2}.

The above expressions require the knowledge of three shape factors: F_{1-2}, F_{1-r}, F_{2-r}. These may be interrelated for some special cases as noted below.

Two Gray Convex Surfaces and a Single Adiabatic Surface

If the two active surfaces of the preceding case are plane or convex so that they do not "see" themselves and $F_{1-1} = F_{2-2} = 0$, then the two shape factors F_{1-r} and F_{2-r} may be eliminated by using $F_{1-r} = 1 - F_{1-2}$, $F_{2-r} = 1 - F_{2-1}$ and $A_1 F_{1-2} = A_2 F_{2-1}$. The resulting heat flow and equilibrium temperature become, after some algebra,

$$q_1 = -q_2 = \frac{E_{b1} - E_{b2}}{\dfrac{1 - \epsilon_1}{A_1 \epsilon_1} + \left[\dfrac{A_1 + A_2 - 2A_1 F_{1-2}}{A_1 A_2 - (A_1 F_{1-2})^2}\right] + \dfrac{1 - \epsilon_2}{A_2 \epsilon_2}},$$

$$\sigma T_r^4 = J_r = \frac{A_1(1 - F_{1-2})J_1 + (A_2 - A_1 F_{1-2})J_2}{A_1 + A_2 - 2A_1 F_{1-2}},$$ (11.94)

$$J_1 = E_{b1} - \frac{1 - \epsilon_1}{A_1 \epsilon_1} q_1; \qquad J_2 = E_{b2} + \frac{1 - \epsilon_2}{A_2 \epsilon_2} q_1.$$

In this instance only the single shape factor between the two active surfaces, F_{1-2}, is needed, and the configuration of the adiabatic surface is immaterial as long as it does not obstruct the view of A_1 and A_2. Again, the reduction to the case when the active surfaces are black is obvious.

Any Number of Black Surfaces and a Single Adiabatic Surface

If an enclosure consists of any number of black surfaces, say n_1 (all of specified E_{bi}), plus *one* adiabatic surface, it is left as an exercise to show that the heat flow from any one of the surfaces is

$$q_i = A_i \sum_{j=1}^{n_1} \left(F_{i-j} + \frac{F_{r-j}F_{i-r}}{1 - F_{r-r}} \right)(E_{bi} - E_{bj}), \tag{11.95}$$

and the equilibrium radiosity of the single adiabatic surface is

$$\sigma T_r^4 = J_r = \frac{\sum_{j=1}^{n_1} F_{r-j}E_{bj}}{1 - F_{r-r}}. \tag{11.96}$$

In Eqs. (11.95) and (11.96), F_{r-r} is the shape factor of the adiabatic surface with respect to itself.

Comparison of Eq. (11.95) with Eq. (11.69) for an enclosure of black surfaces only shows that the effect of an adiabatic surface is to enhance the heat flow between a pair of black active ones.

Two Convex Gray Surfaces in a Radiation-Free Space

Two gray surfaces, A_1 and A_2, located in an otherwise radiation-free space may be characterized as an enclosure in which the enclosing surface, A_3, is a black surface maintained with $E_{b3} = 0$. The corresponding network analogy is shown in Fig. 11.29. Current balances at nodes J_1 and J_2, or Eq. (11.79) yields

$$\frac{A_1\epsilon_1}{1 - \epsilon_1}(E_{b1} - J_1) = A_1F_{1-2}(J_1 - J_2) + A_1F_{1-3}(J_2 - 0),$$

$$\frac{A_2\epsilon_2}{1 - \epsilon_2}(E_{b2} - J_2) = A_2F_{2-1}(J_2 - J_1) + A_2F_{2-3}(J_2 - 0).$$

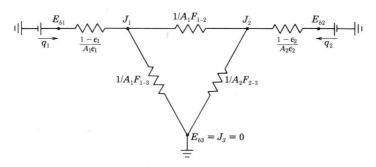

FIGURE 11.29. Network analogy for two gray surfaces in a radiation free space.

If the two active surfaces are convex so that $F_{1-1} = F_{2-2} = 0$, $F_{1-3} = 1 - F_{1-2}$, $F_{2-3} = 1 - F_{2-1}$, and if one uses $A_1F_{1-2} = A_2F_{2-1}$, solution of the above two equations will yield

$$J_1 = \frac{\epsilon_1 E_{b1} + \epsilon_2(1 - \epsilon_1)F_{1-2}E_{b2}}{1 - (1 - \epsilon_1)(1 - \epsilon_2)(A_1/A_2)F_{1-2}^2}.$$

Thus, since

$$q_1 = \frac{A_1\epsilon_1}{1 - \epsilon_1}(E_{b1} - J_1),$$

one has

$$q_1 = A_1\epsilon_1 \frac{[1 - (1 - \epsilon_2)(A_1/A_2)F_{1-2}^2]E_{b1} - \epsilon_2 F_{1-2}E_{b2}}{1 - (1 - \epsilon_1)(1 - \epsilon_2)(A_1/A_2)F_{1-2}^2}. \quad (11.97)$$

Note that in this instance, since energy is lost to space,

$$q_1 \neq -q_2.$$

The heat flow from surface A_2 may be obtained by interchanging the subscripts 1 and 2 throughout Eq. (11.97).

Two Gray Surfaces Only

If, as suggested in Fig. 11.30(a), an enclosure consists of only two gray active surfaces and nothing else, the network analogy readily yields

$$q_1 = -q_2 = \frac{E_{b1} - E_{b2}}{\dfrac{1 - \epsilon_1}{A_1\epsilon_1} + \dfrac{1}{A_1F_{1-2}} + \dfrac{1 - \epsilon_2}{A_2\epsilon_2}}. \quad (11.98)$$

If the surfaces are infinite parallel gray planes, $A_1 = A_2$, $F_{1-2} = 1$, and the above result reduces to that discussed in Sec. 11.6.

One Gray Surface Completely Enclosing a Second Convex Gray Surface

A case of considerable practical value is that in which one gray surface, A_2, completely encloses a second surface, A_1, which does not see itself, as depicted in Fig. 11.30(b). Concentric pipes, concentric spheres, objects in rooms, etc., are examples. In this instance one may use Eq. (11.98) with $F_{1-1} = 0$, $F_{1-2} = 1$:

$$\frac{q_1}{A_1} = \frac{E_{b1} - E_{b2}}{\dfrac{1}{\epsilon_1} + \dfrac{A_1}{A_2}\left(\dfrac{1 - \epsilon_2}{\epsilon_2}\right)}. \quad (11.99)$$

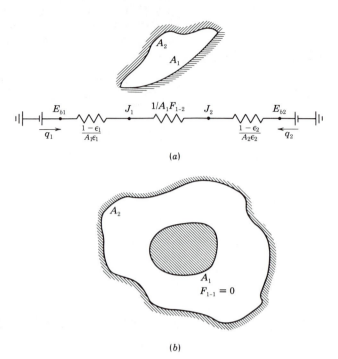

FIGURE 11.30. Enclosure of only two gray surfaces.

If, as may often be the case for heated bodies in a very *large* room, $A_1/A_2 \to 0$, then

$$\frac{q_1}{A_1} = \epsilon_1(E_{b1} - E_{b2}). \qquad (11.100)$$

The heat loss from the small body is seen to be independent of the properties of the large room.

EXAMPLE 11.14

Repeat Examples 11.10 and 11.11 using the closed-form solutions.

Solution

(a) Example 11.10 consisted of an enclosure of two black surfaces ($A_1 = 120$ m², $E_{b1} = 579.8$ W/m², $A_2 = 18$ m², $E_{b2} = 363.7$ W/m²) enclosed by a single adiabatic surface. The shape factor was known to be $F_{1-2} = 0.06$. Since the two active surfaces do not see themselves, Eqs. (11.94) apply with $\epsilon_1 = \epsilon_2 = 1.0$:

$$
\begin{aligned}
q_1 = -q_2 &= \frac{A_1 A_2 - (A_1 F_{1-2})^2}{A_1 + A_2 - 2A_1 F_{1-2}}(E_{b1} - E_{b2}) \\[2mm]
&= \frac{120 \times 18 - (120 \times 0.058)^2}{120 + 18 - 2(120 \times 0.058)}(579.8 - 363.7) \\[2mm]
&= 3678 \text{ W } (12{,}550 \text{ Btu/h-ft}^2),
\end{aligned}
$$

$$J_r = \frac{A_1(1 - F_{1-2})E_{b1} + (A_2 - A_1F_{1-2})E_{b2}}{A_1 + A_2 - 2A_1F_{1-2}}$$

$$= \frac{120 \times 0.942 \times 579.8 + (18 - 120 \times 0.058)363.7}{120 + 18 - 2 \times 120 \times 0.058}$$

$$= = 560.57 \text{ W/m}^2.$$

Thus,

$$T_r = 315°\text{K} = 42°\text{C} (108°\text{F}).$$

(b) Example 11.11 involved two active gray surfaces with $A_1 = A_2 = 50 \text{ ft}^2$, $\epsilon_1 = 0.7$, $\epsilon_2 = 0.9$, $E_{b1} = 1456 \text{ Btu/h-ft}^2$, $E_{b2} = 5864 \text{ Btu/h-ft}^2$. The surfaces did not see themselves and $F_{1-2} = F_{2-1} = 0.285$.

In part (a) of Example 11.11 the surfaces were located in a radiation-free space. Thus, Eq. (11.97) applies. For surface A_1:

$$q_1 = A_1\epsilon_1 \frac{[1 - (1 - \epsilon_2)(A_1/A_2)F_{1-2}^2]E_{b1} - \epsilon_2F_{1-2}E_{b2}}{1 - (1 - \epsilon_1)(1 - \epsilon_2)(A_1/A_2)F_{1-2}^2}$$

$$= 50 \times 0.7 \times \frac{[1 - 0.1 \times (0.285)^2]1456 - 0.9 \times 0.285 \times 5864}{1 - 0.3 \times 0.1 \times (0.285)^2}$$

$$= 2103 \text{ Btu/h} (-616 \text{ W}).$$

For surface A_2, interchange 1's and 2's in the above:

$$q_2 = A_2\epsilon_2 \frac{[1 - (1 - \epsilon_1)(A_2/A_1)F_{2-1}^2]E_{b2} - \epsilon_1F_{2-1}E_{b1}}{1 - (1 - \epsilon_2)(1 - \epsilon_1)(A_2/A_1)F_{2-1}^2}$$

$$= 50 \times 0.9 \frac{[1 - 0.3 \times (0.285)^2]5864 - 0.7 \times 0.285 \times 1456}{1 - 0.1 \times 0.3 \times (0.285)^2}$$

$$= 244,980 \text{ Btu/h} (71,800 \text{ W}).$$

In part (b) of Example 11.11 the surfaces were enclosed by a single adiabatic surface. Thus, Eq. (11.94) gives the heat flows to be

$$q_1 = -q_2 = \frac{E_{b1} - E_{b2}}{\dfrac{1 - \epsilon_1}{A_1\epsilon_1} + \left[\dfrac{A_1 + A_2 - 2A_1F_{1-2}}{A_1A_2 - (A_1F_{1-2})^2}\right] + \dfrac{1 - \epsilon_2}{A_2\epsilon_2}}$$

$$= \frac{1456 - 5864}{\dfrac{0.3}{50 \times 0.7} + \left[\dfrac{50 + 50 - 2 \times 50 \times 0.285}{50 \times 50 - (50 \times 0.285)^2}\right] + \dfrac{0.1}{50 \times 0.9}}$$

$$= -105,100 \text{ Btu/h} (-30,800 \text{ W}).$$

Thus, the radiosities of the surfaces are

$$J_1 = E_{b1} - \frac{1 - \epsilon_1}{A_1\epsilon_1} q_1 = 1456 + \frac{0.3}{50 \times 0.7}(105,100)$$

$$= 2357 \text{ Btu/h-ft}^2,$$

$$J_2 = E_{b2} + \frac{1 - \epsilon_2}{A_2\epsilon_2} = 5864 - \frac{0.1}{50 \times 0.9}(105,100)$$

$$= 5630 \text{ Btu/h-ft}^2,$$

$$J_r = \frac{A_1(1 - F_{1-2})J_1 + (A_2 - A_1F_{1-2})J_2}{A_1 + A_2 - 2A_1F_{1-2}}$$

$$= \frac{50(1 - 0.285)2356 + (50 - 50 \times 0.285)5630}{50 + 50 - 2 \times 50 \times 0.285}$$

$$= 3993 \text{ Btu/h-ft}^2.$$

Thus,

$$T_r = 100\left(\frac{3993}{0.1714}\right)^{1/4} = 1235°R = 775°F \ (413°C). \quad \blacksquare$$

11.11

RADIATION IN THE PRESENCE OF ABSORBING AND EMITTING GASES

For other than the very broad definitions given in Sec. 11.2, all discussions in this chapter have been devoted to the characteristics of solid surfaces (emitting, absorbing and reflecting characteristics) and the exchange of radiation between solid surfaces separated by nonabsorbing media. The solid surfaces considered were taken as opaque to thermal radiation ($\tau = 0$) while the media between surfaces were taken as transparent and nonemitting ($\tau = 1$, $\epsilon = \alpha = 0$).

Elementary gases with symmetrical molecules are, indeed, transparent to thermal radiation. Many gases with more complex molecules, however, do emit and absorb thermal radiation—at least in certain wavelength bands. Water vapor, carbon dioxide, ammonia, and most hydrocarbons are examples of the latter.

Absorptivity and Emissivity of Gases

Figure 11.31 shows, as an example, the absorption characteristics of carbon dioxide. Similar data are available for other gases of engineering significance (Ref. 12). Examination of Fig. 11.31 reveals that the absorptivity depends upon the thickness of the gas layer as well as upon the wavelength of the incident radiation. The thermodynamic state of the gas is also a determining factor for the absorptivity. The dependence of the gas absorption upon wavelength and thickness is described by the equation

$$dI_\lambda = -I_\lambda a_\lambda \, ds,$$

where I_λ represents the monochromatic intensity of the incident beam, s the path length of the beam, and a_λ the *absorption coefficient*. The absorption coefficient is dependent upon wavelength and the thermodynamic state of the gas. To a first approximation, a_λ varies linearly with pressure, at constant tem-

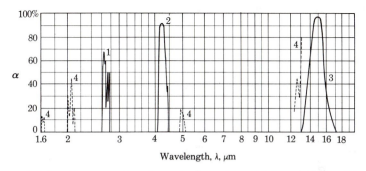

FIGURE 11.31. Absorptivity of carbon dioxide: (1) 5 cm thick; (2) 3 cm thick; (3) 6.3 cm thick; and (4) 100 cm thick. (From E. R. G. Eckert and R. M. Drake, _Heat and Mass Transfer_, New York, McGraw-Hill, 1959. Used by permission.)

perature. Thus, in a mixture of an absorbing and nonabsorbing gas the absorption coefficient should be proportional to the partial pressure of the absorbing gas.

An integration of the above equation yields

$$I_\lambda = I_{\lambda 0} e^{-a_\lambda s},$$

where $I_{\lambda 0}$ represents the intensity of the radiant beam as it enters the gas at $s = 0$. Since the reflection of thermal radiation at a gas-to-gas interface is generally negligible, the above result yields the following for the monochromatic transmissivity, emissivity, and absorptivity of a gas:

$$\tau_\lambda = e^{-a_\lambda s}, \tag{11.101}$$

$$\alpha_\lambda = 1 - e^{-a_\lambda s}.$$

If Kirchhoff's law holds, then a monochromatic emissivity of a gas is

$$\epsilon_\lambda = 1 - e^{-a_\lambda s} = \frac{I_\lambda}{I_{b\lambda}}. \tag{11.102}$$

As indicated in Eq. (11.102) the emissivity of a gas has to be interpreted as the intensity of radiation arriving at a point, divided by the equivalent black body intensity. As such, the emissivity given above is associated with the radiation arriving at a point from a given direction and through a given thickness of gas. In order to account for radiant exchange between a gas mass and an element of its boundary surface, consideration must be made for _all_ the radiation arriving at the element from _all_ directions. The geometry of such a situation is depicted in Fig. 11.32.

Application of Eqs. (11.14), (11.15), and (11.28) shows that the hemispherical monochromatic emissivity of the entire gas mass with respect to dA is

$$\bar{\epsilon}_\lambda = \frac{1}{\pi} \int_0^{2\pi} \int_0^{\pi/2} (1 - e^{-a_\lambda s}) \sin\theta \cos\theta \, d\theta \, d\varphi.$$

The integration in the above equation can be performed only when s is related to θ and φ—meaning that the geometry of the gas mass must be known. Then a second integration must be performed, over the finite boundary surface.

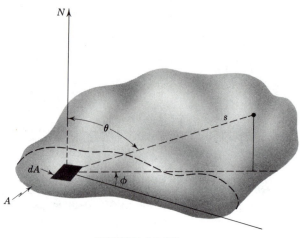

FIGURE 11.32

In general, calculations like those just described are quite complex. However, in the case of a hemispherical mass of gas, the emissivity for radiation to the center of the base may be found easily. Hottel and Sarofim (Ref. 10) have shown that other shapes of practical interest may be reduced to equivalent hemispherical masses. The radii of these equivalent hemispheres are referred to as the *mean beam length,* customarily denoted by L_e. Table 11.2 gives recommended mean beam lengths for some shapes of interest. Since for the hemispherical shape, the mean beam length is constant, the associated value of $\bar{\epsilon}_\lambda$ may be integrated over all wavelengths, to obtain finally a total gas emissivity. Thus, the mean beam length becomes a useful engineering concept, giving the following simpler relation for emissive power of a given gas shape:

$$W = \bar{\epsilon}(L_e)\sigma T^4. \tag{11.103}$$

In Eq. (11.103), $\bar{\epsilon}$ represents the total gas emissivity described above. Quite apparently, $\bar{\epsilon}$ is a function of the gas composition, its thermodynamic state, *and* the mean beam length, L_e.

TABLE 11.2   Mean Beam Lengths for Various Gas Shapes

Shape	L_e
Sphere (radiation to surface)	0.65 × diameter
Circular cylinder, infinite length (radiation to curved surface)	0.95 × diameter
Circular cylinder, length = diameter (radiation to base center)	0.77 × diameter
Infinite planes (radiation to surfaces)	1.80 × distance between planes
Cube (radiation to surfaces)	0.66 × edge length
Arbitrary shape (radiation to surface)	≈ 3.6 × (volume/surface)

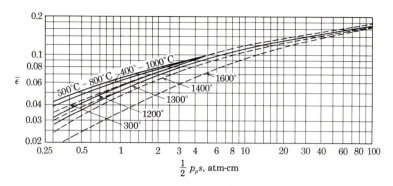

FIGURE 11.33. Emissivity of carbon dioxide in a mixture with air or nitrogen. The total pressure is 1 atm, and p_p represents the partial pressure of the carbon dioxide. (From E. R. G. Eckert and R. M. Drake, *Heat and Mass Transfer*, New York, McGraw-Hill, 1959. Used by permission.)

Values of $\bar{\epsilon}(L_e)$ may be found from experimental data such as those shown in Figs. 11.33 and 11.34 for carbon dioxide and water vapor mixed with air. As noted in the earlier discussion, the gas emissivity is dependent upon density, and in mixtures of emitting and nonemitting gases, $\bar{\epsilon}$ should depend upon the partial pressure of the absorbing component—as Figs. 11.33 and 11.34 show.

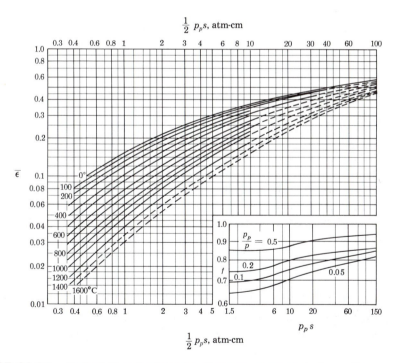

FIGURE 11.34. Emissivity of mixtures of water vapor and air or nitrogen for a total pressure of 1 atm, p_p representing the partial pressure of the water vapor. When p_p differs from 1, values from the large chart must be multiplied by f from the small chart. (From E. R. G. Eckert and R. M. Drake, *Heat and Mass Transfer*, New York, McGraw-Hill, 1959. Used by permission.)

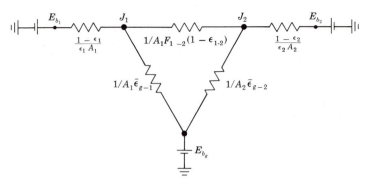

FIGURE 11.35. Network analogy of two gray surfaces separated by an absorbing, emitting gas.

EXAMPLE 11.15

Find the emissivity to be used for calculating the radiant exchange between a spherical mass of water vapor–air mixture and its surface if the mass has the diameter of 3 m, a temperature of 800°C, and a total pressure of 1 atm. The partial pressure of the water vapor is 0.2 atm.

Solution. For a diameter of 3 m, Table 11.2 gives the mean beam length to be $L_e = 0.65 \times 3 = 1.95$ m. Thus, $p_p L_e = 0.2 \times 1.95 \times 100 = 39$ atm-cm. At 800°C, $p_p/p = 0.2$, Fig. 11.34 gives

$$\bar{\epsilon} = 0.32 \times 0.83 = 0.27. \qquad \blacksquare$$

Radiation Exchange in Gas-Filled Enclosures

In the event that the enclosures in Secs. 11.8 and 11.9 are filled with an emitting and absorbing gas, the procedures given in those sections must be modified. The radiating gas must be treated as an individual "surface" itself. Since Eq. (11.103) gives the emissive power of the gas relative to surface i as $\bar{\epsilon}_{g-i}E_{bg}$, the gas may be represented by a node maintained at a potential E_{bg} connected to each active radiating surface through a resistor of $1/A_i\bar{\epsilon}_{g-i}$. The direct radiant exchange between active surfaces, $A_iF_{i-j}(J_i - J_j)$, when no gas is present, must be decreased by the amount absorbed by the intervening gas. Thus, the resistors connecting active nodes have to be of the form $1/A_iF_{i-j}(1 - \bar{\epsilon}_{i-j})$, in which $\bar{\epsilon}_{i-j}$ is the gas absorptivity evaluated for all rays traveling between A_i and A_j. This amounts to the evaluation of the equation for $\bar{\epsilon}$ over the geometrical angles admitted by F_{i-j}. This calculation can become quite complex. References 11 and 12 may be consulted in this respect. Figure 11.35 illustrates the equivalent network for a two-surface enclosure filled with an emitting and absorbing gas.

REFERENCES

1. SIEGEL, R., and J. R. HOWELL, *Thermal Radiation Heat Transfer*, 2nd ed., New York, McGraw-Hill, 1981.

2. DUNKLE, R. V., "Thermal Radiation Tables and Applications" *Trans. ASME*, Vol. 65, 1954, p. 549.

3. KREITH, FRANK, *Radiation Heat Transfer,* Scranton, Pa., International Textbook, 1962.

4. EDWARDS, D. K., D. E. NELSON, R. D. RODDICK, and J. T. GIER, "Basic Studies on the Use and Control of Solar Energy," *Univ. Calif. Dept. Eng. Rep. 60-93,* Los Angeles, 1960.

5. DUFFIE, J. A., and W. A. BECKMAN, *Solar Energy Thermal Processes,* New York, Wiley, 1974.

6. KREITH, FRANK, and J. F. KREIDER, *Principles of Solar Engineering,* Washington, D.C., Hemisphere, 1978.

7. SPARROW, E. M., "A New and Simpler Formulation for Radiative Angle Factors," *J. Heat Transfer, Trans. ASME,* Vol. 85, 1963, p. 81.

8. TOUPS, K. A., "Confac II, A General Computer Program for the Determination of Radiant Interchange Configuration and Form Factors," *Tech. Doc. Rep. FDL-TDR-64-43,* Air Force Flight Dynamics Laboratory, Wright-Patterson Air Force Base, Ohio, 1964.

9. ROHSENOW, W. M., and J. P. HARTNETT, eds., *Handbook of Heat Transfer,* New York, McGraw-Hill, 1973.

10. HOTTEL, H. C., and A. F. SAROFIM, *Radiative Heat Transfer,* New York, McGraw-Hill, 1967.

11. OPPENHEIM, A. K., "Radiation Analysis by the Network Method," *Trans. ASME,* Vol. 78, 1956, p. 725.

12. McADAMS, W. H., *Heat Transmission,* 3rd ed., New York, McGraw-Hill, 1954.

PROBLEMS

11.1 Find the fraction of the total hemispherical emissive power leaving a diffuse emitter that is contained in the directions (a) $0 \leq \theta \leq \pi/6$, $0 \leq \varphi \leq 2\pi$; and (b) $0 \leq \theta \leq \pi/4$, $0 \leq \varphi \leq 2\pi$.

11.2 A hollow spherical cavity, 1 m in diameter, is evacuated and heated to an interior surface temperature of 500°K. How many watts of radiant energy is emitted from a hole 0.2 cm in diameter in the wall of the cavity if the cavity surface has an emissivity of (a) 0.9, and (b) 0.4?

11.3 Using Eqs. (11.20), (11.22) and (11.23), derive the relation given in Eq. (11.24).

11.4 The filament of an ordinary 100-watt light globe may be approximated as a blackbody at 2900°K. What is the wavelength of the maximum monochromatic emission? What fraction of the emission is in the visible portion of the spectrum?

11.5 A blackbody is maintained at a temperature of 200°C. Find (a) the wavelength at which the maximum monochromatic emissive power occurs, (b) the value of the maximum monochromatic emissive power, (c) the total emissive power, and (d) the fraction of the radiation emitted between the wavelengths of 1.0 and 4.0 μm.

11.6 Repeat Prob. 11.5 if the blackbody temperature is 1100°C.

11.7 Verify the data given in Table 11.1 for the sun at a blackbody temperature of 5800°K.

11.8 A blackbody surface is heated to 2000°K. Find the fraction of the emission that occurs in wavelength bands (a) 1 to 5 μm, (b) 5 to 10 μm, (c) 10 to 15 μm, and (d) 15 to 20 μm.

11.9 For the surface of Example 11.2 and described in Fig. 11.10, find the total emissivity for a surface temperature of 1000°K. What is its total absorptivity with respect to a blackbody source at 1000°K and at 3000°K?

11.10 A surface has a monochromatic emissivity of 0.15 for all wavelengths less than or equal to 3.0 μm and 0.8 for all wavelengths greater than or equal to 3.0 μm. Find (a) the total emissivity of the surface for a surface temperature of 1000°K, and (b) the solar absorptivity of the surface.

11.11 If the surface in Prob. 11.10 is allowed to come to thermal equilibrium with the sun, find the equilibrium temperature if the sun's rays are (a) normal to the surface, and (b) inclined at an angle of 45° with the surface normal.

11.12 A surface has a monochromatic emissivity of 0.1 for all wavelengths below a "cutoff" value λ_c, and 0.9 for all wavelengths above λ_c. Find the solar absorptivity of the surface if λ_c is (a) 0 μm (b) 1 μm, (c) 2 μm, (d) 4 μm, and (e) 10 μm.

11.13 If the surface in Prob. 11.12 is allowed to come to equilibrium with the sun's rays at normal incidence, find the equilibrium temperature of the surface for each of the cases.

11.14 Using the data of Fig. 11.7, make an approximate calculation of the solar absorptivity of white epoxy paint on aluminum.

11.15 The air space in the wall of a house is sufficiently thin that the bounding surfaces may be treated as infinite parallel planes. If the inner surfaces of the outside brick ($\epsilon = 0.93$) is at 40°C and the opposite surface of the air space is building paper ($\epsilon = 0.90$) at 25°C, find the radiant heat flux through the space.

11.16 If in Prob. 11.15 a layer of aluminum foil ($\epsilon = 0.09$) is placed on top of the building paper, to what value is the radiant flux reduced if the foil assumes the temperature of the paper?

11.17 Two very large parallel planes are maintained at 150°F and 600°F, respectively. Find the radiant flux exchanged if (a) each plane is black; (b) the 150°F plane is gray with $\epsilon = 0.9$, and the 600°F plane is gray with $\epsilon = 0.6$; (c) the 150°F plane is gray with $\epsilon = 0.6$, and the 600°F plane is gray with $\epsilon = 0.9$; and (d) the 150°F plane is gray with $\epsilon = 0.9$, and the 600°F plane is gray with $\epsilon = 0.1$.

11.18 Two parallel infinite planes each have $\epsilon = 0.095$. One plane is maintained at 90°C and the other at 350°C.

    (a) What is the radiant heat exchanged between the planes?

    (b) If a third plane of the same material is placed between the original two and allowed to come to equilibrium, what is the temperature of this third plane, and what is the radiant exchange between the two original surfaces?

11.19 Two infinite parallel planes, one at 400°C with $\epsilon = 0.8$ and the other at 150°C with $\epsilon = 0.3$ are to be shieded by placing a third plane ($\epsilon = 0.9$, each side) between them and allowing it to come to thermal equilibrium. Find the radiant flux between the planes before and after the insertion of the third plane, and find the equilibrium temperature of the shield.

11.20– Find the shape factors, F_{1-2}, for the configurations shown in the accompanying
11.27 figures.

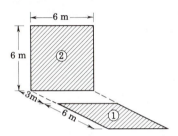

PROBLEM 11.20

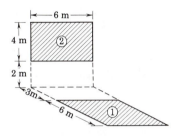

PROBLEM 11.21

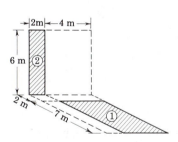

PROBLEM 11.22

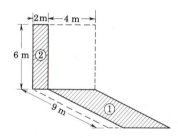

PROBLEM 11.23

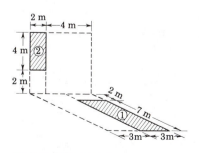

PROBLEM 11.24

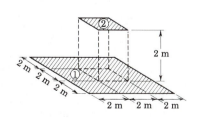

PROBLEM 11.25

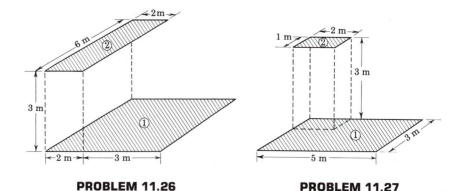

PROBLEM 11.26   **PROBLEM 11.27**

11.28   An enclosure consists of the inner surface of a 10-cm-I.D. cylinder 10 cm long (call it A_2), the outer surface of a 5-cm-O.D. cylinder of the same length and placed coaxially with the first (call it A_1), and the annular surfaces at each end (call them A_3 and A_4). Find the 16 shape factors: F_{i-j}, $i = 1, 2, 3, 4$; $j = 1, 2, 3, 4$.

11.29   Derive Eq. (E.13) of Appendix E for the shape factor between two parallel circular disks as shown in Fig. 11.17.

11.30   Two black rectangles, 2 m × 3 m, are parallel and directly opposed—spaced 2 m apart. If their temperatures are 250°C and 150°C, respectively, find the rate of radiant exchange between them and the rate at which the 250°C rectangle is losing energy. Assume that the surfaces are in a radiation-free environment.

11.31   Two black squares, 5 ft × 5 ft, are spaced 10 ft apart, parallel and directly opposed and placed in a radiation-free space. If their temperatures are 100°F and 300°F, respectively, find the rate of radiant exchange between them and the rate at which each square is losing energy.

11.32   A black rectangle, 2.5 m × 3.5 m, is maintained at 250°C. It is located normal to, and shares a common edge with, another black rectangle, 1.8 m × 3.5 m, at 500°C. What is the rate at which each rectangle is losing radiant energy if they are located in a radiation-free space?

11.33   For the geometry given in Prob. 11.21, let $T_1 = 300°K$ and $T_2 = 400°K$. If both surfaces are black, what is the radiant heat flow away from each surface, assuming that no other surfaces are present?

11.34   A cubical room, 10 ft on a side, is composed of black surfaces. One wall is at 70°F and the opposite wall is at 90°F. All the other walls are at 80°F. What is the net radiant heat flow away from the 90°F wall?

11.35   The accompanying figure depicts an artist's studio with a skylight and door in one wall. The flow is to be used as a radiant heating panel and acts as a black

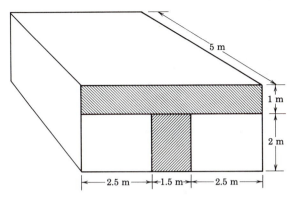

PROBLEM 11.35

plane at 70°C. The skylight and door act as black planes at 10°C, whereas the rest of the walls and ceiling act as black planes at 30°C. Find the net radiant energy given up by the floor.

11.36 If for the geometry of Prob. 11.35, the floor acts as a black surface at 90°C, the door as a black surface at 40°C, and the skylight as a black plane at 5°C, find the net heat flow at each of these surfaces if the remaining walls act as a single adiabatic surface. What is the equilibrium temperature of the adiabatic surface?

11.37 Imagine a system of three black active surfaces. Surface 1 has A_1 = 40 m², T_1 = 500°C; surface 2 has A_2 = 20 m², T_2 = 1000°C; surface 3 has A_3 = 15 m², T_3 = 1600°C. The following shape factors are known: F_{1-1} = 0.0, F_{1-2} = 0.15, F_{1-3} = 0.10, F_{2-2} = 0.0, F_{2-3} = 0.0, F_{3-3} = 0.0.
  (a) If the three surfaces are located in a radiation-free space, find the net heat flow at each surface.
  (b) If the three surfaces are enclosed by a fourth, adiabatic surface, find the equilibrium temperature of this surface and the heat flow at each active surface.

11.38 An enclosure consists of a rectangular parallelpiped 2 m × 4 m × 6 m. One of the 2 m × 4 m surfaces acts as a black surface at 100°C and the other acts as a black surface at 200°C. The other surfaces act as adiabatic surfaces; however, symmetry permits the treatment of the two 2 m × 6 m surfaces as one adiabatic surface and the two 4 m × 6 m surfaces as another. Find the equilibrium temperature of the two adiabatic surfaces and the heat flow at the two active surfaces.

11.39 The results of Prob. 11.38 should show that the two adiabatic surfaces have the same temperature. Show, perhaps with use of the electrical network analogy, that in *any* rectangular parallelepiped with two opposing active surfaces, *gray* or *black,* no error is incurred in finding the heat flow at the active surfaces if the other surfaces are taken to be a single adiabatic one as compared to treating them as individual adiabatic surfaces.

11.40   Verify the shape factors quoted in Example 11.12.

11.41   An enclosure consists of a parallelepiped 6 ft × 12 ft × 24 ft. One of the 12 × 24 surfaces (call it A_1) acts as a black surface at 500°F, and one of the 6 × 24 surfaces (call it A_2) acts as a black surface at 1000°F. The two 6 × 12 surfaces act as a single adiabatic surface (call it A_3) and the remaining 12 × 24 and 6 × 24 surfaces act as a second adiabatic surface (call it A_4). Find the equilibrium temperature of the two adiabatic surfaces and the net heat flow at the two active surfaces.

11.42   Repeat Prob. 11.32 if the 250°C rectangle is gray with $\epsilon = 0.8$ and the 500°C rectangle is gray with $\epsilon = 0.4$. All other data remain the same.

11.43   For the geometry of Prob. 11.21, let A_1 be gray with $\epsilon_1 = 0.2$, $T_1 = 300$°K and let A_2 be gray with $\epsilon_2 = 0.9$, $T_2 = 400$°K. The surfaces are in a radiation-free environment. Find the net heat flow at each surface.

11.44   For the geometry of Prob. 11.35, the floor acts as a black plane at 90°C, the skylight acts as a gray plane at 5°C with $\epsilon = 0.8$, and the door acts as a gray plane at 40°C with $\epsilon = 0.4$. The remaining walls act as a single adiabatic surface. Find the temperature of the adiabatic surface and the heat flow at each of the active surfaces.

11.45   Repeat Prob. 11.44, but let the floor be a gray plane with $\epsilon = 0.6$, all other data remaining the same.

11.46   An enclosure consists of an equilateral tetrahedron, each of the four faces of which is an equilateral triangle, 1 m on each side. One surface is gray with $\epsilon_1 = 0.2$, $T_1 = 200$°K; another is gray with $\epsilon_2 = 0.4$. $T_2 = 400$°K; a third is gray with $\epsilon_3 = 0.7$, $T_3 = 700$°K; the fourth surface is an adiabatic surface. Find the equilibrium temperature of the adiabatic surface.

11.47   Repeat Prob. 11.46 if the three active surfaces are black, all other data remaining unchanged.

11.48   For the data of Prob. 11.37, let A_1 remain black, but let A_2 be gray with $\epsilon_2 = 0.7$ and let A_3 be gray with $\epsilon_3 = 0.3$. all other data remain the same. Find the same information requested in parts (a) and (b).

11.49   The floor of a furnace acts as a plane at $T_1 = 1500$°F, $\epsilon_1 = 0.6$ and the ceiling acts as a plane at $T_2 = 800$°F, $\epsilon_2 = 0.8$. The furnace is 10 ft wide, 12 ft long, and 15 ft high. All the other surfaces act as a single adiabatic surface. What is the heat flow at the two active surfaces and what is the equilibrium temperature of the walls?

11.50   Two 1 m × 1 m squares are parallel, directly opposed, and located 2 m apart. One plane is at $T_1 = 100$°C and the other $T_2 = 300$°C. Find the net heat flow at each surface if they are (a) black, unenclosed in a radiation-free space; (b) black, enclosed by a single adiabatic surface; (c) gray ($\epsilon_1 = 0.6$, $\epsilon_2 = 0.8$), in a radiation-free space; and (d) gray as in part (c), enclosed by a single adiabatic surface.

11.51 Two disks, each 3 ft in diameter, directly opposed and spaced 6 ft apart, are maintained at $T_1 = 300°F$ and $T_2 = 600°F$. Find the net heat flow at each disk if they are (a) black, unenclosed and in radiation-free space; (b) black, enclosed by a single adiabatic surface; (c) gray ($\epsilon_1 = 0.4$, $\epsilon_2 = 0.7$), in a radiation-free space; and (d) gray as in part (c), enclosed by a single adiabatic surface.

11.52 Two parallel, directly opposed, square planes (2 m $\times$ 2 m) are spaced 2 m apart. One plane ($\epsilon_1 = 0.6$) is at $T_1 = 200°C$ and the other ($\epsilon_2 = 0.7$) is at $T_2 = 400°C$. What is the heat flow at each plane if they are (a) in a radiation-free environment; and (b) enclosed by a single adiabatic surface?

11.53 Two parallel, directly opposed square planes (3 m $\times$ 3 m) are spaced 2 m apart. One plane is maintained at $T_1 = 400°C$, $\epsilon_1 = 0.8$ and the other at $T_2 = 150°C$, $\epsilon_2 = 0.3$. A third plane of the same size with $\epsilon_3 = 0.9$ on each side is placed equidistant between the first two and allowed to come to thermal equilibrium, with the same temperature on each side. The entire system is in a radiation-free space. Find the net heat flow from each of the two original planes before and after the third is inserted. The above data are the same as those given for infinite planes in Prob. 11.19. Compare the results, on a heat *flux* basis, and explain any differences.

11.54 A cylindrical cavity is 0.5 m in diameter and 0.5 m long. The ends are closed by circular disks. One end acts as a gray plane with $T_1 = 100°C$, $\epsilon_1 = 0.8$; the other end acts as a gray plane with $T_2 = 300°C$, $\epsilon_2 = 0.4$.
   (a) If the wall of the cavity is an adiabatic surface, find its temperature and the net heat flow at each active surface.
   (b) If the wall of the cavity is a third gray surface at $T_3 = 200°C$, $\epsilon_3 = 0.6$, find the net heat flow at each surface.

11.55 Two infinitely long cylinders are placed coaxially. The outer cylinder has an inside diameter of 6 in., an emissivity of 0.4, and is maintained at 400°F. The inner cylinder has an outside diameter of 2 in., an emissivity of 0.2, and is maintained at 800°F. A third cylindrical shield (very thin so that its temperature is the same on both sides) is placed concentrically midway between the first two and allowed to come to thermal equilibrium. It has an emissivity of 0.6 on both sides. Find (a) the net heat flow, per foot of length, from each cylinder before the shield is put in place; (b) the net heat flow, per foot of length, from each cylinder after the shield is put in place; and (c) the equilibrium temperature of the shield.

11.56 A cylindrical cavity has a diameter of 10 cm and a length of 15 cm. It is closed on one end by a circular disk. The other end is capped with a circular disk provided with a 5-cm-diameter concentric hole. The hole acts as a black plane at 0°K. Find the radiant loss out of the hole if (a) all interior surfaces are black

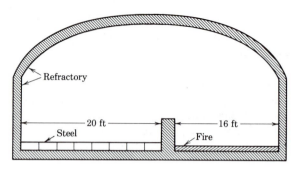

PROBLEM 11.59

at 600°K; (b) the base of the cavity (i.e., the 10-cm-diameter end) is gray at 600°K, $\epsilon = 0.5$, and the other surfaces are adiabatic; and (c) all interior surfaces are gray at 600°K, $\epsilon = 0.5$.

11.57 A piece of steel (30 cm × 100 cm × 300 cm) is heated to a temperature of 1000°C and then placed in a large room with walls at 40°C. The emissivity of the steel is 0.5. What is the initial rate of radiant heat loss from the steel as it cools?

11.58 A 3-in.-nominal bare wrought iron pipe with a surface temperature of 300°F passes through a large furnace at 800°F. What is the radiant loss from the pipe, per foot of length?

11.59 The accompanying figure depicts an annealing furnace in which the "fire" acts as a gray plane ($\epsilon = 0.7$, 2500°F). The steel, shielded so that it does not "see" the fire, acts as a gray surface at 1500°F with $\epsilon = 0.8$. All other surfaces are refractory (adiabatic) surfaces. The furnace is 10 ft deep into the plane of the paper. Even though the fire and steel do not see each other, find the rate of radiant heat transfer between them.

11.60 Verify the algebra leading from Eq. (11.83) to Eq. (11.84).

11.61 Verify the algebra leading from Eqs. (11.91), (11.92), and (11.93) to Eq. (11.94).

11.62 Derive Eqs. (11.95) and (11.96).

11.63 Derive Eq. (11.97).

CHAPTER 12

Heat Transfer by Combined Conduction and Convection— Heat Exchangers

12.1

INTRODUCTORY REMARKS

The title of this chapter would seem to imply that all the foregoing chapters have dealt with situations in which only one mode of heat transfer was taking place. Strictly speaking, this is not true. The previous chapters on the various aspects of conduction and convection certainly emphasized one or the other of these modes; however, in many instances both mechanisms were occurring simultaneously.

In much of the work concerning conduction, particularly Chapters 3 through 5, convection was included as a boundary condition in the solution of the conduction equation. In these cases it was presumed that the surface heat transfer coefficient, h, was known; and from this, and other given facts, much could be deduced about the conduction process within a given body—including the determination of the temperature of the solid surface in contact with the convecting fluid. On the other hand, all of Chapters 6 through 10, which considerd the hydrodynamical and thermodynamical aspects of convection, presumed knowledge of the surface temperature for determination of the heat transfer coefficient.

In real problems encountered in engineering practice, however, conduction may be taking place within the interior of a solid body while convection takes place with ambient fluids in contact with one or more of its surfaces. In such instances neither the surface heat transfer coefficients nor the surface temperatures are known, and it becomes necessary to solve the conduction and convection problems simultaneously. Imagine, for instance, a hot fluid of given temperature and hydrodynamic conditions separated by a wall (of known geometry and composition) from a cold fluid of known conditions. Based on the fluid temperatures, the heat flow through the wall (and, hence, the surface temperatures of the wall) can be found only if the heat transfer coefficients at the two surfaces are known along with the wall geometry and thermal properties. However, the heat transfer coefficients can generally be found only when the surface temperatures are known. Thus, a simultaneous, perhaps iterative solution must be carried out to find the surface temperature *and* the heat transfer coefficients.

The situation cited in the preceding paragraph, that of a wall separating two convecting fluids of different temperatures, is encountered frequently in engineering applications. The wall may be of any geometry (i.e., plane, cylindrical, etc.) and may consist of composite layers. In particular, the analysis of the performance of a heat exchanger (a device to transfer heat between two fluids without mixing) presents such a problem and is considered in some detail in later sections of this chapter.

However, other situations also occur wherein a simultaneous solution of conduction and convection must be carried out—such as the analysis of heat flow in an extended surface on which the surface heat transfer coefficient is not known a priori. It is clearly impossible to generalize sufficiently in order to establish a solution technique that will apply in all instances. Hence, the next section presents the approach that may be used in two particular cases with the hope that the reader may gain sufficient experience to be able to handle other instances requiring simultaneous solution of conduction and convection problems.

12.2

EXAMPLES OF ITERATIVE SOLUTION OF COMBINED CONDUCTION AND CONVECTION

This section will illustrate by the use of two examples the methodology of the simultaneous, iterative solution that may be applied in cases of combined convection and conduction. In both examples the problem of heat transfer between two convecting fluids separated by a wall will be considered, and in both cases a cylindrical-shaped wall will be used. In the first case forced convection will occur inside the cylinder and free convection will exist at the outer surface. In the second case forced convection and condensation occur at the surfaces. These cases are simply examples; any of the convective processes of Chapters 7 through 10 can be occurring at the boundaries—forced convection, free convection, dissipative flow, etc. Similarly, the separating wall may be of any geometry for which a conduction solution can be obtained (e.g., a plane wall), and the exposed surfaces may or may not be equipped with an array of extended surfaces.

In general, then, one must ascertain two surface heat transfer coefficients and two surface temperatures. In the first example, however, a special case will be considered in which one of the surface heat transfer coefficients is independent of temperature.

One Coefficient Independent of Temperature. As a first example, then, consider the case of a pipe (perhaps insulated) through which a hot fluid is flowing at a specified bulk temperature and velocity. Let the pipe be horizontal and exposed to still, atmospheric air at a known temperature so that there is heat loss by free convection at the outer surface. Both the inside pipe surface and the outer exposed surface temperatures are unknown. However, let it be assumed that the flow inside the pipe is turbulent and that the inner surface temperture is such that the conditions specified for determination of the inside heat transfer coefficient by Eq. (8.27) are satisfied. In this fortunate instance, the heat transfer coefficient can be found directly from the known, bulk, fluid temperature and no iterative solution for it is required. At the outer surface, however, the free convection coefficient is dependent on the unknown exposed surface temperature. This temperature is dependent on the heat flow through the pipe, which, in turn, is dependent on the heat transfer coefficient. The dependence of h on the surface temperature is implied through the complex dependence of the air thermal properties on temperature as well as through the correlations of Chapter 9, which involve the difference between the surface and the air temperature. Hence, one cannot, generally, solve directly for the temperature, and some iterative procedure must be established.

First, a trial value of the outer surface temperature, t_s, is assumed. Based on this assumption and the ambient air temperature, t_o, the free convection coefficient at the outside surface, h_o, can be calculated. The heat flow, per unit of outside surface area A_o, is then

$$\frac{q}{A_o} = h_o(t_s - t_o).$$

Since the heat transfer coefficient at the inner surface is known, as is the geometry and thermal conductivity of the pipe and insulation, the overall coefficient, U_o, based on the outer surface area can be calculated in accordance with the methods of Sec. 3.10. Then the heat flow may also be expressed in terms of the overall temperature difference between the outside fluid at t_o and the inside fluid at t_i:

$$\frac{q}{A_o} = U_o(t_i - t_o).$$

Equating these last two expressions gives

$$h_o(t_s - t_o) = U_o(t_i - t_o) \tag{12.1}$$

from which the surface temperature t_s may be calculated. This computed result is compared with the initial assumption, a revised value of t_s is chosen, and the calculations are repeated. This process is continued until a satisfactory agreement between assumed value and the derived values of t_s is achieved. The extent to which this iterative process is carried depends on the accuracy desired or the accuracy that can be justified by the given data.

The revised value of t_s used on the second, and subsequent, iterations need not necessarily be the derived value from the previous iteration. Indeed, depending on the problem at hand, the convergence to an acceptable answer may be hastened by choosing a new value of t_s between the original one and the derived one. Because of the complex way in which the surface temperature enters the calculation of h_o through the dependence of the fluid properties on t_s, it is difficult to formulate a general rule for choosing t_s in subsequent iterations, and the experience gained by working examples is essential. Obviously, the iterative process is shortened if the initial assumption of the surface temperature is close to the correct value. Again, previous experience is invaluable in this regard. In instances in which free convection into atmospheric air is occurring it is useful to remember that the heat transfer coefficient is often very close to 1 Btu/h-ft^2-°F $\approx$ 6 W/m^2-°C. These values may be used as an initial assumption (rather than t_s) to start the iterative process. These points are best illustrated by example.

The following examples are presented in some detail to illustrate the methodology involved. However, the reader should note that modern computer methods may be applied to facilitate considerably the solution of such iterative problems—provided that a capability of generating the fluid thermophysical properties is available.

EXAMPLE 12.1

Steam at 2000 kN/m^2 pressure and 300°C flows with a velocity of 5 m/s in a 6-in. schedule 40 pipe (k = 45 W/m-°C) which is covered with 3.5 cm of magnesia insulation (k = 0.075 W/m-°C). The pipe is positioned horizontally in a room in which the ambient air is at atmospheric pressure and 20°C. Find the temperature of the outer insulation surface, the overall heat transfer coefficient, and the rate of heat loss per unit length of pipe.

Solution. The notation to be used is shown in Fig. 12.1. Table B.1 yields the following diameters:

$$D_2 = 15.41 \text{ cm}, \quad D_3 = 16.83 \text{ cm}, \quad D_4 = 23.83 \text{ cm}.$$

The given data also yield

$$t_1 = 300°C, \quad t_5 = 20°C,$$

$$k_{23} = 45 \text{ W/m-°C}, \quad k_{34} = 0.075 \text{ W/m-°C}.$$

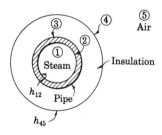

FIGURE 12.1

First the inside heat transfer coefficient, h_{12}, is determined. For steam at 2000 kN/m^2, 300°C, Table A.5 gives

$$\rho = 7.9681 \text{ kg/m}^3, \qquad \mu = 20.10 \times 10^{-6} \text{ kg/m-s}$$

$$k = 49.95 \times 10^{-3} \text{ W/m-°C}, \qquad \text{Pr} = 1.02.$$

Thus, the Reynolds number is

$$\text{Re}_D = \frac{vD_2\rho}{\mu} = \frac{5 \times 0.1541 \times 7.9681}{20.10 \times 10^{-6}} = 3.05 \times 10^5.$$

So the flow is turbulent and Eq. (8.27) may be applied if it is assumed that $|t_1 - t_2| < 60°C$. The validity of this assumption will have to be verified later. Thus, since the steam is being cooled, Eq. (8.27) gives

$$\text{Nu}_D = 0.023\text{Re}_D^{0.8}\text{Pr}^{0.3}$$

$$= 0.023 \times (3.05 \times 10^5)^{0.8} \times (1.02)^{0.3} = 564.9,$$

$$h_{12} = \text{Nu}_D \frac{k}{D_2} = 564.9 \times \frac{0.04595}{0.1541} = 168.5 \text{ W/m}^2\text{-°C}.$$

The coefficient h_{12} is not dependent on the unknown surface temperature t_2 and may thus be taken as constant in the subsequent calculations. Note that one might have used Eq. (8.29) instead of Eq. (8.27) to find h_{12} since the surface temperature is not needed in that equation for gases.

To obtain a starting estimate for the outside surface temperature, t_4, let h_{45} (the heat transfer coefficient at that surface) be taken to be approximately 6 W/m^2-°C, a value typical for free convection to atmospheric air. Thus, according to Eq. (3.38), the overall heat transfer coefficient for the pipe, based on the outside surface area, U_4, is

$$U_4 = \left[\frac{r_4}{r_2h_{12}} + \frac{r_4 \ln (r_3/r_2)}{k_{23}} + \frac{r_4 \ln (r_4/r_3)}{k_{34}} + \frac{1}{h_{45}}\right]^{-1}$$

$$= \left[\frac{0.2383}{0.1541 \times 168.5} + \frac{0.2383}{2 \times 45} \ln \left(\frac{0.1683}{0.1541}\right)\right.$$

$$\left. + \frac{0.2383}{2 \times 0.075} \ln \left(\frac{0.2383}{0.1683}\right) + \frac{1}{6}\right]^{-1}$$

$$= (0.00918 + 0.00023 + 0.5525 + 0.1667)^{-1} = 1.373 \text{ W/m}^2\text{-°C}.$$

It is immediately noticed that the controlling resistances in this case are those of the insulation and the outside coefficient—the resistance of the inside coefficient and the pipe wall being rather small. In some instances it may be justified to neglect these latter two resistances, but they are retained here for the sake of completeness.

Now, the heat flow, per unit of exposed surface, may be written in two ways— based on the overall temperature difference $(t_1 - t_5)$ and the surface-to-air difference $(t_4 - t_5)$, as in Eq. (12.1), to find an estimate of t_4:

$$\frac{q}{A_4} = h_{45}(t_4 - t_5) = U_4(t_1 - t_5)$$

$$6.0(t_4 - 20) = 1.373(300 - 20)$$

$$t_4 = 84°C.$$

Using this value of t_4, one may compute the heat transfer coefficient h_{45}, by use of the free convection correlations of Chapter 9. For $t_4 = 84°C$, $t_5 = 20°C$, one has

$$\Delta t = t_4 - t_5 = 64°C, \qquad t_m = \frac{t_4 + t_5}{2} = 52°C, \qquad \beta = \frac{1}{T_5} = \frac{1}{293.15°K}.$$

Table A.6 gives for air at $t_m = 52°C$:

$$\nu = 18.12 \times 10^{-6} \text{ m}^2/\text{s}, \quad k = 27.95 \times 10^{-3} \text{ W/m-°C}, \quad \text{Pr} = 0.709.$$

Thus, the Grashof and Rayleigh numbers are

$$\text{Gr}_D = \frac{D_{48}^3 g \beta \ \Delta t}{\nu^2} = \frac{(0.2383)^3 \times 9.8 \times (1/293.15) \times 64}{(18.12 \times 10^{-6})^2} = 8.818 \times 10^7,$$

$$\text{Ra}_D = \text{Gr}_D \times \text{Pr} = 6.252 \times 10^7.$$

For a horizontal cylinder, Eq. (9.42) gives

$$\text{Nu}_D = \left\{ 0.60 + 0.387 \ \text{Ra}_D^{1/6} \left[1 + \left(\frac{0.559}{\text{Pr}} \right)^{9/16} \right]^{-8/27} \right\}^2 = 49.00,$$

$$h_{45} = \text{Nu}_D \frac{k}{D_4} = 49.00 \times \frac{27.95 \times 10^{-3}}{0.2383} = 5.75 \text{ W/m}^2\text{-°C}.$$

This value of h_{45} is close to the originally assumed value of 6 W/m²-°C. Now, a new value of U_4 may be found, as above, to give

$$U_4 = \left(0.00918 + 0.00023 + 0.5525 + \frac{1}{5.75} \right)^{-1} = 1.355 \text{ W/m}^2\text{-°C},$$

and the associated value of the surface temperature is

$$h_{45}(t_4 - t_5) = U_4(t_1 - t_5)$$

$$5.75(t_4 - 20) = 1.355(300 - 20)$$

$$t_4 = 86°C.$$

This value of t_4 is probably sufficiently close to the former value of 84°C for most engineering calculations. However, just to complete the process, one may carry out one additional iteration for $t_4 = 85.5°C$, and the results are:

$$t_4 = 85.5°C, \qquad t_m = 52.8°C, \qquad \Delta t = 65.5°C,$$

$$\text{Gr}_D = 8.955 \times 10^7,$$

$$\text{Ra}_D = 6.349 \times 10^7,$$

$$\text{Nu}_D = 49.23,$$

$$h_{45} = 5.78 \ \text{W/m}^2\text{-°C},$$

$$U_4 = 1.357 \ \text{W/m}^2\text{-°C},$$

$$t_4 = 85.7\text{°C}.$$

Thus, the desired results may be taken to be

$$t_4 = 85.6\text{°C} \ (186.1\text{°F}),$$

$$U_4 = 1.357 \ \text{W/m}^2\text{-°C} \ (0.239 \ \text{Btu/h-ft}^2\text{-°F}),$$

$$\frac{q}{L} = \pi D_4 U_4 (t_1 - t_5),$$

$$= 284.5 \ \text{W/m} \ (295.8 \ \text{Btu/h-ft}).$$

The validity of the use of Eq. (8.27) for h_{12} should be determined by finding the inside pipe surface temperature t_2:

$$A_2 h_{12} (t_1 - t_2) = A_4 U_4 (t_1 - t_5),$$

$$\pi D_2 h_{12} (t_1 - t_2) = \pi D_4 U_4 (t_1 - t_5),$$

$$0.1541 \times 168.5(300 - t_2) = 0.2383 \times 1.357(300 - 20)$$

$$t_2 = 297\text{°C}.$$

Thus, $|t_1 - t_2| = 3\text{°C}$, which is less than 60°C, as required by Eq. (8.27) for gases. Thus, the use of Eq. (8.27) is justified. ∎

As was pointed out earlier, the above example is just that—an *example*. The hydrodynamic conditions existing at the two surfaces could be any of those described in Chapters 7 through 10 and the separating wall could be of some geometry other than cylindrical. The basic procedure, however, remains the same—a surface temperature is assumed to find the unknown h, and this h is used to find the surface temperature, which is then revised. The agreement between the assumed and derived temperatures required to terminate the calculation depends on the situation at hand. Usually, the accuracy of the given data does not justify the carrying of the solution as far as was done in the above example.

Two Coefficients Dependent on Surface Temperatures. Example 12.1 was simplified in that the inside surface temperature did not directly influence the calculation of the heat transfer coefficient there. Had the flow in the pipe been laminar or had the conditions on Eq. (8.27) not been met so that Eq. (8.28) must be applied, then knowledge of t_2 would be necessary for the determination of h_{12}. Then a double iterative process would have to be performed, as noted in the next example.

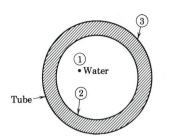

FIGURE 12.2. Example 12.2.

Condensing Steam

Water

Tube

EXAMPLE 12.2

A horizontal condenser tube ($\frac{3}{4}$ in. O.D., 16 gage) has steam at 5 psia condensing on its outer surface. Water at 85°F flows through the tube with a velocity of 5 ft/s. If the tube is made of stainless steel with $k = 10$ Btu/h-ft-°F, find the temperature of the tube surface and the overall heat transfer coefficient.

Solution. The Steam Tables show the saturation temperature and latent heat for steam at 5 psia to be 162°F and 1000.9 Btu/lb$_m$, respectively. These data, along with the problem specifications and Table B.2, yield the following known information when the notation of Fig. 12.2 is used:

$$t_1 = 85°F, \qquad v_1 = 5 \text{ ft/s},$$

$$t_4 = 162°F, \qquad h_{fg} = 1000.9 \text{ Btu/lb}_m,$$

$$D_2 = 0.620 \text{ in.}, \qquad D_3 = 0.750 \text{ in.}, \qquad k_{23} = 10 \text{ Btu/h-ft-°F}.$$

The procedure to be followed here is not vastly different from that in Example 12.1. An initial trial value of the inside heat transfer coefficient, h_{12}, is obtained. With this value held fixed, an iterative calculation is performed to determine the outside surface temperature and coefficient, t_3 and h_{34}. Then the value of t_2, the inside surface temperature, is found, and the trial value of h_{12} is revised. Using the revised h_{12}, the iterative determination of t_3 and h_{34} is repeated. The process is continued, alternating between revisions of h_{12} and the iterative determination of h_{34}, until a satisfactory agreement of all assumed and calculated temperatures is obtained.

1. *Initial estimate of h_{12}:* First it will be assumed that $|t_1 - t_2| < 6°C$ so that Eq. (8.27) may be used for h_{12}. This turns out not to be true, and a different correlation must be used; however, it provides a good starting point. Thus, with $t_1 = 85°F$, Table A.3 gives

$$v = 0.03150 \text{ ft}^2/\text{h}, \qquad k = 0.3549 \text{ Btu/h-ft-°F}, \qquad \text{Pr} = 5.52.$$

Thus,

$$\text{Re}_D = \frac{v_1 D_2}{v} = \frac{5 \times 3600 \times (0.62/12)}{0.03150} = 2.95 \times 10^4.$$

The flow in the tube is turbulent and Eq. (8.27) gives

**COMBINED CONDUCTION AND CONVECTION—
HEAT EXCHANGERS**

$$\text{Nu}_D = 0.023\text{Re}_D^{0.8}\text{Pr}^{0.4} = 171.65,$$

$$h_{12} = \text{Nu}_D \frac{k}{D_2} = 171.65 \times \frac{0.3549}{(0.62/12)},$$

$$= 1179.1 \text{ Btu/h-ft}^2\text{-}°\text{F}.$$

2. *Iteratively find t_3 and h_{34}:*
   (a) *Assume* that $t_3 = 135°\text{F}$. Then $t_m = (135 + 162)/2 = 148.5°\text{F}$ and Table A.3 gives

$$c_{pl} = 1.00 \text{ Btu/lb}_m\text{-}°\text{F}, \qquad \rho_l = 61.22 \text{ lb}_m/\text{ft}^3,$$

$$\mu_l = 1.054 \text{ lb}_m/\text{ft-h}, \qquad k_l = 0.3785 \text{ Btu/h-ft-}°\text{F},$$

   Thus, application of Eq. (10.17) for condensation on a horizontal cylinder, with $\Delta t = 162 - 135 = 27°\text{F}$ and $\rho_l(\rho_l - \rho_v) \approx \rho_l^2$, gives

$$\text{Ja} = \frac{1.00 \times 27}{1000.9} = 0.0270$$

$$h'_{fg} = h_{fg}(1 + 0.68\text{Ja}) = 1019.3 \text{ Btu/lb}_m.$$

$$\text{Nu}_D = 0.728\left(\frac{g\rho_l^2 h'_{fg} D^3}{\mu_l k_l \Delta t}\right)^{1/4}$$

$$= 0.728\left[\frac{(32.2)(3600)^2(61.22)^2(1019.3)(0.75/12)^3}{1.054 \times 0.3785 \times 27}\right]^{1/4}$$

$$= 317.4$$

$$h_{34} = \text{Nu}_D \frac{k}{D_3} = 317.4 \times \frac{0.3785}{0.75/12}$$

$$= 1922.2 \text{ Btu/h-ft}^2\text{-}°\text{F}.$$

   (b) *Find the overall coefficient.* For the geometry of Fig. 12.2, the overall heat transfer coefficient based on the outside surface area is

$$U_3 = \left[\frac{r_3}{r_2 h_{12}} + \frac{r_3 \ln (r_3/r_2)}{k_{23}} + \frac{1}{h_{34}}\right]^{-1}$$

$$= \left[\frac{0.75}{0.62 \times 1179.1} + \frac{0.75 \ln (0.75/0.62)}{2 \times 12 \times 10} + \frac{1}{1922.2}\right]^{-1}$$

$$= 467.1 \text{ Btu/h-ft}^2\text{-}°\text{F}.$$

   (c) *Verify the assumed t_3.* The value of t_3 compatible with the current values of h_{12} and h_{34} can be calculated from

$$\frac{q}{A_4} = h_{34}(t_3 - t_4) = U_3(t_1 - t_4)$$

$$1922.2(t_3 - 162) = 467.1(85 - 162)$$

$$t_3 = 143°\text{F}.$$

This calculated value of t_3 is significantly different from the assumed 135°F. Thus, a new assumption and calculation must be performed.

(d) *Assume a new* $t_3 = 144°F$. Repeat the calculations of steps (a) through (c) above. The results are summarized below:

$$t_3 = 144°F, \qquad t_m = 153°F, \qquad \Delta t = 18°F,$$

$$c_{pl} = 1.00 \text{ Btu/lb}_m\text{-°F}, \qquad \rho_l = 61.13 \text{ lb}_m/\text{ft}^2,$$

$$\mu_l = 1.018 \text{ lb}_m/\text{ft-h}, \qquad k_l = 0.3797 \text{ Btu/h-ft-°F},$$

$$\text{Ja} = 0.0180,$$

$$h'_{fg} = 1013.1 \text{ Btu/lb}_m,$$

$$\text{Nu}_D = 353.2,$$

$$h_{34} = 2145.8 \text{ Btu/h-ft}^2\text{-°F},$$

$$U_3 = 479.2 \text{ Btu/h-ft}^2\text{-°F},$$

$$t_3 = 144.8°F.$$

This latter value is sufficiently close to the assumed $t_3 = 144°F$ to proceed.

3. *Verify the trial value of* h_{12}. Now that values of h_{34} and t_3 consistent with the value of h_{12} found in part 1 have been determined, a value of t_2 may be calculated and h_{12} re-evaluated. Find the inside tube surface temperature from

$$A_2 h_{12}(t_1 - t_2) = A_4 h_{34}(t_3 - t_4)$$

$$D_2 h_{12}(t_1 - t_2) = D_4 h_{34}(t_3 - t_4)$$

$$0.62 \times 1179.1(85 - t_2) = 0.75 \times 2145.8 \times (144 - 162)$$

$$t_2 = 124°F.$$

This value of t_2 yields $|t_1 - t_2| = 40°F = 22°C, > 6°C$, which exceeds the limits imposed on Eq. (8.27). So one must start again, now with a more rational value of t_2.

4. *Assume* t_2 *and find* h_{12}. Based on the above, now assume that $t_2 = 124°F$. In this case $|t_1 - t_2| > 6°C$, so Eq. (8.28) must be used. [Equation (8.29) could also be used.] At $t_1 = 85°F$, the properties and Re_D found in step 1 still apply. However, one needs the viscosity at t_1 and t_2, and Table A.3 gives

$$\text{At } t_1 = 85°F, \ \mu = 1.959 \text{ lb}_m/\text{ft-h},$$

$$\text{At } t_2 = 124°F, \ \mu_s = 1.301 \text{ lb}_m/\text{ft-h}.$$

Equation (8.28) then gives

$$\text{Nu}_D = 0.027\text{Re}_D^{0.8}\text{Pr}^{1/3}\left(\frac{\mu}{\mu_s}\right)^{0.14}$$

$$= 0.027(2.95 \times 10^4)^{0.8}(5.52)^{1/3}\left(\frac{1.959}{1.301}\right)^{0.14}$$

$$= 190.3,$$

$$h_{12} = \text{Nu}_D\frac{k}{D_2} = 1307.1 \text{ Btu/h-ft}^2\text{-}°\text{F}.$$

5. *Repeat step 2.* One must now repeat the iterative calculation for t_3 and h_{34}. However, a very good starting point is the last iteration given in part (d) of step 2—in fact, no iteration is necessary. For an assumed $t_3 = 144°\text{F}$ (as found last), one has the identical calculations for h_{34}. Thus, $h_{34} = 2145.8$ Btu/h-ft²-°F. However, U_3 must be recalculated since h_{12} has changed. Thus,

$$h_{12} = 1307.1 \text{ Btu/h-ft}^2\text{-}°\text{F},$$

$$h_{34} = 2145.8 \text{ Btu/h-ft}^2\text{-}°\text{F},$$

$$U_3 = 504.7 \text{ Btu/h-ft}^2\text{-}°\text{F},$$

and the verification of t_3 yields

$$h_{34}(t_3 - t_4) = U_3(t_1 - t_4)$$

$$t_3 = 144°\text{F, as assumed.}$$

6. *Verify the assumed t_2.* Now, one repeats step 3 to check the assumed t_2 of 124°F. Thus,

$$D_2h_{12}(t_1 - t_2) = D_4h_{34}(t_3 - t_4)$$

$$0.62 \times 1307.1(85 - t_2) = 0.75 \times 2145.8(144 - 162)$$

$$t_2 = 121°\text{F}.$$

This value may be judged to be sufficiently close to the assumed 124°F to terminate the calculation. If, however, one were to repeat the process of steps 5 and 6, including another iterative process as in 2, the following results would be obtained:

$$t_2 = 121°\text{F, assumed,}$$

$$h_{12} = 1302.3 \text{ Btu/h-ft}^2\text{-}°\text{F},$$

$$t_3 = 144°\text{F, assumed,}$$

$$h_{34} = 2146 \text{ Btu/h-ft}^2\text{-}°\text{F},$$

$$U_3 = 503.8 \text{ Btu/h-ft}^2\text{-}°\text{F},$$

$$t_3 = 144°\text{F, calculated,}$$

$$t_2 = 121°\text{F, calculated.} \qquad \blacksquare$$

The procedures outlined in the two preceding examples will be used later in this chapter. The two fluids in a heat exchanger are separated by a wall (plane or cylindrical, with or without fins) and thus calculations of the foregoing sort must be carried out. These calculations may be further complicated by the fact that the local fluid temperatures also change as the fluids move through the heat exchanger.

12.3

HEAT EXCHANGERS—THE VARIOUS TYPES AND SOME OF THEIR GENERAL CHARACTERISTICS

The process of exchanging heat between two different fluid streams is one of the most important and frequently encountered processes found in engineering practice. Boilers, condensers, water heaters, automobile radiators, air heating or cooling coils, etc., are examples of processes in which heat is exchanged between a "hot" and a "cold" fluid. (The terms "hot" and "cold" are used in a relative sense.) The modern petrochemical industry, energy generating plants, etc., are based on innumerable processes involving the use of devices to exchange heat between two fluid streams without physically mixing them. Such devices are generally termed *heat exchangers*. This section will be devoted to descriptions of the more commonly encountered forms of heat exchangers and some of their general characteristics.

Ordinary heat exchangers may be divided into two general classes depending on the relative orientation of the flow direction of the two fluid streams. If the two streams cross one another in space—usually at right angles—the exchanger is termed a *cross-flow* heat exchanger. An automobile radiator or the cooling unit in an air conditioning duct are simple examples of cross-flow heat exchangers.

The second general class of heat exchangers comprise those in which the two streams move in parallel directions in space. The usual *shell-and-tube* heat exchanger is the most frequently encountered form of such an exchanger, and the concentric pipe exchanger (or *double-tube* exchanger) is also of this type. The term *parallel-flow* is not applied to this general class but is reserved for a special subclass as described in the following section.

Shell-and-Tube or Concentric Pipe Heat Exchangers

The simple form of the concentric pipe heat exchanger is depicted in Fig. 12.3. In the form shown in Fig. 12.3(a), the two fluids enter at the same end of the device and flow in the same direction to the other end. Such an arrangement is termed a *parallel-flow* heat exchanger. Figure 12.3(b) depicts an alternative arrangement for the same device in which the flow direction of the fluid in the inner pipe has been reversed. In this instance the two fluid streams travel in parallel paths but in opposite directions, and the arrangement is termed a *counterflow* heat exchanger. A concentric pipe heat exchanger may be limited in the amount of available surface area for the transfer of heat between the two fluid

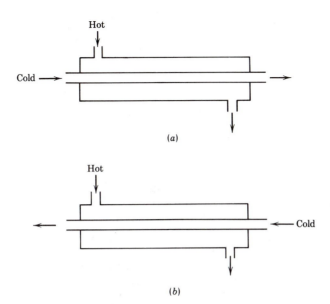

FIGURE 12.3. Schematic drawing of a concentric pipe heat exchanger; (a) parallel-flow; and (b) counterflow.

streams. The transfer area may be made greater by increasing the number of inner pipes as is done in the shell and tube arrangement described next.

The "classical" form of shell-and-tube heat exchanger is indicated by the simplified drawing in Fig. 12.4(a). The exchanger consists of a bundle of tubes (usually many more than shown in Fig. 12.4) secured at either end in tube sheets which are large drilled plates into which the tubes are either rolled or soldered.

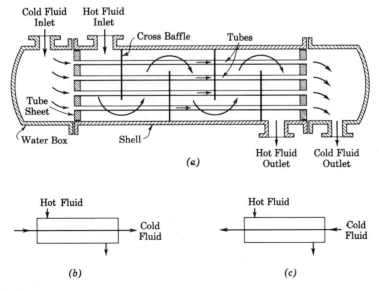

FIGURE 12.4. Schematic drawing of (a) one-shell pass, one-tube-pass heat exchanger; (b) parallel-flow; and (c) counterflow.

The entire tube bundle is placed inside a closed shell, which seals around the tube sheets to form the two domains for the hot and cold fluids. One fluid circulates through the tubes—entering and leaving via the water boxes formed between the tube sheets and shell heads. The other fluid flows around the outside of the tubes—in the space between the tube sheets and enclosed by the outer shell. Baffle plates are located normal to the tube bundle to ensure thorough distribution of the shell side fluid.

Figure 12.4(a) is to be interpreted only as a simplified illustration of the principal components of a typical shell-and-tube exchanger. There are many refinements in the design and construction of a heat exchanger which are not shown. Such factors as thermal expansion stresses, tube fouling due to contaminated fluids, ease of assembly and disassembly, size, weight, etc., must be provided for in a detailed design of a heat exchanger.

The heat exchanger shown in Fig. 12.4(a) is designated as having one shell pass and one tube pass since both the shell side fluid and the tube side fluid make a single traverse through the heat exchanger. Schematically, a one-shell-pass, one-tube-pass heat exchanger will be indicated as shown in Figs. 12.4(b) and 12.4(c). Figure 12.4(b) indicates the flow arrangement for a parallel-flow, one-shell-pass, one-tube-pass heat exchanger. As in the concentric pipe exchanger, the fluids enter and leave at the same ends of the exchanger.

Figure 12.4(c) indicates, schematically, a counterflow, one-shell-pass, one-tube-pass heat exchanger in which the two fluid streams flow in opposite directions through the exchanger.

Although they are formally quite similar, the parallel-flow and counterflow heat exchangers differ greatly in the manner in which the fluid temperatures vary as the fluids pass through. This difference is illustrated in the diagrams in Fig. 12.5, which plot the temperature of the fluid streams against the distance along the heat exchanger and apply whether the exchangers are of the concentric pipe or shell and tube configuration.

As indicated in Fig. 12.5, the parallel-flow exchanger brings the hottest portion of the hot fluid into communication with the coldest portion of the cold fluid. This provides a large thermal *potential difference* initially, but as the fluids move through the exchanger and transfer heat, one to the other, the temperature difference drops rapidly. The temperatures of the two streams approach one another asymptotically, and it would theoretically require an infinitely long heat exchanger to raise the cold fluid to the temperature of the hot fluid.

On the other hand, the counterflow exchanger places the hottest portion of the hot fluid in communication with the hottest portion of the cold fluid—and the coldest portion of the hot fluid in communication with the coldest portion of the cold fluid. This provides a more nearly constant temperature difference between the two fluids all along the heat exchanger and allows the possibility of heating the cold fluid to a temperature greater than the outlet temperature of the hot fluid. In general, the counterflow arrangement results in a greater rate of heat exchange than in the corresponding parallel-flow case.

Figure 12.5 also illustrates the importance of another parameter—the product of the mass flow rate and the specific heat of each of the fluids. If $\dot{m}$ is the mass flow rate and c_p is the specific heat of the fluids, the product, $\dot{m}c_p$, is called the *heat capacity rate*.

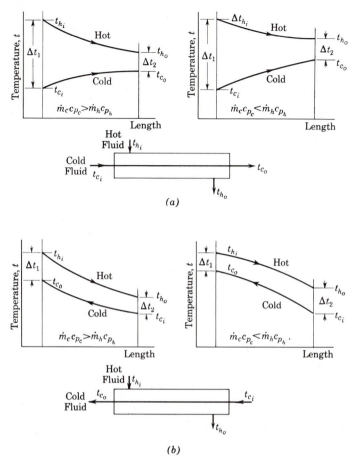

FIGURE 12.5. Temperature variations in (a) parallel-flow; and (b) counterflow heat exchangers.

The subscript c will be used to denote the cold fluid, and the subscript h will be used to denote the hot fluid. The subscript i will be used to indicate the inlet fluid conditions and o will be used to indicate the outlet fluid conditions. If one uses these subscripts and the definition of the heat capacity rate, an overall energy balance of the heat exchanger gives the total heat transferred between the fluids, q, expressed in two ways:

$$q = \dot{m}_c c_{p_c}(t_{c_o} - t_{c_i}) = \dot{m}_h c_{p_h}(t_{h_i} - t_{h_o}),$$

$$\frac{\dot{m}_c c_{p_c}}{\dot{m}_h c_{p_h}} = \frac{t_{h_i} - t_{h_o}}{t_{c_o} - t_{c_i}}.$$

The sketches of Fig. 12.5 show how the relative variation of the two fluid temperatures through the heat exchanger is influenced by whether $\dot{m}_c c_{p_c}$ is greater or less than $\dot{m}_h c_{p_h}$.

An important aspect of the particular flow arrangment chosen is the maximum possible heat that could be transferred, expressed in terms of the input parameters: $\dot{m}_c c_{p_c}$, $\dot{m}_h c_{p_h}$, t_{c_i}, t_{h_i}. This maximum possible heat transfer is that which would result if the exchanger is infinitely long, or has an infinite surface area for heat exchange. In particular, for counterflow, examination of the sketches in Fig. 12.5 shows that this limiting condition is determined by whether $\dot{m}_c c_{p_c}$ is greater or less than $\dot{m}_h c_{p_h}$. When $\dot{m}_c c_{p_c} > \dot{m}_h c_{p_h}$, the maximum possible heat transfer is determined by the fact that the hot fluid can be cooled to the temperature of the cold fluid inlet. That is,

For $\dot{m}_c c_{p_c} > \dot{m}_h c_{p_h}$:

$$t_{h_o} \rightarrow t_{c_i},$$

$$q_{max} = \dot{m}_h c_{p_h}(t_{h_i} - t_{h_o}),$$

$$= \dot{m}_h c_{p_h}(t_{h_i} - t_{c_i}).$$

For the reverse case, however, the limit is determined as the cold fluid is heated to the inlet temperature of the hot fluid:

For $\dot{m}_c c_{p_c} < \dot{m}_h c_{p_h}$:

$$t_{c_o} \rightarrow t_{h_i}$$

$$q_{max} = \dot{m}_c c_{p_c}(t_{c_o} - t_{c_i})$$

$$= \dot{m}_c c_{p_c}(t_{h_i} - t_{c_i}).$$

Thus, for the counterflow exchanger, the above two sets of equations show that in terms of the inlet parameters, the maximum possible heat exchange is determoned by the stream with the smaller heat capacity rate:

$$q_{max,ctr} = (\dot{m}c_p)_{min}(t_{h_i} - t_{c_i}). \tag{12.2}$$

The limiting case is, however, quite different for the parallel-flow arrangement. Regardless of the relative sizes of the two streams, the limiting heat transfer is determined by the fact that the two fluid streams approach the same outlet temperature, the weighted average of the inlet streams:

$$t_{h_o} \rightarrow t_{c_o} \rightarrow \frac{\dot{m}_c c_{p_c} t_{c_i} + \dot{m}_h c_{p_h} t_{h_i}}{\dot{m}_c c_{p_c} + \dot{m}_h c_{p_h}}. \tag{12.3}$$

Thus, the maximum possible heat transfer is, after some algebra:

$$q_{max} = \dot{m}_c c_{p_c}(t_{c_o} - t_{c_i}),$$

$$= \frac{1}{\dfrac{1}{\dot{m}_c c_{p_c}} + \dfrac{1}{\dot{m}_h c_{p_h}}} (t_{h_i} - t_{c_i}). \tag{12.4}$$

So, for parallel flow, the above may be written in the following way regardless of which stream has the larger heat capacity rate:

$$q_{max,par} = \frac{(\dot{m}c_p)_{min} (t_{h_i} - t_{c_i})}{1 + [(\dot{m}c_p)_{min}/(\dot{m}c_p)_{max}]}. \tag{12.5}$$

Comparison of Eqs. (12.2) and (12.5) shows that for given inlet conditions the counterflow arrangement always has a greater potential for heat transfer than does the parallel-flow arrangement. For this reason the parallel-flow configuration is rarely used in practice unless special conditions demand it or unless multiple passes are used, as described later. Also, since counterflow provides the best potential for heat transfer, its limiting case, as expressed in Eq. (12.2), is used for the basis of comparison (or nondimensionalization) of heat exchanger performance as described later.

It is not necessary for all shell-and-tube heat exchangers to be of the one-shell-pass, one-tube-pass form just discussed. In fact, space limitations frequently demand other arrangements, such as shown in Fig. 12.6. In Fig. 12.6(a) a simplified drawing of another heat exchanger is shown. This is the same in all respects as the exchanger depicted in Fig. 12.4(a) except that a pass partition has been added in the water box at one end and the tube fluid outlet has been relocated. This arrangement causes the tube fluid to make one pass through half of the tubes, reverse its direction of flow, and make a second pass through the remaining half of the tubes. This type of exchanger is referred to as a *one-shell-pass, two-tube-pass* heat exchanger and is typical of many steam condensers. This arrangement is shown schematically in Fig. 12.6(b) and it is apparent that the exchanger is neither a parallel-flow nor counterflow heat exchanger—it is often called a parallel-counterflow heat exchanger.

Many other possible flow arrangements exist and are used. By including longitudinal baffles within the heat exchanger shell, it is possible to cause the shell side fluid to make two or more passes through the shell. Proper positions

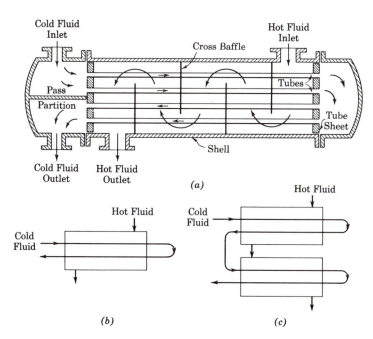

FIGURE 12.6. Multiple pass heat exchangers: (a), (b) one shell pass and two tube passes; (c) two shell passes, four tube passes.

of pass partitions in the tube fluid circuit will provide as many tube passes per shell pass as may be desired. A two-shell-pass, four-tube-pass heat exchanger is depicted schematically in Fig. 12.6(c).

In any of the multiple pass arrangements, as in Fig. 12.6, some portion of the exchanger is in parallel flow. Thus, such exchangers have a heat transfer potential less than that of pure counterflow—leaving the counterflow exchanger as the ideal maximum.

Cross-Flow Heat Exchangers

As defined earlier, a cross-flow heat exchanger is one in which the two fluid streams are oriented at an angle to one another. For purposes of discussion, Fig. 12.7 depicts a simplified plate type of cross-flow heat exchanger. As is often the case in the cross-flow heat exchangers used in aircraft or spacecraft applications, Fig. 12.7 shows the separating plate, or wall, to be provided with fins. Although simple in principle, the analysis of this type of heat exchanger is extremely complex, since the fluid temperatures may vary both in the direction of flow and normal to the direction of flow if no fluid mixing is allowed in the normal direction. This produces inlet and outlet fluid temperature distributions as indicated in Fig. 12.7—resulting in the fact that the fluid temperature difference varies with the two coordinate fluid stream directions. The designations "mixed" and "unmixed" are used to denote whether or not mixing of either fluid is allowed in the direction transverse to the direction of the flow stream. The exchanger illustrated in Fig. 12.7 is, as noted, one in which there is no mixing in either fluid stream. Other cross-flow heat exchangers may allow for mixing in one or both of the fluid streams.

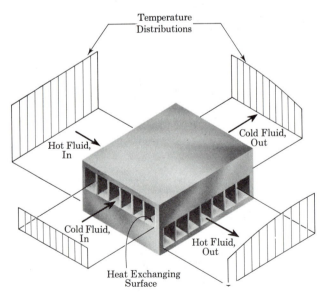

FIGURE 12.7. Temperature variation of unmixed fluids in a cross-flow heat exchanger.

THE OVERALL HEAT TRANSFER COEFFICIENT
OF A HEAT EXCHANGER

One of the basic parameters used to rate a heat exchanger and used to calculate the performance of a given heat exchanger is the overall heat transfer coefficient. This coefficient depends on the geometric configuration of the wall separating the two fluids and the convective heat transfer coefficients on each side. The separating surface may, or may not, be equipped with a fin array to enhance the exchange of heat. Figure 12.8 depicts a separating wall in which t_c and t_h denote the local cold and hot fluid temperatures, h_c and h_h the heat transfer coefficients on each side, A_c and A_h the total exposed surface area of each side, and $\mathscr{R}_w$ the total resistance of the wall exclusive of the fins. The overall coefficient for this geometry, U, is then defined as that quantity which yields the total rate of local heat transfer between the cold and hot fluids according to

$$q = UA(t_h - t_c).$$

Clearly, the value of U depends on which of the surface areas, A_c or A_h, is used. According to the discussion of Secs. 3.10 and 3.18, the UA product in the above equation is

$$\frac{1}{UA} = \frac{1}{A_c\eta_c h_c} + \mathscr{R}_w + \frac{1}{A_h\eta_h h_h}. \tag{12.6}$$

In Eq. (12.6) η_c and η_h represent the total surface effectiveness given in Eq. (3.64):

$$\eta_c = 1 - \frac{A_{f,c}}{A_c}(1 - \kappa_c),$$

$$\eta_h = 1 - \frac{A_{f,h}}{A_h}(1 - \kappa_h),$$

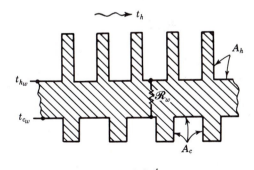

FIGURE 12.8

where $A_{f,c}$ and $A_{f,h}$ are the areas of *fins only* on the cold and hot sides, respectively. In the above expressions the κ's represent the fin effectiveness which can be found from the fin geometry (i.e., rectangular, triangular, etc.) according to the methods of Sec. 3.17. Thus, given the fin shapes and the surface geometries, the η's in Eq. (12.6) can be found, and are presumed known in what follows.

For shell-and-tube heat exchangers, the separating wall has a cylindrical geometry (outside radius r_o, inside radius r_i) and the wall resistance is known to be $\mathcal{R}_w = \ln(r_o/r_i)/2\pi Lk$, with L being the tube length and k its thermal conductivity. In shell and tube exchangers, the overall U is invariably based on the outside surface area, which is also almost always the hot surface, A_h. Thus, for a shell-and-tube exchanger, the overall U is found from

$$\frac{1}{U} = \frac{A/A_c}{\eta_c h_c} + \frac{A/2\pi L}{k} \ln\left(\frac{r_o}{r_i}\right) + \frac{1}{\eta_h h_h}. \tag{12.7}$$

The A without a subscript is the *total* exposed surface area on the hot, outer surface, including fins, if any. This total exposed surface is another important performance parameter for a heat exchanger, as will be seen subsequently. Most shell-and-tube exchangers do not have fins at the inner tube surface, so that $A_c = 2\pi L r_i$ and $\eta_c = 1$:

$$\frac{1}{U} = \frac{A/2\pi L r_i}{h_c} + \frac{A/2\pi L}{k} \ln\left(\frac{r_o}{r_i}\right) + \frac{1}{\eta_h h_h}. \tag{12.8}$$

In many instances the tubes are bare on the outside as well, so that the familiar relation results:

$$\frac{1}{U} = \frac{r_o/r_i}{h_c} + \frac{r_o}{k} \ln\left(\frac{r_o}{r_i}\right) + \frac{1}{h_h}. \tag{12.9}$$

For cross-flow heat exchangers, the situation is somewhat different. The wall resistance is $\mathcal{R}_w = \Delta x/A_w k$, in which Δx and A_w represent the thickness and the bare, unfinned area of the plane separating wall. Since there is no obvious preferred surface, such as the outside surface of a shell and tube exchanger, one may base the overall coefficient on the total exposed surface at the cold side *or* the hot side:

$$\frac{1}{U_c} = \frac{1}{\eta_c h_c} + \left(\frac{A_c}{A_w}\right)\frac{\Delta x}{k} + \frac{A_c/A_h}{\eta_h h_h},$$

$$\frac{1}{U_h} = \frac{A_h/A_c}{\eta_c h_c} + \left(\frac{A_h}{A_w}\right)\frac{\Delta x}{k} + \frac{1}{\eta_h h_h}. \tag{12.10}$$

Depending on which U is used, the exchanger size is specified by the corresponding total area, A_c or A_h. Most cross-flow exchangers will have fins on both the hot and cold sides.

The preceding discussion presumed the heat transfer surfaces involved to be clean. In practice, however, it is not uncommon for the surface to become contaminated with deposits due to impurities in the fluids, chemical reaction,

TABLE 12.1 Selected Values of Heat Exchanger Fouling Factors*

Fluid	R_f m²-°C/W	R_f h-ft²-°F/Btu
Seawater, below 50°C (125°F)	0.00009	0.0005
Seawater, above 50°C (125F)	0.00018	0.001
Treated boiler feedwater	0.00018	0.001
Steam, non-oil bearing	0.00009	0.0005
Refrigerant vapor, oil bearing	0.00035	0.002
Refrigerant liquid	0.00018	0.001
Natural gas	0.00018	0.001

*From Ref. 2.

or the buildup of scale by biological action. Such surface contamination can significantly influence the overall heat transfer coefficient by the introduction of additional thermal resistances. The magnitude of the additional resistances due to such fouling is, of course, dependent on a great number of factors such as the nature of the fluid, the length of service, etc. For general purpose use, experimental values of the "fouling factor" have been determined for a number of situations and some are quoted in Table 12.1. Additional data are available in Ref. 2. These fouling factors, R_f, are *unit* thermal resistances and thus their application must account for the size of the surface area to which they apply. Since either the hot or cold surface may become fouled, Eq. (12.6) must be modified in the following way:

$$\frac{1}{(UA)_{\text{dirty}}} = \frac{1}{(UA)_{\text{clean}}} + \frac{1}{A_c(1/R_{fc})} + \frac{1}{A_h(1/R_{fh})},$$

in which R_{fc} and R_{fh} are the fouling factors for the cold and hot surfaces, respectively. Thus, the U_c or U_h expressions in the foregoing should be modified according to

$$\frac{1}{(U_h)_{\text{dirty}}} = \frac{1}{(U_h)_{\text{clean}}} + \frac{A_h}{A_c} R_{fc} + R_{fh},$$

$$\frac{1}{(U_c)_{\text{dirty}}} = \frac{1}{(U_c)_{\text{clean}}} + R_{fc} + \frac{A_c}{A_h} R_{fh}.$$

(12.11)

The above additions may be made to any of Eqs. (12.7) through (12.10); however, the corresponding special forms are not written here for the sake of brevity.

In what follows, it is presumed that the expressions of this section have been applied to evaluate the overall U for a given heat exchanger—based on whichever total surface area has been taken to characterize the size of the exchanger. As a guide to the reader, Table 12.2 shows typical values encountered in practice for the overall heat transfer coefficient. A wide range of values is apparent, depending on the particular application.

TABLE 12.2 Representative Values of Heat Exchanger Overall Heat Transfer Coefficients

Situation	U W/m²-°C	U Btu/h-ft²-°F
Air to air, cross-flow	2–10	10–50
Air to water, cross-flow	5–15	25–80
Water to oil, shell and tube	20–60	110–350
Water to water, shell and tube	150–300	850–1700
Steam condenser, shell and tube	200–1000	1100–5600

12.5

A SIMPLE HEAT EXCHANGER— FLOW PAST AN ISOTHERMAL SURFACE

The analysis of the performance of a heat exchanger is customarily done in one of two formulations—the "mean-temperature-difference" formulation and the "effectiveness" formulation. These two approaches are treated in some detail in the following sections; however, it is instructive to see the motivation behind each approach and their equivalence by first examining a rather simple special case.

Consider, then, the flow of a fluid past an isothermal surface of known temperature and surface area. Figure 12.9 depicts such a case in which a fluid of known inlet temperature, t_{f_i}, flows at a known mass flow rate, $\dot{m}$, in a tube of known circumference, C, and length, L. The specific heat of the fluid, c_p, is known as is the temperature of the tube surface, t_s, and the heat transfer coefficient at that surface, h. Let it be desired to find the outlet temperature of the fluid, t_{f_o}, and the total heat transferred to it, q. For convenience of discussion the surface temperature is taken as greater than that of the fluid. The fluid temperature increases with distance along the tube, x, and at $x = L$, the exit fluid temperature is t_{f_o}. In any incremental portion of the tube the heat transfer to the fluid is expressed in two ways, in terms of a heat balance on the fluid and in terms of the exchange between the fluid and the surface:

$$dq = \dot{m}c_p dt_f,$$

$$dq = hC\, dx(t_s - t_f).$$

For the entire tube length where $A = LC$ is the total surface area, integrated forms of the above equations are

$$q = \dot{m}c_p(t_{f_o} - t_{f_i}), \qquad (12.12)$$

$$q = hLC(t_s - t_f)_m = hA(t_s - t_f)_m. \qquad (12.13)$$

The term $(t_s - t_f)_m$ represents some suitable mean value of the surface-to-fluid temperature difference which varies from $t_s - t_{f_i}$ at the tube entrance to

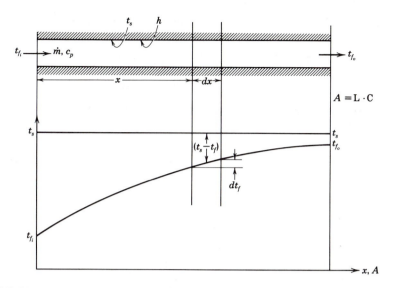

FIGURE 12.9. Temperature variation in an isothermal passage.

$t_s - t_{f_o}$ at the exit. It is t_{f_o} and q which are sought, given t_{f_i}, $\dot{m}$, c_p, h, A, and t_s.

The situation just described and depicted in Fig. 12.9 is identical to that discussed for flow in a pipe at the end of Sec. 8.4 and depicted in Fig. 8.2. Hence, those results will simply be quoted here, with appropriate notational changes. In Sec. 8.4 it was shown that the fluid temperature increases with x according to Eq. (8.37) and the mean value of the temperature difference is the logarithmic mean given in Eq. (8.39):

$$\frac{t_f - t_s}{t_{f_i} - t_s} = \exp\left(-\frac{hCx}{\dot{m}c_p}\right), \tag{12.14}$$

$$\frac{t_{f_o} - t_s}{t_{f_i} - t_s} = \exp\left(-\frac{hA}{\dot{m}c_p}\right), \tag{12.15}$$

$$(t_s - t_f)_m = \frac{(t_s - t_{f_i}) - (t_s - t_{f_o})}{\ln\left(\dfrac{t_s - t_{f_i}}{t_s - t_{f_o}}\right)}. \tag{12.16}$$

Given t_{f_i}, t_s, $\dot{m}$, c_p, h, and A (i.e., given the geometry of the "exchanger," the surface temperature, and the fluid inlet conditions) the *mean-temperature-difference* method determines the performance parameters as follows:

1. Find t_{f_o} from Eq. (12.15).
2. Find the mean temperature difference from Eq. (12.16).
3. Find q from Eq. (12.12).

An exactly equivalent method can be formulated by considering the maximum possible heat that could be transferred to the fluid if the tube (i.e., "exchanger") were infinitely long. In such an instance it is obvious that the maximum heat transfer possible occurs when the fluid is heated to the surface temperature, that is, $t_{f_o} \rightarrow t_s$, so that Eq. (12.12) gives

$$q_{max} = \dot{m}c_p(t_s - t_{f_i}), \qquad (12.17)$$

a quantity calculable from the given data. Then, let the "effectiveness" of a finite tube, ϵ, be defined

$$\epsilon = \frac{q}{q_{max}}$$

$$= \frac{\dot{m}c_p(t_{f_o} - t_{f_i})}{\dot{m}c_p(t_s - t_{f_i})},$$

$$= 1 - \frac{t_s - t_{f_o}}{t_s - t_{f_i}}.$$

From Eq. (12.15), the effectiveness is then

$$\epsilon = 1 - \exp\left(-\frac{hA}{\dot{m}c_p}\right).$$

The quantity $hA/\dot{m}c_p$ is a special case of what is termed "number of transfer units," or NTU:

$$\text{NTU} = \frac{hA}{\dot{m}c_p},$$

$$\epsilon = 1 - \exp(-\text{NTU}). \qquad (12.18)$$

The *effectiveness* method, then, determines q and t_{f_o} from the same input data as follows:

1. Find NTU $= hA/\dot{m}c_p$.
2. Find the effectiveness, ϵ, from Eq. (12.18).
3. Find q_{max} from Eq. (12.17).
4. Find $q = \epsilon q_{max}$ and t_{f_o} from Eq. (12.12).

Clearly the two approaches, the mean-temperature-difference approach and the effectiveness approach, are exactly equivalent. The same equations are used, but cast in different forms. The first method concentrates on expressing performance in terms of the mean difference between the fluid and the surface as the fluid proceeds through the "exchanger." The second method expresses performance in terms of the maximum heat that *could* be transferred under the stated conditions. The equivalence of the two formulations is easy to see because of the simple case chosen—that of a single fluid and a surface of uniform, known, temperature. In the general case of the heat exchangers described in Sec. 12.4, there are *two* fluids, both of varying temperature, and the separating surface varies in temperature along the flow path. Thus, the problem of describing the heat exchanger performance will become more complicated; however two basic representations are possible, in terms of a mean temperature difference or in terms of an effectiveness based on maximum possible performance. These two representations are described in detail for the more complex exchangers in the next two sections.

HEAT EXCHANGER MEAN TEMPERATURE DIFFERENCES

For the case of a heat exchanger involving two fluids the concept of the mean temperature difference introduced in the preceding section must be broadened. Let it be assumed that the geometry of a heat exchanger is completely known. That is, if it is a shell and tube exchanger the dimensions of tubes, the number of tubes, the arrangement of the tubes and the shell passes, and the configuration of any fins are all known. If it is a cross-flow exchanger, comparable data are known. Then, in any incremental portion of the exchanger in which an increment of transfer surface area, dA, is exposed between the two fluids, the incremental heat transferred is

$$dq = U \, dA \, \Delta t. \tag{12.19}$$

In Eq. (12.19) Δt is the temperature difference between the two fluids in the increment dA, and U is the overall heat transfer coefficient (based on the area being used) between the two fluids. One seeks to integrate Eq. (12.19) to give the total heat transferred in the entire heat exchanger in the form

$$q = (U \, \Delta t)_m A,$$

in which A is the heat exchanger total surface area and $(U \, \Delta t)_m$ is a suitable mean value of the $U \, \Delta t$ product. For most cases to be treated here, let it be assumed that an appropriate average value for U can be found that applies throughout the exchanger. Then one seeks a mean temperature difference, Δt_m, so that

$$q = UA \, \Delta t_m. \tag{12.20}$$

Equations (12.19) and (12.20) imply the existence of a mean temperature difference defined as

$$\Delta t_m = \frac{1}{A} \int_0^A (\Delta t) \, dA. \tag{12.21}$$

To determine Δt_m, then, one needs to determine how the temperature of each fluid varies as it flows through the exchanger, so that the difference, Δt, may be expressed as a function of position (i.e., as a function of surface area) and Eq. (12.21) evaluated.

In the case of the elementary exchanger considered in the preceding section, this temperature difference was easily determined. The determination of the more complicated tube-pass and shell-pass configurations will be described in the next few sections.

Parallel-Flow and Counterflow Heat Exchangers

First consider the one-tube-pass, one-shell-pass heat exchangers described in Sec. 12.3. The heat exchanger may be of either the parallel-flow or counterflow arrangement. The discussion which follows considers the parallel flow case, but the discussion should make it apparent that the results apply equally well to the counterflow case. The following assumptions are made:

1. The average overall U is constant throughout the exchanger.
2. If either fluid undergoes a phase change (e.g., as in a condenser), then this phase change occurs throughout the heat exchanger—not just part of it—resulting in a constant fluid temperature.
3. The specific heat and mass flow rate, and hence the heat capacity rate, of each fluid are constant.
4. No heat is lost to the surroundings.
5. There is no conduction in the direction of flow in either the fluids or the tube walls.
6. At any cross section in the heat exchanger, each of the fluids may be characterized by a single temperature—i.e., perfect transverse mixing in each fluid is presumed.

The notation used in Sec. 12.3 will be used here. That is, the subscripts h and c refer to the hot and cold fluids, respectively, whereas the subscripts i and o represent inlet and outlet conditions. The subscripts 1 and 2 denote the left and right ends of the exchanger. Thus, for the parallel flow device shown in Fig. 12.5(a) the temperature differences at the two ends are

$$\Delta t_1 = t_{h_i} - t_{c_i},$$

$$\Delta t_2 = t_{h_o} - t_{c_o};$$

whereas for a counterflow exchanger, as in Fig. 12.5(b),

$$\Delta t_1 = t_{h_i} - t_{c_o},$$

$$\Delta t_2 = t_{h_o} - t_{c_i}.$$

The four fluid terminal temperatures t_{c_i}, t_{c_o}, t_{h_i}, t_{h_o} are presumed known.

Now, for the parallel-flow heat exchanger depicted in Fig. 12.10, the heat exchanged in an incremental length of the exchanger (of surface area dA) may be expressed in the following three ways:

$$dq = U \, dA \, \Delta t, \tag{12.22}$$

$$dq = \dot{m}_c c_{p_c} \, dt_c, \tag{12.23}$$

$$dq = -\dot{m}_h c_{p_h} \, dt_h. \tag{12.24}$$

The difference between the two fluid temperatures in the increment dA is denoted by $\Delta t = t_h - t_c$, and dt_c and dt_h represent the changes in the cold and hot streams as they pass through the increment.

Equations (12.23) and (12.24) show that if one were to plot the fluid temperatures against the amount of heat transferred up to the section under consideration, the resulting curve would be a straight line as illustrated in Fig. 12.10.

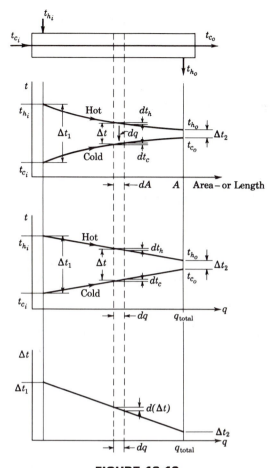

FIGURE 12.10

Thus, the temperature difference, Δt, is also a straight line when plotted against q—also shown in Fig. 12.10. This fact is true whether or not the exchanger is a parallel-flow one or a counterflow one. Thus, Δt is a linear function of q—varying from Δt_1 where $q = 0$ at the left end to Δt_2 where $q = q_{\text{total}}$ at the right end. Thus,

$$\frac{d(\Delta t)}{dq} = \frac{\Delta t_2 - \Delta t_1}{q_{\text{total}}}.$$

Since Eq. (12.22) gives $dq = U\, dA\, \Delta t$,

$$\frac{d(\Delta t)}{\Delta t} = \frac{U(\Delta t_2 - \Delta t_1)}{q_{\text{total}}}\, dA. \tag{12.25}$$

Integration of the above between $A = 0$, $\Delta t = \Delta t_1$ and $A = A$, $\Delta t = \Delta t_2$ gives

$$\ln \Delta t_2 - \ln \Delta t_1 = U \frac{\Delta t_2 - \Delta t_1}{q_{\text{total}}} A,$$

or

$$q_{total} = UA \frac{\Delta t_1 - \Delta t_2}{\ln (\Delta t_1/\Delta t_2)},$$

with A representing the total surface area. This latter expression, when compared with Eq. (12.20), yields the desired mean temperature difference:

$$\Delta t_m = \Delta t_{lm} = \frac{\Delta t_1 - \Delta t_2}{\ln (\Delta t_1/\Delta t_2)}, \qquad (12.26)$$

$$q = UA \, \Delta t_{lm}. \qquad (12.27)$$

Again, the logarithmic mean temperature difference is seen to apply, hence the symbol Δt_{lm}. The logarithmic mean is the mean of the temperature differences at the two ends of the heat exchanger, and while the same relation appears to apply to both the parallel-flow and counterflow cases, it is important to note that Δt_1 and Δt_2 are different in the two instances. In terms of the four fluid terminal temperatures, the log-mean temperature differences for the parallel-flow and counterflow cases are

$$\Delta t_{lm,par} = \frac{(t_{h_i} - t_{c_i}) - (t_{h_o} - t_{c_o})}{\ln \left(\dfrac{t_{h_i} - t_{c_i}}{t_{h_o} - t_{c_o}} \right)}, \qquad (12.28)$$

$$\Delta t_{lm,ctr} = \frac{(t_{h_i} - t_{c_o}) - (t_{h_o} - t_{c_i})}{\ln \left(\dfrac{t_{h_i} - t_{c_o}}{t_{h_o} - t_{c_i}} \right)}. \qquad (12.29)$$

For a condenser, or any other heat exchanger in which the hot fluid maintains a constant temperature $t_h = t_{h_o} = t_{h_i}$, the above two mean temperature differences become the same, as one would expect, and become equal to the mean temperature difference deduced for the case of an isothermal surface in Sec. 12.5.

EXAMPLE 12.3

In a one-tube-pass, one-shell-pass exchanger, the hot fluid enters at $t_{h_i} = 425°C$ and leaves $t_{h_o} = 260°C$, while the cold fluid enters at $t_{c_i} = 40°C$ and leaves at $t_{c_o} = 150°C$. Find the log-mean temperature difference if the heat exchanger is arranged for (a) parallel flow; (b) counterflow.

Solution

(a) For the parallel-flow case the temperature differences at each end are

$$\Delta t_1 = 425 - 40 = 385°C,$$

$$\Delta t_2 = 260 - 250 = 110°C,$$

and the mean is

$$\Delta t_{lm} = \frac{385 - 110}{\ln\left(\dfrac{385}{110}\right)} = 219.5°C \ (395.1°F).$$

(b) For counterflow,

$$\Delta t_1 = 425 - 150 = 275°C,$$

$$\Delta t_2 = 260 - 40 = 220°C,$$

$$\Delta t_{lm} = \frac{275 - 220}{\ln\left(\dfrac{275}{220}\right)} = 246.5°C \ (443.7°F).$$

Note that for equal values of U and A, the counterflow exchanger would transfer 12% more heat than the parallel-flow one. ∎

In the event that the overall heat transfer coefficient, U, varies greatly from one end of a heat exchanger to the other, it may not be possible to represent U by an average, constant value. If one presumes that the variation of U may be represented as a linear function of the temperature difference, then

$$U = a + b \, \Delta t.$$

The symbols a and b denote constants. Equation (12.25) then gives

$$\frac{d(\Delta t)}{\Delta t(a + b \, \Delta t)} = \frac{dA(\Delta t_2 - \Delta t_1)}{q_{total}}.$$

Upon integration and evaluation at the limits, this equation gives

$$(\Delta t_2 - \Delta t_1)\frac{A}{q_{total}} = \frac{1}{a} \ln \left(\frac{\Delta t}{a + b \, \Delta t}\right)^{\Delta t_2}_{\Delta t_1}$$

$$= \frac{1}{a} \ln \left[\frac{\Delta t_2(a + b \, \Delta t_1)}{\Delta t_1(a + b \, \Delta t_2)}\right].$$

Since $U_1 = a + b \, \Delta t_1$ and $U_2 = a + b \, \Delta t_2$, the constant, a may be found to be

$$a = \frac{U_1 \, \Delta t_2 - U_2 \, \Delta t_1}{\Delta t_2 - \Delta t_1}, \tag{12.30}$$

$$q = A\left[\frac{U_2 \, \Delta t_1 - U_1 \, \Delta t_2}{\ln\left(\dfrac{U_2 \, \Delta t_1}{U_1 \, \Delta t_2}\right)}\right].$$

This equation would then replace Eqs. (12.26) and (12.27).

Multiple Pass Shell-and-Tube Heat Exchangers. Cross-Flow Heat Exchangers

The above analysis and the derived log-mean temperature difference applies only to heat exchangers having just one shell pass and one tube pass. In the event that a multiple pass heat exchanger is under consideration, it is necessary to obtain a different expression for the mean temperature difference, dependent on the arrangement of the shell and tube passes.

Consider, for instance, the one-shell-pass, two-tube-pass heat exchanger discussed earlier and shown again in Fig. 12.11. The subscripts h and c are used as in the above example, and again the subscripts i and o will denote inlet and outlet fluid conditions. The subscripts a and b will be used to denote the first and second tube passes, respectively.

For an incremental length of the heat exchanger—involving a total tube surface area of dA—the following expression may be written for the heat transferred:

$$dq = \dot{m}_h c_{p_h}\, dt_h,$$

$$dq = \dot{m}_c c_{p_c}(dt_{c_a} - dt_{c_b}),$$

$$dq = U\, dA[(t_h - t_{c_a}) + (t_h - t_{c_b})].$$

These three equations may be combined to eliminate any two of the three unknown temperatures—say t_{c_a} and t_{c_b}—resulting in a differential equation in t_h. This differential equation may be solved along with the following fact, derived from an overall heat balance of the exchanger:

$$\dot{m}_h c_{p_h}(t_{h_i} - t_{h_o}) = \dot{m}_c c_{p_c}(t_{c_o} - t_{c_i}).$$

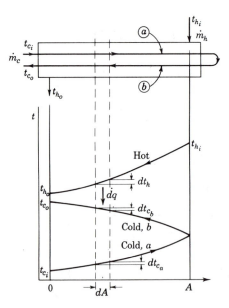

FIGURE 12.11. Temperature variation in a one-shell-pass, two-tube-pass heat exchanger.

The following expression for the total heat transfer then results:

$$q = UA\, \Delta t_m,$$

in which the mean temperature difference, Δt_m, is

$$\Delta t_m = \cfrac{\sqrt{(t_{h_i} - t_{h_o})^2 + (t_{c_o} - t_{c_i})^2}}{\ln\left[\cfrac{(t_{h_i} + t_{h_o}) - (t_{c_i} + t_{c_o}) + \sqrt{(t_{h_i} - t_{h_o})^2 + (t_{c_o} - t_{c_i})^2}}{(t_{h_i} + t_{h_o}) - (t_{c_i} + t_{c_o}) - \sqrt{(t_{h_i} - t_{h_o})^2 + (t_{c_o} - t_{c_i})^2}}\right]}.$$

$$(12.31)$$

The actual algebra leading to this expression has been omitted due to its length and complexity.

Similar expressions may be developed for each individual shell and tube pass arrangement—or, in a much more complex analysis, for cross-flow heat exchangers. This is done in some detail in Ref. 1. In each case the mean temperature difference is expressed in terms of the four terminal fluid temperatures: t_{h_i}, t_{h_o}, t_{c_i}, and t_{c_o}. The expressions for the mean temperature difference may be simplifed (see Ref. 2) by introducing the following dimensionless ratios:

$$\text{Capacity ratio:} \quad R = \frac{\dot{m}_c c_{pc}}{\dot{m}_h c_{ph}} = \frac{t_{h_i} - t_{h_o}}{t_{c_o} - t_{c_i}}. \qquad (12.32)$$

$$\text{Effectiveness:} \quad P = \frac{\dot{m}_c c_{pc}(t_{c_o} - t_{c_i})}{\dot{m}_c c_{pc}(t_{h_i} - t_{c_i})} = \frac{t_{c_o} - t_{c_i}}{t_{h_i} - t_{c_i}}. \qquad (12.33)$$

The capacity ratio, R, is so named since it is the ratio of $\dot{m}_c c_{pc}$ to $\dot{m}_h c_{ph}$, above. The effectiveness is the ratio of the total heat transferred to the cold fluid to the heat that would be transferred to the cold fluid if it were raised to the inlet temperature of the hot fluid. With these definitions it is possible to relate the mean temperature difference to P, R, and the log-mean temperature difference calculated for *counterflow*. That is, one may define a function of P and R, call it $F(P, R)$, so that

$$\Delta t_m = F(P, R)\, \Delta t_{\text{lm,ctr}}. \qquad (12.34)$$

Here $\Delta t_{\text{lm,ctr}}$ is given by eq. (12.29), and the function $F(R, P)$ is then determined by the shell and tube pass arrangement of the particular heat exchanger.

For instance, in the case just discussed of a one-shell-pass, two-tube-pass heat exchanger, Eqs. (12.31) through (12.34) may be combined with Eq. (12.29) to obtain

$$F(P, R) = \frac{\sqrt{R^2 + 1}}{R - 1} \cdot \cfrac{\ln\left(\cfrac{1 - P}{1 - PR}\right)}{\ln\left[\cfrac{2 - P(R + 1 - \sqrt{R^2 + 1})}{2 - P(R + 1 + \sqrt{R^2 + 1})}\right]}. \qquad (12.35)$$

With F thus defined, it may be considered to be a *correction factor* to be applied to the usual, and readily calculated, log-mean temperature difference for counterflow according to Eq. (12.29). It can be noted in Eq. (12.35) that a solution for F does not exist for all possible combinations of R and P. The reason that there are certain areas of impossible solutions is due to the definition of the effectiveness P. The effectiveness is defined as the ratio of $\dot{m}_c c_{p_c}(t_{c_o} - t_{c_i})$ to $\dot{m}_c c_{p_c}(t_{h_i} - t_{c_i})$. In Sec. 12.3 it was noted that the *maximum* possible heat that could be transferred in a *counterflow* heat exchanger with *infinite* surface area was equal to $\dot{m}_c c_{p_c}(t_{h_i} - t_{c_i})$ for $R < 1$. This quantity is the denominator in the definition of the effectiveness [see Eq. (12.33)]. The two-tube-pass heat exchanger described by Eq. (12.35) exchanges some heat in parallel flow as well as counterflow. For this reason there are certain values of the effectiveness which cannot be achieved—even for an exchanger of infinite area. The maximum effectiveness would, of course, be dependent on the capacity ratio R.

In addition to the limitations just described, it is economically poor practice to employ a heat exchanger for conditions under which $F < 0.75$. This criterion further restricts the maximum effectiveness of a heat exchanger. For applications in which the specified values of the terminal temperatures are such that the associated values of P and R give no solution for F, or if F is less than 0.75, some other shell and tube arrangement must be employed. Some other arrangements will be discussed below.

Values of F for a one-shell-pass, two-tube-pass heat exchanger have been computed from Eq. (12.35) by Bowman et al. (Ref. 3). The results are shown in Fig. 12.12. Similar results have been obtained by similar analyses for a number of shell and tube arrangements and for various cross-flow heat exchangers. These results may be found in Ref. 2, and some of them are presented

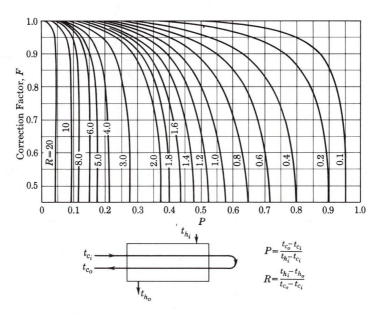

FIGURE 12.12. Correction factor F for a one-shell-pass, two-tube-pass heat exchanger. (From *Standards of Tubular Exchanger Manufacturers Association*, 3rd ed., New York, Tubular Exchanger Manufacturers Association, Inc., 1952. Used by permission.)

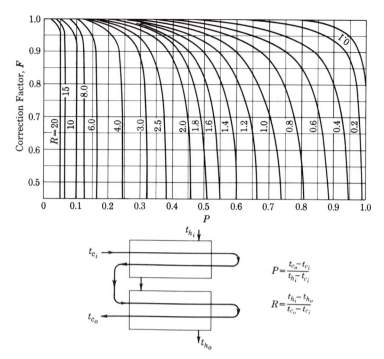

$$P = \frac{t_{c_o} - t_{c_i}}{t_{h_i} - t_{c_i}}$$

$$R = \frac{t_{h_i} - t_{h_o}}{t_{c_o} - t_{c_i}}$$

FIGURE 12.13. Correction factor *F* for a two-shell-pass, four-tube pass heat exchanger. (From *Standards of Tubular Exchanger Manufacturers Association*, 3rd ed., New York, Tubular Exchanger Manufacturers Association, Inc., 1952. Used by permission.)

here. Figure 12.13 gives *F* for a heat exchanger with two shell passes and two tube passes per shell pass. Figure 12.14 presents the results of Bowman et al. (Ref. 3) for a cross-flow heat exchanger in which there is no mixing of either fluid stream in a direction transverse to their direction of flow. The outlet fluid temperature in this case is interpreted as the "mixed mean" outlet temperature of the fluids—i.e., transverse mixing being allowed after exit from the heat exchanger.

Another important item to be noted about this manner of presentation of the mean temperature difference of a multiple pass heat exchanger is the unfortunate shape of the curves shown in Figs. 12.12 through 12.14. As is readily apparent, the curves are difficult to read with accuracy as *P* increases (depending on *R*). This is particularly true for $R > 1$ as $P \to 1/R$, since *F* becomes undefined. However, as has been recently pointed out by Shamsundar (Ref. 4), half the *F* versus *P* curves are redundant since it may be shown from the definitions of *P* and *R* that the following reciprocity relation holds

$$F(P, R) = F(PR, 1/R). \tag{12.36}$$

Thus, the curves for $R > 1$ are uniquely related to those for $R < 1$ and only one set is actually needed. Consequently, in the regions of the charts of Figs. 12.12 through 12.14 for $R > 1$ where good accuracy is difficult to obtain, the above reciprocity relation may be applied and the curves for $R < 1$ used instead.

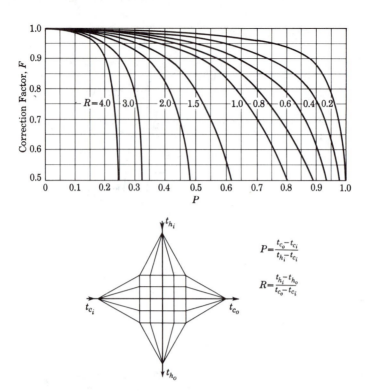

$$P = \frac{t_{c_o} - t_{c_i}}{t_{h_i} - t_{c_i}}$$

$$R = \frac{t_{h_i} - t_{h_o}}{t_{c_o} - t_{c_i}}$$

FIGURE 12.14. Correction factor F for a cross-flow heat exchanger with both fluids unmixed. (From R. A. Bowman, A. C. Mueller, and W. M. Nagle, "Mean Temperature Difference in Design," *Trans. ASME,* Vol. 62, 1940, p. 288. Used by permission.)

One final note concerning the F factor might be made. The above presentation concentrated on heat exchangers with multiple tube and shell passes. It could be applied as well to the one-shell-pass, one-tube-pass exchangers discussed earlier. Clearly, from the definition of F, it is equal to 1 for the simple counterflow exchanger; however, this is not so for the parallel-flow case. One could derive the $F(P, R)$ relation for a simple parallel-flow exchanger (this is left as an exercise in one of the problems at the end of the chapter), but this is usually not done since Δt_m for this case is so easily calculated from Eq. (12.26) or (12.28).

EXAMPLE 12.4

A one-shell-pass, two-tube-pass exchanger is used to heat 20,000 lb_m/h of water from 100°F to 250°F by the use of 10,000 lb_m/h of water entering the exchanger at 600°F. The total heat exchange surface area is 50 ft^2. Find the average over-all heat transfer coefficient of the heat exchanger.

Solution. The given data include $t_{c_i} = 100°F$, $t_{c_o} = 250°F$, $t_{h_i} = 600°F$. Thus, an overall heat balance gives $(600 - t_{h_o}) = (250 - 100)(20,000/10,000)$, so the hot water leaves the exchanger at $t_{h_o} = 300°F$. The log-mean temperature difference for counterflow is, by Eq. (12.29),

$$\Delta t_{\text{lm,ctr}} = \frac{(600 - 250) - (300 - 100)}{\ln\left(\dfrac{600 - 250}{300 - 100}\right)} = 268°F.$$

The capacity ratio and effectiveness defined in Eqs. (12.32) and (12.33) are

$$R = \frac{600 - 300}{250 - 100} = 2,$$

$$P = \frac{250 - 100}{600 - 100} = 0.30.$$

Figure 12.12 or Eq. (12.35) gives $F = 0.883$, so the mean temperature difference is given by

$$\Delta t_m = 268 \times 0.883 = 237°F.$$

Since

$$q = UA \, \Delta t_m = \dot{m}_h c_{ph}(t_{h_i} - t_{h_o}),$$

$$U = \frac{10,000 \times 1 \times (600 - 300)}{237 \times 50} = 253 \text{ Btu/h-ft}^2\text{-}°F \ (1440 \text{ W/m}^2\text{-}°C). \blacksquare$$

12.7

HEAT EXCHANGER EFFECTIVENESS AND THE NUMBER OF TRANSFER UNITS

One difficulty associated with the mean-temperature-difference formulation for heat exchanger performance is that Δt_m cannot be found unless all four fluid terminal temperatures (t_{c_i}, t_{h_i}, t_{c_o}, t_{h_o}) are known. In certain design problems these temperatures are known and one may proceed as in Sec. 12.6. However, if one wishes to estimate the performance (i.e., q) of a given exchanger subject to given *inlet* conditions, then the outlet temperatures are not known until q is determined. Thus, an iterative calculation would appear to be necessary.

If the performance equations are recast in the terms of an effectiveness parameter as suggested in Sec. 12.5, this difficulty can be circumvented. While this is not the only reason for use of the effectiveness formulation, it is an important one. The concept of heat exchange "effectiveness" and the associated "number of transfer units" was first suggested by Nusselt and has been developed extensively by Kays and London (Ref. 5).

The effectiveness of a heat exchanger, consistent with the concept introduced in Sec. 12.5, is defined as the actual heat transferred by the exchanger divided by the maximum possible heat that could be transferred, i.e.,

$$\epsilon = \frac{q}{q_{max}}.$$

The discussion of Sec. 12.3 showed that for an infinite transfer area the most heat would be transferred in counterflow and, further, the maximum q was limited by the fluid stream with the least heat capacity rate. That is, according to Eq. (12.2),

$$q_{max} = \dot{m}_c c_{p_c}(t_{h_i} - t_{c_i}), \quad \dot{m}_c c_{p_c} < \dot{m}_h c_{p_h},$$

$$q_{max} = \dot{m}_h c_{p_h}(t_{h_i} - t_{c_i}), \quad \dot{m}_c c_{p_c} > \dot{m}_h c_{p_h}.$$

The actual heat transfer is found from a heat balance on either the cold or the hot stream: $q = \dot{m}_c c_{p_c}(t_{c_o} - t_{c_i}) = \dot{m}_h c_{p_h}(t_{h_i} - t_{h_o})$. In order to make the effectiveness dependent on only the terminal temperatures, the first form for q is used when $\dot{m}_c c_{p_c} < \dot{m}_h c_{p_h}$ and the second when the reverse is true. Thus, the following definition is used for heat exchanger effectiveness:

$$\epsilon = \begin{cases} \dfrac{t_{c_o} - t_{c_i}}{t_{h_i} - t_{c_i}}, & \dot{m}_c c_{p_c} < \dot{m}_h c_{p_h}, \\[3mm] \dfrac{t_{h_i} - t_{h_o}}{t_{h_i} - t_{c_i}}, & \dot{m}_c c_{p_c} > \dot{m}_h c_{p_h}. \end{cases} \tag{12.37}$$

The relative thermal "size" of the two fluid streams is obviously an important parameter, and in order to make it a quantity that is always less than unity, the *capacity ratio* is defined as

$$C_R = \frac{(\dot{m}c_p)_{min}}{(\dot{m}c_p)_{max}}. \tag{12.38}$$

The product UA represents the heat exchanging capacity of an exchanger, per degree of temperature difference. This thermal "size" of an exchanger can be nondimensionalized by referring it to the storage capacity of one of the fluid streams. Since the stream with the smaller heat capacity rate is the one which limits the maximum heat transfer, it is used in the nondimensionalization. Thus, similar to the definition made for the elementary case in Sec. 12.5, the *number of transfer units* of a heat exchanger, NTU, is defined:

$$\text{NTU} = \frac{UA}{(\dot{m}c_p)_{min}}. \tag{12.39}$$

The three parameters just defined (ϵ, C_R, and NTU) do not involve any new quantities not used in the mean-temperature-difference formulation but simply constitute a different combination of the same ones. The utility of the ϵ–C_R–NTU formulation lies in the fact that the effectiveness, ϵ, can be written as an explicit function of C_R and NTU for a given heat exchanger. Both C_R and NTU involve only the fluid flow rates, heat capacities, the heat exchanger size, and the overall coefficient. Thus, the effectiveness, ϵ, can be found without knowing any temperatures, and since ϵ involves only *one* of the outlet temperatures, that temperature may be found explicitly when the inlet temperatures are given.

The following sections give the $\epsilon = \epsilon(C_R, \text{NTU})$ functions for several heat exchanger configurations. The actual determination of this function is discussed for a parallel-flow heat exchanger.

The Parallel-Flow Exchanger

For any heat exchanger one may write the heat transferred in the following three ways:

$$q = UA \, \Delta t_m,$$

$$q = \dot{m}_c c_{p_c}(t_{c_o} - t_{c_i}),$$

$$q = \dot{m}_h c_{p_h}(t_{h_i} - t_{h_o}),$$

or, equivalently,

$$\dot{m}_c c_{p_c}(t_{c_o} - t_{c_i}) = \dot{m}_h c_{p_h}(t_{h_i} - t_{h_o}), \tag{12.40}$$

$$UA \, \Delta t_m = \dot{m}_h c_{p_h}(t_{h_i} - t_{h_o}). \tag{12.41}$$

For the parallel-flow heat exchanger, Eq. (12.28) for Δt_m may be introduced into Eq. (12.41) to yield

$$\ln\left(\frac{t_{h_i} - t_{c_i}}{t_{h_o} - t_{c_o}}\right) = \left[\frac{(t_{h_i} - t_{c_i}) - (t_{h_o} - t_{c_o})}{t_{h_i} - t_{h_o}}\right]\frac{UA}{\dot{m}_h c_{p_h}}$$

$$= \left(1 + \frac{t_{c_o} - t_{c_i}}{t_{h_i} - t_{h_o}}\right)\frac{UA}{\dot{m}_h c_{p_h}}.$$

Equation (12.40) may be introduced to give

$$\ln\left(\frac{t_{h_i} - t_{c_i}}{t_{h_o} - t_{c_o}}\right) = \left(1 + \frac{\dot{m}_h c_{p_h}}{\dot{m}_c c_{p_c}}\right)\frac{UA}{\dot{m}_h c_{p_h}},$$

or $\hfill (12.42)$

$$\frac{t_{h_o} - t_{c_o}}{t_{h_i} - t_{c_i}} = \exp\left[-\left(\frac{1}{\dot{m}_c c_{p_c}} + \frac{1}{\dot{m}_h c_{p_h}}\right)UA\right].$$

This latter equation may be solved with Eq. (12.40) for either of the fluid outlet temperatures, t_{c_o} or t_{h_o}.

Solving first for t_{c_o}, Eqs. (12.40) and (12.42) yield

$$1 - \left(1 + \frac{\dot{m}_c c_{p_c}}{\dot{m}_h c_{p_h}}\right)\frac{t_{c_o} - t_{c_i}}{t_{h_i} - t_{c_i}} = \exp\left[-\left(1 + \frac{\dot{m}_c c_{p_c}}{\dot{m}_h c_{p_h}}\right)\frac{UA}{\dot{m}_c c_{p_c}}\right].$$

In this form, one sees that if $\dot{m}_c c_{p_c} < \dot{m}_h c_{p_h}$, then the temperature difference ratio on the left side of the equation is the effectiveness defined in Eq. (12.37), $\dot{m}_c c_{p_c}/\dot{m}_h c_{p_h}$ is C_R in Eq. (12.38), and $UA/\dot{m}_c c_{p_c}$ is the NTU of Eq. (12.37). Thus,

$$1 - (C_R + 1)\,\epsilon = \exp\left[-(C_R + 1)\text{NTU}\right]. \tag{12.43}$$

If, on the other hand, Eqs. (12.40) and (12.42) are solved for t_{h_o}, the hot fluid outlet, one has

$$1 - \left(1 + \frac{\dot{m}_h c_{p_h}}{\dot{m}_c c_{p_c}}\right) \frac{t_{h_i} - t_{h_o}}{t_{h_i} - t_{c_i}} = \exp\left[-\left(1 + \frac{\dot{m}_h c_{p_h}}{\dot{m}_c c_{p_c}}\right) \frac{UA}{\dot{m}_h c_{p_h}}\right],$$

so that if $\dot{m}_h c_{p_h} < \dot{m}_c c_{p_c}$ there results

$$1 - (C_R + 1)\epsilon = \exp\left[-(C_R + 1)\text{NTU}\right].$$

This latter expression is identical with Eq. (12.43), which is precisely why the parameters ϵ, C_R, and NTU were defined in the way they were. For either relative size of the hot and cold streams, $\dot{m}_c c_{p_c} < \dot{m}_h c_{p_h}$ or $\dot{m}_c c_{p_c} > \dot{m}_h c_{p_h}$, then

$$\epsilon = \frac{1 - \exp\left[-(C_R + 1)\text{NTU}\right]}{C_R + 1},$$

$$\text{NTU} = \frac{-\ln\left[1 - (C_R + 1)\epsilon\right]}{C_R + 1}. \tag{12.44}$$

Equation (12.44) is the sought-for relation giving ϵ as a function of C_R and NTU. Given the heat exchanger size and overall U, the fluid flow rate and heat capacities, and the fluid inlet temperatures, Eq. (12.44) yields whichever fluid outlet temperature is involved in ϵ—depending on which form of Eq. (12.37) applies. The remaining fluid outlet temperature is then simply obtained from the heat balance of Eq. (12.40). Also quoted in Eq. (12.44) is its inversion, expressing NTU as a function of C_R and ϵ. This form is sometimes useful for design calculations.

Equation (12.44) is easy to evaluate; however, it is also presented graphically in Fig. 12.15. It is useful to note that for the case in which one fluid changes

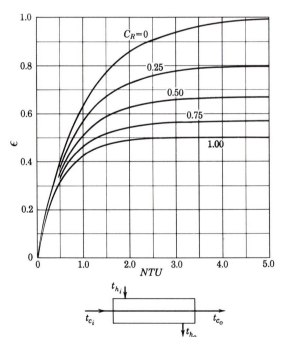

FIGURE 12.15. Effectiveness—NTU relationship for a parallel-flow heat exchanger. (Adapted from W. M. Kays and A. L. London, *Compact Heat Exchangers*, New York, McGraw-Hill, 1958. Used by permission.)

COMBINED CONDUCTION AND CONVECTION—
HEAT EXCHANGERS

phase, as in a condenser, then $t_{h_o} = t_{h_i} = t_h$ is constant. Thus, $\dot{m}_h c_{p_h} \to \infty$ and $C_R \to 0$. In this case the first form for ϵ in Eq. (12.37) applies and

$$\epsilon = \frac{t_{c_o} - t_{c_i}}{t_{h_i} - t_{c_i}} = 1 - \exp(-\text{NTU}),$$

$$\text{NTU} = -\ln(1 - \epsilon). \tag{12.45}$$

The Counterflow Exchanger

An analogous derivation may be carried out for the counterflow heat exchanger to show that

$$\epsilon = \frac{1 - \exp[-(1 - C_R)\text{NTU}]}{1 - C_R \exp[-(1 - C_R)\text{NTU}]}, \tag{12.46}$$

$$\text{NTU} = \frac{\ln[(1 - \epsilon)/(1 - \epsilon C_R)]}{C_R - 1}.$$

Equation (12.46) is shown graphically in Fig. 12.16. In the case of a condenser with $C_R \to 0$, Eq. (12.45) is again obtained. This must be so since the distinction between a parallel-flow and a counterflow heat exchanger does not apply when one fluid remains at a fixed temperature. The form of Eq. (12.45) is, for the same reason, identical to that found in Sec. 12.5, Eq. (12.18), for the elementary isothermal wall case.

FIGURE 12.16. Effectiveness–NTU relationship for a counterflow heat exchanger. (Adapted from W. M. Kays and A. L. London, *Compact Heat Exchangers*, New York, McGraw-Hill, 1958. Used by permission.)

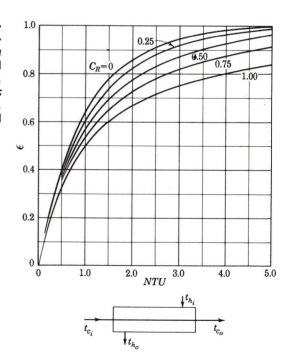

Equation (12.46) for the counterflow exchanger becomes indeterminate when the two fluid streams are the same "size" thermally, or as $C_R \to 1$. One may readily show that in this case Eq. (12.46) becomes

$$\epsilon = \frac{\text{NTU}}{1 + \text{NTU}},$$

(12.47)

$$\text{NTU} = \frac{\epsilon}{1 - \epsilon}.$$

The One-Shell-Pass, Two-Tube-Pass Exchanger

For the one-shell-pass, two-tube-pass heat exchanger depicted in Fig. 12.6, Ref. 3 shows the effectiveness to be

$$\epsilon = 2 \left[(1 + C_R) + (1 + C_R^2)^{1/2} \frac{1 + \exp\left[-(1 + C_R^2)^{1/2}\text{NTU}\right]}{1 - \exp\left[-(1 + C_R^2)^{1/2}\text{NTU}\right]} \right]^{-1},$$

(12.48)

$$\text{NTU} = -(1 + C_R^2)^{-1/2} \ln \left[\frac{2/\epsilon - 1 - C_R - (1 + C_R^2)^{1/2}}{2/\epsilon - 1 - C_R + (1 + C_R^2)^{1/2}} \right].$$

Equation (12.48) is awkward to use, so the graphical display in Fig. 12.17 may be used.

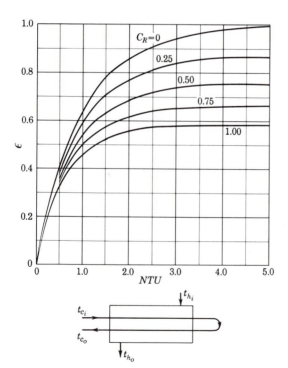

FIGURE 12.17. Effectiveness–NTU relationship for a one-shell-pass, two-tube-pass heat exchanger. (Adapted from W. M. Kays and A. L. London, *Compact Heat Exchangers*, New York, McGraw-Hill, 1958. Used by permission.)

The Cross-Flow Exchanger

Determination of the ϵ–C_R–NTU relation in cross flow is complex since the fluid temperatures are dependent on two spatial variables. An analytic result for the case in which both fluids are unmixed may be found in Ref. 6; however this result is a complicated double infinite series. The results of this analysis are displayed in Fig. 12.18. An approximation that may be used when C_R is close to unity is:

$$\epsilon \cong 1 - \exp\{C_R \text{NTU}^{0.22}[\exp(-C_R \text{NTU}^{0.78}) - 1]\}.$$

Other Heat Exchanger Configurations

Relations and graphical plots of the $\epsilon = \epsilon(C_R, \text{NTU})$ function for a number of other heat exchanger configurations, particularly multiple pass cases for both shell-and-tube and cross-flow exchangers have been determined and reported in the literature. Reference 5 should be consulted for these results.

FIGURE 12.18. Effectiveness–NTU relationship for a cross-flow heat exchanger with fluids unmixed. (Adapted from W. M. Kays and A. L. London, *Compact Heat Exchangers*, New York, McGraw-Hill, 1958. Used by permission.)

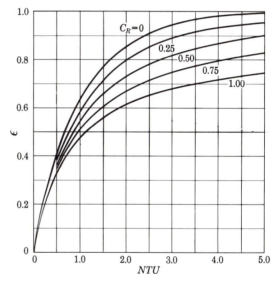

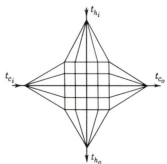

CALCULATION OF THE PERFORMANCE
OF A GIVEN HEAT EXCHANGER

Two general types of problems are normally encountered in dealing with heat exchanger calculations. The first consists of the determination of the *performance* of a given heat exchanger. That is, one wishes to find the heat exchanged between two fluid streams (and their outlet temperatures) when given the complete physical description of the exchanger and the inlet conditions of the streams. The *design* problem consists of determining the physical description of a heat exchanger that will accomplish a desired amount of heat transfer between the two given streams. Clearly, the design problem is more complex than the performance problem and has no unique answer. Most design procedures consist of making a great number of performance calculations; hence the performance problem will be discussed first.

The performance problem, then, presumes that the complete physical description of the heat exchanger is known. That is, one knows the flow arrangement (i.e., whether it is a cross-flow exchanger, a shell-and-tube exchanger, etc.; how many passes of each fluid there are; whether it is parallel-flow, counterflow, etc.) and one also knows the physical dimensions involved (i.e., tube size and length, the size of the separating wall and any fins if cross-flow, etc.). The question then is: Given the above and given the inlet fluid conditions (i.e., $\dot{m}_c$, c_{p_c}, t_{c_i}, $\dot{m}_h$, c_{p_h}, t_{h_i}), what are the rate of heat transfer and the fluid outlet temperatures (i.e., q, t_{c_o}, t_{h_o})? Three relations are needed to find these three performance quantities. Two relations come immediately from the heat exchanger heat balance:

$$q = \dot{m}_c c_{p_c}(t_{c_o} - t_{c_i}),$$ (12.49)

$$q = \dot{m}_h c_{p_h}(t_{h_i} - t_{h_o}).$$

The third relation to be used to complete the problem depends on whether the mean temperature difference or the effectiveness formulation is to be used. In terms of the mean temperature difference,

$$q = UA \, \Delta t_m,$$ (12.50)

in which Δt_m depends on the four terminal fluid temperatures and the fluid flow arrangement. In terms of the effectiveness one has

$$\epsilon = \epsilon(C_R, \text{NTU}),$$ (12.51)

in which the form of the effectiveness function depends on the flow arrangement and the definitions in Eqs. (12.37) through (12.39).

In either formulation one needs the UA product for the exchanger. The surface area is known from the given geometry. However, in general the overall U depends on the fluid temperatures and the hydrodynamic conditions on each side of the exchanger. Thus, the determination of U has to be done, iteratively perhaps, simultaneous with the determination of the fluid outlet temperatures.

In order to separate the question of the determination of t_{co} and t_{ho} and the determination of U, a special subcase of the performance problem will be discussed first in which it is presumed U is known. Such knowledge might be the result of extensive experience or it might be the result of some previous calculation in a longer iterative scheme to be described later.

Performance Calculation—Known U

If one presumes U to be known, then the only unknowns of the performance problem are q, t_{co}, and t_{ho}. Once any one of these three are known, Eqs. (12.49) readily yield the other two. In the use of the mean-temperature-difference formulation, the Δt_m is a function (often complex) of the *known* t_{ci}, t_{hi}, and the unknown t_{co}, t_{ho}. Thus, an iterative calculation would be necessary. However, the definition of the effectiveness in Eq. (12.37) involves only one of the unknown outlets and since ϵ can be expressed as a function of C_R and NTU, one may solve directly for t_{co} or t_{ho} without iteration. Both approaches are illustrated in the following example for the sake of completeness. Clearly, the effectiveness method would be preferred; however the advantage of this method vanishes in the general problem when U is not known.

EXAMPLE 12.5

A one-shell-pass, two-tube-pass heat exchanger has a total surface area of 5 m^2, and its overall heat transfer coefficient based on that area is known to be 1400 W/m^2-°C. If 4500 kg/h of water enters the shell side at 315°C while 9000 kg/h of water enters the tube side at 40°C, find the outlet temperatures using (a) the mean temperature difference; (b) the effectiveness. Take the specific heat of both fluid streams to be 4.187 kJ/kg-°C.

Solution

(a) The capacity ratio R, by Eq. (12.32) is

$$R = \frac{\dot{m}_c c_{pc}}{\dot{m}_h c_{ph}} = \frac{9000}{4500} = 2.0.$$

As a first assumption, take $t_{ho} = 175$°C. Then Eq. (12.49) gives the outlet cold fluid temperature to be

$$9000(t_{co} - 40) = 4500(315 - 175)$$

$$t_{co} = 110\text{°C}.$$

Thus, the parameter P in Eq. (12.33) is

$$P = \frac{t_{co} - t_{ci}}{t_{hi} - t_{ci}} = \frac{110 - 40}{315 - 40} = 0.255,$$

and the log mean Δt for counterflow is, by Eq. (12.29),

$$\Delta t_{\text{lm,ctr}} = \frac{(175 - 40) - (315 - 110)}{\ln\left(\dfrac{175 - 40}{315 - 110}\right)} = 167.6°\text{C}.$$

Figure 12.13 or Eq. (12.35) gives $F = 0.938$ for $R = 2.0$, $P = 0.255$. Thus, the exchanger Δt_m is

$$\Delta t_m = 167.6 \times 0.938 = 157.2°\text{C}.$$

Equating Eqs. (12.49) and (12.50), one may calculate t_{h_o}.

$$q = UA\,\Delta t_m = \dot{m}_h c_{p_h}(t_{h_i} - t_{h_o})$$

$$1400 \times 5 \times 157.2 = \frac{4500}{3600} \times 4.187 \times 1000 \times (315 - t_{h_o})$$

$$t_{h_o} = 104.7°\text{C}.$$

This calculated value is considerably lower than the assumed 175°C, so another calculation must be performed with a revised estimate of t_{h_o}. The final results, after repeating the above, are

$$t_{h_o} = 146.5°\text{C, assumed,}$$

$$t_{c_o} = 124.3°\text{C (255.7°F),}$$

$$R = 2.0, P = 0.306, F = 0.872,$$

$$\Delta t_{\text{lm,ctr}} = 144.5°\text{C,}$$

$$\Delta t_m = 126.0°\text{C,}$$

$$t_{h_o} = 146.5°\text{C (295.7°F), calculated.}$$

(b) The ϵ–NTU approach requires no iteration. From the given data:

$$\dot{m}_c c_{p_c} = 9000 \times 4.187 \times \frac{1000}{3600} = 10{,}468 \text{ W/°C} = (\dot{m}c_p)_{\text{max}},$$

$$\dot{m}_h c_{p_h} = 4500 \times 4.187 \times \frac{1000}{3600} = 5234 \text{ W/°C} = (\dot{m}c_p)_{\text{min}}.$$

Thus,

$$C_R = \frac{(\dot{m}c_p)_{\text{min}}}{(\dot{m}c_p)_{\text{max}}} = 0.5,$$

$$\text{NTU} = \frac{UA}{(\dot{m}c_p)_{\text{min}}} = \frac{1400 \times 5}{5234} = 1.337.$$

Figure 12.17 or Eq. (12.48) gives the effectiveness to be $\epsilon = 0.613$. Since $\dot{m}_c c_{p_c} > \dot{m}_h c_{p_h}$, the second definition of Eq. (12.37) applies, and the hot fluid outlet can be found:

$$\epsilon = \frac{t_{h_i} - t_{h_o}}{t_{h_i} - t_{c_i}},$$

$$0.613 = \frac{315 - t_{h_o}}{315 - 40},$$

$$t_{h_o} = 146.4°C \ (295.6°F).$$

Then the cold fluid outlet is

$$\dot{m}_c c_{p_c}(t_{c_o} - t_{c_i}) = \dot{m}_h c_{p_h}(t_{h_i} - t_{h_o})$$

$$10{,}468(t_{c_o} - 40) = 5234(315 - 146.4)$$

$$t_{c_o} = 124.2°C \ (255.5°F).$$

Parts (a) and (b) differ slightly due to round-off errors. ■

Performance Calculation—Unknown U

As noted earlier, the overall coefficient, U, is generally not known. In reality, U depends on the temperatures of the two fluids. As shown in Examples 12.1 and 12.2, an iterative calculation for U is often required even when the two fluid temperatures are known. In addition, in the case of a heat exchanger only the *inlet* temperatures of the fluids are known. For a parallel-flow heat exchanger one might estimate U at one end of the exchanger; however, if one is dealing with, say, a counterflow exchanger or if, as is more often the case, one wishes to find a representative average U based on the *mean* temperatures of the two fluids, one needs knowledge of the fluid *outlet* temperature before U can be found. Thus, a doubly iterative calculation must be made in which fluid outlet temperatures are assumed, an overall U calculated (iteratively, perhaps) based on these assumptions, the outlet temperatures calculated by either of the methods of Example 12.5, and then these results are compared with the initial assumptions and revised accordingly. Since assumptions must be made for t_{c_o} and t_{c_i} to find U, there is no particular advantage of the ϵ–NTU method over the Δt_m method.

Even with some knowledge of a representative mean temperature for each fluid stream, determination of the overall U involves finding the heat transfer coefficients between each stream and the separating wall. For shell-and-tube exchangers, the relations of Chapter 8 can be used to find the h to be applied at the inside tube surface. The method of estimating h on the outside tube surface depends on the mechanism taking place there (forced convection, condensing, etc.) and many of the relations of Chapters 8 or 10 may be applied. If the tubes are equipped with fins, or if one is dealing with a cross-flow heat exchanger which almost invariably involves a complicated finned surface, then some knowledge of the heat transfer correlation applicable for the particular geometry is necessary. In order to establish the method an example will first be shown in which the correlations of Chapters 8 and 10 are applied. Then, typical data for more complex surface geometries and an associated second example will be presented.

EXAMPLE 12.6

A steam condenser is made of $\frac{5}{8}$-in. brass ($k = 110$ W/m-°C) tubes, 18 gage. The exchanger has one shell pass and two tube passes with 129 tubes per pass. The length of the tubes is 1.85 m. Estimate the overall heat transfer coefficient and the mass rate of condensation of saturated steam at a pressure of 15 kN/m² (approximately 4.5 in. of mercury) that occurs when cooling water enters the tubes at 20°C and flows with an average velocity of 1.4 m/s.

Solution. The notation to be used in reference to the tube geometry is shown in Fig. 12.19. For a $\frac{5}{8}$-in. 18-gage tube, Table B.2 gives the following geometric data:

$$D_2 = 1.399 \text{ cm,}$$

$$D_3 = 1.588 \text{ cm,}$$

$$\text{surface area} = 0.04989 \text{ m}^2/\text{m,}$$

$$\text{cross-sectional area} = 0.0001408 \text{ m}^2.$$

Thus, the total exposed tube surface area is

$$A = 0.04989 \times 1.85 \times 2 \times 129 = 23.81 \text{ m}^2,$$

and the total cross-sectional area for cooling water flow, per tube pass, is

$$A_x = 0.0001408 \times 129 = 0.0182 \text{ m}^2.$$

The Steam Tables give the saturation temperature and latent heat of steam at 15 kN/m² to be: $t_{\text{sat}} = 53.97 \approx 54$°C, $h_{\text{fg}} = 2373.1$ kJ/kg.

The tube side fluid (i.e., cold) increases in temperature along its path. To estimate the inside tube heat transfer coefficient (and to find an average tube surface temperature for the determination of the outside heat transfer coefficient) one must assume an outlet cold fluid temperature. Likewise, a trial value of the outlet hot fluid temperature would generally also be needed, and could be found from the assumed cold fluid outlet and an overall heat balance on the exchanger. In the present special case of a condenser, however, it is known that the hot fluid is of constant temperature, so such a calculation need not be made and one takes $t_4 = t_{\text{sat}} = t_{h_i} = t_{h_o} = 54$°C. Based then on an assumed average cold

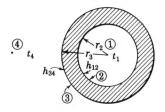

FIGURE 12.19

fluid temperature and the hot fluid temperature (known in ths case), one calculates the resulting h's, finds the overall U, and then verifies (or modifies) the assumed cold fluid outlet temperature. Thus, one proceeds as follows:

1. *Assume that $t_{c_o} = 35°C$. This assumption is based on the experience that cooling water temperature increases in steam condensers are typically of the order of 15 to 25°C.*

   (a) *Determine the inside tube coefficient h_{12}. This coefficient is based, in part, on the average cold fluid temperature:* $t_1 = (t_{c_o} + t_{c_i})/2 = (35 + 20)/2 = 27.5°C$. *Subsequent calculations reveal (as in Example 12.2) that the inside tube surface temperature is such that the conditions of Eq. (8.27) are not met and one must use Eq. (8.28) which involves knowledge of t_2. However, subsequent calculations also show that the temperature drop through the tube wall is negligibly small so that one may take $t_2 \approx t_3$. Thus, one need not perform iterative calculations for both t_2 and t_3 as was done in Example 12.2. So, anticipating the above results, and showing the* last *iteration only,* assume *that $t_2 = 46°C$. With $t_1 = 27.5°C$, as above, then Table A.3 gives $\mu = 0.8443 \times 10^{-3}$ kg/m-s, $\mu_s = 0.5864 \times 10^{-3}$ kg/m-s, $\nu = 0.8473 \times 10^{-6}$ m²/s, $k = 0.6113$ W/m-°C, Pr $= 5.78$. Thus, Eq. (8.28) gives*

   $$Re_D = \frac{1.4 \times 0.01399}{0.8473 \times 10^{-6}} = 2.312 \times 10^4,$$

   $$Nu_D = 0.027 Re_D^{0.8} Pr^{1/3} \left(\frac{\mu}{\mu_s}\right)^{0.14} = 158.0$$

   $$h_{12} = 158.0 \times \frac{0.6113}{0.01399} = 6903 \text{ W/m}^2\text{-°C}.$$

   (b) *Determine h_{34}. Consistent with the above assumption of $t_2 = 46°C$, assume that $t_3 = 46°C$. Thus, with $t_4 = 54°C$, $\Delta t = 54 - 46 = 8°C$, $t_m = (54 + 46)/2 = 50°C$, Table A.3 gives $c_{pl} = 4.182$ kJ/kg-°C, $\rho_l = 988.0$ kg/m³, $\mu_l = 0.5471 \times 10^{-3}$ kg/m-s, $k_l = 0.6405$ W/m-°C. Thus, Eq. (10.17) gives*

   $$Ja = \frac{c_{pl} \Delta t}{h_{fg}} = 0.0141,$$

   $$h'_{fg} = h_{fg}(1 + 0.68Ja) = 2396 \text{ kJ/kg},$$

   $$Nu_D = 0.728 \left[\frac{9.8 \times 1000 \times (988)^2 \times 2396 \times (0.01588)^3}{0.5471 \times 10^{-3}(0.6405 \times 8)}\right]^{1/4}$$
   $$= 309.7,$$

   $$h_{34} = 309.7 \times \frac{0.6405}{0.01588} = 12,490 \text{ W/m}^2\text{-°C}.$$

(c) *Determine U.* For the geometry in question, the overall U is

$$U = \left[\frac{r_3}{r_2 h_{12}} + \frac{r_3 \ln (r_3/r_2)}{k_{23}} + \frac{1}{h_{34}} \right]^{-1}$$

$$= \left[\frac{0.01588}{0.01399 \times 6903} + \frac{0.01588 \ln (0.01588/0.01399)}{2 \times 110} \right.$$

$$\left. + \frac{1}{12,490} \right]^{-1},$$

$$= 3942 \text{ W/m}^2\text{-°C}.$$

(d) *Verify t_2 and t_3.* The trial values of t_2 and t_3 may now be checked:

$$D_2 h_{12}(t_1 - t_2) = D_3 h_{34}(t_3 - t_4) = D_3 U(t_1 - t_4),$$

$$0.01399 \times 6903(27.5 - t_2) = 0.01588 \times 12,490(t_3 - 54)$$

$$= 0.01588 \times 3942(27.5 - 54),$$

$$t_2 = 44.7°C, \qquad t_3 = 45.6°C.$$

These values are sufficiently close to the assumed value of 46°C and verify the presumption that the tube wall temperature drop is negligible. Had these computed values differed significantly from the assumed, steps (a), (b), and (c) should be repeated. Only the last iteration is shown here.

2. *Verify the assumed t_{co}.* The verification of the assumed cold fluid outlet may be made using either the Δt_m or ϵ method. Both are shown for sake of completeness.

Based on the cold fluid average temperature, $t_1 = 27.5°C$, $\rho = 996.4$ kg/m³, $c_{p_c} = 4.180$ kJ/kg-°C. Thus,

$$\dot{m}_c = A_x \rho v = 0.0182 \times 996.4 \times 1.4 = 25.39 \text{ kg/s}$$

$$\dot{m}_c c_{p_c} = 25.39 \times 4.180 = 106.1 \text{ kW/°C}.$$

It is not necessary to find the R, P, and F factors since the hot fluid is of constant temperature, and either Eq. (12.28) or (12.29) gives, with $t_{ci} = 20°C$, $t_{co} = 35°C$, $t_{ho} = t_{hi} = 54°C$:

$$\Delta t_m = \frac{(54 - 20) - (54 - 35)}{\ln \left(\dfrac{54 - 20}{54 - 35} \right)} = 25.78°C.$$

Thus, $q = UA \Delta t_m = 3492 \times 23.81 \times 25.78 = 2143$ kW. Then t_{co} is found:

$$q = \dot{m}_c c_{p_c}(t_{co} - t_{ci}),$$

$$2143 = 106.1(t_{co} - 20),$$

$$t_{co} = 40.2°C.$$

Using the ϵ–NTU method, one has $\dot{m}_c c_{p_c} = 106.1 \text{ kW}/°\text{C}$, $\dot{m}_c c_{p_h} \to \infty$, so that

$$C_R = 0,$$

$$\text{NTU} = \frac{UA}{\dot{m}_c c_{p_c}} = \frac{3492 \times 23.81}{1000 \times 106.1} = 0.783.$$

Then Eq. (12.44), (12.45), or (12.46) gives

$$\epsilon = 1 - \exp(-\text{NTU}) = 0.543,$$

or either Fig. (12.15) or (12.16) could be used. From Eq. (12.37),

$$\epsilon = \frac{t_{c_o} - t_{c_i}}{t_{h_i} - t_{c_i}} = 0.543,$$

$$t_{c_o} = 38.5°\text{C}.$$

This result differs slightly from that given by the Δt_m method since Δt_m is based on an incorrect value of t_{c_o}. When a verified t_{c_o} is achieved, this difference disappears.

3. *Modify the assumed it$_{c_o}$, and repeat steps 1 and 2.* The calculated t_{c_o}, above, differs significantly from the assumed value of 35°C. Hence, revised calculations are in order. The process is not as extensive as one might fear since the values of $t_3 \approx t_2$ found in the first iteration usually work well in the second. If one assumes a new value of $t_{c_o} = 40°\text{C}$, a repetition of steps 1 and 2 yields the following results:

$$t_{c_o} = 40°\text{C, assumed,}$$

$$t_1 = 30°\text{C,}$$

$$h_{12} = 7053 \text{ W/m}^2\text{-}°\text{C,}$$

$$h_{34} = 12,490 \text{ W/m}^2\text{-}°\text{C,}$$

$$t_2 = t_3 = 46°\text{C,}$$

$$U = 3998 \text{ W/m}^2\text{-}°\text{C,}$$

$$\Delta t_m = 22.5°\text{C,}$$

$$\dot{m}_c c_{p_c} = 106.1 \text{ kW}/°\text{C,}$$

$$C_R = 0,$$

$$\text{NTU} = 0.897,$$

$$\epsilon = 0.592,$$

$$t_{c_o} = 40.1°\text{C, calculated.}$$

4. The above agreement between the assumed and calculated t_{c_o} is adequate, so the heat transfer is

$$q = \dot{m}_c c_{p_c}(t_{c_o} - t_{c_i}) = 2122 \text{ kW } (4.17 \times 10^6 \text{ Btu/h}),$$

and the condensate produced is

$$\dot{m} = \frac{q}{h_{fg}} = 0.894 \text{ kg/s} = 3219 \text{ kg/h} (7095 \text{ lb}_m/\text{h}).$$ ∎

The above example utilized heat transfer correlations developed in Chapters 8 and 10 for the calculation of the heat transfer coefficients at the two surfaces of the heat exchanger tubes. For other surface configurations or other heat transfer processes it is necessary to have appropriate correlations for the particular situations at hand. This is particularly true for finned surfaces for which analytical solutions are generally not available. Data of this sort are available in various sources in the heat transfer literature, and the reader is referred to sources such as Refs. 2, 5, and 7.

As *examples* of the kind of information available, Figs. 12.20 and 12.21,

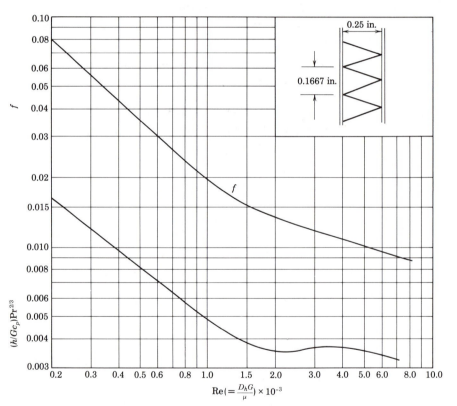

Fin pitch = 12.0 per in.; Fin thickness = 0.006 in.
Plate spacing = 0.250 in.; Flow hydraulic diamter = 0.000235 ft
Fin area/total surface area = 0.773
Total surface area/volume between plates = 392.7 ft²/ft³
Free flow area/normal area between plates = 0.9420

FIGURE 12.20. Heat transfer and friction data for a plate-fin heat exchanger surface. (From W. M. Kays and A. L. London, *Compact Heat Exchangers*, 2nd ed., New York, McGraw-Hill, 1964. Used by permission.)

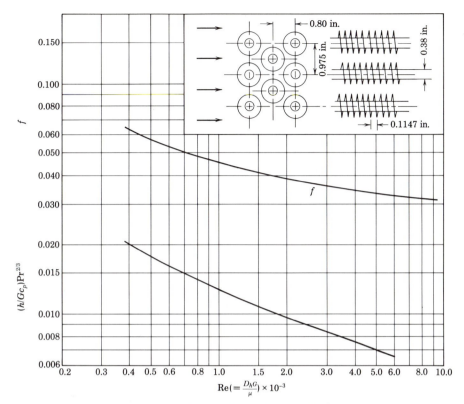

FIGURE 12.21. Heat transfer and friction data for a finned-tube heat exchanger core. (From W. M. Kays and A. L. London, *Compact Heat Exchangers*, 2nd ed., New York, McGraw-Hill, 1964. Used by permission.)

taken from Ref. 5, are shown. Figure 12.20 shows information for plates with straight fins (in a triangular pitch for ease of manufacture) and Fig. 12.21 shows data for a bank of finned circular tubes. Similar data are available for a vast variety of configurations and these are shown simply to indicate typical presentations. Shown are correlations for both the surface heat transfer coefficients and the friction factor for use in estimating pressure losses. Also, various geometric parameters (such as fin area, total area, etc.) are given. Because of the complexity of the geometry involved, the dimensionless parameters used are modified in form somewhat from those used in Chapters 7 through 10. The Reynolds number is based on the *hydraulic diameter* introduced in Eq. (8.40):

$$D_h = \frac{4 \times \text{flow area}}{\text{wetted perimeter}}$$

and the *mass* velocity, G, introduced in Eq. (10.21):

$$G = \frac{\text{mass flow rate}}{\text{flow area}}. \tag{12.52}$$

Thus, the Reynolds number is expressed

$$\mathrm{Re} = \frac{U\rho D_h}{\mu} \tag{12.53}$$

$$= D_h \frac{G}{\mu}.$$

The available flow area for the calculation of G is shown in the figures presented as is the hydraulic diameter.

Instead of expressing the heat transfer coefficient in terms of the Nusselt number, it is given in terms of

$$\left(\frac{\mathrm{Nu}}{\mathrm{Re}}\right)\mathrm{Pr}^{-1/3} = \left(\frac{hD_h}{k}\frac{\mu}{D_h G}\right)\mathrm{Pr}^{-1/3}, \tag{12.54}$$

$$= \left(\frac{h}{Gc_p}\right)\mathrm{Pr}^{2/3}.$$

Likewise, the friction factor is modified to the form

$$f = \frac{\tau_0}{\frac{1}{2}\rho U^2} = \frac{2\rho^2\tau_0}{G^2}. \tag{12.55}$$

Use of these parameters is no different from those used previously as shown in the next example. This example is that of a typical cross-flow heat exchanger using an array of plate-fin surfaces.

EXAMPLE 12.7

A compact, cross-flow, aircraft heat exchanger is depicted in Fig. 12.22. The overall dimensions are shown. The basic construction of the exchanger is of the plate-fin type. Plates 0.012 in. thick, spaced 0.25 in. apart, form the separating walls between the two fluids. The surface of each plate is covered with an array of longitudinal fins, 0.006 in. thick, placed in a triangular pitch with 12 fins per inch. The material of both the fins and the plates has a $k = 15$ Btu/h-ft-°F. By alternating the direction of the fins on successive plates, a multiple cross-flow pattern is created as shown. The heat exchanger is 6 in. wide normal to the cold fluid flow direction and 4 in. wide normal to the hot fluid flow direction. Appropriate manifolding (not shown) distributes the two fluid streams into the exchanger and collects the exiting streams. The fin array need not be the same in the two flow directions as assumed here, but heat transfer data for each type would be needed.

Let there be 12 flow passages in each flow direction; hence, the total height of the stack is 6.30 in. when the plate thickness is accounted for. (Actually, there probably would be one more passage in one direction than the other to provide symmetry at the ends of the stack; however, for simplicity of discussion imagine there to be an equal number in each direction, representing the last passage on each end as a half-passage.)

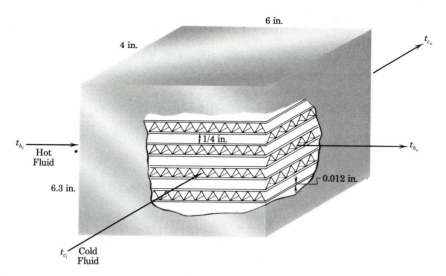

FIGURE 12.22. Compact cross-flow heat exchanger.

The cold fluid is air entering at $t_{c_i} = 0°F$ and flowing at the rate of $\dot{m}_c = 800\ lb_m/h$. The hot fluid consists of combustion gases (to be approximated with the transport properties of air) entering at $t_{h_i} = 800°F$ and flowing at the rate of $\dot{m}_h = 200\ lb_m/h$. Estimate the mixed outlet temperature of the two fluid streams and the rate of heat transfer between them.

Solution. The plate-fin configuration conforms to that in Fig. 12.20. Thus, the following geometric parameters apply:

$$D_h = 0.009412\ \text{ft},$$

$$\frac{\text{total area}}{\text{volume between plates}} = 392.7\ \text{ft}^2/\text{ft}^3,$$

$$\frac{\text{fin area}}{\text{total area}} = 0.773$$

$$\frac{\text{flow area}}{\text{normal area between plates}} = 0.9240.$$

For either fluid stream (with 12 passages) the volume between plates is $(12 \times 0.25 \times 4 \times 6)/1728 = 0.04167\ \text{ft}^3$. Thus, the total surface area on both the hot and cold sides is

$$A_c = A_h = 0.04167 \times 392.7 = 16.325\ \text{ft}^2.$$

The free-flow area for each of the streams is

$$A_{x_c} = \frac{12 \times (0.25 \times 6) \times 0.9240}{144} = 0.1155\ \text{ft}^2,$$

$$A_{x_h} = \frac{12 \times (0.25 \times 4) \times 0.9240}{144} = 0.0770\ \text{ft}^2.$$

Thus, the mass velocity in each stream is

$$G_c = \frac{\dot{m}_c}{A_{x_c}} = \frac{800}{0.1155} = 6926.4 \ \text{lb}_m/\text{h-ft}^2,$$

$$G_h = \frac{\dot{m}_h}{A_{x_h}} = \frac{200}{0.077} = 2597.4 \ \text{lb}_m/\text{h-ft}^2.$$

The other known data are

$$t_{c_i} = 0°\text{F}, \qquad \dot{m}_c = 800 \ \text{lb}_m/\text{h},$$

$$t_{h_i} = 800°\text{F}, \qquad \dot{m}_h = 200 \ \text{lb}_m/\text{h}.$$

The same basic approach is used as in the preceding example. Trial values of t_{c_o} and t_{h_o} are chosen. Based on average cold and hot fluid temperatures, values of the heat transfer coefficients on each fluid surface, h_c and h_h, are then found. Then an overall U and the outlet fluid temperatures can be calculated and compared with the trial values. Repetition of this process eventually yields acceptable values for t_{c_o} and t_{h_o}. Only the last, verified iteration is given here:

1. *Assume that* $t_{c_o} = 140°\text{F}$. Since $t_{c_i} = 0°\text{F}$, an average cold fluid temperature of $70°\text{F}$ results for which $c_{p_c} = 0.2413 \ \text{Btu/lb}_m\text{-}°\text{F}$. If, for the moment, one takes $c_{p_h} \approx c_{p_c}$, then a basic heat balance gives $800 (140 - 0) = 200(800 - t_{h_o})$, or $t_{h_o} = 240°\text{F}$. This leads to an average t_h of $520°\text{F}$ for which $c_{p_h} \cong 0.248 \ \text{Btu/lb}_m\text{-}°\text{F}$. Repeating this calculation, the following consistent assumptions of t_{c_o} and t_{h_o} are made:

$$t_{c_o} = 140°\text{F},$$

$$t_{h_o} = 256°\text{F}.$$

2. *Determine* h_c. Based on the assumed t_{c_o} the cold side heat transfer coefficient can be found using $t_c = (t_{c_i} + t_{c_o})/2 = 70°\text{F}$. Table A.6 gives $c_{p_c} = 0.2403 \ \text{Btu/lb}_m\text{-}°\text{F}$, Pr $= 0.713$, $\mu = 4.409 \times 10^{-2} \ \text{lb}_m/\text{ft-h}$. Equation (12.53) for Re gives

$$\text{Re} = \frac{D_h G_c}{\mu} = \frac{0.009412 \times 6926.4}{4.409 \times 10^{-2}} = 1478.$$

Figure 12.20 presents the heat transfer correlation for the plate-fin geometry in question. Thus, Fig. 12.20 at Re $= 1478$ gives

$$\frac{h_c}{G_c c_{p_c}} \ \text{Pr}^{2/3} = 0.0038,$$

so that

$$h_c = 7.92 \ \text{Btu/h-ft}^2\text{-}°\text{F}.$$

3. *Determine* h_h. With $t_{h_i} = 800°\text{F}$ and t_{h_o} assumed to be $256°\text{F}$, an average hot fluid temperature of $528°\text{F}$ results for which $c_{p_h} = 0.2483 \ \text{Btu/lb}_m\text{-}°\text{F}$, Pr $= 0.698$, $\mu = 6.905 \times 10^{-2} \ \text{lb}_m/\text{ft-h}$. With $G_h = 2597.4 \ \text{lb}_m/\text{h-ft}^2$,

calculations similar to the above give

$$\text{Re} = \frac{D_h G_h}{\mu} = 354,$$

$$\frac{h_h}{G_h c_{p_h}} \text{Pr}^{2/3} = 0.0105 \quad \text{(Fig. 12.20)},$$

$$h_h = 8.61 \text{ Btu/h-ft}^2\text{-}°\text{F}.$$

4. *Determine the effectiveness of both surfaces.* The plate fins have a thickness of $w = 0.006$ in. Their length (half-length between plates) is $L = [(0.25)^2 + (0.1667/2)^2]^{1/2}/2 = 0.1318$ in. According to Eq. (3.60), the fin effectiveness is $\kappa = (\tanh mL)/mL$ with $mL = L\sqrt{2h/kw}$. The surface effectiveness is, then, according to Eq. (3.64): $\eta = 1 - (A_f/A)(1 - \kappa)$. The ratio A_f/A is 0.773, for both sides, as given by Fig. 12.20. Thus, one has

Cold side: $mL = (0.1318/12)\sqrt{2 \times 7.92/15 \times (0.006/12)} = 0.5048,$

         $\kappa_c = 0.9228,$

         $\eta_c = 0.940.$

Hot side: $mL = (0.1318/12)\sqrt{2 \times 8.61/15 \times (0.006/12)} = 0.5263,$

         $\kappa_h = 0.9169,$

         $\eta_h = 0.936.$

5. *Determine the overall U.* The overall U based on the hot side surface is, by Eq. (12.10),

$$U_h = \left[\frac{A_h/A_c}{\eta_c h_c} + (A_h/A_w)\frac{\Delta x}{k} + \frac{1}{\eta_h h_h} \right]^{-1}.$$

The unfinned plate area (for 23 separating plates) is $A_w = (4 \times 6 \times 23)/144 = 3.833 \text{ ft}^2$. Thus,

$$U_h = \left(\frac{16.325/16.325}{0.940 \times 7.92} + \frac{16.325}{3.833} \times \frac{0.012/12}{15} + \frac{1}{0.936 \times 8.61} \right)^{-1}$$

$$= 3.87 \text{ Btu/h-ft}^2\text{-}°\text{F}.$$

Since $A_c = A_h$ in this instance, then $U_c = U_h$.

6. *Verify the fluid outlet temperatures.* From the given mass flow rates and the specific heats found in steps 2 and 3 above:

$$\dot{m}_c c_{p_c} = 800 \times 0.2403 = 192.24 \text{ Btu/h-}°\text{F},$$

$$\dot{m}_h c_{p_h} = 200 \times 0.2483 = 49.66 \text{ Btu/h-}°\text{F}.$$

Thus,

$$C_R = \frac{49.66}{192.24} = 0.258,$$

$$\text{NTU} = \frac{UA}{(\dot{m}c_p)_{\min}} = \frac{3.87 \times 16.325}{49.66} = 1.27,$$

so that Fig. 12.18 gives $\epsilon = 0.68$. Then by Eq. (12.37) with $\dot{m}_h c_{p_h} < \dot{m}_c c_{p_c}$,

$$\epsilon = \frac{t_{h_i} - t_{h_o}}{t_{h_i} - t_{c_i}},$$

$$0.68 = \frac{800 - t_{h_o}}{800 - 0},$$

$$t_{h_o} = 256°F.$$

This agrees with the assumption, so no further iteration is necessary and $t_{c_o} = 140°F$ as assumed, also. Then since $q = \dot{m}_h c_{p_h}(t_{h_i} - t_{h_o})$, the final desired results are

$$t_{c_o} = 140°F \ (60°C),$$

$$t_{h_o} = 256°F \ (124°C),$$

$$q = 27{,}000 \ \text{Btu/h} \ (7910 \ \text{W}). \qquad \blacksquare$$

12.9

HEAT EXCHANGER DESIGN

The discussion of the preceding section was limited to the calculation of the performance of a given heat exchanger of known geometry. That is, the total surface area and the flow pass arrangement were known, and it was desired to determine the outlet temperatures of the fluids for various inlet conditions.

The design problem, that of selecting the geometry of a heat exchanger to accomplish a desired transfer of heat between two fluids of specified terminal temperatures, is more complex inasmuch as there is no single answer to the problem. Several different heat exchangers may be able to satisfy stated conditions equally well from the thermal point of view. The ultimate decision as to which of several possible designs should be used usually depends on many other factors—such as cost, space requirements, operating expenses, personal taste of the designer, etc. Designs are also often limited by the desirability of adhering to certain standard practices—such as the use of standard tube sizes, standard fin arrays, etc.

If the flow rates, specific heats, and terminal temperatures of two heat exchanging fluids are specified, one may easily calculate the total heat to be transferred, the capacity ratio (R or C_R), and the effectiveness (P or ϵ). Then selection of the basic flow pattern readily yields Δt_m or the NTU from which the necessary UA product is found. The question becomes, then, what heat exchanger will have the proper area and overall coefficient to satisfy this need?

For a shell-and-tube exchanger, the tube size and the fluid velocity determine the number of tubes, per pass, needed to accommodate the desired flow. The number of tube passes and their length determine the total surface, A. These same dimensions, and the fluid velocities, and temperatures, serve to establish U. Thus one seeks a synthesis of all these factors to find the combination to give the desired UA. A similar reasoning applies to compact cross-flow exchangers.

In order to separate some of these parameters, two sets of examples will be shown. First it will be assumed that U is known, and the effect of flow pass arrangement and tube size will be shown. Then a case in which U is simultaneously found will be discussed.

EXAMPLE 12.8

A shell-and-tube heat exchanger is to be constructed with 1-in. 12-gage tubes. The cold fluid, flowing in the tubes, is water flowing at the rate of 18,000 kg/h and is to be heated from $t_{c_i} = 35°C$ to $t_{c_o} = 65°C$. The hot shell side fluid is also water and flows at the rate of 12,800 kg/h, entering at 100°C. The tube water flows at an average velocity of 0.3 m/s and the overall heat transfer coefficient of the exchanger is known to be 1600 W/m²-°C. For simplicity, assume that the specific heat of each fluid is constant at 4.18 kJ/kg-°C. Determine the number of tubes, per pass, and the required length of the tubes—examining various tube-pass possibilities but limiting the design to one shell pass.

Solution. Table B.2 gives the following data for 1-in. 12-gage tubes:

$$\frac{\text{surface area}}{\text{unit length}} = 0.0798 \text{ m}^2/\text{m, per tube,}$$

$$\text{cross-sectional area} = 0.0003098 \text{ m}^2, \text{ per tube.}$$

For the given $t_{c_i} = 35°C$, $t_{c_o} = 65°C$, an average of 50°C gives $\rho = 988.0$ kg/m³ (Table A.3), so that the number of tubes needed, per pass, to accommodate the 18,000-kg/h flow is

$$\dot{m} = 18,000 = N\rho A_x v = N \times 988.0 \times 0.0003098 \times 0.3 \times 3600,$$

$$N = 54.5.$$

Thus, 55 tubes per pass should be used. The capacity rates of the two streams are

$$\dot{m}_c c_{p_c} = 18,000 \times \frac{4.18}{3600} = 20.90 \text{ kW/°C,}$$

$$\dot{m}_h c_{p_h} = 12,800 \times \frac{4.18}{3600} = 14.86 \text{ kW/°C.}$$

The hot fluid outlet is given by

$$q = \dot{m}_c c_{p_c}(t_{c_o} - t_{c_i}) = \dot{m}_h c_{p_h}(t_{h_i} - t_{h_o}),$$

$$t_{h_o} = 57.8°C.$$

The problem specifications indicate that only one shell pass should be used. Investigate, first, the use of one tube pass. In this instance the use of a parallel-flow exchanger is ruled out since $t_{h_o} < t_{c_o}$. Thus, the only one-shell-pass, one-tube-pass arrangement possible is that for counterflow. Since $\dot{m}_h c_{p_h} < \dot{m}_c c_{p_c}$, the definition of the effectiveness in Eq. (12.37) gives

$$\epsilon = \frac{t_{h_i} - t_{h_o}}{t_{h_i} - t_{c_i}} = \frac{100 - 57.8}{100 - 35} = 0.649.$$

Also, $C_R = 14.86/20.90 = 0.711$. Figure 12.16 may be used to find the required NTU, but Eq. (12.46) is more accurate:

$$NTU = \frac{\ln\,[(1 - \epsilon)/(1 - \epsilon C_R)]}{C_R - 1} = 1.481,$$

and since NTU $= UA/(\dot{m}c_p)_{min}$,

$$NTU = 1.481 = \frac{UA}{14.86},$$

$$UA = 22.01 \text{ kW/°C}.$$

One could, as well, find Δt_m from Eq. (12.29):

$$\Delta t_m = \Delta t_{lm,ctr} = \frac{(t_{h_i} - t_{c_o}) - (t_{h_o} - t_{c_i})}{\ln\left(\dfrac{t_{h_i} - t_{c_o}}{t_{h_o} - t_{c_i}}\right)} = 28.47°C,$$

and UA from

$$q = UA\,\Delta t_m = \dot{m}_c c_{p_c}(t_{c_o} - t_{c_i}),$$

$$UA \times 28.47 = 20.90(65 - 35),$$

$$UA = 22.01 \text{ kW/°C}.$$

In this example U is known to be 1600 W/m²-°C $= 1.6$ kW/m²-°C, so the necessary exchanger surface area is

$$A = \frac{22.01}{1.6} = 13.76 \text{ m}^2.$$

The total surface area is determined by the tube length, L, so that for N tubes

$$A = N \times L \times \text{(surface area/unit length/tube)},$$

$$13.76 = 55 \times L \times 0.0798,$$

$$L = 3.13 \text{ m}.$$

Thus, a counterflow, one-shell-pass, one-tube-pass exchanger with 55 tubes approximately 3 m long will satisfy the stated requirements.

If, because of space limitations or other constraints, the above calculated tube length results in a heat exchanger considered to be too long, one might examine the possibility of using a one-shell-pass, two-tube-pass arrangement. In this instance, the effectiveness and capacity ratio remain $\epsilon = 0.649$, $C_R = 0.711$. Figure 12.17 is now applicable, but it is difficult to read the required NTU with much accuracy—a value of NTU of about 2.5 seeming to be indicated. Using the Δt_m–P–R formulation does not give much better results, the definitions of R and P giving $R = 1.41$, $P = 0.462$. Figure 12.12 for the F correction to be applied to $\Delta t_{lm,ctr}$ is likewise difficult to read with accuracy. The implication of either of these facts is that such an arrangement is not an effective use of the exchanger surface area and should probably be avoided. Proceeding, however, to show the method, one may apply Eq. (12.48) to find NTU with better accuracy than use of Fig. 12.17. Thus, for $\epsilon = 0.649$, $C_R = 0.711$, this equation yields.

$$NTU = 2.359 = \frac{UA}{\dot{m}_h c_{p_h}},$$

$$UA = 35.05 \text{ kW/°C},$$

$$A = 21.9 \text{ m}^2.$$

(Similarly, Eq. (12.35) will give $F = 0.620$, an undesirably low value). While the surface area required is greater than the one-tube-pass case, there are twice as many tubes to provide the area:

$$A = 21.9 = 2 \times 55 \times L \times 0.0798,$$
$$L = 2.5 \text{ m}.$$

The reduction in length is not much, and the increased number of tubes (i.e., twice) probably results in a higher cost of manufacture. Thus, this second option should probably not be used. The marginal improvement of the second design is the result of the fact, as noted earlier, that the surface area is not being utilized very effectively as evidenced by the low value of the temperature difference correction factor F. ■

The above example emphasized the effect of the flow pass arrangement on heat exchanger design. Altering other parameters will also affect the solution. The fluid velocity in the tubes affects the number of tubes needed to pass the tube side fluid and altering it will significantly modify the results. The fluid velocity also determines to some degree the overall U, assumed known above, as well as the pumping requirements to move the fluid. Likewise, the tube size may be varied to alter performance. It is not possible to quantify all these effects to produce general rules for design. However, the following example illustrates some of these effects, as well as showing the procedure involved when, as is actually the case, the overall U is not known. This example, as the above, is presented for a shell-and-tube exchanger. Analogous procedures would apply for compact cross-flow devices.

EXAMPLE 12.9

It is desired to design a feedwater heater to supply a boiler. The inlet water is at a temperature of $t_{c_i} = 70°F$, flows at the rate of 20,000 lb_m/h, and is to be heated to 150°F. The feedwater is to be heated in a horizontal shell-and-tube heat exchanger by condensing saturated steam at 20 psia. Because of space limitations, the tubes should not exceed 6 ft in length. Determine the configuration of a heat exchanger to accomplish these needs.

Solution. The Steam Tables give, at 20 psia, $t_{sat} = 228$°F. Using 1 Btu/lb$_m$-°F for the specific heat of the tube water, then the stated data give

$$\dot{m}_c c_{p_c} = 20{,}000 \text{ Btu/h-°F}, \quad t_{c_i} = 70\text{°F}, \quad t_{c_o} = 150\text{°F},$$

$$\dot{m}_h c_{p_h} \rightarrow \infty, \quad\quad\quad\quad t_{h_i} = 228\text{°F}, \quad t_{h_o} = 228\text{°F}.$$

Since the shell fluid is of constant temperature, the number of tube passes does not affect the mean temperature difference or the ϵ–NTU–C_R relation to be used (since $C_R \rightarrow 0$). Thus, the required UA product can be found in advance, and one does not have to be concerned with the effect of the flow pass arrangement as in the preceding example. For the above data,

$$\Delta t_m = \frac{(228 - 70) - (228 - 150)}{\ln\left(\dfrac{228 - 70}{228 - 150}\right)} = 113.3\text{°F},$$

and $q = \dot{m}_c c_{p_c}(t_{c_o} - t_{c_i}) = UA \, \Delta t_m$ yields

$$UA = \frac{20{,}000(150 - 70)}{113.3} = 14{,}120 \text{ Btu/h-°F}.$$

Use of the definition of $\epsilon = (t_{c_o} - t_{c_i})/(t_{h_i} - t_{c_i}) = 0.5063$ and Eq. (12.45) for $C_R = 0$ gives NTU $= 0.7059$ and the same result for UA. Thus, the design process devolves into finding configurations and flow velocities that will yield the above UA.

As a starting assumption, pick 1-in. 16-gage brass ($k = 64$ Btu/h-ft-°F) tubes. In order to keep pressure losses to a minimum, select a low water velocity in the tubes, say $v = 1$ ft/s. Using the notation of Fig. 12.19, or Example 12.6, Table B.2 gives the geometric data for the tubes:

$$D_2 = 0.870 \text{ in.}, \quad\quad D_3 = 1.00 \text{ in.},$$

$$\text{cross section} = 0.004128 \text{ ft}^2,$$

$$\text{surface area} = 0.2618 \text{ ft}^2/\text{ft}.$$

The number of tubes per pass is established by the need to pass 20,000 lb$_m$/h at a velocity of 1 ft/s. For a mean tube water temperature of 110°F, $\rho = 61.86$ lb$_m$/ft^3. Thus,

$$N = \frac{20{,}000}{61.86 \times 1 \times 3600 \times 0.004128} = 21.8,$$

so 22 tubes per pass must be provided.

Determination of the overall U will enable one to establish the total area needed and, thus, the required tube length. The overall U is found by obtaining the inside and outside heat transfer coefficients, h_{12} and h_{34}. These coefficients are found by the methods outlined in some detail in both Examples 12.2 and 12.6. The tube surface temperatures t_2 and t_3 need to be assumed, h_{12} found by application of the forced convection relation of Eq. (8.28), and h_{34} found by use of the condensing formula of Eq. (10.17). Then the assumed values of t_2 and t_3 are verified, modified appropriately, and the process repeated until satisfactory agreement is obtained. As in Example 12.6, one can probably neglect the tube wall resistance and let $t_2 \approx t_3$, so that an iteration on only one temperature need be made. Since these calculations are identical to those in Example 12.6 (or 12.2), they are not shown here. The results of this iterative calculation are

$$t_2 \approx t_3 = 212°F,$$

$$h_{12} = 410 \text{ Btu/h-ft}^2\text{-°F},$$

$$h_{34} = 2302 \text{ Btu/h-ft}^2\text{-°F}.$$

Thus, the overall U may be found:

$$U = \left[\frac{r_3}{r_2 h_{12}} + \frac{r_3 \ln (r_3/r_2)}{k_{23}} + \frac{1}{h_{34}} \right]^{-1},$$

$$= \left[\frac{1.0}{0.87 \times 410} + \frac{1.0 \ln (1.0/0.87)}{2 \times 12 \times 64} + \frac{1}{2303} \right]^{-1} = 300.4 \text{ Btu/h-ft}^2\text{-°F}.$$

Then, the required surface area is

$$A = \frac{14,120}{300.4} = 47.0 \text{ ft}^2.$$

For one tube pass, the required tube length will be, for 22 tubes per pass,

$$L = \frac{47.0}{22 \times 0.2618} = 8.16 \text{ ft}.$$

This exceeds the 6-ft maximum imposed in the specifications, so two tube passes must be provided. Since in this case the Δt_m does not change when two tube passes are used, the length is $L = 8.16/2 = 4.08 \approx 4$ ft. As a summary, then, one has for a first design:

$$\left. \begin{array}{l} \text{1 shell pass} \\ \text{2 tube passes} \\ \text{22 tubes per pass, 1-in.-O.D. 16-gage 4-ft-long tubes} \\ \text{1 ft/s water velocity} \\ h_{12} = 410 \text{ Btu/h-ft}^2\text{-°F} \\ h_{34} = 2303 \text{ Btu/h-ft}^2\text{-°F} \\ U = 300 \text{ Btu/h-ft}^2\text{-°F} \\ \Delta t_m = 113.3°F \\ A = 47.0 \text{ ft}^2 \end{array} \right\} \text{Design 1}$$

The above design does not allow for other factors such as tube fouling, shell side fluid distribution, etc. Nonetheless, these calculations illustrate the method and the effect of certain parameters.

Since the length of a single pass in the above calculation was not too much greater than the imposed maximum of 6 ft, one might hope to avoid the use of two passes by increasing U and, thus, reducing the required surface area. An increased tube water velocity should serve to increase U, and indeed it does, but the number of tubes required per pass is reduced—thus giving fewer tubes to provide the area.

To show these effects, increase the water velocity to 1.5 ft/s, leaving all other quantities the same. With the same size tubes, the final results are summarized below:

> 1 shell pass
> 2 tube passes
> 15 tubes per pass, 1-in.-O.D. 16-gage 4.6-ft-long tubes
> 1.5 ft/s water velocity
> $h_{12} = 567$ Btu/h-ft^2-°F
> $h_{34} = 2173$ Btu/h-ft^2-°F
> $U = 389$ Btu/h-ft^2-°F
> $\Delta t_m = 113.3$°F
> $A = 36.3$ ft^2

Design 2

The overall U is, indeed, greater and A is smaller. However, only 15 tubes are required per pass, to allow the tube water to flow at the rate of 20,000 lb$_m$/h. Thus, the tube length in a single pass becomes greater, 9.2 ft, so that two tube passes are still required. The second design requires fewer total tubes than the first, so its cost of manufacture would be less. However, the pressure loss required to pump the tube water in the second design would be of the order of 2.5 times that in the first design since this loss varies, approximately, as the square of the water velocity. Thus, operating costs associated with the second design would be expected to be greater than the first.

The effect of tube size can be seen by returning to a water velocity of 1 ft/s, as in the first design, but using instead $\frac{3}{4}$-in. 16-gage tubes. In this case, calculations yield

> 1 shell pass
> 1 tube pass
> 43 tubes per pass, $\frac{3}{4}$-in.-O.D. 16-gage 5.4-ft-long tubes
> 1 ft/s water velocity
> $h_{12} = 440$ Btu/h-ft^2-°F
> $h_{34} = 2610$ Btu/h-ft^2-°F
> $U = 310$ Btu/h-ft^2-°F
> $\Delta t_m = 113.3$°F
> $A = 45.5$ ft^2

Design 3

This design requires only one tube pass, but the total number of tubes and their length are comparable to those required in the first design. The pressure loss in the tubes would be of the same order as that in the first case, so one may take Design 1 and Design 3 as approximately equivalent.

The three designs given above are certainly not the only ones possible for the stated specifications. Others exist, and many should be examined and then compared through, probably, a detailed economic analysis to ascertain which should be adopted as the final design. More details of typical design procedures are available in sources such as Refs. 5, 8, and 9. ∎

CLOSURE

The emphasis of this chapter has been on the *methodology* of the treatment of problems of combined conduction and convection. Section 12.2 concentrated on the simultaneous solution of these two heat transfer mechanisms. Iterative, and sometimes double-iterative, solutions were found to be necessary. In Sections 12.8 and 12.9, in which heat exchanger performance analysis and heat exchanger design were discussed, multiple iterative solutions were again found to be necessary. The solution techniques were discussed in some detail in order to emphasize the fundamental origin of the conduction solutions and the convective correlations being used.

Quite apparently, the actual implementation of the solution methods described in this chapter can be greatly facilitated by the use of modern digital computing equipment. As noted earlier, the successful implementation of any computing system to the solution of these problems requires the capability of generating the necessary physical property data for the large number of fluids and solids that might be involved. Many modern engineering design offices have available computing systems with such capability.

REFERENCES

1. JAKOB, M., *Heat Transfer,* Vol. 2., New York, Wiley, 1957.
2. *Standards of Tubular Exchange Manufacturers Association,* 6th ed., New York, Tubular Exchanger Manufacturers Association, Inc. 1978.
3. BOWMAN, R. A., A. C. MUELLER, and W. M. NAGLE, "Mean Temperature Difference in Design," *Trans. ASME,* Vol. 63, 1940, p. 283.
4. SHAMSUNDAR, N., "A Property of the Log-Mean Temperature-Difference Correction Factor," *Mech. Eng. News,* Vol. 19, No. 3, 1982, p. 14.
5. KAYS, W. M., and A. L. LONDON, *Compact Heat Exchangers,* 2nd ed., New York, McGraw-Hill, 1964.
6. MASON, J. L., "Heat Transfer in Cross-flow," *Proc. Appl. Mech.,* 2nd U.S. Natl. Congr. 1954, p. 801.
7. ROHSENOW, W. M., and J. P. HARTNET, eds., *Handbook of Heat Transfer,* New York, McGraw-Hill, 1975.
8. KERN, D. Q., *Process Heat Transfer,* New York, McGraw-Hill, 1950.
9. KERN, D. Q., and A. D. KRAUS, *Extended Surface Heat Transfer,* New York, McGraw-Hill, 1972.

PROBLEMS

12.1 Steam at 4000 kN/m², 450°C, flows at a velocity of 7.5 m/s through a 6-in. schedule 40 pipe which is covered with 3.5 cm of 85% magnesia insulation.

The pipe is located horizontally in a room in which the ambient air temperature is 25°C. Find the temperature of the outer surface of the insulation, the overall heat transfer coefficient, and the heat loss, per meter of length.

12.2 Steam at 600°F, 800 psia, flows at 20 ft/s through a horizontal 5-in. schedule 40 pipe covered with 1 in. of insulation (k = 0.5 Btu/h-ft-°F). The pipe is exposed to atmospheric stagnant air at 100°F. Find the temperature of the outer surface of the insulation and the overall heat transfer coefficient.

12.3 A 1-in. 16-gage condenser tube is made of brass. Steam at 20 kN/m² is condensing on the tube surface, in a horizontal position, and water at 20°C flows with a velocity of 1.2 m/s through the tube. Find the tube surface temperature and the overall heat transfer coefficient.

12.4 A ¾-in. 18-gage horizontal condenser tube has steam at 5 psia condensing on its outer surface and cooling water at 80°F flowing through it with a velocity of 2 ft/s. Find the overall heat transfer coefficient if the tube is (a) brass with k = 64 Btu/h-ft-°F, and (b) stainless steel with k = 10 Btu/h-ft-°F.

12.5 A bare 10-in. schedule 40 steel pipe passes horizontally through a room of still air at 85°C. The pipe carries steam at 3000 kN/m², 550°C, flowing with a velocity of 100 m/s. Find the overall heat transfer coefficient and the pipe surface temperature.

12.6 Steam at 500 kN/m², 450°C, flows at a velocity of 5 m/s through a horizontal 5-in. schedule 80 pipe covered with 7.5 cm of 85% magnesia insulation. The pipe is exposed to still atmospheric air at 35°C. Find the overall heat transfer coefficient and the outer insulation temperature.

12.7 The lubricating oil listed in Table A.4 flows at 0.3 m/s through a 1-in. 16-gage brass tube 2 m long. The oil is at 50°C and the outer tube surface is exposed, horizontally, to atmospheric air at 20°C by free convection. Find the outer tube surface temperature.

12.8 The lubricating oil listed in Table A.4 flows at 1.5 ft/s through a horizontal 1-in. 16-gage brass tube 7 ft long. The oil temperature is 120°F and the tube is exposed to atmospheric air at 70°F. Find the tube wall temperature.

12.9 Superheated steam at 2000 kN/m², 300°C, flows with a velocity of 6 m/s through a 6-in. schedule 40 steel pipe. Atmospheric air at 20°C flows normal to the cylinder with a velocity of 12 m/s. Estimate the outer pipe surface temperature.

12.10 A cast iron rod, 2 cm in diameter and 25 cm long, protrudes horizontally from a heat source at 150°C into still atmospheric air at 30°C. Estimate the heat loss from the rod.

12.11 In a one-shell-pass, one-tube-pass heat exchanger, the cold fluid (water with c_{p_c} = 4.18 kJ/kg-°C) enters at 50°C and flows at the rate of 900 kg/h. The hot fluid (also water with $c_{p_h} = 4.18$ kJ/kg-°C) enters at 370°C and flows at the rate of 1260 kg/h. If the heat exchanger becomes infinitely large, find the maximum heat that can be transferred and the outlet temperatures of the fluids for (a) parallel flow, and (b) counterflow.

12.12 Repeat Prob. 12.11 if the cold fluid flow rate is 1260 kg/h and the hot fluid flow rate is 900 kg/h, the other data remaining the same.

12.13 In a one-shell-pass, one-tube-pass heat exchanger, the hot fluid enters at 425°C and leaves at 315°C. The cold fluid enters at 40°C and leaves at 260°C. Find the mean temperature difference for (a) parallel flow, and (b) counterflow.

12.14 Find the mean temperature difference in a one-shell-pass, two-tube-pass heat exchanger for the temperatures noted in Prob. 12.13.

12.15 In a heat exchanger, the cold fluid enters at 110°F and leaves at 500°F while the hot fluid enters at 750°F and leaves at 600°F. Find the mean temperature difference for (a) parallel-flow, (b) counterflow, and (c) one-shell-pass, two-tube-pass.

12.16 In a one-shell-pass, one-tube-pass heat exchanger the cold fluid enters at 40°C and leaves at 200°C while the hot fluid enters a 370°C and leaves at 150°C. Find the mean temperature difference.

12.17 In a cross-flow heat exchanger, the cold fluid enters at 40°C and leaves at 150°C while the hot fluid enters at 425°C and leaves at 260°C. Find the mean temperature difference.

12.18 In terms of the parameters R and P defined in Eqs. (12.32) and (12.33), show that the F correction factor defined in Eq. (12.34) for a parallel-flow heat exchanger is

$$F = \frac{R+1}{R-1} \frac{\ln\left(\dfrac{1-P}{1-PR}\right)}{\ln\left[\dfrac{1}{1-P(R+1)}\right]}.$$

12.19 Equation (12.44) for the effectiveness of a parallel-flow exchanger was derived from Eqs. (12.40) and (12.41) by introduction of Eq. (12.28). Follow the same procedure to derive Eq. (12.46) from Eqs. (12.40) and (12.41) by introducing Eq. (12.29).

12.20 Show that Eq. (12.47) is the limiting form for Eq. (12.46) as $C_R \to 1$.

12.21 A 1-in. 14-gage heat exchanger tube is equipped with 20 equally spaced straight fins of uniform thickness placed longitudinally along the tube. The fins are 2.5

cm long in the radial direction and are 0.16 cm thick. Both the tube and the fins are made of steel with $k = 45$ W/m-°C. The inside and outside surface heat transfer coefficients are 1130 and 255 W/m²-°C, respectively. What is the overall heat transfer coefficient for the exchanger, based on the outer exposed surface area?

12.22 The separating surface of a cross-flow heat exchanger is a plate 0.03 cm thick. On the hot fluid side, there is an array of straight uniform thickness fins 0.3 cm long, 0.015 cm thick, spaced 0.2 cm on centers, and the heat transfer coefficient there is 45 W/m²-°C. On the cold fluid side of the plate there is also an array of straight uniform thickness fins of the same dimensions but spaced 0.13 cm on centers and the heat transfer coefficient is 54 W/m²-°C. What is the overall heat transfer coefficient based on the exposed surface area where $h = 45$ W/m²-°C if the fins and the plate have a thermal conductivity of 40 W/m°C?

12.23 The overall heat transfer coefficient for a one-shell-pass, two-tube-pass heat exchanger is known to be 1700 W/m²-°C. The shell side fluid ($c_p \cong 4.18$ kJ/kg-°C) flows at the rate of 9100 kg/h, entering at 260°C. The tube side fluid ($c_p = 3.55$ kJ/kg-°C) flows at the rate of 27,300 kg/h, entering at 30°C. For a total surface area of 9.5 m², find the rate of heat transfer between the two fluids and their outlet temperatures using (a) the Δt_m formulation, and (b) the ϵ–NTU formulation.

12.24 A one-shell-pass, one-tube-pass, counterflow heat exchanger uses 6800 kg/h of water ($c_p = 4.18$ kJ/kg-°C) entering at 25°C to cool 18,000 kg/h of oil ($c_p = 2.9$ kJ/kg-°C) entering at 100°C. The total exchanger surface area is 14 m² and the overall coefficient is $U = 370$ W/m²-°C. Find the rate of heat exchange and the outlet temperature of the two fluids by either the Δt_m or ϵ–NTU formulation.

12.25 A one-shell-pass, two-tube-pass heat exchanger has a known overall heat transfer coefficient of 280 Btu/h-ft²-°F. The shell side fluid is water ($c_p \cong 1.0$ Btu/lb$_m$-°F) flowing at the rate of 18,000 lb$_m$/h and entering at 500°F. The tube side fluid is oil ($c_p = 0.85$ Btu/lb$_m$-°F) flowing at the rate of 50,000 lb$_m$/h and entering at 80°F. For a total surface area of 125 ft², find the outlet fluid temperatures using (a) the Δt_m formulation, and (b) the ϵ–NTU formulation.

12.26 A one-shell-pass, one-tube-pass heat exchanger operates in counterflow. The exchanger uses 6500 kg/h of water ($c_p = 4.18$ kJ/kg-°C) entering at 25°C to cool 13,000 kg/h of oil ($c_p = 2.09$ kJ/kg-°C) entering at 100°C. The total surface area of the exchanger is 16 m² and the overall heat transfer coefficient is 350 W/m²-°C. Find the outlet temperatures of the two fluids using both the Δt_m and ϵ–NTU formulations.

12.27 A steam condenser is made of $\frac{3}{4}$-in. 14-gage brass tubes, 5 ft long. There are two tube passes with 115 tubes per pass, and the cooling water enters at 70°F with a velocity of 5 ft/s. The inside tube heat transfer coefficient is known to be 1140 Btu/h-ft^2-°F and the outside tube surface condensing heat transfer coefficient is known to be 1250 Btu/h-ft^2-°F. The steam is being condensed at 5 psia. Find (a) the outlet temperature of the cooling water, and (b) the lb$_m$/h of steam condensed.

12.28 A steam condenser is made of $\frac{5}{8}$-in. 14-gage brass tubes, 1.8 m long. There are two tube passes with 125 tubes per pass. Cooling water enters at 24°C and flows with an average tube velocity of 1.37 m/s. The inside tube heat transfer coefficient is known to be 6735 W/m^2-°C and the outside tube surface condensing heat transfer coefficient is known to be 11,340 W/m^2-°C. Find the outlet water temperatuure and the kg/h of steam condensed if the steam is saturated at 14 kN/m^2 pressure.

12.29 A cross-flow heat exchanger has a surface configuration with heat transfer coefficients as described in Prob. 12.22. The fluid passages are 0.6 cm wide, so the fin length, by symmetry, is 0.3 cm as given. The hot fluid flows at a rate of 45 kg/h (c_p = 1 kJ/kg-°C), entering at 300°C while the cold fluid flows at the rate of 55 kg/h (c_p = 1.2 kJ/kg-°C), entering at 100°C. The total exposed surface area on the hot side is 1.5 m^2. Find the fluid outlet temperatures.

12.30 Water flows at an average velocity of 0.3 m/s through a 1-in. schedule 40 steel pipe. The pipe is bare and passes horizontally through a large room where it losses heat by free convection to the ambient air at 20°C. If the water enters the pipe at 95°C and the pipe is 150 m long, estimate the outlet temperature of the water.

12.31 Water flows at the rate of 2.5 gpm through a 1-in. schedule 40 steel pipe. The pipe is bare and passes horizontally through a large room where it loses heat by free convection to the ambient air at 70°F. If the water enters the pipe at 190°F and is 500 ft long, estimate the outlet temperature of the water.

12.32 A horizontal steam condenser is condensing steam at a pressure of 14 kN/m^2. There are two tube passes with 180 tubes per pass, 2.4 m long. The tubes are brass, $\frac{5}{8}$-in., 16-gage, and the tube water velocity is 1 m/s. If the cooling water enters at 20°C, find the outlet water temperature and the kg/h of steam condensed.

12.33 A horizontal steam condenser has two tube passes, 210 tubes per pass, of $\frac{3}{4}$-in. 16-gage brass tubes 2.75 m long. Cooling water is supplied at 25°C with an average tube velocity of 1.5 m/s. What is the outlet water temperature if the steam is being condensed at a pressure of 10 kN/m^2?

12.34 A one-shell-pass, two-tube-pass steam condenser is made of $\frac{3}{4}$-in. 18-gage brass tubes, 8 ft long. There are 220 tubes per pass and the average tube water velocity is 5 ft/s. If the condenser is in a horizontal position, the water inlet at 70°F, and the condensing pressure is 1.5 psia, find the outlet water temperature and the lb$_m$/h of steam condensed.

12.35 A compact cross-flow heat exchanger is constructed of a plate-fin array with the same plate thickness, plate spacing, fin spacing, and fin dimensions as that considered in Example 12.7. The plates are, however, 8 in. × 8 in. square, and there are 20 flow passages in each direction. The cold fluid is gaseous carbon dioxide at atmospheric pressure entering at 50°F and flowing at the rate of 1200 lb_m/h. The hot fluid is air, at atmospheric pressure, entering at 600°F and flowing at the rate of 900 lb_m/h. Determine the outlet temperature of the two streams and the rate of heat transfer between them. Take the plate and fin thermal conductivity to be 15 Btu/h-ft-°F.

12.36 The heating coil in a ventilation system duct consists of a bank of finned tubes of the configuration shown in Fig. 12.21. The duct is 2 ft × 2 ft in cross section. Thus the tubes are 2 ft long and, as the geometry in Fig. 12.21 shows, the 2-ft height allows for a bank of tubes 25 high in the first row, 24 in the second, and 25 in the third (last) row. The tubes have an I.D. of 0.28 in., O.D. of 0.38 in., and are equipped with fins (k = 95 Btu/h-ft°F) of the dimensions shown. According to the dimensions shown, the thickness from the front face of the tube bank to the back face is 2.52 in. Hot water is the heating medium, and enters at one end of the tubes at 120°F. The water velocity is 2 ft/s in in the tubes. Air flows in the duct at a volume rate of 2000 cfm in an unobstructed cross section. If the air approaching the heating coil is at 50°F (1 atm pressure), what is the temperature of the air leaving the coil?

12.37 In a counterflow heat exchanger, 400,000 lb_m/h of water (c_p ≅ 1 Btu/lb_m-°F) is heated from 250°F to 310°F by the use of hot gases entering at 750°F and leaving at 500°F. If the exchanger has a total surface area of 22,000 ft², find the overall U. Use both the Δt_m and ε–NTU formulations.

12.38 A heat exchanger is used to heat 3600 kg/h of water (c_p ≅ 4.18 kJ/kg-°C) from 40°C to 175°C by cooling another stream of water flowing at 4550 kg/h and entering at 315°C. The overall heat transfer coefficient for the exchanger is 1760 W/m²-°C. Using both the Δt_m and ε–NTU formulations, find the required surface area if the heat exchanger is (a) parallel-flow, (b) counterflow, and (c) one-shell-pass, two-tube-pass.

12.39 A one-shell-pass, two-tube-pass heat exchanger is used to heat 27,000 kg/h of water from 85°C to 100°C by the use of steam condensing at 345 kN/m². The overall heat transfer coefficient is known to be 2800 W/m²-°C. If there are 30 1-in.-O.D. tubes per pass, how long must the tubes be?

12.40 A counterflow heat exchanger is used to heat 2160 kg/h of water (c_p = 4.18 kJ/kg-°C) from 35°C to 90°C by the use of oil (c_p = 2.1 kJ/kg-°C) flowing at the rate of 3240 kg/h with an inlet temperature of 175°C. If the overall heat transfer coefficient is 425 W/m²-°C, find the surface area required.

12.41 A cross-flow heat exchanger is used to heat water from 35°C to 85°C, flowing at the rate of 3 kg/s, with atmospheric pressure air being cooled from 225°C to

100°C. If the overall heat transfer coefficient is 210 W/m²-°C, find the surface area required.

12.42 A water-to-water, shell-and-tube heat exchanger is to be made of $\frac{3}{4}$-in. 18-gage tubes. Cold water enters the tubes at a flow rate of 15,000 kg/h and is heated from 30°C to 60°C. The hot fluid, also water, flows at the rate of 10,000 kg/h, entering at 105°C. The tube water velocity is to be 0.4 m/s and the overall heat transfer coefficient for the exchanger is 1750 W/m²-°C. Find the number of tubes per pass and the required length of the tubes if the exchanger is (a) parallel-flow; (b) counterflow; (c) one-shell-pass, one-tube-pass; and (d) two-shell-passes, four-tube-passes. Discuss the relative merits of each design.

12.43 A water-to-water heat exchanger is to be made of $\frac{3}{4}$-in. 16-gage tubes. The tube water enters at 90°F leaves at 120°F and flows at the rate of 30,000 lb$_m$/h. The shell water flows at the rate of 20,000 lb$_m$/h, entering at 200°F. The overall heat transfer coefficient is 250 Btu/h-ft²-°F. The exchanger is to have one shell pass, and the tube water velocity is to be 1 ft/s. If the length of the tubes is not to exceed 8 ft, find (a) the number of tubes per pass, (b) the number of tube passes, and (c) the length of the tubes.

12.44 A one-shell-pass, one-tube-pass heat exchanger is made of 60 tubes ($\frac{5}{8}$ in., 14 gage). Water enters the shell side at 150°C and leaves at 40°C, flowing at the rate of 9000 kg/h. The tube water enters at 25°C and leaves at 65°C. The inside and outside tube surface heat transfer coefficients are 1700 and 8500 W/m²-°C, respectively. Find the required length of the tubes.

12.45 A steam condenser is made of brass tubes, $\frac{3}{4}$ in., 14 gage. There are 150 tubes per pass. The cooling water flows at the rate of 1.52 m/s, entering at 25°C and leaving at 50°C. The condensing heat transfer coefficient is known to be 10,200 W/m²-°C and the inside tube surface coefficient is 6800 W/m²-°C. Steam is being condensed at 77°C. The tubes are not to exceed 3 m in length. Find the number of tube passes and the length of the tubes.

12.46 It is desired to design a steam condenser to condense 10,200 kg/h of saturated steam at 35 kN/m² pressure. The condenser is to be supplied with cooling water at 26°C and the water temperature rise is to be 14°C. The tube water velocity is to be 2.1 m/s. The tubes are selected to be $\frac{7}{8}$-in. 16-gage brass (2.23 cm O.D., 1.89 cm I.D.) and are not to exceed 2.5 m in length. Find the number of tube passes and the required tube length.

12.47 Repeat Prob. 12.46 if the tube water velocity is changed to 1.8 m/s.

12.48 A heat exchanger is to be designed to heat 6800 kg/h of water from 20°C to 38°C by the use of saturated steam condensing at 240 kN/m². The tubes are to be 1-in. 12-gage brass and are not to exceed 2 m in length. The tube water velocity is to be 0.45 m/s. Determine the number of tubes per pass, the number of tube passes, and the length of the tubes.

12.49 Design a steam condenser (i.e., find the number of tubes per pass, number of passes, and tube length) that will condense 20,000 lb_m/h of steam at 5 psia. It is desired to use tubes with 0.875 in. O.D., 0.745 in. I.D. The tube water velocity is 7 ft/s and the tubes are not to exceed 12 ft in length. The tube water is to enter at 75°F and leave at 100°F.

12.50 Design a feedwater heater to heat 1.5 kg/s of water from 20°C to 55°C by the use of steam condensing at 200 kN/m^2. Set the maximum tube length at 2.5 m.

12.51 A cross-flow heat exchanger is to be designed using the plate-fin configuration of Example 12.7. The same plate thickness, spacing, fin dimensions and thermal conductivity are to be used; however, the plate dimensions in the directions transverse to the fluid streams and the number of flow passages for each fluid are to be determined. It is desired to heat a stream of air flowing at the rate of 1000 lb_m/h from 20°F to 160°F by the use of another airstream being cooled from 700°F to 200°F.

CHAPTER 13

Additional Cases of Combined Heat Transfer

13.1

INTRODUCTORY REMARKS

The examples of combined heat transfer considered in Chapter 12 were restricted to cases in which only the conduction and convection modes were present, and primary emphasis was placed on the application of these cases to the analysis and design of heat exchangers. This chapter will be devoted to other examples of combined heat transfer, some of which will include thermal radiation as one of the simultaneous modes taking place. Applications of these principles for the estimation of the thermometric errors will be illustrated. Some recent engineering applications involving multimode heat transfer will be discussed briefly. The analysis of a solar collector involves all of the basic heat transfer modes— conduction, convection, and radiation. The so-called *heat pipe* involves the combined effects of conduction, convection, vaporization, and condensation.

13.2

SIMULTANEOUS CONVECTION AND RADIATION— THE RADIATION COEFFICIENT

The instances of heat transfer from a body surface to its surroundings treated thus far have been limited to cases in which the sole mechanism taking place at the surface was either pure convection (Chapters 7 through 10) or pure radiation (Chapter 11). In many cases of practical interest both the convective and radiative mechanisms occur simultaneously and the conditions may be such that the heat transfer by each of these mechansims may be of the same order of magnitude. When the convecting fluid surrounding a surface is radiatively nonabsorbing and nonemitting, one may treat the convection and radiation mechanisms as independent.

The surface heat transfer by convection, q_c, is represented by the familiar relation

$$q_c = Ah_c(T_s - T_a),$$ (13.1)

in which h_c is the *convective* heat transfer coefficient as determined by the methods of Chapters 7 through 10, T_s is the surface temperature, and T_a represents the temperature of the ambient fluid surroundings. In Chapter 11, the radiant heat transfer from a surface, q_r, was shown to be of the form

$$q_r = A\mathcal{F}(E_{bs} - E_{be}),$$ (13.2)

$$= A\sigma\mathcal{F}(T_s^4 - T_e^4).$$

In Eq. (13.2), T_e is the temperature of the radiation environment seen by the surface and $\mathcal{F}$ is a function (sometimes complex) of the surface radiation properties (ϵ, α, etc.) *and* the geometry of the radiative environment. The exact form of $\mathcal{F}$ depends on the particular situation at hand, but all the results of Chapter 11 (particularly the analytic results of Sec. 11.10) may be reduced to this form.

Often Eq. (13.2) is linearized into the form

$$q_r = Ah_r(T_s - T_a)$$ (13.3)

by introduction of the *radiation coefficient* h_r:

$$h_r = \sigma\mathcal{F}\frac{T_s^4 - T_e^4}{T_s - T_a}.$$

If, in particular, $T_e = T_a$, the above becomes

$$h_r = \sigma\mathcal{F}(T_s^2 + T_a^2)(T_s + T_a).$$ (13.4)

The radiation coefficient was first introduced in Eq. (10.29) in connection with film boiling. The linearization given by Eq. (13.3) is deceptive in that the radiation coefficient, h_r, is a strong function of temperature.

The use of h_r allows one to express, formally, the combined heat flow from the surface as

$$q = q_c + q_r,$$

$$= A(h_c + h_r)(T_s - T_a),$$ (13.5)

$$= Ah_{cr}(T_s - T_a),$$

in which $h_{cr} = h_c + h_r$ is termed the *combined coefficient*.

Some applications of these concepts will now be presented.

Combined Heat Loss from a Completely Enclosed Body

As a first example, imagine a body completely enclosed in a large room. The convective coefficient h_c is determined by the hydrodynamic conditions imposed—free convection, forced convection, etc., and let it be presumed that h_c has been appropriately determined.

For a body with surface temperature T_s and surface emissivity ϵ_s, completely enclosed by a large room at temperature T_a, Eq. (11.100) gives the radiant heat transfer to be

$$q = \epsilon_s(E_{bs} - E_{ba}),$$
$$= \sigma\epsilon_s(T_s^4 - T_a^4),$$

that is, the $\mathscr{F}$ in Eq. (13.2) is simply $\mathscr{F} = \epsilon_s$. Then the associated radiation coefficient is, by Eq. (13.4),

$$h_r = \sigma\epsilon_s(T_s^2 + T_a^2)(T_s + T_a). \tag{13.6}$$

The above relations may be used to estimate the combined convective and radiative losses from bodies in large enclosures as illustrated in the next example.

EXAMPLE 13.1

A horizontal oxidized wrought iron pipe (16 cm O.D.) has a surface temperature of 100°C and is located in a large room in which heat loss occurs to the ambient air by free convection and to the walls by radiation. Both the air and the walls are at 20°C. Find the combined heat loss, per unit surface area, by convection and radiation.

Solution. The pipe diameter is given as $D = 16$ cm, and Table A.10 gives $\epsilon_s = 0.94$.

The convective coefficient is found by the methods of Chapter 9. For the stated conditions the mean film temperature is $(100 + 20)/2 = 60°C$, so that Table A.6 gives $\nu = 18.90 \times 10^{-6}$ m²/s, $k = 28.52 \times 10^{-3}$ W/m-°C, $Pr = 0.708$. Also, $\Delta t = 100 - 20 = 80°C$, $\beta = 1/(273.15 + 20)°K$. Thus, Eq. (9.42) yields

$$Gr_D = \frac{D^3 g\, \beta\, \Delta t}{\nu^2} = 3.07 \times 10^7,$$

$$Ra = 2.17 \times 10^7,$$

$$Nu_D = \left\{0.60 + 0.387 Ra_D^{1/6}\left[1 + \left(\frac{0.559}{Pr}\right)^{9/16}\right]^{-8/27}\right\}^2 = 35.58,$$

$$h_c = 6.34 \text{ W/m}^2\text{-°C}.$$

By Eq. (13.6) the radiation coefficient is with $T_s = 373.15°K$ and $T_a = 293.15°K$:

$$h_r = \sigma\epsilon(T_s^2 + T_a^2)(T_s + T_a) = 8.00 \text{ W/m}^2\text{-°C}.$$

Thus, the combined coefficient and the heat loss are

$$h_{cr} = h_c + h_r = 14.34 \text{ W/m}^2\text{-}°\text{C} \ (2.53 \text{ Btu/h-ft}^2\text{-}°\text{F}),$$

$$\frac{q}{A} = h_{cr}(T_s - T_a) = 1147.4 \text{ W/m}^2 \ (363.7 \text{ Btu/h-ft}^2).$$

Note that even for the modest surface temperature used in this example, the convective and radiative coefficients are of the same order of magnitude. If the surface temperature is increased to, say, 480°C, the convective coefficient increases modestly to 9.62 W/m^2-°C, while the radiative coefficient increases by a factor of nearly 4.5 to 36.42 W/m^2-°C. ∎

The method of calculation shown in Example 13.1 forms the basis for the various tables and charts available (see, for example, Ref. 1) in the literature which give values of the combined coefficient for pipes as a function of the diameter and temperature difference. The results for insulated pipes are substantially the same as those illustrated above for a bare pipe (except that lower surface temperatures are usually involved) because of the fact that the emissivity of the canvas insulation wrapping is about the same as that of oxidized steel.

An instance of heat transfer by a combination of all three modes is encountered, for example, in the case of heat loss from a pipe (or wall) when the surface temperature is not specified as it was in the above example. If, instead, the bulk temperature of the fluid flowing inside the pipe is specified, one must account for the inside convective film coefficient and the thermal resistance of the pipe and insulation (if any) as well as the combined coefficient. The determination of the rate of heat loss from the pipe then necessitates a trial and error solution just like those presented in Sec. 12.2, for cases where radiation was neglected, except that the computation of h_{cr} illustrated above replaces the step of the calculations wherein the outside convective coefficient was computed. The iterative solution of such problems converges less rapidly when the thermal radiation is included than when it is neglected, because the radiation coefficient is much more sensitive to changes in the surface temperature than is the convective coefficient.

Combined Convection and Radiation in Air Spaces

Another example of combined convection and radiation heat transfer is that found in the case of air spaces between two surfaces of different temperature. Such air spaces are found between the walls of a building, between glass layers in windows, or between the absorbing surface of a solar collector and its glass cover plate. The air spaces may be inclined or in a vertical position.

If the two surface temperatures are denoted by T_h and T_c (for "hot" and "cold," respectively) the convective transfer between them is

$$\frac{q_c}{A} = h_c(T_h - T_c)$$

where the convective coefficient is that due to free convection as found from the correlations given in Chapter 9 in Eqs. (9.48) through (9.53). The radiative transfer may be modeled as that between two infinite parallel planes as given by Eq. (11.53):

$$\frac{q_r}{A} = \frac{\sigma(T_h^4 - T_c^4)}{(1/\epsilon_h) + (1/\epsilon_c) - 1}.$$

Even when the surfaces are glass, the above representation is probably valid since glass is nearly opaque to low temperature radiation in the long (infrared) part of the spectrum. If *solar* radiation is present, other considerations must be made, as illustrated later in the chapter.

For the above representation the radiation coefficient is then defined as

$$\frac{q_r}{A} = h_r(T_h - T_c)$$

$$\tag{13.7}$$

$$h_r = \frac{\sigma(T_h^2 + T_c^2)(T_h + T_c)}{(1/\epsilon_h) + (1/\epsilon_c) - 1},$$

so that the combined heat transfer is

$$q = (h_r + h_c)(T_h - T_c)$$

$$\tag{13.8}$$

$$= h_{cr}(T_h - T_c).$$

Given the thermal conditions and the surface emissivities, the combined effects of convection and radiation may be deduced. Calculations such as those given in the next example form the basis of an air space conductance values quoted in sources such as Ref. 2. Representations such as the above are also used in the thermal analysis of solar collectors as shown later in this chapter.

EXAMPLE 13.2

A vertical air space is 1.5 in. wide and 3 ft. high. One of the bounding surfaces is at $T_h = 95°F = 555°R$ and the other is at $T_c = 65°F = 525°R$. Find the combined heat transfer coefficient due to convection and radiation between the surfaces if both their emissivatives are (a) 0.9; (b) 0.333.

Solution.

(a) The convective coefficient is found from the correlations given in Eq. (9.49). From the given data: $H = 3$ ft, $L = 1.5$ in. $= 0.125$ ft, $Ar = H/L = 24$, $\Delta t = 30°F$. For a mean temperature of 80°F, Table A.6 gives $\nu = 0.6086$ ft²/h, $k = 0.01511$ Btu/h-ft-°F, $Pr = 0.711$, $\beta = (1/540°R)$. Thus, Eq. (9.49) gives

$$Ra = \frac{L^3 g \beta \Delta t}{\nu^2} Pr = 8.69 \times 10^4,$$

$$Nu_1 = 0.0605 Ra^{1/3} = 2.68,$$

$$NU_2 = \left\{ 1 + \left[\frac{0.104 Ra^{0.293}}{(1 + (6310/Ra)^{1.36})} \right]^3 \right\}^{1/3} = 2.87,$$

$$\mathrm{Nu}_3 = 0.242 \left(\frac{\mathrm{Ra}}{\mathrm{Ar}} \right)^{0.272} = 2.25,$$

$$h_c = \mathrm{Nu}_2 \frac{k}{L} = 0.347 \; \text{Btu/h-ft}^2\text{-}^\circ\text{F}.$$

For $\epsilon_h = \epsilon_c = 0.9$, Eq. (13.7) gives the radiation coefficient to be

$$h_r = \frac{\sigma(T_h^2 + T_c^2)(T_h + T_c)}{(1/\epsilon_h) + (1/\epsilon_c) - 1} = 0.884 \; \text{Btu/h-ft}^2\text{-}^\circ\text{F},$$

so that

$$h_{cr} = 0.347 + 0.884 = 1.23 \; \text{Btu/h-ft}^2\text{-}^\circ\text{F} \; (6.97 \; \text{W/m}^2\text{-}^\circ\text{C}).$$

Note, once more, that the radiation coefficient is of the same order of magnitude as the convective one.

(b) For $\epsilon_h = \epsilon_c = 0.333$, the radiation coefficient becomes

$$h_r = 0.216 \; \text{Btu/h-ft}^2\text{-}^\circ\text{F}.$$

The convection coefficient is unchanged, so

$$h_{cr} = 0.347 + 0.216 = 0.56 \; \text{Btu/h-ft}^2\text{-}^\circ\text{F} \; (3.18 \; \text{W/m}^2\text{-}^\circ\text{C}). \quad \blacksquare$$

13.3

THERMOCOUPLE LEAD ERROR IN SURFACE TEMPERATURE MEASUREMENTS

Some of the principles of the various modes of heat transfer may be combined to analyze, approximately, the error that may be expected when the temperature of a body or fluid is measured with one of the usual thermometric devices—a thermometer or a thermocouple. Since the introduction of a temperature-measuring device into a system in which heat transfer is taking place will alter the thermal conditions existing there, one must expect that the device will indicate a value of the sought-after temperature that differs from the temperature that would actually exist had no measurement been made. This error is due to the thermal effects of the measuring device itself on the system and is not a calibration error of the thermometer. Some typical cases will be discussed in the following sections.

If a thermocouple is attached to the surface of a solid body in the manner illustrated in Fig. 13.1(a), heat will be conducted along the thermocouple leads and dissipated into the surroundings by convection from the wires. This conduction of heat sets up temperature gradients within the body in the vicinity of the point of attachment—a condition that would not exist had the thermocouple not been attached. The thermocouple will read a value corresponding to the local depression in the surface temperature at the point of attachment.

Since the thermocouple wires are usually long and thin, the conduction of the heat along them—and its eventual dissipation into the surroundings—may be represented as heat flow along a spine of uniform cross section and infinite

FIGURE 13.1. Measurement of surface temperatures by the use of thermocouples.

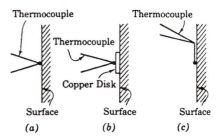

Thermocouple   Thermocouple

Thermocouple

Copper Disk

Surface   Surface   Surface

(a)   (b)   (c)

length. This was treated in Prob. 3.41. To simplify the system under consideration, let the thermocouple be represented as a single wire of radius R and thermal conductivity k_t. The surface film coefficient at the surface of the wire will be denoted by h. Let t_a denote the temperature of the ambient air and let t_t denote the depressed temperature of the surface under the point of attachment of the thermocouple. Then Prob. 3.41 shows that the rate of heat flow from the surface into the wire is

$$q = k_t \sqrt{\frac{hC}{k_t A}} \times A(t_t - t_a).$$

Here C and A are the perimeter and cross-sectional areas of the wire, respectively. Thus,

$$q = \pi \sqrt{2k_t R^3 h} \times (t_t - t_a). \tag{13.9}$$

The heat noted in Eq. (13.9) must be supplied from within the body. The body may be represented as a semi-infinite solid maintained at a temperature t_s at points far removed from the point of the thermocouple attachment. The point of attachment may be approximated as an area of radius R maintained at temperature t_t. Thus, t_s is the true surface temperature, t_t is the temperature indicated by the thermocouple, and $(t_s - t_t)$ is the error introduced by the attachment of the thermocouple.

One of the classical solutions of the mathematical theory of heat conduction is that for a semi-infinite solid of a uniform temperature, say t_s, at points far removed from a heat sink on the surface which has a finite radius, say R, and a temperature t_t. The solution is not given here but may be found in Ref. 3. The rate of heat flow into the sink is given by this solution to be

$$q = 4Rk_s(t_s - t_t). \tag{13.10}$$

In Eq. (13.10) k_s is used to denote the thermal conductivity of the surface material. Combination of Eqs. (13.9) and (13.10) gives the measured error to be

$$t_s - t_t = \frac{\dfrac{\pi}{k_s} \sqrt{\dfrac{k_t R h}{8}}(t_s - t_a)}{1 + \dfrac{\pi}{k_s} \sqrt{\dfrac{k_t R h}{8}}}.$$

Now, h is determined by the free convection relations of Chapter 9. Generally, the diameter of a thermocouple wire is quite small, and the average temperature difference between the wire and the ambient air would not be expected to be very great. Hence, one may estimate h by assuming that the Grashof, or Rayleigh, number for the free convection around the wire is approximately zero. In such an instance, the correlation for free convection on a horizontal cylinder given in Eq. (9.42) yields $\mathrm{Nu}_D = 0.36 = h(2R)/k_a$, k_a being the thermal conductivity of the ambient air. Thus, the above relation gives the estimated error to be

$$t_s - t_t = \frac{0.47(k_a k_t/k_s^2)^{1/2}}{1 + 0.47(k_a k_t/k_s^2)^{1/2}} (t_s - t_a). \tag{13.11}$$

This relation gives an approximation to the expected error in the surface temperature indicated by a thermocouple when it is attached in the manner shown in Fig. 13.1(a). Equation (13.11) is only an approximation in view of the many simplifying assumptions that were made in its derivation, but it is valuable for the purpose of estimating the order of magnitude of the error likely to be encountered. It is interesting to note that the error is independent of the diameter of the wire. Also, the greater the thermal conductivity of the wire, k_t, the greater the error is, since heat will be more readily conducted away. Similarly, the smaller the value of the thermal conductivity of the surface material, k_s, the greater is the error, since the heat conducted away by the wire must be supplied from within the body. In some instances, the error involved may be quite significant, as the following example shows.

EXAMPLE 13.3

A furnace wall of chrome brick has a surface temperature of 200°C. The ambient air is at 40°C. What error may be expected to be indicated by a thermocouple made of constantan (60% Cu, 40% Ni) attached to the surface?

Solution. For the materials given, the tables of Appendix A give

$$k_a = 0.0271, \qquad k_t = 26, \qquad k_s = 2.32 \text{ W/m-°C.}$$

With $t_s = 200°C$ and $t_a = 40°C$, Eq. (13.11) gives the error to be

$$t_s - t_t = \frac{0.47[0.0271 \times 26/(2.32)^2]^{1/2}}{1 + 0.47[0.0271 \times 26/(2.32)^2]^{1/2}} (200 - 40)$$

$$= 23°C \ (42°F).$$

The estimate of error is seen to be rather significant. Materials were chosen to illustrate a "worst case." The wire has a high conductivity which allows it to withdraw heat readily from the surface. The surface has a low conductivity so that it has difficulty supplying the heat and the temperature is locally depressed rather dramatically. Had the surface been made of a better conductor (say iron with $k = 62$ W/m-°C), similar calculations give a much lower error:

$$t_s - t_t = 1.0°C \ (1.8°F). \qquad \blacksquare$$

The error in the reading of a thermocouple attached to a surface may be reduced by one of the schemes indicated in Fig. 13.1(b) or (c). Figure 13.1(b) shows the thermocouple attached to a thin disk of material of high thermal conductivity (say copper). The disk is then placed in contact with the surface. The arrangement is advantageous since it provides a large area at the surface from which the heat conducted away by the thermocouple is withdrawn, producing a less severe local depression in the surface temperature.

If the leads of the thermocouple are placed along the surface for a short distance before being led away, as shown in Fig. 13.1(c), the junction of the thermocouple does not lie at the point of departure of the leads from the surface—the point at which the greatest local depression in temperature is likely to occur.

13.4

THERMOMETER WELL ERRORS DUE TO CONDUCTION

The temperature of a fluid flowing in a closed conduit is often measured by the use of a thermometer or thermocouple placed in a thermometer well inserted into the fluid stream. Such an arrangement is indicated schematically in Fig. 13.2.

If the thermometer well is assumed to have good thermal contact with the pipe wall, at temperature t_s, then it may be treated as a spine of uniform cross section and finite length L. The thermometer or thermocouple placed in the well will be assumed to have perfect contact with the bottom of the well, so the indicated temperature, t_t, may be assumed to be that of the end of the spine.

The finite length spine was treated in Sec. 3.12, and when losses out of the end of the spine are neglected, Eq. (3.45) shows that the temperature of the end is given by

$$t_t - t_f = \frac{t_s - t_f}{\cosh ml},$$

(13.12)

$$m = \sqrt{\frac{hC}{kA}}.$$

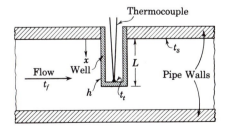
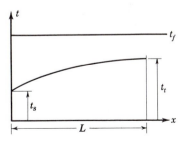

FIGURE 13.2

In Eq. (13.12) t_f is the fluid temperature—the desired quantity—so that $(t_t - t_f)$ is the error in the reading given by the thermocouple. Also, in Eq. (13.12) A is the cross-sectional area for heat flow in the thermometer well wall and k is the thermal conductivity of the well material. If one assumes that little convection takes place at the inner well surface (the well is usually filled with a stagnant liquid), C denotes the perimeter of the outer well surface, and h is the surface film coefficient there. Usually, a thermometer well is of a cylindrical shape, and with the pipe fluid flowing normal to it, correlations of the form given in Eqs. (8.41) and (8.42) of Sec. 8.6 may be used to estimate the outer surface heat transfer coefficient, h. These correlations are based on the mean film temperature (i.e., the mean between the fluid at t_f and the well surface). Thus, some representative average surface temperature must be found, and the average between t_s and t_t would be appropriate. Thus, an iterative calculation using the above correlations and Eq. (13.12) might be necessary.

The estimate of the error in the temperature reading given by Eq. (13.12) involves knowledge of the pipe surface temperature t_s. The error is directly proportional to the difference $(t_s - t_f)$, so the error may be reduced by insulating the outside pipe surface in the vicinity of the well in order to reduce this difference. Also, any means of making the product mL as great as possible, such as increasing L, will reduce the error.

13.5

RADIATION EFFECTS IN THE MEASUREMENT OF GAS TEMPERATURES

If the temperature of a high-temperature gas stream is to be measured by the insertion of a thermometer or thermocouple, the effects of the radiant exchange between the pipe walls and the temperature-sensing element may be of such an order of magnitude as to influence markedly the value indicated.

Conduction Effects Negligible. If the thermocouple (perhaps contained in a housing) is inserted into the gas stream in the manner illustrated in Fig. 13.3— with a significant length of the thermocouple parallel to the flow—then one might be able to neglect the effects of the heat which is conducted along the probe to the pipe wall. If this is so, any loss in heat by radiation from the thermocouple-sensing element is balanced by the gain in heat due to convective heat transfer between the fluid and the element. On a unit area basis, this balance

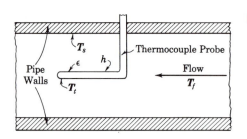

FIGURE 13.3

of heat gives

$$h_c(T_f - T_t) = \sigma\epsilon(T_t^4 - T_s^4). \qquad (13.13)$$

where T_f = fluid temperature, T_t = thermocouple temperature, T_s = pipe surface temperature, h_c = convective film coefficient, and ϵ = emissivity of thermocouple housing. The expression on the right side of Eq. (13.13) is based on Eq. (11.100) for the radiant exchange between a surface (the thermocouple) completely enclosed by a second, much longer, surface (the pipe walls).

The coefficient of convective heat transfer, h_c, is to be found by the methods of Chapters 7 through 9, depending on the geometry of the thermocouple housing. Since the correlations for convection are generally based on an average film temperature $[(T_t + T_f)/2$ in this case], it is necessary to perform a trial and error solution of Eq. (13.13) in order to obtain the error $(T_f - T_s)$.

EXAMPLE 13.4

High temperature air flows in a large duct with a velocity of 20 ft/s. The duct wall temperature is 300°F. A thermocouple probe (diameter = 0.5 in., ϵ = 0.5) is placed normal to the flow. If conduction effects are negligible, estimate the air temperature when the thermocouple reads 1000°F.

Solution. An iterative solution is necessary, but for the sake of brevity, only the final iteration is given here. From the given data, T_t = 1000°F = 1460°R, T_s = 300°F = 760°R.

If one *assumes* that T_f = 1300°F = 1760°R, the mean film temperature is $(1000 + 1300)/2$ = 1150°F and Table A.6 gives v = 3.840 ft²/hr, k = 0.03586 Btu/h-ft²-°F, Pr = 0.706. For v = 20 ft/s, D = 0.5 in.,

$$\mathrm{Re}_D = \frac{vD}{v} = 781.$$

Using Eq. (8.42) for flow normal to a cylinder,

$$\mathrm{Nu}_D = 0.3 + \frac{0.62\mathrm{Re}_D^{1/2}\mathrm{Pr}^{1/3}}{[1 + (0.4/\mathrm{Pr})^{2/3}]^{1/4}}\left[1 + \left(\frac{\mathrm{Re}_D}{2.82 \times 10^5}\right)^{5/8}\right]^{4/5}$$

$$= 13.8 = \frac{h_c D}{k}$$

$$h_c = 11.9 \text{ Btu/h-ft}^2\text{-}°F.$$

Equation (13.13) may be used to verify the assumed value of T_f:

$$11.9(T_f - 1000) = 0.1714 \times 0.5\left[\left(\frac{1460}{100}\right)^4 + \left(\frac{760}{100}\right)^4\right],$$

$$T_f = 1303°F \ (706°C).$$

This calculated value is sufficiently close to the assumed and may be taken as an estimate of the fluid temperature. It is seen that an error of about 300°F is observed to occur in the measurement of the gas temperature! ∎

The radiation error involved in the measurement of the temperature of a very hot gas may be reduced considerably by the use of radiation shields. The use of plane radiation shields was discussed in Sec. 11.6, and it was noted in that case that the use of one shield reduced the radiant energy exchange to one-half of its original, two shields reduced it to one-third, etc. The same sort of advantage can be deduced for a radiation shield around a thermocouple housing. If t_{sh} denotes the equilibrium temperature of a radiation shield placed around a thermocouple, and if one assumes that the shield is large compared to the thermocouple but small compared to the duct walls, the equilibrium temperature is given by

$$\sigma\epsilon_t(T_t^4 - T_{sh}^4) = \sigma\epsilon_{sh}(T_{sh}^4 - T_s^4).$$

If one can say that the emissivities of the thermocouple and pipe walls are practically the same,

$$T_{sh}^4 = \frac{T_s^4 + T_t^4}{2}.$$

Thus, Eq. (13.13) may be replaced by

$$h_c(T_f - T_t) = \sigma\epsilon(T_{sh}^4 - T_s^4), \tag{13.14}$$

$$= \frac{\sigma\epsilon}{2}(T_t^4 - T_s^4).$$

The error is reduced to one-half of its former value. It is not hard to show, as in the case of plane radiation shields, that if N shields are placed around the thermocouple probe, the error will be reduced to $1/(N + 1)$ of its value when no shield is used.

Conduction Effects Not Negligible. Section 13.4 considered the error involved in the use of a thermometer well in a gas stream when the effects of radiation were neglected. The discussion above considered the case in which conduction errors were negligible and radiant effects were dominant. This section will consider the case of the error likely to be involved in the measurement of the temperature of a fluid stream by means of a thermometer well when all three modes of heat transfer must be taken into account.

Figure 13.4 illustrates the simplified system assumed to represent the ther-

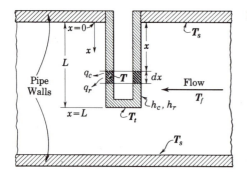

FIGURE 13.4

mometer well. The temperatures of the pipe surface, flowing gas, and thermometer well end are denoted by T_s, T_f, and T_t, respectively. By assuming no losses in the interior of the well, T_t will be the temperature indicated by the thermometer or thermocouple. The symbol h_c denotes the convective film coefficient on the surface of the well. If the well surface is modeled, for radiation purposes, as a small surface enclosed by a large one, the radiation coefficient at the surface may be taken from Eq. (11.100) as was done earlier in Sec. 13.2 and Eq. (13.6):

$$h_r = \sigma\epsilon(T_s^2 + T_f^2)(T_s + T_f). \tag{13.15}$$

Thus, the heat flow, per unit area of the well surface, is a combination of radiation and convection:

$$\frac{q_c + q_r}{A} = h_c(T - T_f) + h_r(T - T_s). \tag{13.16}$$

The symbol T denotes the local surface temperature at some point on the well surface. Here the well surface "sees" two environments for the transfer of heat: a fluid at T_f through the coefficient h_c, and the radiative environment at T_s through the coefficient h_r.

The above surface heat flux may be expressed in terms of a weighted equivalent environment temperature, T_e, defined as

$$T_e = \frac{h_r T_s + h_c T_f}{h_r + h_c}, \tag{13.17}$$

so that Eq. (13.16) becomes

$$\frac{q_c + q_r}{A} = (h_c + h_r)(T - T_e). \tag{13.18}$$

With these definitions, the thermometer well may be characterized as a spine of uniform cross section, A, and uniform perimeter, C, maintained at a base temperature T_s and exposed to an environment at T_e through a linear coefficient $h_c + h_r$. These conditions are identical to those considered in Sec. 3.12 for a convecting spine, so the following solution is immediately available from Eq. (3.45):

$$m = \sqrt{\frac{(h_c + h_r)C}{kA}}, \tag{13.19}$$

$$\frac{T - T_e}{T_s - T_e} = \frac{\cosh m(L - x)}{\cosh mL}. \tag{13.20}$$

In Eq. (13.20), L is the length of the well and x is the local distance away from the pipe wall. By taking $x = L$, one obtains T_t, the temperature observed by the thermocouple or thermometer inserted into the well:

$$\frac{T_t - T_e}{T_s - T_e} = \frac{1}{\cosh mL}. \tag{13.21}$$

Equation (13.21) may be used as it stands to estimate T_t and how this value differs from the correct fluid temperature, T_f, or one may introduce Eq. (13.17) into Eq. (13.21) to find the indicated error, $T_f - T_t$:

$$\frac{T_f - T_t}{T_f - T_s} = 1 - \frac{h_c}{h_r + h_c}\left(1 - \frac{1}{\cosh mL}\right). \qquad (13.22)$$

Given the well geometry and thermal properties, the flow conditions of the fluid, the fluid and surface temperatures (T_f and T_s) one may use Eq. (13.22) to estimate the error. It must be remembered that the parameter m has a different definition in this case as given by Eq. (13.19). An iterative solution, similar to that given in the preceding example, is usually required.

13.6

HEAT TRANSFER FROM RADIATING FINS

One of the most important problems of modern space technology is that of *thermal control*—the maintenance of a desired temperature level within a spacecraft (manned or unmanned) or satellite. Such spacecraft are subjected to heat inputs of both internal and external sources. Internal sources consist of electrically generated heat and metabolically generated heat, while the external sources consist of thermal irradiation from the sun, planetary bodies, etc. If a net gain in heat is experienced, some means of heat rejection must be provided in order that thermal control be maintained.

In outer space, the only means of rejecting heat is that of thermal radiation. For this purpose *heat rejection space radiators* have been developed. Many varieties of such systems have been developed (see Ref. 4). A typical space radiator configuration is illustrated in Fig. 13.5. Internally generated heat within

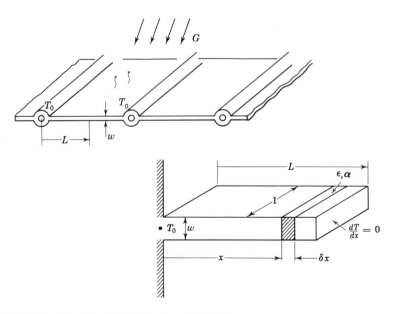

FIGURE 13.5. Heat rejection space radiator.

the spacecraft is conveyed to the radiator by a circulating coolant (e.g., a water-glycol mixture). The coolant circulates in tubes (usually connected in a parallel circuit) which are joined by finned surfaces. Heat from the coolant conducts into the fin and is dissipated by radiation to the surroudings. The finned surface may be subjected to an external irradiation from a source such as the sun. The basic question to be answered is: Given a radiating fin of known geometry, maintained at a known base (i.e., tube temperature) and subjected to a certain external radiation, what is the rate at which heat is dissipated by the fin to the environment?

The solution for one such situation will be given here. If the fluid in adjacent radiator tubes is at the same temperature, the fin may be represented by the model suggested in Fig. 13.5—a straight fin of length L and uniform thickness w maintained, at one end, at a known temperature T_0. The other end of the fin has the condition $dT/dx = 0$ imposed by symmetry. The fin material has thermal conductivity k, and the exposed surface has a total emissivity ϵ. The surface is exposed to an external irradiation G (perhaps solar) to which it exhibits a total absorptivity α. Since such radiators are often made an integral part of the space-craft skin, only one surface of the fin is considered to be radiatively active.

An approach similar to that developed in Sec. 3.12 for a fin with a convective boundary condition may be used here. If the conduction in the fin is taken to be one-dimensional, a heat balance taken on an element dx in length yields the following relation (for a unit depth):

$$-kw\frac{dT}{dx} = -kw\frac{dT}{dx} - kw\frac{d^2T}{dx^2}\,\delta x + \cdots + \epsilon\sigma T^4\,\delta x - \alpha G\,\delta x.$$

No convective loss at the fin surface is included. For $\delta x \to 0$, the following differential equation must be satisfied at each point:

$$\frac{d^2T}{dx^2} - \frac{\epsilon\sigma}{kw}\left(T^4 - \frac{\alpha G}{\epsilon\sigma}\right) = 0.$$

The nonhomogeneous term is recognized as the equilibrium temperature, or equivalent sink temperature, defined in Eq. (11.47). This temperature is the equilibrium temperature the surface would achieve if isolated in the irradiation G. Thus, with

$$T_s^4 = \frac{\alpha G}{\epsilon\sigma}, \tag{13.23}$$

the differential equation for the temperature distribution in the fin is

$$\frac{d^2T}{dx^2} - \frac{\epsilon\sigma}{kw}(T^4 - T_s^4) = 0. \tag{13.24}$$

The solution of Eq. (13.24) is best discussed in terms of the following di-mensionless variables:

$$\theta = \frac{T}{T_0},$$

$$\theta_s = \frac{T_s}{T_0},$$

$$\xi = \frac{x}{L},$$

(13.25)

$$\lambda = \frac{\epsilon\sigma T_0^3 L^2}{kw}.$$

Introduction of these definitions into Eq. (13.24) yields

$$\frac{d^2\theta}{d\xi^2} - \lambda(\theta^4 - \theta_s^4) = 0.$$

(13.26)

The boundary conditions to be satisfied are

At $\xi = 0 (x = 0)$: $T = T_0, \theta = 1.$

At $\xi = 1 (x = L)$: $dT/dx = d\theta/d\xi = 0.$

(13.27)

The nonlinearity of Eq. (13.26) makes its solution in closed form impossible; although the first integral may readily be written

$$\frac{d\theta}{d\xi} = -\sqrt{\tfrac{2}{5}\lambda(\theta^5 - 5\theta\theta_s^4) + C},$$

(13.28)

C being a constant of integration.

A complete solution of the above system, Eqs. (13.27) and (13.28), may be accomplished only by numerical means. Rather than seek the solution for the temperature distribution, the principal item of interest is the heat dissipated by the fin:

$$q = -kw\left(\frac{dT}{dx}\right)_{x=0} = -\frac{kwT_0}{L}\left(\frac{d\theta}{d\xi}\right)_{\xi=0}.$$

(13.29)

As in the case of the convective fins of Chapter 3, the heat dissipation is best expressed in terms of a fin effectiveness (see Sec. 3.17), which is the ratio of q to the heat that *would* be dissipated if the whole fin was maintained at T_0:

$$\kappa = \frac{q}{(\epsilon\sigma T_0^4 - \alpha G)L}$$

(13.30)

$$= \frac{q}{\epsilon\sigma L T_0^4(1 - \theta_s^4)}.$$

Thus, when eqs. (13.29) and (13.30) are combined,

$$\kappa = -\frac{1}{\lambda(1 - \theta_s^4)}\left(\frac{d\theta}{d\xi}\right)_{\xi=0}.$$

(13.31)

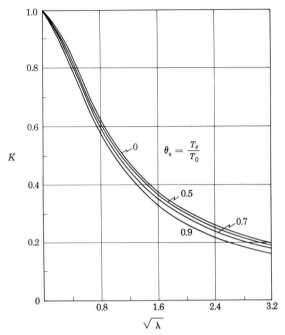

When Eq. (13.28) is solved, subject to the conditions of Eqs. (13.27), the results may be applied to Eq. (13.31) to yield the fin effectiveness in the following form:

$$\kappa = fn(\theta_s, \lambda). \tag{13.32}$$

The parameter θ_s measures the relative magnitude of the fin base temperature and the irradiation of the environment, while λ represents a combination of the thermal parameters of the fin. Lieblein, in Ref. 5, performed the numerical integration referred to above. The results are shown graphically in Fig. 13.6. By use of Fig. 13.6, along with the definitions of κ, θ_s, and λ given in Eqs. (13.23), (13.25), and (13.30), the heat rejected by the radiator may be determined.

EXAMPLE 13.5

A space radiator is constructed of fins of 3-in. half-width and thickness 0.0150 in. The fin material is aluminum alloy with $k = 65$ Btu/h-ft-°F. A surface coating on the fin has $\epsilon = 0.5$, $\alpha = 0.1$. If the fin base is maintained at 120°F, find the heat rejected per foot of width when the external irradiation is 100 Btu/h-ft^2.

Solution. For the data given,

$$T_s = \left(\frac{0.1 \times 100}{0.5 \times 0.1714 \times 10^8} \right)^{1/4}$$

$$= 328.67°R,$$

$$\theta_s = \frac{328.67}{580} = 0.567,$$

$$\lambda = \frac{0.5 \times 0.1714 \times 10^{-8}(120 + 460)^3(3/12)^2}{65 \times 0.150/12}$$

$$= 0.1286.$$

For $\theta_0 = 0.567$ and $\lambda = 0.1286$, Fig. 13.6 yields

$$\kappa = 0.82.$$

Thus, Eq. (13.30) gives, for a 1-ft width:

$$q = 0.82 \times 0.5 \times 0.1714 \times 10^{-8} \times (3/12) \times (580)^4[1 - (0.567)^4]$$
$$= 17.83 \ \text{Btu/h-ft} \ (17.14 \ \text{W/m}). \qquad \blacksquare$$

13.7

THE FLAT PLATE SOLAR COLLECTOR

An application of considerable current interest which involves all three modes of heat transfer is the collection of solar energy for commercial or domestic uses. One form of device used for solar collection is the flat plate solar collector such as depicted in Fig. 13.7. Such devices usually consist of a flat absorber

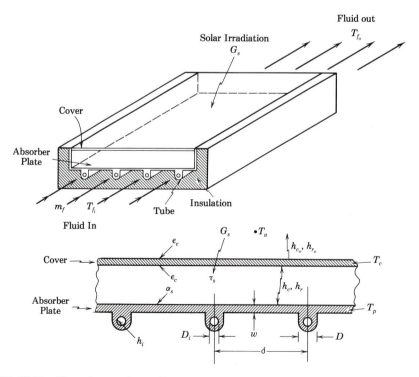

FIGURE 13.7. Flat plate solar collector.

plate on which solar energy falls and is partly absorbed. Some of the absorbed energy is reradiated by the absorber plate and that which is not is removed from the surface by a coolant fluid in tubes attached to the plate.

Some energy is also lost from the absorbing surface by convection to whatever fluid surrounds the plate, usually the ambient air. In order to reduce the energy lost both by reradiation from the absorber and by convection to the ambient, most solar collectors are also provided with a transparent cover, often glass, which is opaque to the long-wave re-emitted radiation, thus returning it to the absorber. The cover also serves to protect the absorbing surface from air movement, thus reducing the convective loss there to that by free convection. Some advanced designs provide an evacuated space between the absorbing plate and the cover to further reduce convective losses.

Heat removal from the absorbing surface is accomplished by circulating a fluid in tubes attached to the surface as suggested in Fig. 13.7. The tubes may be attached to the back side of the plate, as shown, or may be made integrally with the plate in a form quite analogous to that illustrated in Fig. 13.6 for a space radiator. The analysis differs only slightly for the two configurations, so that shown in Fig. 13.7 will be used here. In this instance, then, the heat is transferred to the fluid by conduction in the plate surface between the tubes. For the geometry shown in Fig. 13.7, the tubes act as heat sinks with fins attached. The fin length is $(\delta - D)/2$ if δ is the on-center tube spacing and D is the outside tube diameter.

Let the solar irradiation falling on the collector be denoted by G_s. This amount depends on the orientation of the collector with respect to the sun, the time of year and day, the longitude and latitude on the earth's surface where the collector is located, etc. Let it be assumed that this quantity has been determined and is known. The cover of the collector should be as transparent to the incident solar radiation as possible; although some energy may be lost to the environment by reflection from the cover. The solar radiation that passes through the cover is mostly absorbed at the absorbing surface, but some may be reflected there also. Let $(\tau_s \alpha_s)$ represent the apparent solar transmissivity–absorbtivity product for the cover–plate array that accounts for all these effects. Typically, $\tau_s \approx 0.9$ and α_s is close to the solar absorbtivity of the collector plate. In any event $(\tau_s \alpha_s)G_s$ represents the solar energy absorbed at the absorbing surface.

In addition to the solar beam just discussed, the absorber plate and cover exchange radiation which results from their own emission. The cover radiates to the environment, and both the cover and the plate are subjected to convective heat transfer. All of these energy flows interact to establish equilibrium temperatures for the cover and the absorbing plate when the tube fluid temperature is known. The tube fluid, of course, varies in temperature as it proceeds through the collector and one wishes to find, among other things, the temperature of the fluid leaving the collector (given its inlet) so that the amount of useful energy absorbed can be found. The variation of the fluid temperature along the length of the collector implies a variation in the temperatures of the cover and the absorber plate, also in the longitudinal direction.

The analysis to follow will attempt to determine the useful heat collected (i.e., find the outlet fluid temperature) when one is given the inlet fluid temperature, the fluid flow rate, the various geometrical and thermal parameters for the collector (i.e., cover, absorber, tubes), and the ambient environment and solar irradiation. In this analysis it will be assumed that the ambient air is at a temperature, T_a, and that the radiative environment to which the cover radiates is at the same temperature. Further, it will be assumed that there exist representative average temperatures for the cover and the absorber plate, T_c and T_p, respectively, which take into account any longitudinal variation such as discussed above. These representative temperatures, T_c and T_p, will be assumed known for the present; however, later it will be pointed out how they may be determined. The analysis here will also assume that the collector is well insulated on the back side of the absorber plate so that no heat is lost to the environment there.

There are two convective exchanges to be accounted for—that at the outer cover surface and that between the inner cover surface and the absorber plate. The coefficient at the outer cover surface will be denoted by h_{co}, and may be determined when T_a, T_c, and the airflow conditions past the cover are known. The coefficient between the inner cover surface and the absorber plate will be denoted by h_c and is found by application of the air-space correlations of Eqs. (11.48) through (11.53), involving knowledge of T_p and T_c. Often solar collectors are tilted toward the sun, so the inclined air-space correlations are particularly important.

Radiation coefficients must also be found for the outer cover and between the cover and the absorber. For the outer cover one may once again use the result of Eq. (11.100) to obtain the radiation coefficient there, h_{ro}, as

$$h_{ro} = \sigma \epsilon_c (T_c^2 + T_a^2)(T_c + T_a), \tag{13.33}$$

in which ϵ_c is the emissivity of the cover plate material. The radiation exchange, due to surface emission, between the inner surface of the cover and the absorber plate is normally in the low temperature long wave portion of the spectrum to which the cover plate is thermally opaque. Hence, the radiation coefficient used in Eq. (13.7) may be applied here:

$$h_r = \frac{\sigma (T_p^2 + T_c^2)(T_p + T_c)}{(1/\epsilon_c) + (1/\epsilon_p) - 1}. \tag{13.34}$$

In Eq. (13.33) ϵ_p represents the emissivity of the absorber plate and is generally different from its solar absorptivity, α_s above, if the surface is not gray.

With the above four coefficients known (h_{co}, h_c, h_{ro}, h_r), an effective cover–absorber overall coefficient, U_c, may be found,

$$\frac{1}{U_c} = \frac{1}{h_{co} + h_{ro}} + \frac{1}{h_c + h_r}, \tag{13.35}$$

and expresses the combined conductance due to convection and long-wave radiation between the absorber plate at T_p and the ambient environment at T_a. The total heat flow, per unit of collector surface area, is the sum of that due to the mechanisms just discussed and that absorbed from the solar beam. If q_p is the *net* heat flow to the absorber plate and A_p is the collector surface area (cover or absorber plate), then

$$\frac{q_p}{A_p} = U_c(T_a - T_p) + (\tau_s \alpha_s) G_s. \tag{13.36}$$

Equation (13.36) may be put into a more standard form by defining an *equivalent environment temperature* T_a' so that

$$T_a' = T_a + \frac{(\tau_s \alpha_s) G_s}{U_c}, \tag{13.37}$$

$$\frac{q_p}{A_p} = U_c(T_a' - T_p) \tag{13.38}$$

The heat flow in Eq. (13.36) or (13.38) must eventually go into the fluid circulating in the tubes. The solar collector may be characterized as a heat exchanger in which the "cold" fluid is the collecting fluid entering at T_{fi} and leaving at T_{fo}, and the "hot" fluid is one which remains at a constant temperature equal to the equivalent environment temperature, T_a'. Thus, Eq. (12.45) gives the fluid temperature rise to be

$$\frac{T_{fo} - T_{fi}}{T_a' - T_{fi}} = 1 - \exp\left(-\frac{\overline{U}A_p}{\dot{m}_f c_{pf}}\right) \tag{13.39}$$

in which $\overline{U}$ is some overall coefficient for the collector (between the fluid and the environment) to be discussed shortly and $\dot{m}_f$ and c_{pf} are the flow rate and specific heat of the collector fluid.

In Eq. (13.39), $\overline{U}$ is the overall coefficient for the collector between the fluid and the environment at T_a'. If h_i is the heat transfer coefficient applying at the inside tube surface (as given by the appropriate forced convection correlation of Chapter 8) and A_i is the inside tube surface area ($A_i = \pi D_i L$, L being the total length of the tubes), then the principles of Sec. 3.18 give

$$\frac{1}{\overline{U}A_p} = \frac{1}{h_i A_i} + \frac{1}{A_p U_c \eta} \tag{13.40}$$

if the tube wall resistance is neglected and η represents the effectiveness of the plate surface (i.e., "fins") between adjacent tubes. In order to avoid interrupting the current line of analysis, detailed discussion of the effectiveness η will be delayed until later, so let it be treated as known for the present.

With $\overline{U}$ determined from Eq. (13.40), the heat gained by the fluid (i.e., the total useful heat collected) may be determined from

$$q = \dot{m}_f c_{pf}(T_{fo} - T_{fi}). \tag{13.41}$$

Substitution of Eqs. (13.39) and (13.37) into (13.41) gives

$$q = \frac{\dot{m}_f c_{pf}}{U_c}[(\tau_s \alpha_s) G_s - U_c(T_{fi} - T_a)]\left[1 - \exp\left(-\frac{\overline{U}A_p}{\dot{m}_f c_{pf}}\right)\right] \tag{13.42}$$

Equation (13.42) constitutes the desired result. It expresses the useful heat absorbed in a collector by the circulating fluid when the fluid inlet conditions are known, the collector geometry is specified, and the environment (solar and ambient) is known. One proceeds as follows: Values for the average cover and absorber plate temperatures are assumed, i.e., T_c and T_p. The convective coefficients h_{co} and h_c are found using these temperatures and the applicable convection correlations. The radiation coefficients are found from Eqs. (13.33) and (13.34) so that the cover–absorber coefficient U_c can be found from Eq. (13.35). Knowledge of the flow conditions in the collector tubes then permits calculation of the overall coefficient $\overline{U}$ of the collector from Eq. (13.40), and Eq. (13.42) then yields their desired heat collected. The assumed value of T_p is verified by substituting the just-found q into Eq. (13.38). Then the assumed value of T_c is verified by use of the identity $(h_{co} + h_{ro})(T_c - T_a) = (h_c + h_r)(T_p - T_c)$. If necessary T_p and T_c can be revised and the entire calculation repeated.

The above calculation requires the determination of the effectiveness of the absorber plate surface. This is found by first finding the fin effectiveness for the portion of the plate between the tubes. Equation (3.60) shows this to be

$$\kappa = \frac{\tanh ml'}{ml'}, \qquad (13.43)$$

where l' is the apparent fin length between tubes and $m = \sqrt{hC/kA}$. For the geometry of Fig. 13.7, and the fact that only one side of the plate is subjected to heat transfer, one has for Eq. (13.43):

$$ml' = \sqrt{\left(\frac{U_c}{kw}\right)} \frac{\delta - D}{2}. \qquad (13.44)$$

In Eq. (13.44) δ is the tube spacing, D the tube outside diameter, and w the plate thickness. The surface effectiveness, η, is then by Eq. (3.64):

$$\eta = 1 - \frac{A_f}{A_p}(1 - \kappa),$$

$$= 1 - \left(1 - \frac{D}{\delta}\right)(1 - \kappa), \qquad (13.45)$$

$$= \frac{D}{\delta} + \kappa\left(1 - \frac{D}{\delta}\right).$$

Practitioners in solar energy collectors often rewrite the formulation given in the foregoing by defining the *collector efficiency factor*, F', and the *collector heat removal factor*, F_R, as follows:

$$F' = \frac{\overline{U}}{U_c} = \frac{1/U_c}{\dfrac{A_p}{A_i h_i} + \dfrac{1}{U_c \eta}} = \frac{1/U_c}{\delta\left[\dfrac{1}{\pi D_i h_i} + \dfrac{1}{U_c[D + \kappa(\delta - D)]}\right]}, \qquad (13.46)$$

$$F_R = \frac{\dot{m}_f c_{p_f}}{A_p U_c}\left[1 - \exp\left(-\frac{A_p U_c F'}{\dot{m}_f c_{p_f}}\right)\right] \qquad (13.47)$$

In terms of these expressions, Eq. (13.42) for the useful heat collected is

$$q = A_p F_R[(\tau_s \alpha_s) G_s - U_c(T_{fi} - T_a)]. \qquad (13.48)$$

The formulation given by Eqs. (13.46) through (13.48) is exactly the same as that leading to Eq. (13.42).

13.8

THE HEAT PIPE

The heat pipe is a recent (Refs. 6 and 7) development which utilizes the high latent heat of evaporation, or condensation, together with the phenomenon of capillary pumping to transfer very high heat fluxes without the addition of external work. As a result, the device, seemingly inert, can transport heat fluxes many times that which can be conducted by solid metal bars of equal cross section.

The basic schematic diagram of a typical heat pipe is shown in Fig. 13.8. Structurally, the device consists of a closed vessel, or pipe, of circular cross section, a capillary wick, and a transport fluid. The wick materials are usually wire screen, woven cloth, ceramic materials, or narrow grooves machined into the pipe wall. At one end, the evaporator, heat is added at some temperature,

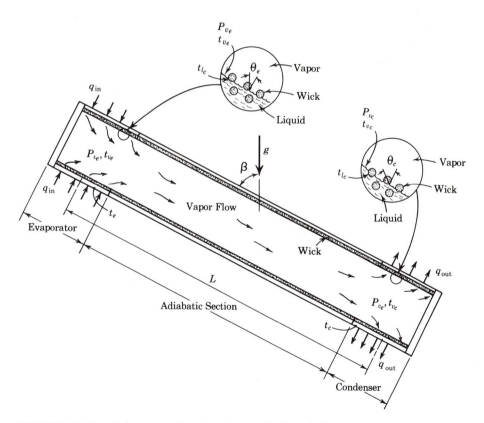

FIGURE 13.8.   Schematic drawing of a typical heat pipe.

vaporizing the liquid fluid from the wick material. The vaporized fluid flows down the central core of the pipe to the somewhat-cooler, lower-pressure end, termed the *condenser*. At the condenser end, the vapor is condensed back to a liquid with the release of the associated latent heat. The condensed liquid is ''pumped'' back to the evaporator section by the action of surface tension in the capillary structure of the wick material. The evaporator and condenser sections are normally separated by an adiabatic section. Because of the relatively low pressure drop in the flowing vapor, the difference between the evaporator temperature (where the heat is added) and then condensor temperature (when the heat is rejected) is relatively small—at least compared with the latent heat transported. These facts lead to a very high apparent conductivity between the two ends of the pipe.

Apparently, the nature of the working fluid has a significant effect on the performance of a heat pipe, as do the properties of the wick material and the relative orientation of the pipe in the gravity field. Working fluids such as water, ammonia, alcohol, molten metals, and nitrogen are used—the choice being based, in part, on the desired working temperature of the device.

In space applications in which no gravity forces act, the heat pipe may be used to transport heat over relatively large distances by lengthening the adiabatic section.

Some of the governing parameters of heat pipe performance may be obtained in the following simplified analysis. The basic transport mechanism involved is that of capillary pumping in the wick. The driving potential for this process is the capillary pressure difference which results from the surface tension at a vapor–liquid interface. Figure 13.9 shows the basic physical mechanism. By assuming that the vapor–liquid interface occurs in a circular tube of radius, r; that the liquid wets the tube wall with a contact angle θ; and that the surface

FIGURE 13.9

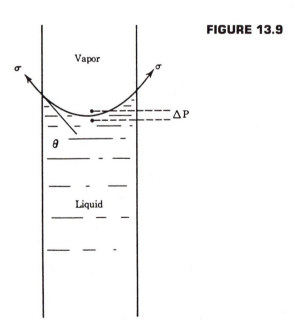

tension at the interface is σ, a simple balance of forces shows that the pressure of the vapor exceeds that of the liquid by an an amount

$$\Delta P = \frac{2\pi r\sigma \cos \theta}{\pi r^2}$$

(13.49)

$$= \frac{2\sigma \cos \theta}{r}.$$

Using the notation of Fig. 13.8, one finds that the pressure of the liquid at the condensing interface in the condenser, P_{l_c}, exceeds that at the evaporative interface in the evaporator, P_{l_e}, by an amount determined by the difference in the vapor pressure (P_{v_c} and P_{v_e}), the difference in the interfacial pressures, and the hydrostatic head imposed by gravity (if any):

$$P_{l_c} - P_{l_e} = (P_{v_c} - P_{v_e}) - (\Delta P_c - \Delta P_e) - g\rho_l L \cos \beta$$

(13.50)

$$= (P_{v_c} - P_{v_e}) + 2\sigma \left(\frac{\cos \theta_e}{r} - \frac{\cos \theta_c}{r} \right) - g\rho_l L \cos \beta.$$

In Eq. (13.50) the subscripts l and v denote liquid and vapor and the subscripts c and e denote condenser and evaporator. The angle β specifies the orientation of the pipe relative to the gravity vector and, for simplicity, L denotes a mean distance between the condenser and evaporator sections. It has been presumed that the wick material can be represented as being composed of circular capillaries of radius r. The symbols θ_e and θ_c represent the liquid contact angles in the evaporator and condenser. Also in Eq. (13.50) it has been presumed that the evaporator is located higher in the gravity field than the condenser—hence the negative sign on the last term.

The pressure difference given by Eq. (13.50) is the driving force to transport the liquid from the condenser back to the evaporator through the wick material. The wick, naturally, offers a resistance to this flow. Darcy's law may be used in this instance:

$$\dot{m} = (P_{l_c} - P_{l_e}) \frac{\rho_l A K}{\mu_l L},$$

in which $\dot{m}$ is the mass flow rate of liquid, A is the cross-sectional area of wick material available for the liquid flow, and K is the permeability of the wick. Thus, the flow rate in the heat pipe is

$$\dot{m} = \frac{AK}{L} \frac{\sigma\rho_l}{\mu_l} \left[\frac{P_{v_c} - P_{v_e}}{\sigma} + \frac{2}{r} (\cos \theta_e - \cos \theta_c) - \frac{g\rho_l L\cos \beta}{\sigma} \right].$$

For many applications, certainly adequate for this simple analysis, one may neglect the difference between the vapor pressures at each end of the pipe ($P_{v_c} \approx P_{v_e}$). Thus, the mass flow is

$$\dot{m} = \frac{AK}{L} \frac{\sigma\rho_l}{\mu_l} \left[\frac{2}{r} (\cos \theta_e - \cos \theta_c) - \frac{g\rho_l L}{\sigma} \cos \beta \right],$$

(13.51)

and the heat flux between the condenser and the evaporator is

$$q = \frac{AK}{L} \frac{\sigma \rho_l h_{fg}}{\mu_l} \left[\frac{2}{r} (\cos \theta_e - \cos \theta_c) - \frac{g\rho_l L}{\sigma} \cos \beta \right], \qquad (13.52)$$

where h_{fg} is the latent heat of vaporization.

In order to produce a positive flow rate, and thus heat flux, $(\cos \theta_e - \cos \theta_c)$ must be positive. As the power input, q, to the heat pipe increases and the rates of vaporization and condensation increase, the apparent contact angles in the evaporator and condenser must change so that $(\cos \theta_e - \cos \theta_c)$ becomes more positive. The maximum flow rate and maximum heat flux are given when $\cos \theta_e \rightarrow 1$ and $\cos \theta_c \rightarrow 0$:

$$q_{max} = \frac{2AK}{rL} \frac{\sigma \rho_l h_{fg}}{\mu_l} \left(1 - \frac{g\rho_l Lr}{2\sigma} \cos \beta \right). \qquad (13.53)$$

Although Eq. (13.53) may be lacking in some respects for the prediction of actual heat pipe performance, it does permit the identification of significant parameters. It is seen that q_{max} is the product of three factors. The first, $2AK/rL$, is a grouping of wick properties. The third, $g\rho_l Lr \cos \beta / 2\sigma$, is a grouping of wick and liquid properties which represents the relative importance of gravitational and surface tension forces. In space, where $g = 0$, this is 0. The second term is quite significant in that it is a grouping of liquid properties called the *liquid transport factor:*

$$N_l = \frac{\rho_l h_{fg} \sigma}{\mu_l}.$$

The designer wishes to choose this factor to be as large as possible. It is limited on one extreme by the liquid freezing point, where $\mu_l \rightarrow \infty$, and the other extreme by the critical point, where $h_{fg} \rightarrow 0$. Figure 13.10 presents values of N_l for some typical heat pipe fluids. It is apparent that the desired operating range temperature has a profound effect in the choice of fluid to be used. Also, it may be noticed that the heat pipes operating at high temperatures using, say,

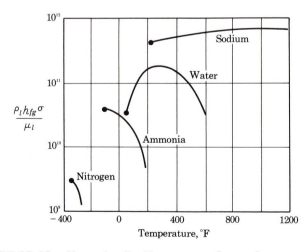

FIGURE 13.10. Heat pipe liquid transport factor for several fluids.

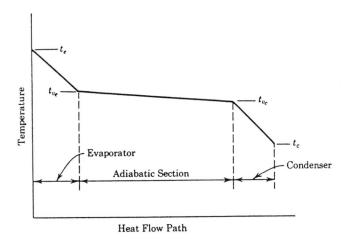

FIGURE 13.11.   Temperature profile of a typical heat pipe.

liquid metals, may transport heat fluxes of several orders of magnitude greater than those at lower temperatures.

The maximum limitation on heat flux given by Eq. (13.53) was imposed by the maximum pumping capability of the wick material. Other considerations may override this limitation. One is associated with the fact that in practice a finite temperature difference is required between the evaporator, T_{v_e}, and the outside evaporator wall, T_e. This difference is required to drive the heat from the heat source into the heat pipe. A typical temperature distribution through a heat pipe is shown in Fig. 13.11. The overall temperature drop is seen to exceed considerably the vapor temperature drop. For high heat loads the required difference, $t_e - t_{v_e}$, may be large enough that local superheating of the liquid will occur within the wick material. If this superheating is great enough, nucleate boiling may be initiated within the wick with a subsequent degradation of its pumping capability. Thus, the heat pipe may become "boiling-limited" rather than "pumping-limited," as discussed above.

REFERENCES

1. MARKS, L. S., ed., *Mechanical Engineers' Handbook,* 8th ed., New York, McGraw-Hill, 1978.

2. *ASHRAE Handbook of Fundamentals,* Atlanta, Am. Soc. Heating, Refrig., Air-Cond. Engrs., 1981.

3. GRÖBER, H., and S. ERK, *Die Grundgesetze der Wärmeübertragung,* 3rd ed., rev. by U. Grigull, Berlin, Springer-Verlag, 1961.

4. MACKAY, D. B., *Design of Space Power Plants,* Englewood Cliffs, N.J., Prentice-Hall, 1963.

5. LIEBLEIN, S., "Analysis of Temperature Distribution and Radiant Heat Transfer Along a Rectangular Fin of Uniform Thickness," *NASA Tech. Note D-196,* Washington, D.C., Nov. 1959.

6. COTTER, T. P., "Theory of Heat Pipes," *Los Alamos Sci. Lab. Rep. LA-3246-MS,* Los Alamos, N.M., 1965.

7. Feldman, K. T., and G. H. Whiting, "The Heat Pipe," *Mech. Eng.*, Feb. 1967, p. 30.

PROBLEMS

13.1 A bare, oxidized, wrought iron pipe (8-in. schedule 40) is placed horizontally in still air at 20°C. Its surface temperature is 95°C. Find the combined heat transfer coefficient due to convection and radiation, and find the heat loss from 30 m of the pipe.

13.2 Repeat Prob. 13.1 if the pipe temperature raised to 465°C.

13.3 Steam at 800°F, 500 psia flows through a 6-in. schedule 40 horizontal steel pipe with a velocity of 10 ft/s. The pipe is covered with 1 in. of 85% magnesia insulation. The emissivity of the outer insulation surface is 0.75. The pipe is located in a large room where the ambient temperature of the walls and still air is 70°F. Accounting for both convective and radiative losses, find the temperature of the outer insulation surface and the overall heat transfer coefficient.

13.4 Steam at 4000 kN/m², 450°C, flows with a velocity of 7.5 m/s through a 6-in. schedule 40 pipe which is covered with 3.5 cm of 85% magnesia insulation (emissivity = 0.8). The pipe is located horizontally in a large room in which the ambient air and walls are at 25°C. Accounting for both convection and radiation, find the temperature of the outer insulation surface, the overall heat transfer coefficient, and the heat loss, per meter of pipe. Compare the results with those found in Prob. 12.1, in which radiation was neglected.

13.5 For the oil flowing in a tube described in Prob. 12.8, assume that the brass tube is oxidized with $\epsilon = 0.3$ and that the radiative environment is also at 70°F. Repeat the problem, accounting for the combined effects of convection and radiation, and compare the results.

13.6 A 100-watt light bulb delivers about 10% of its energy in the visible spectrum. This energy and about 2% of the remaining radiation from the filament passes through the glass of the bulb. Make an engineering estimate of the surface temperature of such a light bulb.

13.7 A vertical air space is 2 m high and 6 cm wide. One bounding surface is maintained at 35°C and the other at 15°C. The emissivity of the two surfaces is 0.65. Find the heat flux through the air space if (a) radiation is neglected, and all heat transfer is by convection; (b) the space is evacuated and *only* radiation takes place; and (c) both the effects of radiation and convection are accounted for.

13.8 The air space in the wall of a house is approximately 3.75 in. wide and 8 ft high. The outer surface of the air space is at 95°F and has an emissivity of 0.8. The inner surface is at 70°F. Find the heat flux through the wall by combined convection and radiation if the emissivity of the inner surface is (a) 0.8 or (b) 0.2.

13.9   The temperature of a surface at 300°C is to be measured with a thermocouple extending horizontally into ambient air at 25°C. Find the magnitude of the error of the reading indicated by the thermocouple if (a) the surface is made of iron and the thermocouple is made of (1) constantan or (2) iron, and (b) the surface is made of magnisite brick and the thermocouple is made of (1) constantan or (2) iron.

13.10   A thermocouple is attached to a steel surface and extends horizontally into still air at 75°F. If the thermocouple is made of constantan and indicates a temperature of 350°F, what is the temperature of the surface?

13.11   A thermometer well (0.6 cm O.D., 0.3 cm I.D.) is made of brass and extends into a pipe normal to the direction of flow. The pipe has an inside diameter of 15 cm and carries air at 315°C, 700 kN/m², flowing with a velocity of 4.6 m/s. The thermometer well is 10 cm long and the pipe wall temperature is 175°C. Neglecting the effects of radiation, find the temperature indicated by a thermometer in contact with the bottom of the well.

13.12   A thermometer well (0.25 in O.D., 0.125 in. I.D.) is made of brass and extends a length of 6 in. normal to the direction of flow in an 8-in. schedule 40 steel pipe. The pipe carries air at 20 psia, flowing at the rate of 1000 lb$_m$/h. The pipe is bare and is located horizontally in a room of atmospheric air at 100°F. Neglecting the effects of radiation, estimate the temperature of the air flowing in the pipe if a thermometer inserted in the well reads 700°F.

13.13   A bare thermocouple ($\epsilon = 0.8$) is 0.25 cm in diameter, and it is placed in a flue duct carrying combustion gases (approximated as air) at 340°C. The duct walls are at 120°C. The thermocouple is placed normal to the stream in the duct, but is sufficiently long that conduction may be neglected. The gas in the duct is at atmospheric pressure and flows with a velocity of 7.6 m/s. What error in the thermocouple reading results?

13.14   A glass thermometer 0.25 in. in diameter is inserted normal to the flow of air in a duct. The thermometer extends for a distance of 6 in. into the duct, which is 1 ft in diameter and has a wall surface temperature of 100°F. The air in the duct is at 14.7 psia and flows at a velocity of 50 ft/s. If the thermometer reads 350°F, estimate the air temperature.

13.15   A thermocouple well (0.6 cm O.D., 0.3 cm I.D.) is made of steel ($\epsilon = 0.5$) and extends 15 cm into a 30-cm-diameter duct, normal to the flow. The duct carries atmospheric air flowing with a velocity of 6 m/s. The duct is also steel with $\epsilon = 0.5$ and has a surface temperature of 95°C. The thermocouple inserted into the well reads 315°C. Accounting for conduction effects in the well, estimate the air temperature if (a) radiation is neglected, and (b) radiation is not neglected.

13.16 Verify the algebra leading from Eq. (13.16) to Eq. (13.18) and the algebra leading from Eq. (13.21) to Eq. (13.22).

13.17 Verify that Eq. (13.28) is a first integral of Eq. (13.26).

13.18 Repeat Example 13.5 for the case of a zero-irradiation environment, all other data remaining the same.

13.19 For the radiating fin described in Sec. 13.6, if the length L becomes very great, one might assume that $\theta_L \to \theta_s$ as well as $(d\theta/d\xi)_L = 0$. Under this circumstance, show that the effectiveness is

$$ K_\infty = \frac{\left[\dfrac{2}{5\lambda}(1 - 5\theta_s^4 + 4\theta_s^5) \right]^{1/2}}{1 - \theta_s^4}. $$

13.20 A flat plate solar collector panel, 2 m × 1 m, consists of a lower absorbing plate at 70°C with a glass cover plate at 30°C. The air in the space between the surfaces is at atmospheric pressure and the surfaces are spaced 5 cm apart. The 2 m dimension of the collector is inclined at an angle of 60° with the horizontal. The long-wave emissivity of the glass cover plate is 0.9 on both sides and that for the absorber surface is 0.2. The ambient air is at 20°C and this is also the temperature of the radiative environment. Wind blowing over the collector cover produces a convective heat transfer coefficient of 17 W/m²-°C on the outside of that surface. Find the cover–absorber overall coefficient, U_c.

13.21 A flat solar collector, such as depicted in Fig. 13.7, is 1.05 m wide and 3 m long in the direction of flow. The absorber plate and tubes are made of aluminum. The tubes are 1.2 cm O.D., 1.1 cm I.D., spaced 15 cm on centers, and the absorber plate is 0.05 cm thick. The cover–absorber overall heat transfer coefficient is known to be $U_c = 6$ W/m²-°C. The cover–absorber combination has an apparent solar transmissivity–absorptivity product of $(\tau_s\alpha_s) = 0.81$. Water enters the tubes at 57°C and flows at the rate of 0.02 kg/s with an inside tube heat transfer coefficient of 1200 W/m²-°C. The solar irradiation incident on the collector is $G_s = 730$ W/m² and the ambient air temperature is 20°C. Find (a) the water outlet temperature; (b) the useful energy collected; (c) the collector efficiency factor, F'; and (d) the heat removal factor, F_R.

Tables and Charts of Thermal Properties of Substances

TABLE A.1 Thermal Properties of Metals*

Metal	Properties at 20°C				Thermal Conductivity, k, W/m-°C									
	ρ kg/m³	c_p kJ/kg-°C	k W/m-°C	$\alpha \times 10^5$ m²/s	−100°C	0°C	100°C	200°C	300°C	400°C	600°C	800°C	1000°C	1200°C
Aluminum														
Pure	2,707	0.896	204	8.418	215	202	206	215	228	249				
Al-Cu (Duralumin) 94–96 Al, 3–5 Cu, trace Mg	2,787	0.883	164	6.676	126	159	182	194						
Al-Mg (Hydronalium) 91–95 Al, 5–9 Mg	2,611	0.904	112	4.764	93	109	125	142						
Al-Si (Silumin) 87 Al, 13 Si	2,659	0.871	164	7.099	149	163	175	185						
Al-Si (Silumin, copper bearing) 86.5 Al, 12.5 Si, 1 Cu	2,659	0.867	137	5.933	119	137	144	152	161					
Al-Si (Alusil) 78–80 Al, 20–22 Si	2,627	0.854	161	7.172	144	157	168	175	178					
Al-Mg-Si, 97 Al, 1 Mg, 1 Si, 1 Mn	2,707	0.892	177	7.311		175	189	204						
Lead	11,373	0.130	35	2.343	36.9	35.1	33.4	31.5	29.8					
Iron														
Pure	7,897	0.452	73	2.026	87	73	67	62	55	48	40	36	35	36
Wrought iron (C < 0.5%)	7,849	0.46	59	1.626		59	57	52	48	45	36	33	33	33
Cast iron (C ≈ 4%)	7,272	0.42	52	1.703										
Steel (C max. ≈ 1.5%)														
Carbon steel, C ≈ 0.5%	7,833	0.465	54	1.474		55	52	48	45	42	35	31	29	31
1.0%	7,801	0.473	43	1.172		43	43	42	40	36	33	29	28	29
1.5%	7,753	0.486	36	0.970		36	36	36	35	33	31	28	28	29
Nickel steel, Ni ≈ 0%	7,897	0.452	73	2.026	87	73	67	62	55	48	40	36	35	36
10%	7,945	0.46	26	0.720										
20%	7,993	0.46	19	0.526										
30%	8,073	0.46	12	0.325										

Material	ρ	c	λ	a										
40%	8,169	0.46	10	0.279										
50%	8,266	0.46	14	0.361										
60%	8,378	0.46	19	0.493										
70%	8,506	0.46	26	0.666										
80%	8,618	0.46	35	0.872										
90%	8,762	0.46	47	1.156										
100%	8,906	0.448	90	2.276										
Invar, Ni = 36%	8,137	0.46	10.7	0.286										
Chrome steel, Cr = 0%	7,897	0.452	73	2.026	87	73	67	62	55	48	40	36	35	36
1%	7,865	0.46	61	1.665		62	55	52	47	42	36	33	33	
2%	7,865	0.46	52	1.443		54	48	45	42	38	33	31	31	
5%	7,833	0.46	40	1.110		40	38	36	36	33	29	29	29	
10%	7,785	0.46	31	0.867		31	31	31	29	29	28	28	29	
20%	7,689	0.46	22	0.635		22	22	22	22	24	24	26	29	
30%	7,625	0.46	19	0.542										
Cr–Ni (chrome–nickel); 15 Cr, 10 Ni	7,865	0.46	19	0.526										
18 Cr, 8 Ni (V2A)	7,817	0.46	16.3	0.444		16.3	17	17	19	19	22	26	31	
20 Cr, 15 Ni	7,833	0.46	15.1	0.415										
25 Cr, 20 Ni	7,865	0.46	12.8	0.361										
Ni–Cr (nickel–chrome); 80 Ni, 15 Cr	8,522	0.46	17	0.444										
60 Ni, 15 Cr	8,266	0.46	12.8	0.333										
40 Ni, 15 Cr	8,073	0.46	11.6	0.305										
20 Ni, 15 Cr	7,865	0.46	14.0	0.390		14.0	15.1	15.1	16.3	17	19	22		
Cr–Ni–Al; 6 Cr, 1.5 Al, 0.55 Si (Sicromal 8)	7,721	0.490	22	0.594										
24 Cr, 2.5 Al, 0.55 Si (Sicromal 12)	7,673	0.494	19	0.501										
Manganese steel, Mn = 0%	7,897	0.452	73	2.026	87	73	67	62	55	48	40	36	35	36
1%	7,865	0.46	50	1.388		36	36	36	36	35	33			
2%	7,865	0.46	38	1.050										
5%	7,849	0.46	22	0.637										
10%	7,801	0.46	17	0.483										

TABLE A.1 (Cont.)

Metal	Properties at 20°C				Thermal Conductivity, k, W/m-°C									
	ρ kg/m³	c_p kJ/kg-°C	k W/m-°C	$\alpha \times 10^5$ m²/s	−100°C	0°C	100°C	200°C	300°C	400°C	600°C	800°C	1000°C	1200°C
Tungsten steel, W = 0%	7,897	0.452	73	2.026	87	73	67	62	55	48	40	36	35	36
1%	7,913	0.448	66	1.858		62	59	54	48	45	36			
2%	7,961	0.444	62	1.763										
5%	8,073	0.435	54	1.525										
10%	8,314	0.419	48	1.391										
20%	8,826	0.389	43	1.249										
Silicon steel, Si = 0%	7,897	0.452	73	2.026	87	73	67	62	55	48	40	36	35	36
1%	7,769	0.46	42											
2%	7,673	0.46	31											
5%	7,417	0.46	19											
Copper Pure	8,954	0.3831	386	11.234	407	386	379	374	369	363	353			
Aluminum bronze; 95 Cu, 5 Al	8,666	0.410	83	2.330										
Bronze; 75 Cu, 25 Sn	8,666	0.343	26	0.859										
Red brass; 85 Cu, 9 Sn, 6 Zn	8,714	0.385	61	1.804		59	71							

Brass; 70 Cu, 30 Zn	8,522	0.385	111	3.412	88		128	144	147	147				
German silver, 62 Cu, 15 Ni, 22 Zn	8,618	0.394	24.9	0.733	19.2		31	40	45	48				
Constantan; 60 Cu, 40 Ni	8,922	0.410	22.7	0.612	21		22.2	26						
Magnesium Pure	1,746	1.013	171	9.708	178	171	168	163	157					
Mg–Al (electrolytic); 6–8 Al, 1–2 Zn	1,810	1.00	66	3.605		52	62	74	83					
Mg–Mn; 2% Mn	1,778	1.00	114	6.382	93	111	125	130						
Molybdenum	10,220	0.251	123	4.790	138	125	118	114	111	109	106	102	99	92
Nickel Pure (99.9%)	8,906	0.4459	90	2.266	104	93	83	73	64	59	55	62	67	69
Impure (99.2%)	8,906	0.444	69	1.747		69	64	59	55	52				
Ni–Cr; 90 Ni, 10 Cr	8,666	0.444	17	0.444		17.1	18.9	20.9	22.8	24.6				
80 Ni, 20 Cr	8,314	0.444	12.6	0.343		12.3	13.8	15.6	17.1	18.9	22.5			
Silver Purest	10,524	0.2340	419	17.004	419	417	415	412						
Pure (99.9%)	10,524	0.2340	407	16.563	419	410	415	374	362	360				
Tin pure	7,304	0.2265	64	3.884	74	65.9	59	57						
Tungsten	19,350	0.1344	163	6.271		166	151	142	133	126	112	76		
Zinc, pure	7,144	0.3843	112.2	4.106	114	112	109	106	100	93				

*From E. R. G. Eckert and R. M. Drake, *Analysis of Heat and Mass Transfer*, New York, copyright McGraw-Hill, 1972. Used by permission.

TABLE A.2 Thermal Properties of Some Nonmetals*

Substance	c_p kJ/kg·°C		ρ kg/m³		t °C	k W/m·°C		α × 10⁷ m²/s
			Structural Materials					
Asphalt	1.59	(b)	1273	(b)	20	0.74	(a)	1.14
Bakelite					20	0.232	(b)	
Bricks	0.84	(d)						
Common			1602	(d)	20	0.69	(a)	5.2
Face			2050	(d)	20	1.32	(a)	
Carborundum brick					600	18.5	(a)	
					1400	11.1	(a)	
Chrome brick	0.84	(d)	3011	(d)	200	2.32	(a)	9.2
					550	2.47	(a)	9.8
					900	1.99	(a)	7.9
Diatomaceous earth (fired)					204	0.24	(a)	
					871	0.31	(a)	
Fire-clay brick (burnt 1330°C)	0.96	(d)	2050	(d)	500	1.04	(a)	5.3
					800	1.07	(a)	5.4
					1100	1.09	(a)	5.5
Fire-clay brick (burnt 1450°C)	0.96	(d)	2323	(d)	500	1.28	(a)	5.7
					800	1.37	(a)	6.1
					1100	1.40	(a)	6.3
Fire-clay brick (Missouri)	0.96	(d)	2643	(d)	200	1.00	(a)	3.9
					600	1.47	(a)	5.8
					1400	1.77	(a)	7.0
Magnesite	1.13	(d)	1500	(a)	204	3.81	(a)	
					650	2.77	(a)	
					1204	1.90	(a)	
Cement, portland					24	0.29	(a)	
Cement, mortar					20	1.16	(a)	
Concrete			1900–2300	(b)	24	0.81–1.40	(b)	4.0–8.4
Concrete, cinder	0.88	(b)			24	0.76	(b)	
Glass, plate	0.84	(b)	2700	(b)	20	0.76	(b)	3.2

Material	c		ρ		t	k		
Glass, borosilicute	0.84	(d)	2225	(b)	30	1.09	(b)	4.0
Plaster, gypsum			1440	(d)	21	0.48	(a)	
Plaster, metal lath					21	0.47	(a)	
Plaster, wood lath					21	0.28	(a)	
Stone								
Granite	0.82	(d)	2640	(d)	100–300	1.73–3.98	(a)	8.0–18.3
Limestone	0.91	(d)	2480	(d)		1.26–1.33	(a)	5.6–5.9
Marble	0.81	(b)	2500–2700	(b)	20	2.77	(b)	12.7–13.7
Sandstone	0.71	(b)	2160–2300	(b)	20	1.63–2.08	(b)	10.0–13.6
Wood, cross grain								
Balsa			140	(a)	30	0.055	(a)	
Cypress			465	(d)	30	0.097	(a)	
Fir	2.72	(d)	417	(b)	24	0.11	(a)	0.96
Oak	2.39	(d)	480–600	(b)	30	0.17	(a)	1.3
Yellow pine	2.81	(d)	640	(d)	24	0.16	(a)	0.91
White pine			430	(d)	30	0.11	(a)	
Wood, radial								
Fir	2.72	(b)	417–421	(b)	20	0.14	(b)	1.2
Oak	2.39	(b)	480–600	(b)	20	0.17–0.21	(b)	1.2–1.8

Insulating Materials

Material	ρ		t	k	
Asbestos	469	(b)	−200	0.074	(b)
			0	0.155	(b)
Asbestos	577	(b)	0	0.151	(b)
			100	0.192	(b)
			200	0.208	(b)
			400	0.223	(b)
Asbestos	697	(b)	−200	0.156	(b)
			0	0.234	(b)
Asbestos cement			20	2.08	(a)
Asbestos cement board			51	0.744	(a)
Asbestos sheet			38	0.166	(a)
Asbestos felt			38	0.057	(a)
(40 laminations			149	0.069	(a)
per inch)			260	0.083	(a)

TABLE A.2 (Cont.)

Substance	c_p kJ/kg·°C	ρ kg/m³	t °C	k W/m·°C	$\alpha \times 10^7$ m²/s
Asbestos felt (20 laminations per inch)	1.88 (b)		⎰ 38 ⎱ 149 ⎰ 260	0.078 (a) 0.095 (a) 0.113 (a)	
Asbestos, corrugated (4 plies per inch)			⎰ 38 ⎱ 93 ⎰ 149	0.087 (a) 0.100 (a) 0.119 (a)	
Balsam wool		32 (a)	32	0.040 (a)	1.6–4.3
Cardboard, corrugated				0.064 (a)	
Celotex			32	0.048 (a)	
Corkboard		160 (b)	30	0.043 (b)	
Cork, expanded scrap		45–119 (b)	20	0.036 (b)	
Cork, ground		151 (b)	30	0.043 (b)	
Diatomaceous earth (powdered)		160 (e)	⎰ 93 ⎱ 204 ⎰ 316	0.050 (e) 0.066 (e) 0.083 (e)	
Diatomaceous earth (powdered)		224 (e)	⎰ 93 ⎱ 204 ⎰ 316	0.057 (e) 0.068 (e) 0.080 (e)	
Diatomaceous earth (powdered)		288 (e)	⎰ 93 ⎱ 204 ⎰ 316	0.069 (e) 0.078 (e) 0.085 (e)	
Felt, hair		131 (c)	⎰ −7 ⎱ 38 ⎰ 93	0.0410 (c) 0.0466 (c) 0.0537 (c)	
Felt, hair		183 (c)	⎰ −7 ⎱ 38 ⎰ 93	0.0367 (c) 0.0440 (c) 0.0518 (c)	
Felt, hair		205 (c)	⎰ −7 ⎱ 38 ⎰ 93	0.0403 (c) 0.0454 (c) 0.0511 (c)	
Fiber insulating board		237 (b)	21	0.049 (b)	
Glass wool		24 (c)	⎰ −7 ⎱ 38 ⎰ 93	0.0376 (c) 0.0542 (c) 0.0753 (c)	

Material	c_p	ρ	T	k	
Glass wool		64 (c)	−7	0.0310 (c)	
			38	0.0414 (c)	
			93	0.0549 (c)	
Glass wool		96 (c)	−7	0.0282 (c)	
			38	0.0377 (c)	
			93	0.0499 (c)	
Kapok			30	0.035 (a)	
			38	0.068 (a)	
			93	0.071 (a)	
Magnesia, 85%		270 (c)	149	0.074 (a)	
			204	0.080 (a)	
Rock wool		64 (c)	−7	0.0260 (c)	
			38	0.0388 (c)	
			93	0.0549 (c)	
Rock wool		128 (c)	−7	0.0296 (c)	
			38	0.0345 (c)	
			93	0.0518 (c)	
Rock wool		192 (c)	−7	0.0317 (c)	
			38	0.0391 (c)	
			93	0.0486 (c)	

Miscellaneous Materials

Material	c_p	ρ	T	k	
Aerogel, silica		136 (b)	120	0.023 (b)	10.0
Clay	0.88 (b)	1458 (b)	20	1.28 (b)	1.4–1.7
Coal, anthracite	1.26 (b)	1200–1500 (b)	20	0.26 (b)	1.2
Coal, powdered	1.30 (b)	737 (b)	30	0.116 (b)	5.7
Cotton	1.30 (b)	80 (b)	20	0.059 (b)	1.4
Earth, coarse	1.84 (b)	2050 (b)	20	0.52 (b)	
Ice	1.93 (b)	913 (b)	0	2.22 (b)	12.6
Rubber, hard		2000 (b)	0	0.151 (b)	
Sawdust			24	0.059 (a)	
Silk	1.38 (b)	58 (b)	20	0.036 (b)	4.5

*Adapted from (a) A. I. Brown and S. M. Marco, *Introduction to Heat Transfer*, 3rd ed., New York, McGraw-Hill, 1958; (b) E. R. G. Eckert, *Introduction to the Transfer of Heat and Mass*, New York, McGraw-Hill, 1950; (c) R. H. Heilman, *Ind. Eng. Chem.*, Vol. 28, 1936, p. 782; (d) L. S. Marks, *Mechanical Engineers' Handbook*, 6th ed., New York, McGraw-Hill, 1958; (e) R. Calvert, *Diatomaceous Earth*, Chemical Catalog Company, Inc., 1930; (f) H. F. Norton, *J. Am. Ceramic Soc.*, Vol. 10, 1957, p. 30.

TABLE A.3-SI Properties of Saturated Water (SI Units)*

t °C	c_p kJ/kg-°C	ρ kg/m³	$\mu \times 10^3$ kg/m-s	$\nu \times 10^6$ m²/s	k W/m-°C	$\alpha \times 10^7$ m²/s	$\beta \times 10^3$ 1/°K	Pr
0	4.218	999.8	1.791	1.792	0.5619	1.332	−0.0853	13.45
5	4.203	1000.0	1.520	1.520	0.5723	1.362	0.0052	11.16
10	4.193	999.8	1.308	1.308	0.5820	1.389	0.0821	9.42
15	4.187	999.2	1.139	1.140	0.5911	1.413	0.148	8.07
20	4.182	998.3	1.003	1.004	0.5996	1.436	0.207	6.99
25	4.180	997.1	0.8908	0.8933	0.6076	1.458	0.259	6.13
30	4.180	995.7	0.7978	0.8012	0.6150	1.478	0.306	5.42
35	4.179	994.1	0.7196	0.7238	0.6221	1.497	0.349	4.83
40	4.179	992.3	0.6531	0.6582	0.6286	1.516	0.389	4.34
45	4.182	990.2	0.5962	0.6021	0.6347	1.533	0.427	3.93
50	4.182	998.0	0.5471	0.5537	0.6405	1.550	0.462	3.57
55	4.184	985.7	0.5043	0.5116	0.6458	1.566	0.496	3.27
60	4.186	983.1	0.4668	0.4748	0.6507	1.581	0.529	3.00
65	4.187	980.5	0.4338	0.4424	0.6553	1.596	0.560	2.77
70	4.191	977.7	0.4044	0.4137	0.6594	1.609	0.590	2.57
75	4.191	974.7	0.3783	0.3881	0.6633	1.624	0.619	2.39
80	4.195	971.6	0.3550	0.3653	0.6668	1.636	0.647	2.23
85	4.201	968.4	0.3339	0.3448	0.6699	1.647	0.675	2.09
90	4.203	965.1	0.3150	0.3264	0.6727	1.659	0.702	1.97
95	4.210	961.7	0.2978	0.3097	0.6753	1.668	0.728	1.86
100	4.215	958.1	0.2822	0.2945	0.6775	1.677	0.755	1.76
120	4.246	942.8	0.2321	0.2461	0.6833	1.707	0.859	1.44
140	4.282	925.9	0.1961	0.2118	0.6845	1.727	0.966	1.23
160	4.339	907.3	0.1695	0.1869	0.6815	1.731	1.084	1.08
180	4.411	886.9	0.1494	0.1684	0.6745	1.724	1.216	0.98
200	4.498	864.7	0.1336	0.1545	0.6634	1.706	1.372	0.91
220	4.608	840.4	0.1210	0.1439	0.6483	1.674	1.563	0.86
240	4.770	813.6	0.1105	0.1358	0.6292	1.622	1.806	0.84
260	4.991	783.9	0.1015	0.1295	0.6059	1.549	2.130	0.84
280	5.294	750.5	0.0934	0.1245	0.5780	1.455	2.589	0.86
300	5.758	712.2	0.0858	0.1205	0.5450	1.329	3.293	0.91
320	6.566	666.9	0.0783	0.1174	0.5063	1.156	4.511	1.02
340	8.234	610.2	0.0702	0.1151	0.4611	0.918	7.170	1.25
360	16.138	526.2	0.0600	0.1139	0.4115	0.485	21.28	2.35

*c_p, ρ, μ, β computed from equations recommended in *ASME Steam Tables*, 3rd ed., New York, Am. Soc. Mech. Engrs., 1977. k computed from equation recommended in J. Kestin, "Thermal Conductivity of Water and Steam," *Mech. Eng.*, Aug. 1978, p. 47.

TABLE A.3-ENGL   Properties of Saturated Water (English Units)*

t °F	c_p Btu/lb$_m$-°F	ρ lb$_m$/ft^3	μ lb$_m$/ft-h	$\nu \times 10^2$ ft^2/h	k Btu/h-ft-°F	$\alpha \times 10^3$ ft^2/h	$\beta \times 10^3$ 1/°R	Pr
32	1.008	62.41	4.333	6.943	0.3247	5.163	−0.0474	13.45
40	1.004	62.42	3.742	5.994	0.3300	5.264	−0.0023	11.39
50	1.001	62.41	3.163	5.069	0.3363	5.381	0.0456	9.42
60	1.000	62.37	2.175	4.354	0.3421	5.485	0.0862	7.94
70	0.999	63.31	2.361	3.789	0.3475	5.584	0.121	6.79
80	0.998	62.22	2.075	3.335	0.3525	5.677	0.153	5.88
90	0.998	62.12	1.842	2.965	0.3572	5.762	0.181	5.15
100	0.998	62.00	1.648	2.659	0.3616	5.844	0.206	4.55
110	0.998	61.86	1.486	2.402	0.3656	5.921	0.230	4.06
120	0.998	61.71	1.348	2.185	0.3693	5.994	0.253	3.65
130	0.999	61.55	1.231	2.000	0.3728	6.062	0.274	3.30
140	1.000	61.38	1.129	1.840	0.3760	6.127	0.294	3.00
150	1.000	61.19	1.041	1.701	0.3789	6.190	0.313	2.75
160	1.001	60.99	0.964	1.580	0.3815	6.251	0.331	2.53
170	1.002	60.79	0.896	1.474	0.3839	6.303	0.349	2.34
180	1.003	60.57	0.835	1.379	0.3861	6.356	0.366	2.17
190	1.004	60.34	0.782	1.296	0.3880	6.407	0.383	2.02
200	1.006	60.11	0.734	1.221	0.3897	6.448	0.400	1.89
210	1.007	59.86	0.691	1.154	0.3912	6.487	0.416	1.78
220	1.009	59.61	0.652	1.094	0.3924	6.527	0.432	1.68
230	1.010	59.35	0.617	1.039	0.3934	6.567	0.448	1.58
240	1.012	59.08	0.585	0.990	0.3943	6.592	0.464	1.50
250	1.014	58.80	0.556	0.945	0.3949	6.624	0.480	1.43
260	1.016	58.52	0.529	0.904	0.3953	6.648	0.497	1.36
270	1.019	58.22	0.505	0.867	0.3955	6.667	0.513	1.30
280	1.022	57.92	0.483	0.834	0.3956	6.680	0.530	1.25
290	1.025	57.61	0.462	0.803	0.3954	6.697	0.547	1.20
300	1.027	57.30	0.444	0.774	0.3950	6.711	0.565	1.15
350	1.050	55.59	0.369	0.663	0.3905	6.693	0.663	0.99
400	1.078	53.66	0.316	0.589	0.3816	6.596	0.784	0.89
450	1.125	51.46	0.277	0.538	0.3681	6.360	0.946	0.85
500	1.192	48.94	0.246	0.502	0.3501	6.001	1.183	0.84
550	1.302	45.97	0.219	0.476	0.3269	5.463	1.569	0.87
600	1.516	42.31	0.193	0.457	0.2978	4.643	2.316	0.99

*c_p, ρ, μ, β computed from equations recommended in *ASME Steam Tables*, 3rd ed., New York, Am. Soc. Mech. Engrs., 1977. k computed from equation recommended in J. Kestin, "Thermal Conductivity of Water and Steam," *Mech. Eng.*, Aug. 1978, p. 47

TABLE A.4 Properties of Saturated Liquids*

t °C	ρ kg/m³	c_p kJ/kg-°C	$\nu \times 10^6$ m²/s	k W/m-°C	$\alpha \times 10^7$ m²/s	Pr	$\beta \times 10^3$ 1/°K
			Sulfur Dioxide, SO_2				
−50	1560.84	1.3595	0.484	0.242	1.141	4.24	
−40	1536.81	1.3607	0.424	0.235	1.130	3.74	
−30	1520.64	1.3616	0.371	0.230	1.117	3.31	
−20	1488.60	1.3624	0.324	0.225	1.107	2.93	
−10	1463.61	1.3628	0.288	0.218	1.097	2.62	
0	1438.46	1.3636	0.257	0.211	1.081	2.38	
10	1412.51	1.3645	0.232	0.204	1.066	2.18	
20	1386.40	1.3653	0.210	0.199	1.050	2.00	1.94
30	1359.33	1.3662	0.190	0.192	1.035	1.83	
40	1329.22	1.3674	0.173	0.185	1.019	1.70	
50	1299.10	1.3683	0.162	0.177	0.999	1.61	
			Methyl Chloride, CH_3Cl				
−50	1052.58	1.4759	0.320	0.215	1.388	2.31	
−40	1033.35	1.4826	0.318	0.209	1.368	2.32	
−30	1016.53	1.4922	0.314	0.202	1.337	2.35	
−20	999.39	1.5043	0.309	0.196	1.301	2.38	
−10	981.45	1.5194	0.306	0.187	1.257	2.43	
0	962.39	1.5378	0.302	0.178	1.213	2.49	
10	942.36	1.5600	0.297	0.171	1.166	2.55	
20	923.31	1.5860	0.293	0.163	1.112	2.63	
30	903.12	1.6161	0.288	0.154	1.058	2.72	
40	883.10	1.6504	0.281	0.144	0.996	2.83	
50	861.15	1.6890	0.274	0.133	0.921	2.97	
			Ammonia, NH_3				
−50	703.69	4.463	0.435	0.547	1.742	2.60	
−40	691.68	4.467	0.406	0.547	1.775	2.28	
−30	679.34	4.476	0.387	0.549	1.801	2.15	
−20	666.69	4.509	0.381	0.547	1.819	2.09	
−10	653.55	4.564	0.378	0.543	1.825	2.07	
0	640.10	4.635	0.373	0.540	1.819	2.05	
10	626.16	4.714	0.368	0.531	1.801	2.04	
20	611.75	4.798	0.359	0.521	1.775	2.02	2.45
30	596.37	4.890	0.349	0.507	1.742	2.01	
40	580.99	4.999	0.340	0.493	1.701	2.00	
50	564.33	5.116	0.330	0.476	1.654	1.99	
			Carbon Dioxide, CO_2				
−50	1156.34	1.84	0.119	0.0855	0.4021	2.96	
−40	1117.77	1.88	0.118	0.1011	0.4810	2.46	
−30	1076.76	1.97	0.117	0.1116	0.5272	2.22	
−20	1032.39	2.05	0.115	0.1151	0.5445	2.12	
−10	983.38	2.18	0.113	0.1099	0.5133	2.20	
0	926.99	2.47	0.108	0.1045	0.4578	2.38	
10	860.03	3.14	0.101	0.0971	0.3608	2.80	
20	772.57	5.0	0.091	0.0872	0.2219	4.10	14.00
30	597.81	36.4	0.080	0.0703	0.0279	28.7	

TABLE A.4 Properties of Saturated Liquids*

t °C	ρ kg/m³	c_p kJ/kg-°C	$\nu \times 10^6$ m²/s	k W/m-°C	$\alpha \times 10^7$ m²/s	Pr	$\beta \times 10^3$ 1/°K
Dichlorodifluoromethane (Freon-12), CCl_2F_2							
−50	1546.75	0.8750	0.310	0.067	0.501	6.2	2.63
−40	1518.71	0.8847	0.279	0.069	0.514	5.4	
−30	1489.56	0.8956	0.253	0.069	0.526	4.8	
−20	1460.57	0.9073	0.235	0.071	0.539	4.4	
−10	1429.49	0.9203	0.221	0.073	0.550	4.0	
0	1397.45	0.9345	0.214	0.073	0.557	3.8	
10	1364.30	0.9496	0.203	0.073	0.560	3.6	
20	1330.18	0.9659	0.198	0.073	0.560	3.5	
30	1295.10	0.9835	0.194	0.071	0.560	3.5	
40	1257.13	1.0019	0.191	0.069	0.555	3.5	
50	1215.96	1.0216	0.190	0.067	0.545	3.5	
Lubricating Oil (Approx. SAE 50)							
0	899.12	1.796	4280	0.147	0.911	47100	
20	888.23	1.880	900	0.145	0.872	10400	0.70
40	876.05	1.964	240	0.144	0.834	2870	
60	864.04	2.047	83.9	0.140	0.800	1050	
80	852.02	2.131	37.5	0.138	0.769	490	
100	840.01	2.219	20.3	0.137	0.738	276	
120	828.96	2.307	12.4	0.135	0.710	175	
140	816.94	2.395	8.0	0.133	0.686	116	
160	805.89	2.483	5.6	0.132	0.663	84	
Glycerine, $C_3H_5(OH)_3$							
0	1276.03	2.261	8310	0.282	0.983	84700	
10	1270.11	2.319	3000	0.284	0.965	31000	
20	1264.02	2.386	1180	0.286	0.947	12500	0.50
30	1258.09	2.445	500	0.286	0.929	5380	
40	1252.01	2.512	220	0.286	0.914	2450	
50	1244.96	2.583	150	0.287	0.893	1630	
Ethylene Glycol, $C_2H_4(OH_2)$							
0	1130.75	2.294	57.53	0.242	0.934	615	
20	1116.65	2.382	19.18	0.249	0.939	204	0.65
40	1101.43	2.474	8.69	0.256	0.939	93	
60	1087.66	2.562	4.75	0.260	0.932	51	
80	1077.56	2.650	2.98	0.261	0.921	32.4	
100	1058.50	2.742	2.03	0.263	0.908	22.4	

*From E. R. G. Eckert and R. M. Drake, *Analysis of Heat and Mass Transfer,* New York, copyright McGraw-Hill, 1972 used by permission.

TABLE A.5-SI Properties of Steam (SI Units)*

Pressure, kN/m²(kPa)
(Sat. Temp., °C)

Temperature, °C		100 (99.63)	500 (151.84)	1000 (179.88)	1500 (198.29)	2000 (212.37)	4000 (250.33)	6000 (275.55)	8000 (294.97)	10,000 (310.96)	12,000 (324.65)	14,000 (336.64)	16,000 (347.33)	18,000 (356.96)
Saturated vapor	c_p	2.027	2.329	2.595	2.820	3.025	3.788	4.613	5.600	6.770	8.158	10.15	13.46	18.22
	ρ	0.5904	2.6690	5.1469	7.5955	10.047	20.101	30.828	42.507	55.432	70.035	87.106	107.61	132.70
	$\mu \times 10^6$	12.26	14.08	15.07	15.72	16.22	17.61	18.61	19.49	20.33	21.21	22.20	23.37	24.87
	$k \times 10^3$	24.75	31.03	35.40	38.76	41.65	51.28	59.99	68.96	78.96	90.89	106.3	127.6	159.5
	Pr	1.00	1.06	1.10	1.14	1.18	1.30	1.43	1.58	1.74	1.90	2.12	2.47	2.84
150°C	c_p	1.987												
	ρ	0.5165												
	$\mu \times 10^6$	14.19												
	$k \times 10^3$	28.80												
	Pr	0.98												
200°C	c_p	1.979	2.161	2.446	2.800									
	ρ	0.4603	2.3532	4.8563	7.5538									
	$\mu \times 10^6$	16.18	16.07	15.93	15.79									
	$k \times 10^3$	33.37	34.24	36.06	38.75									
	Pr	0.96	1.01	1.08	1.14									
250°C	c_p	1.990	2.088	2.233	2.406	2.608								
	ρ	0.4156	2.1078	4.2965	6.5793	8.9727								
	$\mu \times 10^6$	18.22	18.15	18.07	17.99	17.91								
	$k \times 10^3$	38.28	38.81	39.69	40.82	42.22								
	Pr	0.95	0.98	1.02	1.06	1.11								
300°C	c_p	2.010	2.066	2.145	2.235	2.337	2.876	3.715	5.232					
	ρ	0.3790	1.9136	3.8763	5.8927	7.9681	16.997	27.666	41.214					
	$\mu \times 10^6$	20.29	20.25	20.20	20.15	20.10	19.93	19.80	19.74					
	$k \times 10^3$	43.49	43.89	44.49	45.17	45.95	50.14	56.67	67.27					
	Pr	0.94	0.95	0.97	1.00	1.02	1.14	1.30	1.53					
350°C	c_p	2.037	2.071	2.118	2.169	2.225	2.504	2.883	3.404	4.175	5.413	7.420	11.06	
	ρ	0.3483	1.7542	3.5407	5.3610	7.2172	15.050	23.684	33.391	44.601	58.116	75.682	102.34	
	$\mu \times 10^6$	22.37	22.35	22.32	22.29	22.27	22.18	22.14	22.13	22.18	22.33	22.64	23.35	

$k \times 10^3$	48.97	49.32	49.79	50.31	50.86	53.55	57.14	61.96	68.57	78.05	92.89	120.8	
Pr	0.93	0.94	0.95	0.96	0.97	1.04	1.12	1.22	1.35	1.55	1.81	2.14	
400°C													
c_p	2.067	2.090	2.120	2.152	2.186	2.346	2.550	2.806	3.127	3.545	4.116	4.933	6.131
ρ	0.3223	1.6203	3.2628	4.9281	6.6170	13.629	21.107	29.146	37.867	47.429	58.050	70.055	83.948
$\mu \times 10^6$	24.45	24.44	24.43	24.41	24.40	24.38	24.38	24.42	24.49	24.61	24.79	25.05	25.43
$k \times 10^3$	54.71	55.02	55.43	55.87	56.33	58.41	60.96	64.08	67.94	72.75	78.84	86.69	97.09
Pr	0.92	0.93	0.93	0.94	0.95	0.98	1.02	1.07	1.13	1.20	1.29	1.43	1.61
450°C													
c_p	2.099	2.116	2.138	2.160	2.183	2.286	2.408	2.554	2.728	2.935	3.184	3.490	3.877
ρ	0.2999	1.5059	3.0275	4.5651	6.1191	12.507	19.193	26.216	33.623	41.467	49.813	58.736	68.327
$\mu \times 10^6$	26.52	26.52	26.52	26.51	26.52	26.53	26.57	26.63	26.72	26.85	27.01	27.21	27.46
$k \times 10^3$	60.69	60.98	61.35	61.73	62.14	63.92	65.97	68.36	71.12	74.33	78.06	82.43	87.56
Pr	0.92	0.92	0.92	0.93	0.93	0.95	0.97	1.00	1.02	1.06	1.10	1.15	1.22
500°C													
c_p	2.132	2.146	2.162	2.179	2.197	2.270	2.353	2.446	2.551	2.672	2.810	2.968	3.151
ρ	0.2805	1.4069	2.8251	4.2549	5.6962	11.582	17.670	23.978	30.525	37.333	44.426	51.831	59.578
$\mu \times 10^6$	28.57	28.58	28.58	28.59	28.60	28.64	28.71	28.79	28.90	29.03	29.18	29.36	29.57
$k \times 10^3$	66.89	67.16	67.50	67.86	68.22	69.80	71.56	73.54	75.75	78.23	81.01	84.11	87.59
Pr	0.91	0.91	0.92	0.92	0.92	0.93	0.94	0.96	0.97	0.99	1.01	1.04	1.06
550°C													
c_p	2.166	2.177	2.191	2.205	2.219	2.276	2.337	2.403	2.475	2.554	2.640	2.735	2.839
ρ	0.2634	1.3203	2.6489	3.9859	5.3313	10.799	16.411	22.172	28.092	34.181	40.448	46.904	53.561
$\mu \times 10^6$	30.61	30.62	30.63	30.64	30.65	30.72	30.80	30.90	31.02	31.15	31.30	31.47	31.67
$k \times 10^3$	73.30	73.55	73.87	74.20	74.53	75.96	77.54	79.26	81.15	83.22	85.47	87.94	90.64
Pr	0.91	0.91	0.91	0.91	0.91	0.92	0.93	0.94	0.95	0.96	0.97	0.98	0.99
600°C													
c_p	2.201	2.210	2.221	2.233	2.245	2.292	2.341	2.392	2.446	2.503	2.563	2.627	2.694
ρ	0.2483	1.2439	2.4939	3.7501	5.0125	10.125	15.341	20.664	26.096	31.643	37.308	43.097	49.013
$\mu \times 10^6$	32.62	32.63	32.64	32.66	32.68	32.76	32.86	32.96	33.09	33.22	33.37	33.54	33.73
$k \times 10^3$	79.89	80.13	80.43	80.73	81.05	82.37	83.80	85.34	87.01	88.81	90.75	92.83	95.07
Pr	0.90	0.90	0.90	0.90	0.91	0.91	0.92	0.92	0.93	0.94	0.94	0.95	0.96
650°C													
c_p	2.235	2.243	2.253	2.262	2.272	2.313	2.354	2.396	2.439	2.484	2.530	2.577	2.626
ρ	0.2348	1.1759	2.3564	3.5414	4.7312	9.5366	14.418	19.357	24.412	29.529	34.728	40.010	45.379
$\mu \times 10^6$	34.60	34.61	34.63	34.65	34.68	34.77	34.87	34.99	35.11	35.25	35.40	35.56	35.74
$k \times 10^3$	86.65	86.88	87.16	87.45	87.74	88.97	90.28	91.69	93.20	94.81	96.52	98.34	100.28
Pr	0.89	0.89	0.90	0.90	0.90	0.90	0.91	0.92	0.92	0.93	0.93	0.94	

Note: c_p in kJ/kg·°C, ρ in kg/m³, μ in kg/m·s, k in W/m·°C.
*c_p, ρ, μ computed from equations recommended in *ASME Steam Tables*, 3rd ed., New York. Am. Soc. Mech. Engrs., 1977. k computed from equation recommended in J. Kestin, "Thermal Conductivity of Water and Steam," *Mech. Eng.*, Aug. 1978, p. 47.

TABLE A.5-ENGL Properties of Steam (English Units)*

Temperature, °F		14.696 (212.00)	100 (327.82)	200 (381.80)	400 (444.60)	600 (486.20)	800 (518.21)	1000 (544.58)	1200 (567.19)	1400 (587.07)	1600 (604.87)	1800 (621.02)	2000 (635.80)	2500 (668.11)
							Pressure, psia (Sat. Temp., °F)							
Saturated vapor	c_p	0.4844	0.5820	0.6610	0.7924	0.9172	1.051	1.202	1.373	1.566	1.779	2.029	2.360	3.910
	ρ	0.03732	0.2257	0.4372	0.8614	1.2991	1.7576	2.2424	2.7590	3.3137	3.9146	4.5757	5.3119	7.6711
	$\mu \times 10^2$	2.970	3.513	3.768	4.070	4.277	4.447	4.599	4.742	4.882	5.026	5.177	5.342	5.871
	$k \times 10^2$	1.432	1.900	2.196	2.632	2.998	3.345	3.693	4.059	4.456	4.899	5.414	6.029	8.442
	Pr	1.00	1.08	1.13	1.23	1.31	1.40	1.50	1.60	1.72	1.83	1.94	2.09	2.72
300°F	c_p	0.4748												
	ρ	0.03276												
	$\mu \times 10^2$	3.421												
	$k \times 10^2$	1.659												
	Pr	0.98												
400°F	c_p	0.4729	0.5365	0.6364										
	ρ	0.02884	0.2026	0.4238										
	$\mu \times 10^2$	3.957	3.920	3.876										
	$k \times 10^2$	1.952	2.032	2.200										
	Pr	0.96	1.04	1.12										
500°F	c_p	0.4761	0.5078	0.5542	0.6806	0.8664								
	ρ	0.02579	0.1790	0.3670	0.7787	1.2589								
	$\mu \times 10^2$	4.508	4.486	4.461	4.413	4.367								
	$k \times 10^2$	2.271	2.315	2.388	2.605	2.949								
	Pr	0.95	0.98	1.04	1.15	1.28								
600°F	c_p	0.4820	0.4992	0.5225	0.5817	0.6594	0.7601	0.8964	1.094	1.396				
	ρ	0.02333	0.1609	0.3270	0.6774	1.0575	1.4763	1.9465	2.4902	3.1487				
	$\mu \times 10^2$	5.065	5.052	5.039	5.013	4.991	4.974	4.962	4.958	4.966				
	$k \times 10^2$	2.610	2.643	2.690	2.807	2.961	3.162	3.424	3.775	4.267				
	Pr	0.94	0.95	0.98	1.04	1.11	1.20	1.30	1.44	1.62				

	700°F												
c_p	0.4945	0.4997	0.5129	0.5439	0.5821	0.6284	0.6838	0.7507	0.8335	0.9398	1.081	1.273	2.133
ρ	0.02131	0.1464	0.2961	0.6061	0.9323	1.2775	1.6448	2.0386	2.4640	2.9282	3.4416	4.0197	5.9542
$\mu \times 10^2$	5.624	5.618	5.611	5.600	5.592	5.587	5.586	5.591	5.601	5.618	5.644	5.682	5.870
$k \times 10^2$	2.968	2.996	3.033	3.116	3.212	3.326	3.458	3.615	3.802	4.030	4.311	4.669	6.248
Pr	0.93	0.94	0.95	0.98	1.01	1.06	1.10	1.16	1.23	1.31	1.42	1.55	2.00
800°F													
c_p	0.4977	0.5046	0.5130	0.5316	0.5531	0.5778	0.6063	0.6389	0.6761	0.7189	0.7685	0.8269	1.031
ρ	0.01961	0.1344	0.2709	0.5509	0.8409	1.1417	1.4546	1.7810	2.1222	2.4802	2.8569	3.2547	4.3616
$\mu \times 10^2$	6.182	6.180	6.178	6.176	6.176	6.180	6.187	6.197	6.210	6.228	6.251	6.279	6.378
$k \times 10^2$	3.344	3.369	3.401	3.471	3.550	3.638	3.737	3.849	3.975	4.117	4.278	4.460	5.037
Pr	0.92	0.93	0.93	0.95	0.96	0.98	1.00	1.03	1.05	1.09	1.12	1.16	1.31
900°F													
c_p	0.5065	0.5115	0.5176	0.5304	0.5442	0.5594	0.5762	0.5947	0.6152	0.6378	0.6628	0.6905	0.7749
ρ	0.01816	0.1242	0.2500	0.5061	0.7688	1.0384	1.3153	1.6001	1.8933	2.1956	2.5075	2.8299	3.6868
$\mu \times 10^2$	6.736	6.737	6.738	6.743	6.749	6.758	6.769	6.783	6.800	6.820	6.843	6.869	6.951
$k \times 10^2$	3.736	3.760	3.788	3.850	3.917	3.990	4.069	4.156	4.250	4.353	4.465	4.587	4.942
Pr	0.91	0.92	0.92	0.93	0.94	0.95	0.96	0.97	0.98	1.00	1.02	1.03	1.09
1000°F													
c_p	0.5155	0.5195	0.5242	0.5339	0.5440	0.5547	0.5660	0.5780	0.5909	0.6047	0.6194	0.6352	0.6801
ρ	0.01691	0.1155	0.2321	0.4686	0.7096	0.9551	1.2055	1.4609	1.7215	1.9875	2.2593	2.5371	3.2595
$\mu \times 10^2$	7.284	7.287	7.291	7.300	7.311	7.323	7.338	7.355	7.374	7.395	7.418	7.444	7.519
$k \times 10^2$	4.144	4.165	4.192	4.247	4.307	4.371	4.439	4.511	4.589	4.672	4.761	4.855	5.120
Pr	0.91	0.91	0.91	0.92	0.92	0.93	0.94	0.94	0.95	0.96	0.97	0.97	1.00
1100°F													
c_p	0.5245	0.5278	0.5317	0.5396	0.5476	0.5559	0.5644	0.5732	0.5824	0.5919	0.6018	0.6122	0.6402
ρ	0.01583	0.1080	0.2168	0.4367	0.6597	0.8859	1.1153	1.3481	1.5844	1.8242	2.0677	2.3149	2.9499
$\mu \times 10^2$	7.825	7.830	7.835	7.848	7.861	7.877	7.893	7.912	7.932	7.954	7.978	8.003	8.075
$k \times 10^2$	4.565	4.585	4.609	4.661	4.715	4.772	4.832	4.896	4.963	5.034	5.109	5.188	5.404
Pr	0.90	0.90	0.90	0.91	0.91	0.92	0.92	0.93	0.93	0.94	0.94	0.94	0.96
1200°F													
c_p	0.5335	0.5364	0.5396	0.5463	0.5529	0.5597	0.5666	0.5736	0.5807	0.5880	0.5955	0.6031	0.6230
ρ	0.01487	0.1014	0.2034	0.4090	0.6169	0.8270	1.0393	1.2541	1.4711	1.6906	1.9125	2.1368	2.7087
$\mu \times 10^2$	8.359	8.364	8.371	8.386	8.401	8.418	8.437	8.456	8.477	8.500	8.524	8.549	8.618
$k \times 10^2$	4.998	5.017	5.040	5.087	5.137	5.189	5.244	5.301	5.361	5.424	5.490	5.558	5.743
Pr	0.89	0.89	0.90	0.90	0.91	0.91	0.91	0.92	0.92	0.92	0.92	0.93	0.93

Note: c_p in Btu/lb$_m$-°F, ρ in lb$_m$/ft³, μ in lb$_m$/ft-h, k in Btu/h-ft-°F.
*c_p, ρ, μ computed from equations recommended in *ASME Steam Tables*, 3rd ed., New York, Am. Soc. Mech. Engrs., 1977. k computed from equation recommended in J. Kestin, "Thermal Conductivity of Water and Steam," *Mech. Eng.*, Aug. 1978, p. 47.

TABLE A.6-SI   Properties of Dry Air at Atmospheric Pressure (SI Units)*

t °C	c_p kJ/kg-°C	ρ kg/m³	$\mu \times 10^6$ kg/m-s	$\nu \times 10^6$ m²/s	$k \times 10^3$ W/m-°C	Pr
−50	1.0064	1.5819	14.63	9.25	20.04	0.735
−40	1.0060	1.5141	15.17	10.02	20.86	0.731
−30	1.0058	1.4518	15.69	10.81	21.68	0.728
−20	1.0057	1.3944	16.20	11.62	22.49	0.724
−10	1.0056	1.3414	16.71	12.46	23.29	0.721
0	1.0057	1.2923	17.20	13.31	24.08	0.718
10	1.0058	1.2467	17.69	14.19	24.87	0.716
20	1.0061	1.2042	18.17	15.09	25.64	0.713
30	1.0064	1.1644	18.65	16.01	26.38	0.712
40	1.0068	1.1273	19.11	16.96	27.10	0.710
50	1.0074	1.0924	19.57	17.92	27.81	0.709
60	1.0080	1.0596	20.03	18.90	28.52	0.708
70	1.0087	1.0287	20.47	19.90	29.22	0.707
80	1.0095	0.9996	20.92	20.92	29.91	0.706
90	1.0103	0.9721	21.35	21.96	30.59	0.705
100	1.0113	0.9460	21.78	23.02	31.27	0.704
110	1.0123	0.9213	22.20	24.10	31.94	0.704
120	1.0134	0.8979	22.62	25.19	32.61	0.703
130	1.0146	0.8756	23.03	26.31	33.28	0.702
140	1.0159	0.8544	23.44	27.44	33.94	0.702
150	1.0172	0.8342	23.84	28.58	34.59	0.701
160	1.0186	0.8150	24.24	29.75	35.25	0.701
170	1.0201	0.7966	24.63	30.93	35.89	0.700
180	1.0217	0.7790	25.03	32.13	36.54	0.700
190	1.0233	0.7622	25.41	33.34	37.18	0.699
200	1.0250	0.7461	25.79	34.57	37.81	0.699
210	1.0268	0.7306	26.17	35.82	38.45	0.699
220	1.0286	0.7158	26.54	37.08	39.08	0.699
230	1.0305	0.7016	26.91	38.36	39.71	0.698
240	1.0324	0.6879	27.27	39.65	40.33	0.698
250	1.0344	0.6748	27.64	40.96	40.95	0.698
260	1.0365	0.6621	27.99	42.28	41.57	0.698
270	1.0386	0.6499	28.35	43.62	42.18	0.698
280	1.0407	0.6382	28.70	44.97	42.79	0.698
290	1.0429	0.6268	29.05	46.34	43.40	0.698
300	1.0452	0.6159	29.39	47.72	44.01	0.698
310	1.0475	0.6053	29.73	49.12	44.61	0.698
320	1.0499	0.5951	30.07	50.53	45.21	0.698
330	1.0523	0.5853	30.41	51.95	45.84	0.698
340	1.0544	0.5757	30.74	53.39	46.38	0.699

TABLE A.6-SI   Properties of Dry Air at Atmospheric Pressure (SI Units)*

t °C	c_p kJ/kg-°C	ρ kg/m³	$\mu \times 10^6$ kg/m-s	$\nu \times 10^6$ m²/s	$k \times 10^3$ W/m-°C	Pr
350	1.0568	0.5665	31.07	54.85	46.92	0.700
360	1.0591	0.5575	31.40	56.31	47.47	0.701
370	1.0615	0.5489	31.72	57.79	48.02	0.701
380	1.0639	0.5405	32.04	59.29	48.58	0.702
390	1.0662	0.5323	32.36	60.79	49.15	0.702
400	1.0686	0.5244	32.68	62.31	49.72	0.702
410	1.0710	0.5167	32.99	63.85	50.29	0.703
420	1.0734	0.5093	33.30	65.39	50.86	0.703
430	1.0758	0.5020	33.61	66.95	51.44	0.703
440	1.0782	0.4950	33.92	68.52	52.01	0.703
450	1.0806	0.4882	34.22	70.11	52.59	0.703
460	1.0830	0.4815	34.52	71.70	53.16	0.703
470	1.0854	0.4750	34.82	73.31	53.73	0.703
480	1.0878	0.4687	35.12	74.93	54.31	0.704
490	1.0902	0.4626	35.42	76.57	54.87	0.704
500	1.0926	0.4566	35.71	78.22	55.44	0.704
510	1.0949	0.4508	36.00	79.87	56.01	0.704
520	1.0973	0.4451	36.29	81.54	56.57	0.704
530	1.0996	0.4395	36.58	83.23	57.13	0.704
540	1.1020	0.4341	36.87	84.92	57.68	0.704
550	1.1043	0.4288	37.15	86.63	58.24	0.704
560	1.1066	0.4237	37.43	88.35	58.79	0.705
570	1.1088	0.4187	37.71	90.07	59.33	0.705
580	1.1111	0.4138	37.99	91.82	59.87	0.705
590	1.1133	0.4090	38.27	93.57	60.41	0.705
600	1.1155	0.4043	38.54	95.33	60.94	0.705
610	1.1177	0.3997	38.81	97.11	61.47	0.706
620	1.1198	0.3952	39.09	98.89	62.00	0.706
630	1.1219	0.3908	39.36	100.69	62.52	0.706
640	1.1240	0.3866	39.62	102.50	63.03	0.707
650	1.1260	0.3824	39.89	104.32	63.55	0.707

*ρ computed from ideal gas law. c_p, μ, ν, k computed from equations recommended in *Thermophysical Properties of Refrigerants*, New York, ASHRAE, 1976.

TABLE A.6-ENGL **Properties of Dry Air at Atmospheric Pressure (English Units)***

t °F	c_p Btu/lb$_m$-°F	$\rho \times 10^2$ lb$_m$/ft^3	$\mu \times 10^2$ lb$_m$/ft-h	ν ft^2/h	$k \times 10^2$ Btu/h-ft-°F	Pr
− 100	0.2407	11.029	3.230	0.2929	1.045	0.744
− 80	0.2405	10.448	3.380	0.3235	1.099	0.739
− 60	0.2404	9.925	3.526	0.3552	1.153	0.735
− 40	0.2403	9.452	3.669	0.3882	1.206	0.731
− 20	0.2402	9.022	3.809	0.4222	1.258	0.727
0	0.2402	8.630	3.947	0.4574	1.310	0.724
20	0.2402	8.270	4.082	0.4936	1.361	0.720
40	0.2402	7.939	4.215	0.5309	1.412	0.717
60	0.2403	7.633	4.345	0.5692	1.462	0.714
80	0.2403	7.350	4.473	0.6086	1.511	0.711
100	0.2405	7.088	4.599	0.6489	1.557	0.710
120	0.2406	6.843	4.723	0.6902	1.602	0.709
140	0.2408	6.615	4.845	0.7324	1.648	0.708
160	0.2409	6.401	4.965	0.7756	1.693	0.707
180	0.2412	6.201	5.083	0.8197	1.737	0.706
200	0.2414	6.013	5.199	0.8647	1.781	0.705
220	0.2417	5.836	5.314	0.9105	1.824	0.704
240	0.2419	5.669	5.427	0.9573	1.867	0.703
260	0.2422	5.512	5.539	1.0049	1.910	0.702
280	0.2426	5.363	5.649	1.0533	1.952	0.702
300	0.2429	5.222	5.757	1.1026	1.995	0.701
320	0.2433	5.088	5.864	1.1527	2.036	0.701
340	0.2437	4.960	5.970	1.2036	2.078	0.700
360	0.2441	4.839	6.075	1.2552	2.119	0.700
380	0.2445	4.724	6.178	1.3077	2.160	0.699
400	0.2450	4.614	6.280	1.3609	2.201	0.699
420	0.2455	4.509	6.380	1.4149	2.242	0.699
440	0.2460	4.409	6.480	1.4697	2.282	0.698
460	0.2465	4.313	6.578	1.5252	2.322	0.698
480	0.2470	4.221	6.676	1.5814	2.362	0.698
500	0.2476	4.133	6.772	1.6384	2.402	0.698
520	0.2481	4.049	6.867	1.6960	2.441	0.698
540	0.2487	3.968	6.961	1.7544	2.480	0.698
560	0.2493	3.890	7.055	1.8134	2.519	0.698
580	0.2499	3.815	7.147	1.8732	2.558	0.698

TABLE A.6-ENGL Properties of Dry Air at Atmospheric Pressure (English Units)*

t °F	c_p Btu/lb$_m$-°F	$\rho \times 10^2$ lb$_m$/ft^3	$\mu \times 10^2$ lb$_m$/ft-h	ν ft^2/h	$k \times 10^2$ Btu/h-ft-°F	Pr
600	0.2505	3.743	7.238	1.9336	2.597	0.698
620	0.2511	3.674	7.329	1.9948	2.635	0.698
640	0.2517	3.607	7.418	2.0566	2.623	0.699
660	0.2523	3.543	7.507	2.1190	2.707	0.700
680	0.2530	3.481	7.595	2.1821	2.743	0.701
700	0.2536	3.421	7.682	2.2459	2.778	0.701
720	0.2542	3.363	7.768	2.3103	2.814	0.702
740	0.2549	3.307	7.854	2.3753	2.851	0.702
760	0.2555	3.252	7.939	2.4409	2.887	0.702
780	0.2561	3.200	8.023	2.5072	2.924	0.703
800	0.2568	3.149	8.106	2.5741	2.961	0.703
820	0.2574	3.100	8.189	2.6416	2.998	0.703
840	0.2580	3.052	8.270	2.7098	3.035	0.703
860	0.2587	3.006	8.352	2.7785	3.072	0.703
880	0.2593	2.961	8.432	2.8478	3.108	0.703
900	0.2600	2.917	8.512	2.9177	3.145	0.704
920	0.2606	2.875	8.592	2.9882	3.182	0.704
940	0.2612	2.834	8.670	3.0593	3.218	0.704
960	0.2618	2.794	8.748	3.1310	3.254	0.704
980	0.2625	2.755	8.826	3.2032	3.290	0.704
1000	0.2631	2.718	8.903	3.2760	3.326	0.704
1020	0.2637	2.681	8.979	3.3494	3.361	0.704
1040	0.2643	2.645	9.055	3.4234	3.397	0.705
1060	0.2649	2.610	9.130	3.4978	3.432	0.705
1080	0.2655	2.576	9.205	3.5729	3.466	0.705
1100	0.2661	2.543	9.279	3.6485	3.501	0.705
1120	0.2667	2.511	9.353	3.7246	3.535	0.706
1140	0.2672	2.480	9.426	3.8013	3.569	0.706
1160	0.2678	2.449	9.499	3.8785	3.602	0.706
1180	0.2684	2.419	9.571	3.9562	3.635	0.706
1200	0.2689	2.390	9.643	4.0345	3.668	0.707

*ρ computed from ideal gas law. c_p, μ, ν, k computed from equations recommended in *Thermophysical Properties of Refrigerants*, New York, ASHRAE, 1976.

TABLE A.7 Properties of Gases at Atmospheric Pressure*

t °C	c_p kJ/kg-°C	$\mu \times 10^6$ kg/m-s	$k \times 10^3$ W/m-°C	Pr
Ammonia (NH$_3$)				
−20	2.251	8.64	19.96	0.974
0	2.194	9.32	21.82	0.938
20	2.167	10.02	23.83	0.912
40	2.164	10.73	25.98	0.894
60	2.177	11.45	28.23	0.883
80	2.203	12.18	30.59	0.877
100	2.236	12.91	33.05	0.874
120	2.274	13.65	35.59	0.872
140	2.314	14.39	38.21	0.871
160	2.353	15.13	40.91	0.870
180	2.390	15.87	43.68	0.869
200	2.425	16.62	46.51	0.867
220	2.457	17.37	49.41	0.864
240	2.488	18.13	52.37	0.861
260	2.519	18.88	55.39	0.859
Carbon Dioxide (CO$_2$)				
−50	0.7813	11.34	11.05	0.80
0	0.8277	13.74	14.57	0.78
50	0.8730	16.05	18.58	0.75
100	0.9167	18.27	22.24	0.75
150	0.9581	20.40	26.31	0.74
200	0.9969	22.42	30.25	0.74
250	1.033	24.37	33.96	0.74
300	1.062	26.24	38.16	0.73
400	1.114	29.75	46.59	0.71
500	1.158	33.01	54.14	0.71
600	1.196	36.08	60.60	0.71
Helium (He)				
−100	5.1931	13.75	104.5	0.68
−50	5.1931	16.01	123.5	0.67
0	5.1931	18.03	142.3	0.66
50	5.1931	20.95	160.2	0.68
100	5.1931	23.15	177.7	0.68
150	5.1931	25.23	194.8	0.67
200	5.1931	27.22	211.5	0.67
300	5.1931	31.01	243.3	0.66
400	5.1931	34.58	272.9	0.65
500	5.1931	37.96	300.4	0.65
Nitrogen (N$_2$)				
−100	1.044	11.35	15.89	0.745
−50	1.043	14.07	20.08	0.731
0	1.041	16.58	24.04	0.718
50	1.041	18.89	27.59	0.713
100	1.043	21.05	30.86	0.711

TABLE A.7 (Cont.)

t °C	c_p kJ/kg-°C	$\mu \times 10^6$ kg/m-s	$k \times 10^3$ W/m-°C	Pr
Nitrogen (N$_2$)				
150	1.047	23.07	34.00	0.710
200	1.053	24.97	37.02	0.710
300	1.069	28.40	42.71	0.711
400	1.091	31.55	47.92	0.719
500	1.116	34.46	52.85	0.728
600	1.140	37.16	57.50	0.737
Oxygen (O$_2$)				
0	0.9159	19.15	24.49	0.716
50	0.9239	21.95	28.55	0.710
100	0.9348	24.57	32.26	0.712
150	0.9479	27.02	35.83	0.715
200	0.9627	29.33	39.33	0.718
300	0.9944	33.64	46.23	0.724
400	1.0249	37.55	52.73	0.730
500	1.0492	41.16	58.74	0.735
600	1.0693	44.52	64.54	0.738
700	1.0864	47.68	70.25	0.738
Freon-12 (CCl$_2$F$_2$)				
− 20	0.5525	10.70	7.36	0.804
0	0.5748	11.52	8.34	0.794
20	0.5956	12.33	9.35	0.785
40	0.6150	13.12	10.39	0.777
60	0.6331	13.90	11.44	0.769
80	0.6500	14.66	12.51	0.758
100	0.6656	15.41	13.60	0.754
120	0.6801	16.14	14.70	0.747
140	0.6936	16.86	15.81	0.740
160	0.7060	17.56	16.92	0.733
180	0.7175	18.25	18.05	0.726
200	0.7282	18.93	19.17	0.719
Freon-22 (CHClF$_2$)				
− 40	0.5675	10.13	4.04	1.42
0	0.6156	11.85	9.39	0.777
40	0.6625	13.54	11.79	0.761
80	0.7076	15.20	14.19	0.758
120	0.7503	16.81	16.59	0.760
160	0.7900	18.39	18.99	0.765
200	0.8259	19.92	21.39	0.769
240	0.8575	21.42	23.79	0.772
280	0.8841	22.89	26.19	0.773
320	0.9051	24.31	28.59	0.770

TABLE A.7 (Cont.)

t °C	c_p kJ/kg-°C	$\mu \times 10^6$ kg/m-s	$k \times 10^3$ W/m-°C	Pr
Butane (normal, C_4H_{10})				
20	1.6860	7.41	15.41	0.811
40	1.7781	7.89	17.27	0.812
60	1.8692	8.37	19.25	0.813
80	1.9591	8.85	21.31	0.813
100	2.0478	9.32	23.45	0.814
120	2.1351	9.80	25.64	0.816
140	2.2210	10.27	27.87	0.818
160	2.3052	10.74	30.12	0.822
180	2.3877	11.20	32.38	0.826
200	2.4684	11.66	34.62	0.832
220	2.5472	12.12	36.85	0.838
240	2.6238	12.58	39.04	0.845
260	2.6983	13.03	41.19	0.855
Ethane (C_2H_6)				
−100	1.4252	5.62	8.00	1.00
−50	1.5260	7.14	12.66	0.861
0	1.6791	8.65	18.30	0.794
50	1.8660	10.12	24.87	0.759
100	2.0713	11.53	31.96	0.747
150	2.2825	12.87	39.49	0.744
200	2.4904	14.15	47.36	0.744
250	2.6890	15.40	55.50	0.746
300	2.8750	16.58	63.83	0.747
350	3.0485	17.71	72.44	0.745
400	3.2125	18.80	81.03	0.745
Methane (CH_4)				
0	2.1702	10.32	30.78	0.727
20	2.2193	10.98	33.39	0.721
40	2.2724	11.62	36.06	0.732
60	2.3290	12.25	38.73	0.737
80	2.3889	12.86	41.55	0.740
100	2.4515	13.46	44.38	0.744
120	2.5166	14.05	47.25	0.748
140	2.5839	14.62	50.88	0.742
160	2.6529	15.18	54.61	0.737
180	2.7233	15.72	58.37	0.734
200	2.7948	16.26	62.14	0.731
220	2.8670	16.79	65.91	0.730
240	2.9396	17.30	69.68	0.730
260	3.0121	17.81	73.44	0.730

TABLE A.7 (Cont.)

t °C	c_p kJ/kg-°C	$\mu \times 10^6$ kg/m-s	$k \times 10^3$ W/m-°C	Pr
		Propane (C_3H_8)		
0	1.5956	7.55	15.27	0.789
20	1.6779	8.07	17.41	0.778
40	1.7628	8.60	19.60	0.773
60	1.8496	9.12	21.84	0.773
80	1.9375	9.64	24.12	0.775
100	2.0260	10.16	26.44	0.779
120	2.1144	10.67	28.78	0.784
140	2.2022	11.18	31.15	0.790
160	2.2889	11.68	33.52	0.797
180	2.3742	12.17	35.92	0.805
200	2.4577	12.66	38.31	0.812
220	2.5391	13.14	40.72	0.819
240	2.6183	13.61	43.12	0.826

*Tabulated values computed from equations recommended in *Thermophysical Properties of Refrigerants,* New York, ASHRAE, 1976. The values of c_p for helium, Freon-12, and Freon-22 are for the ideal gas state as $p \rightarrow 0$; they may be used at one atmosphere without serious error.

TABLE A.8 Critical Properties

Substance	Formula	Molecular Weight	T_c °K	p_c kN/m²(kPa)
Air	—	28.96	132.4	3,770
Ammonia	NH_3	17.03	405.5	11,280
Butane	C_4H_{10}	58.124	425.2	3,800
Carbon dioxide	CO_2	44.01	304.2	7,390
Ethane	C_2H_6	30.070	305.5	4,880
Freon-12	CCl_2F_2	120.91	384.7	4,010
Freon-22	$CHClF_2$	86.48	369.2	4,980
Helium	He	4.003	5.3	230
Methane	CH_4	16.043	191.1	4,640
Nitrogen	N_2	28.013	126.2	3,390
Oxygen	O_2	31.999	154.8	5,080
Propane	C_3H_8	44.097	370.0	4,260

TABLE A.9 Properties of Some Liquid Metals*

Metal	t °C	$\rho \times 10^{-3}$ kg/m³	$\mu \times 10^4$ kg/m-s	k W/m-°C	c_p kJ/kg-°C	$\nu \times 10^7$ m²/s	$\alpha \times 10^5$ m²/s	Pr
Bismuth	316	10.011	16.04	16.4	0.1444	1.602	1.134	0.0141
	427	9.867	13.48	15.6	0.1495	1.366	1.058	0.0129
	538	9.739	11.08	15.6	0.1545	1.138	1.037	0.0110
	649	9.611	9.18	15.6	0.1595	0.955	1.018	0.0094
	760	9.467	8.06	15.6	0.1645	0.851	1.002	0.0085
Lead	371	10.540	24.22	16.1	0.1591	2.298	0.960	0.024
	454	10.444	19.92	15.6	0.1549	1.907	0.964	0.020
	538	10.348	16.82	15.4	0.1549	1.625	0.961	0.017
	621	10.236	15.58	15.1	0.1549	1.522	0.952	0.016
	704	10.140		14.9	0.1549			
Mercury	10	13.567	16.12	8.13	0.1382	1.188	0.434	0.027
	93	13.359	12.07	10.4	0.1382	0.904	0.563	0.016
	149	13.231	10.25	11.6	0.1382	0.775	0.634	0.012
	204	13.087	10.13	12.5	0.1340	0.774	0.713	0.011
	316	12.847	8.60	14.0	0.1340	0.669	0.813	0.0082
Potassium	149	0.8073	3.75	45.0	0.795	4.65	7.01	0.0066
	260	0.7801	2.41	42.7	0.795	3.40	6.89	0.0049
	427	0.7417	1.82	39.5	0.754	2.45	7.06	0.0035
	593	0.7016	1.50	35.7	0.754	2.14	6.75	0.0032
	704	0.6744	1.36	33.1	0.754	2.02	6.51	0.0031

Sodium	93	0.9291	6.95	86.2	1.38	7.48	6.72	0.011
	204	0.9018	4.42	80.3	1.34	4.90	6.65	0.0074
	371	0.8602	2.89	72.3	1.30	3.36	6.47	0.0052
	538	0.8201	2.03	65.4	1.26	2.48	6.33	0.0039
	704	0.7785	1.82	59.7	1.26	2.32	6.09	0.0038
Zinc	454	6.904	32.08	58.3	0.498	4.65	1.70	0.027
	538	6.856	26.54	57.5	0.486	3.87	1.72	0.022
	649	6.760	20.92	56.8	0.473	3.09	1.78	0.017
	816	6.536	16.87	56.4	0.448	2.58	1.93	0.013
55% Bi,	149	10.524						
44.5% Pb	288	10.348	17.61	9.05	0.146	1.70	0.589	0.024
(eutectic)	371	10.236	15.34	10.73	0.146	1.50	0.710	0.019
	593	9.931		11.86	0.146		0.794	
	649	9.835						
44% K,	93	0.8874	5.75	25.6	1.130	6.48	2.55	0.025
56% Na	204	0.8618	3.56	26.5	1.093	4.13	2.82	0.015
	371	0.8217	2.31	27.5	1.055	2.81	3.17	0.0089
	538	0.7817	1.78	28.4	1.038	2.28	3.50	0.0065
	704	0.7401	1.61	28.9	1.043	2.18	3.74	0.0058

*Adapted from M. Jakob, *Heat Transfer*, Vol. II, New York, Wiley, 1957, p. 564. Original data are contained in *Liquid Metals Handbook*, 2nd ed. (rev.) NAVEXOS p. 733 (rev.), Atomic Energy Commission, Dept. of the Navy, Washington, D.C., Jan. 1954.

TABLE A.10 Emissivities of Various Surfaces*

Surface Description	Emissivity, ϵ/T, °C
Metals and metal plating	
Aluminum	
Foil, bright side, as received (1)	0.04/371
75S-T alloy, weathered (1)	0.16/816
1100-0, commercially pure (1)	0.05/93; 0.05/204; 0.05/316; 0.05/427
1100-0, commercially pure, oxidized at	
316°C (1)	0.04/93; 0.04/204; 0.05/316; 0.05/427
24S-T81, with chromic acid anodize (1)	0.17/149; 0.17/168; 0.17/185; 0.17/204
24S-T81, with H_2SO_4 anodize	0.85/149; 0.82/168; 0.80/185; 0.78/427
Beryllium (1)	0.16/149; 0.21/371; 0.26/482; 0.30/593
Beryllium, anodized (1)	0.90/149; 0.88/371; 0.85/482; 0.82/593
Brass	
Highly polished (2)	0.030/277
Polished (8)	0.10/38; 0.10/316
Rolled plate (3)	0.06/22
Chromium	
Polished (6)	0.08/38; 0.36/1093
0.1-mil-thick plate on 0.5-mil nickel on	0.12/93; 0.13/171; 0.14/249; 0.15/327;
321 stainless steel (1)	0.15/399
Gold, evaporated on fiber glass (1)	0.05/93; 0.05/149; 0.075/260
Gold, coated on stainless steel (1)	0.09/93; 0.09/171; 0.11/254; 0.15/316;
	0.14/399
Iron	
Cast, polished (5)	0.21/200
Cast, oxidized (5)	0.64/199; 0.78/593
Plate, completely rusted (3)	0.69/19
Wrought, polished (7)	0.28/38; 0.28/249
Wrought, oxidized (7)	0.94/21; 0.94/360
Platinum on polished steel (1)	0.13/93; 0.15/171; 0.14/249; 0.15/327;
	0.15/399
Silver	
Pure, polished (2)	0.020/227; 0.032/627
5-mil silver plate on 0.5-mil nickel on 321	0.06/93; 0.06/171; 0.06/249; 0.07/327;
stainless steel (1)	0.08/399
Silver plate on stainless steel (1)	0.11/93; 0.11/204; 0.11/316; 0.13/427
Steel	
Polished plate (4)	0.066/100
Rough plate (8)	0.94/38; 0.97/371
Stainless, type 301 (1)	0.14/93; 0.15/204; 0.16/316; 0.18/427
Stainless, type 321, oxidized at 260°C (1)	0.27/93; 0.27/204; 0.28/316; 0.32/427
Stainless, type 321, with black oxide (1)	0.66/93; 0.66/204; 0.69/316; 0.76/427
Inconel X, oxidized at 1052°C (1)	0.61/93; 0.79/171; 0.81/249; 0.86/327;
	0.81/399
Tungsten filament (9)	0.39/3316

TABLE A.10   Emissivities of Various Surfaces*

Surface Description	Emissivity, ϵ/T, °C
Paints and Lacquers	
Aluminized silicon resin paint on 321 stainless steel, baked at 316°C (1)	0.20/93; 0.20/204; 0.21/316; 0.22/427
Black lacquer on iron (3)	0.88/24
Black lacquer, flat black (8)	0.96/38; 0.98/93
Black, heat resistant, on 321 stainless (1)	0.81/93; 0.76/204; 0.76/316; 0.80/427
Black, high heat Dixon 208, on stainless (1)	0.93/149; 0.92/260; 0.93/371; 0.93/482; 0.93/593
International orange on 2024-T4A1(1)	0.72/93; 0.68/204; 0.53/316
White enamel, fused on iron (3)	0.90/19
White acrylic resin paint on aluminum (1)	0.90/93; 0.87/204
Miscellaneous	
Asbestos board (3)	0.96/23
Asbestos paper (8)	0.93/38; 0.94/371
Brick	
Red, rough (3)	0.93/21
Building (10)	0.45/1000
Fire clay (10)	0.75/1000
Magnesite refractory (10)	0.38/1000
Concrete tile (10)	0.63/1000
Glass (3)	0.94/22
Graphite, pressed (11)	0.98/249; 0.98/510
Lampblack, thick coat on iron (3)	0.97/20
Plaster, rough (7)	0.91/10; 0.91/88
Roofing paper (3)	0.91/21

*When more than one value is given, linear interpolation is permissible. Compiled from data given in (1) D. K. Edwards, K. E. Nelson, R. D. Roddick, and G. T. Gier, "Basic Studies on the Use and Control of Solar Energy," *Univ. Calif. Dept. Eng. Rep. 60-93,* Los Angeles, 1960; and from data given in W. H. McAdams, *Heat Transmission,* 3rd ed., New York, McGraw-Hill, 1954, as compiled by H. C. Hottel, from (2) H. Schmidt and E. Furthmann, *Mitt. Kaiser-Wilhelm-Inst. Eisenforsch.,* Abhandl., Vol. 109, 1928, p. 225; (3) E. Schmidt, *Gesundh.-Ing.,* Beiheft 20, Reihe 1, 1927; (4) B. T. Barnes, W. E. Forsythe, and E. Q. Adams, *J. Opt. Soc. Am.,* Vol. 37, 1947, p. 804; (5) C. F. Randolph and M. J. Overholtzer, *Phys. Rev.,* Vol. 2, 1913, p. 144; (6) E. O. Hurlbert, *Astrophys. J.,* Vol. 42, 1915, p. 205; (7) F. Wamsler, *Z.V.D.I.,* Vol. 55, 1911, p. 599; *Mitt. Forsch.,* Vol. 98, 1911, p. 1; (8) R. H. Heilmann, *Trans. ASME,* Vol. 32, 1944, p. 239; (9) C. Zwikker, *Arch. Neerl. Sci.,* Vol. 9, 1925, p. 207; (10) M. W. Thring, *Sciences of Flames and Furnaces,* London, Chapman & Hall, 1952; (11) M. Pirani, *J. Sci. Instrum.,* Vol. 16, 1939.

TABLE A.11 Solar Absorptivity of Various Surfaces*

Surface Description	Solar Absorptivity α_s	Infrared Emissivity 250°K ϵ	250°K α_s/ϵ	308°K ϵ	308°K α_s/ϵ	556°K ϵ	556°K α_s/ϵ
Aluminum foil, coated with 10-μm silicon	0.522			0.12	4.35	0.12	4.35
Silicon solar cell, 1 mm thick	0.938			0.316	2.97	0.497	1.89
Chromium plate 0.1 mil thick on 0.5-mil nickel on 321 stainless steel	0.778			0.150	5.18	0.182	4.27
Stainless steel, type 410	0.764			0.130	5.88	0.180	4.24
Titanium 75A, heated 454°C 300 h	0.798			0.211	3.78	0.294	2.72
Titanium C-110, heated 427°C 100 h	0.524			0.162	3.24	0.202	2.59
Titanium, anodized	0.515	0.866	0.59			0.835	0.62
Ebanol C on copper	0.908			0.11	8.25		
Ebanol S on steel	0.848			0.10	8.48		
Aluminum, 6061-T-6, 1 mil anodize	0.923	0.841	1.10			0.847	1.09
Iconel X, oxidized	0.898	0.711	1.26			0.809	1.11
Stainless steel 301, with Armco black oxide	0.891	0.746	1.19			0.756	1.18
Graphite, on sodium silicate on polished aluminum	0.960	0.908	1.06			0.930	1.03
Glass, 3 mils on silicon solar cell	0.925	0.843	1.10			0.877	1.05
Titanox, 2 mils on black paint	0.154	0.885	0.17			0.905	0.17
Flat black epoxy paint on aluminum	0.951	0.888	1.07			0.924	1.03
White epoxy paint on aluminum	0.248	0.882	0.28			0.912	0.27

*Compiled from data given in D. K. Edwards, R. D. Roddick, and J. T. Gier, "Basic Studies on the Use and Control of Solar Energy," *Univ. Calif. Dept. Eng. Rept. 60-93*, Los Angeles, 1960.

TABLE A.12 Radiation Functions

λT (μm·°K)	$\dfrac{E_{b\lambda}}{\sigma T^5} \times 10^5$ (μm·°K)$^{-1}$	$F_{0-\lambda}(T)$	λT (μm·°K)	$\dfrac{E_{b\lambda}}{\sigma T^5} \times 10^5$ (μm·°K)$^{-1}$	$F_{0-\lambda}(T)$	λT (μm·°K)	$\dfrac{E_{b\lambda}}{\sigma T^5} \times 10^5$ (μm·°K)$^{-1}$	$F_{0-\lambda}(T)$
500	0.00000672	0	3000	22.627	0.27323	5500	10.340	0.69088
600	0.0003269	0	3100	22.450	0.29578	5600	9.939	0.70102
700	0.004650	0	3200	22.178	0.31810	5700	9.553	0.71077
800	0.03114	0.000016	3300	21.827	0.34011	5800	9.182	0.72013
900	0.1275	0.000087	3400	21.411	0.36173	5900	8.827	0.72914
1000	0.3723	0.000321	3500	20.942	0.38291	6000	8.486	0.73779
1100	0.8550	0.000911	3600	20.432	0.40360	6100	8.159	0.74611
1200	1.646	0.00213	3700	19.891	0.42376	6200	7.845	0.75411
1300	2.774	0.00432	3800	19.327	0.44337	6300	7.544	0.76181
1400	4.223	0.00779	3900	18.748	0.46241	6400	7.256	0.76920
1500	5.934	0.01285	4000	18.160	0.48087	6500	6.980	0.77632
1600	7.827	0.01972	4100	17.568	0.49873	6600	6.716	0.78317
1700	9.811	0.02853	4200	16.976	0.51600	6700	6.463	0.78976
1800	11.799	0.03934	4300	16.389	0.53268	6800	6.220	0.79610
1900	13.716	0.05211	4400	15.809	0.54878	6900	5.988	0.80220
2000	15.501	0.06673	4500	15.240	0.56430	7000	5.766	0.80808
2100	17.113	0.08305	4600	14.681	0.57926	7100	5.553	0.81373
2200	18.524	0.10089	4700	14.136	0.59367	7200	5.348	0.81918
2300	19.720	0.12003	4800	13.606	0.60754	7300	5.153	0.82443
2400	20.698	0.14026	4900	13.091	0.62089	7400	4.966	0.82949
2500	21.465	0.16136	5000	12.591	0.63373	7500	4.786	0.83437
2600	22.031	0.18312	5100	12.108	0.64608	8000	3.995	0.85625
2700	22.412	0.20536	5200	11.642	0.65795	8500	3.354	0.87457
2800	22.626	0.22789	5300	11.191	0.66936	9000	2.832	0.88999
2900	22.692	0.25056	5400	10.758	0.68034	9500	2.404	0.90304

λT μm-°K	$\dfrac{E_{b\lambda}}{\sigma T^5} \times 10^5$ $(\mu$m-°K$)^{-1}$	$F_{0-\lambda}(T)$	λT μm-°K	$\dfrac{E_{b\lambda}}{\sigma T^5} \times 10^5$ $(\mu$m-°K$)^{-1}$	$F_{0-\lambda}(T)$	λT μm-°K	$\dfrac{E_{b\lambda}}{\sigma T^5} \times 10^5$ $(\mu$m-°K$)^{-1}$	$F_{0-\lambda}(T)$
10000	2.052	0.91416	15000	0.5399	0.96893	35000	0.0247	0.99695
10500	1.761	0.92367	15500	0.4821	0.97149	40000	0.0149	0.99792
11000	1.518	0.93185	16000	0.4317	0.97377	45000	0.0095	0.99852
11500	1.315	0.93892	16500	0.3877	0.97581	50000	0.0063	0.99890
12000	1.145	0.94505	17000	0.3492	0.97765	55000	0.0044	0.99917
12500	1.000	0.95041	18000	0.2853	0.98081	∞	0	1.00000
13000	0.8779	0.95509	19000	0.2353	0.98341			
13500	0.7733	0.95921	20000	0.1958	0.98555			
14000	0.6837	0.96285	25000	0.0869	0.99217			
14500	0.6066	0.96607	30000	0.0441	0.99529			

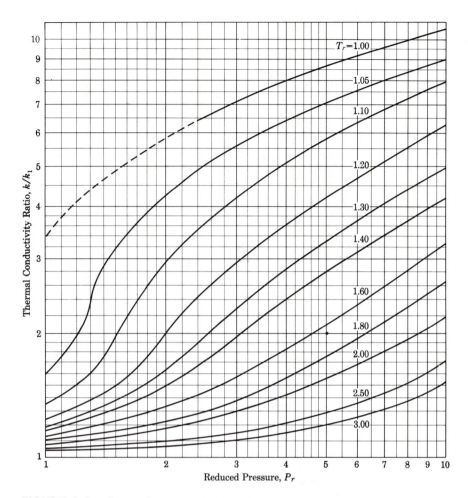

FIGURE A.1. Generalized correlation chart of the thermal conductivity of gases at high pressures: k = thermal conductivity at high pressure, k_1 = thermal conductivity at atmospheric pressure and same temperature. (From E. W. Comings and M. F. Nathan, *Ind. Eng. Chem.* Vol. **39**, 1947, pp. 964–970. Copyright 1947 by the American Chemical Society. Used by permission.)

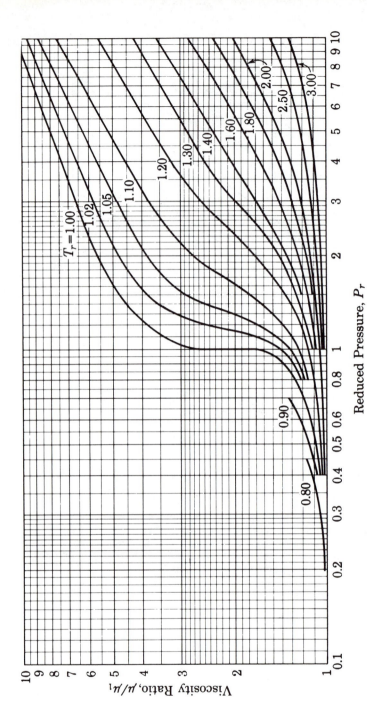

FIGURE A.2. Generalized correlation chart of the dynamic viscosity of gases at high pressures: μ = viscosity at high pressure, μ_1 = viscosity at atmospheric pressure and same temperature. (From E. W. Comings, B. J. Mayland, and R. S. Egly, *Univ. Ill. Eng. Exp. Stat. Bull. 354*, Urbana, Ill., **1944**. Used by permission.)

Miscellaneous Tables

TABLE B.1 Dimensions of Standard Pipe Sizes

Nominal Size in.	Outside Diameter in.	Outside Diameter cm	Outside Surface Area per Length ft²/ft	Outside Surface Area per Length m²/m	Schedule No.	Inside Diameter in.	Inside Diameter cm	Inside Cross-Sectional Area ft²	Inside Cross-Sectional Area m²
$\frac{1}{2}$	0.840	2.134	0.2199	0.06704	40	0.622	1.580	0.002110	0.0001961
					80	0.546	1.387	0.001626	0.0001511
1	1.315	3.340	0.3443	0.1049	40	1.049	2.664	0.006002	0.0005574
					80	0.957	2.431	0.004995	0.0004642
2	2.375	6.033	0.6218	0.1895	40	2.067	5.250	0.02330	0.002165
					80	1.939	4.925	0.02051	0.001905
3	3.500	8.890	0.9163	0.2793	40	3.068	7.793	0.05134	0.004770
					80	2.900	7.366	0.04587	0.004261
4	4.500	11.43	1.178	0.3591	40	4.026	10.226	0.08840	0.008213
					80	3.826	9.718	0.07984	0.007417
5	5.563	14.13	1.456	0.4439	40	5.047	12.82	0.1389	0.01291
					80	4.813	12.23	0.1263	0.01175
6	6.625	16.83	1.734	0.5287	40	6.065	15.41	0.2006	0.01865
					80	5.761	14.63	0.1810	0.01681
8	8.625	21.91	2.258	0.6683	40	7.981	20.27	0.3474	0.03227
					80	7.625	19.37	0.3171	0.02947
10	10.750	27.31	2.814	0.8580	40	10.020	25.45	0.5476	0.05087
					80	9.564	24.29	0.4989	0.04634
12	12.750	32.39	3.338	1.0176	40	11.938	30.32	0.7773	0.07220
					80	11.376	28.90	0.7058	0.06560

TABLE B.2 Dimensions of Standard Tubing

Nominal Size in.	Outside Diameter in.	Outside Diameter cm	Outside Surface Area per Length ft²/ft	Outside Surface Area per Length m²/m	Gage	Inside Diameter in.	Inside Diameter cm	Inside Cross-Sectional Area ft²	Inside Cross-Sectional Area m²
$\frac{1}{2}$	0.500	1.270	0.1309	0.03990	18	0.402	1.021	0.0008814	0.00008187
					16	0.370	0.940	0.0007467	0.00006940
					14	0.334	0.848	0.0006084	0.00005648
$\frac{5}{8}$	0.625	1.588	0.1636	0.04989	18	0.527	1.399	0.001515	0.0001408
					16	0.495	1.257	0.001336	0.0001241
					14	0.459	1.166	0.001149	0.0001068
$\frac{3}{4}$	0.750	1.905	0.1963	0.05985	18	0.652	1.656	0.002319	0.0002154
					16	0.620	1.575	0.002097	0.0001948
					14	0.584	1.483	0.001860	0.0001727
1	1.000	2.540	0.2618	0.07980	18	0.902	2.291	0.004438	0.0004122
					16	0.870	2.210	0.004128	0.0003836
					14	0.834	2.118	0.003794	0.0003523
					12	0.782	1.986	0.003335	0.0003098
					10	0.732	1.589	0.002922	0.0002714
$1\frac{1}{4}$	1.250	3.175	0.3272	0.09975	18	1.152	2.926	0.007238	0.0006724
					16	1.120	2.845	0.006842	0.0006357
					14	1.084	2.753	0.006409	0.0005953
					12	1.032	2.621	0.005809	0.0005395
					10	0.982	2.494	0.005260	0.0004885
$1\frac{1}{2}$	1.500	3.810	0.3927	0.1197	13	1.310	3.327	0.009360	0.0008694
					12	1.282	3.256	0.008964	0.0008326
					11	1.260	3.200	0.008659	0.0008042
					10	1.232	3.129	0.008278	0.0007690
					8	1.170	2.972	0.007460	0.0006937
2	2.000	5.080	0.5236	0.1596	13	1.810	4.597	0.01787	0.001660
					12	1.782	4.526	0.01732	0.001609
					11	1.760	4.470	0.01690	0.001569
					10	1.732	4.399	0.01632	0.001520
					8	1.670	4.242	0.01521	0.001413

A Short Summary of Bessel Functions with Brief Tables

An equation of some importance which is encountered quite frequently in certain phases of applied mathematics (Chapters 3 and 4 of this book) is

$$\frac{d^2y}{dx^2} + \frac{1}{x}\frac{dy}{dx} + \left(1 - \frac{n^2}{x^2}\right)y = 0, \tag{C.1}$$

in which n is constant. This is called *Bessel's equation of order n*. The various forms of the solution to this equation are termed *Bessel functions*.

Much of the difficulty that is often experienced by many students in gaining an understanding of Bessel functions may be avoided if one approaches their definitions as simply the solutions to a particular differential equation. The same approach may be used to define other more familiar functions. For example, the method of Frobenius for expressing the solutions of a differential equation as infinite series shows that the two solutions to

$$\frac{d^2y}{dx^2} + y = 0$$

are

$$y_1 = 1 - \frac{x^2}{2!} + \frac{x^4}{4!} - \frac{x^6}{6!} + \cdots,$$

$$y_2 = x - \frac{x^3}{3!} + \frac{x^5}{5!} - \cdots.$$

For convenience one could define the symbols $\cos x = y_1$ and $\sin x = y_2$ as abbreviations for the above series. Without any reference to the sides of a right triangle, one could then derive all the familiar trigonometric properties and identities from these definitions.

In a similar fashion, the Bessel functions are simply defined as the solutions to Bessel's equation, Eq. (C.1). This equation is a linear, second-order, ordinary, homogeneous differential equation and as such has two linearly independent solutions. The general solution is then obtained by adding these two solutions, each solution being multiplied by an arbitrary constant. The method of Frobenius gives one such solution to be

$$J_n(x) = \sum_{p=0}^{\infty} (-1)^p \frac{1}{\Gamma(n + p + 1)p!} \left(\frac{x}{2}\right)^{n+2p} \tag{C.2}$$

where $\Gamma(n + p + 1)$ represents Euler's gamma function. This series is represented by the symbol $J_n(x)$ for convenience and is called the "Bessel function of the first kind, order n." The second solution is found to be

$$J_{-n}(x) = \sum_{p=0}^{\infty} (-1)^p \frac{1}{\Gamma(-n + p + 1)p!} \left(\frac{x}{2}\right)^{-n+2p} \tag{C.3}$$

Thus, a general solution to Eq. (C.1) is

$$y = AJ_n(x) + BJ_{-n}(x).$$

However, this solution is not always the general one, for if the constant n is an *integer*, an expansion of the two series given in Eqs. (C.2) and (C.3) will show [since $\Gamma(m) = (m - 1)!$ for $m = $ an integer]

$$J_{-n}(x) = (-1)^n J_n(x), \qquad \text{for } n \text{ an integer.}$$

So, when n is an integer (usually the case in physical problems involving Bessel functions), the two solutions given above are *not* linearly independent. The result is a single solution to Bessel's equation, but since it is a second-order equation, two independent solutions are required. Thus, in the case of $n = $ an integer (including 0), a second solution must be sought.

The second solution is

$$Y_n(x) = \frac{2}{\pi} \left[\ln\left(\frac{x}{2} + \gamma\right) J_n(x) - \frac{1}{2} \sum_{p=0}^{n-1} \frac{(n - p - 1)!}{p!} \left(\frac{x}{2}\right)^{2p-n} \right.$$

$$\left. + \frac{1}{2} \sum_{p=0}^{\infty} (-1)^{p+1} \frac{(x/2)^{n+2p}}{(n + p)!p!} [\varphi(n + p) + \varphi(p)] \right]. \tag{C.4}$$

In this expression,

$$\varphi(k) = 1 + \frac{1}{2} + \frac{1}{3} + \cdots + \frac{1}{k} = \sum_{r=1}^{k} \frac{1}{r},$$

$$\gamma = \text{Euler's constant} = \lim_{m \to \infty} [\varphi(m) - \ln m] = 0.5772157.$$

Also, note that

$$Y_{-n}(x) = (-1)^n Y_n(x).$$

For $n = 0$ the second solution is

$$Y_0(x) = \frac{2}{\pi}\left[\ln\left(\frac{x}{2} + \gamma\right)J_0(x) + \sum_{p=0}^{\infty}(-1)^{p+1}\frac{(x/2)^{2p}}{p!p!}\,\varphi(p)\right]. \quad (C.5)$$

Thus, the general solution to Eq. (C.1) is

$$y = AJ_n(x) + BY_n(x), \qquad \text{for } n = 0 \text{ or an integer.}$$

Consider now the following variation of Bessel's equation:

$$\frac{d^2y}{dx^2} + \frac{1}{x}\frac{dy}{dx} + \left(c^2 - \frac{n^2}{x^2}\right)y = 0. \quad (C.6)$$

This may be altered by use of the transformation

$$z = cx,$$

giving

$$\frac{d^2y}{dz^2} + \frac{1}{z}\frac{dy}{dz} + \left(1 - \frac{n^2}{z^2}\right)y = 0.$$

This latter form is the standard form of Bessel's equation as stated in Eq. (C.1). Thus, the solution to Eq. (C.6) is

$$y = AJ_n(cx) + BJ_{-n}(cx), \qquad n \neq \text{integer,}$$

$$y = AJ_n(cx) + BY_n(cx), \qquad n = \text{integer.}$$

Consider the equation

$$\frac{d^2y}{dx^2} + \frac{1}{x}\frac{dy}{dx} + \left(-1 - \frac{n^2}{x^2}\right)y = 0. \quad (C.7)$$

This equation may be transformed by $z = ix$ ($i = \sqrt{-1}$) into the standard form of Bessel's equation:

$$\frac{d^2y}{dz^2} + \frac{1}{z}\frac{dy}{dz} + \left(1 - \frac{n^2}{z^2}\right)y = 0.$$

So the solution to Eq. (C.7) may be written

$$y = AJ_n(ix) + BJ_{-n}(ix), \qquad n \neq \text{integer,}$$

$$y = AJ_n(ix) + BY_n(ix), \qquad n = \text{integer.}$$

$$(C.8)$$

It is conventional to define new functions of real argument rather than leave the solution in the forms given above. These definitions are:

"Modified Bessel function of the first kind": $I_n(x) = i^{-n}J_n(ix).$ (C.9)

"Modified Bessel function of the second kind":

$$K_n(x) = \frac{\pi}{2}i^{n+1}[i^nI_n(x) + iY_n(ix)]. \quad (C.10)$$

Substitution of Eqs. (C.9) and (C.10) into the solution of Eq. (C.8) gives the following way of expressing the solution of Eq. (C.7) (using the case of $n =$ integer as an example):

$$y = Ai^n I_n(x) + B \frac{1}{i}\left[\frac{2}{\pi}\frac{1}{i^{n+1}} K_n(x) - i^n I_n(x)\right]$$

$$= CI_n(x) + DK_n(x),$$

in which C and D are other arbitrary constants involving A, B, and i. For $n \neq$ an integer,

$$y = CI_n(x) + DI_{-n}(x).$$

Summarizing: For

$$\frac{d^2y}{dx^2} + \frac{1}{x}\frac{dy}{dx} + \left(c^2 - \frac{n^2}{x^2}\right)y = 0,$$

the solution is

$$y = AJ_n(cx) + BJ_{-n}(cx), \qquad n \neq \text{integer},$$

$$y = AJ_n(cx) + BY_n(cx), \qquad n = \text{integer}.$$

For

$$\frac{d^2y}{dx^2} + \frac{1}{x}\frac{dy}{dx} + \left(-c^2 - \frac{n^2}{x^2}\right)y = 0,$$

the solution is

$$y = CI_n(cx) + DI_{-n}(cx), \qquad n \neq \text{integer},$$

$$y = CI_n(cx) + DK_n(cx), \qquad n = \text{integer}.$$

Also:

$$\left.\begin{array}{l} J_{-n}(x) = (-1)^n J_n(x) \\ Y_{-n}(x) = (-1)^n Y_n(x) \\ K_{-n}(x) = K_n(x) \\ I_{-n}(x) = I_n(x) \end{array}\right\} \quad n = \text{integer}.$$

Values of J_n, Y_n, I_n, and K_n for $n = 0$ and $n = 1$ are given in Tables C.1 and C.2 at the end of this appendix. These orders, 0 and 1, are those most frequently encountered in physical problems which are described mathematically by Bessel's equation.

From the definition of J_n in Eq. (C.2) one may write

$$u^n J_n(u) = \sum_{p=0}^{\infty} (-1)^p \frac{1}{\Gamma(p + n + 1)p!}\left(\frac{1}{2}\right)^{n+2p} u^{2n+2p}$$

If u is a function of x and if the above function is differentiated with respect to x,

$$\frac{d}{dx}[u^n J_n(u)] = \sum_{p=0}^{\infty} (-1)^p \frac{1}{\Gamma(n+p+1)p!} \left(\frac{1}{2}\right)^{n+2p} 2(n+p)u^{2n+2p-1} \frac{du}{dx}$$

$$= u^n \sum_{p=0}^{\infty} (-1)^p \frac{n+p}{\Gamma(n+p+1)p!} \left(\frac{u}{2}\right)^{n+2p-1} \frac{du}{dx}$$

$$= u^n \sum_{p=0}^{\infty} (-1)^p \frac{1}{\Gamma(n+p)p!} \left(\frac{u}{2}\right)^{n+2p-1} \frac{du}{dx}$$

$$= u^n J_{n-1}(u) \frac{du}{dx}.$$

After expanding the left side of this equation, the following formula for the derivative of the Bessel function is obtained:

$$\frac{d}{dx} J_n(u) = \left[J_{n-1}(u) - \frac{n}{u} J_n(u) \right] \frac{du}{dx}. \tag{C.11}$$

By similar reasoning, the following analogous equations for the other Bessel functions may be obtained:

$$\frac{d}{dx} Y_n(u) = \left[Y_{n-1}(u) - \frac{n}{u} Y_n(u) \right] \frac{du}{dx}, \tag{C.12}$$

$$\frac{d}{dx} I_n(u) = \left[I_{n-1}(u) - \frac{n}{u} I_n(u) \right] \frac{du}{dx}, \tag{C.13}$$

$$\frac{d}{dx} K_n(u) = \left[-K_{n-1}(u) - \frac{n}{u} K_n(u) \right] \frac{du}{dx}. \tag{C.14}$$

The following special cases are of particular use:

$$\frac{d}{dx}[J_0(ax)] = aJ_{-1}(ax) = -aJ_1(ax),$$

$$\frac{d}{dx}[Y_0(ax)] = aY_{-1}(ax) = -aY_1(ax),$$

$$\frac{d}{dx}[I_0(ax)] = aI_{-1}(ax) = aI_1(ax),$$

$$\frac{d}{dx}[K_0(ax)] = aK_{-1}(ax) = -aK_1(ax). \tag{C.15}$$

Two more general differential equations which lead to solutions involving Bessel functions are given below. The solutions are also given and may be verified by substitution.

$$\frac{d^2y}{dx^2} + \frac{1-2a}{x}\frac{dy}{dx} + \left[(bcx^{c-1})^2 + \frac{a^2 - n^2c^2}{x^2} \right] y = 0,$$

$$y = x^a[AJ_n(bx^c) + BY_n^{J-n}(bx^c)],$$

$$\frac{d^2y}{dx^2} + \left[(1 - 2a)\frac{1}{x} + 2\alpha \right] \frac{dy}{dx} + \left[(bcx^{c-1})^2 \right.$$

$$\left. + \frac{\alpha(2a - 1)}{x} + \frac{a^2 - n^2c^2}{x^2} + \alpha^2 \right] y = 0,$$

$$y = x^a \, e^{\alpha x} [AJ_n(bx^c) + B^{J_-n}_{Y_n}(bx^c)].$$

In the above solutions, if the arguments of the Bessel functions are imaginary, the modified functions (I_n, K_n) should replace those shown.

TABLE C.1 Selected Values of the Bessel Functions of the First and Second Kinds, Orders Zero and One

x	$J_0(x)$	$J_1(x)$	$Y_0(x)$	$Y_1(x)$
0.0	1.00000	0.00000	$-\infty$	$-\infty$
0.2	+0.99002	+0.09950	-1.0811	-3.3238
0.4	+0.96039	+0.19603	-0.60602	-1.7809
0.6	+0.91200	+0.28670	-0.30851	-1.2604
0.8	+0.84629	+0.36884	-0.08680	-0.97814
1.0	+0.76520	+0.44005	+0.08825	-0.78121
1.2	+0.67113	+0.49830	+0.22808	-0.62113
1.4	+0.56686	+0.54195	+0.33790	-0.47915
1.6	+0.45540	+0.56990	+0.42043	-0.34758
1.8	+0.33999	+0.58152	+0.47743	-0.22366
2.0	+0.22389	+0.57672	+0.51038	-0.10703
2.2	+0.11036	+0.55596	+0.52078	+0.00149
2.4	+0.00251	+0.52019	+0.51042	+0.10049
2.6	-0.09680	+0.47082	+0.48133	+0.18836
2.8	-0.18503	+0.40971	+0.43591	+0.26355
3.0	-0.26005	+0.33906	+0.37685	+0.32467
3.2	-0.32019	+0.26134	+0.30705	+0.37071
3.4	-0.36430	+0.17923	+0.22962	+0.40101
3.6	-0.39177	+0.09547	+0.14771	+0.41539
3.8	-0.40256	+0.01282	+0.06540	+0.41411
4.0	-0.39715	-0.06604	-0.01694	+0.39792
4.2	-0.37656	-0.13864	-0.09375	+0.36801
4.4	-0.34226	-0.20278	-0.16333	+0.32597
4.6	-0.29614	-0.25655	-0.22345	+0.27374
4.8	-0.24042	-0.29850	-0.27230	+0.21357
5.0	-0.17760	-0.32760	-0.30851	+0.14786
5.2	-0.11029	-0.34322	-0.33125	+0.07919
5.4	-0.04121	-0.34534	-0.34017	+0.01013
5.6	+0.02697	-0.33433	-0.33544	-0.05681
5.8	+0.09170	-0.31103	-0.31775	-0.11923
6.0	+0.15065	-0.27668	-0.28819	-0.17501
6.2	+0.20175	-0.23292	-0.24830	-0.22228
6.4	+0.24331	-0.18164	-0.19995	-0.25955
6.6	+0.27404	-0.12498	-0.14523	-0.28575
6.8	+0.29310	-0.06252	-0.08643	-0.30019
7.0	+0.30007	-0.00468	-0.02595	-0.30267
7.2	+0.29507	+0.05432	+0.03385	-0.29342
7.4	+0.27859	+0.10963	+0.09068	-0.27315
7.6	+0.25160	+0.15921	+0.14243	-0.24280
7.8	+0.25541	+0.20136	+0.18722	-0.20389
8.0	+0.17165	+0.23464	+0.22352	-0.15806
8.2	+0.12222	+0.25800	+0.25011	-0.10724
8.4	+0.06916	+0.27079	+0.26622	-0.05348
8.6	+0.01462	+0.27275	+0.27146	-0.00108
8.8	-0.03923	+0.26407	+0.26587	+0.05436
9.0	-0.09033	+0.24531	+0.24994	+0.10431
9.2	-0.13675	+0.21471	+0.22449	+0.14911
9.4	-0.17677	+0.18163	+0.19074	+0.18714
9.6	-0.20898	+0.13952	+0.15018	+0.21706
9.8	-0.23227	+0.09284	+0.10453	+0.23789
10.0	-0.24594	+0.04347	+0.05567	+0.24902

TABLE C.2 Selected Values of the Modified Bessel Functions of the First and Second Kinds, Orders Zero and One

x	$I_0(x)$	$I_1(x)$	$(2/\pi)K_0(x)$	$(2/\pi)K_1(x)$
0.0	1.000	0.0000	$+\infty$	$+\infty$
0.2	1.0100	0.1005	1.1158	3.0405
0.4	1.0404	0.2040	0.70953	1.3906
0.6	1.0920	0.3137	0.49498	0.82941
0.8	1.1665	0.4329	0.35991	0.54862
1.0	1.2661	0.5652	0.26803	0.38318
1.2	1.3937	0.7147	0.20276	0.27667
1.4	1.5534	0.8861	0.15512	0.20425
1.6	1.7500	1.0848	0.11966	0.15319
1.8	1.9896	1.3172	0.92903×10^{-1}	0.11626
2.0	2.2796	1.5906	0.72507	0.89041×10^{-1}
2.2	2.6291	1.9141	0.56830	0.68689
2.4	3.0493	2.2981	0.44702	0.53301
2.6	3.5533	2.7554	0.35268	0.41561
2.8	4.1573	3.3011	0.27896	0.32539
3.0	4.8808	3.9534	0.22116	0.25564
3.2	5.7472	4.7343	0.17568	0.20144
3.4	6.7848	5.6701	0.13979	0.15915
3.6	8.0277	6.7028	0.11141	0.12602
3.8	9.5169	8.1404	0.8891×10^{-2}	0.9999×10^{-2}
4.0	11.3019	9.7595	0.7105	0.7947
4.2	13.4425	11.7056	0.5684	0.6327
4.4	16.0104	14.0462	0.4551	0.5044
4.6	19.0926	16.8626	0.3648	0.4027
4.8	22.7937	20.2528	0.2927	0.3218
5.0	27.2399	24.3356	0.2350	0.2575
5.2	32.5836	29.2543	0.1888	0.2062
5.4	39.0088	35.1821	0.1518	0.1653
5.6	46.7376	42.3283	0.1221	0.1326
5.8	56.0381	50.9462	0.9832×10^{-3}	0.1064
6.0	67.2344	61.3419	0.7920	0.8556×10^{-3}
6.2	80.7179	73.8859	0.6382	0.6879
6.4	96.9616	89.0261	0.5146	0.5534
6.6	116.537	107.305	0.4151	0.4455
6.8	140.136	129.378	0.3350	0.3588
7.0	168.593	156.039	0.2704	0.2891
7.2	202.921	188.250	0.2184	0.2331
7.4	244.341	227.175	0.1764	0.1880
7.6	294.332	274.222	0.1426	0.1517
7.8	354.685	331.099	0.1153	0.1424
8.0	427.564	399.873	0.9325×10^{-4}	0.9891×10^{-4}
8.2	515.593	483.048	0.7543	0.7991
8.4	621.944	583.657	0.6104	0.6458
8.6	750.461	705.377	0.4941	0.5220
8.8	905.797	852.663	0.4000	0.4221
9.0	1093.59	1030.91	0.3239	0.3415
9.2	1320.66	1246.68	0.2624	0.2763
9.4	1595.28	1507.88	0.2126	0.2236
9.6	1927.48	1824.14	0.1722	0.1810
9.8	2329.39	2207.13	0.1396	0.1465
10.0	2815.72	2670.99	0.1131	0.1187

A Short Summary of the Orthogonal Functions Used in Chapter 4

Given in a countably infinite set of functions: $g_1(x)$, $g_2(x)$, $g_3(x)$, . . . , $g_n(x)$, . . . , $g_m(x)$, . . . , the functions are termed orthogonal in the interval $a \leq x \leq b$ if

$$\int_a^b g_m(x)g_n(x) \, dx = 0 \quad \text{for } m \neq n. \tag{D.1}$$

A set of orthogonal functions has particular value in the possible representation of an arbitrary function as an infinite series of the orthogonal set in the specified interval. If $f(x)$ denotes the arbitrary function, consider the possibility of expressing it as a linear combination of the orthogonal functions:

$$f(x) = C_1 g_1(x) + C_2 g_2(x) + \cdots + C_n g_n(x) + \cdots + C_m g_m(x) + \cdots$$

$$= \sum_{n=1}^{\infty} C_n g_n(x). \tag{D.2}$$

The C's are constants to be determined. If the series of Eq. (D.2) is convergent and integrable after multiplication by one of the functions, say $g_n(x)$, then

$$\int_a^b f(x)g_n(x) \, dx = C_1 \int_a^b g_1(x)g_n(x) \, dx + C_2 \int_a^b g_2(x)g_n(x) \, dx + \cdots$$

$$+ C_n \int_a^b g_n^2(x) \, dx + \cdots + C_m \int_a^b g_m(x)g_n(x) \, dx + \cdots.$$

The orthogonality definition given in Eq. (D.1) makes all the integrals on the right side of the equation above vanish except for the one term when $m = n$. Thus,

$$\int_a^b f(x)g_n(x) \, dx = 0 + 0 + \cdots + C_n \int_a^b g_n^2(x) \, dx + 0 + \cdots,$$

and the constant C_n may be calculated:

$$C_n = \frac{\int_a^b f(x)g_n(x) \, dx}{\int_a^b g_n^2(x) \, dx}. \tag{D.3}$$

Thus, when the function $f(x)$ is given, Eq. (D.3) enables one to calculate the constants, C_n, to be used in the series representation of $f(x)$. These constants are expressed in terms of the given set of orthogonal functions, $g_n(x)$.

A set of functions $[g_1(x), g_2(x), \ldots]$ may form an orthogonal set in the interval $a \leq x \leq b$ with respect to a weighting factor, $p(x)$, if

$$\int_a^b p(x)g_n(x)g_m(x) \, dx = 0 \quad \text{for } m \neq n. \tag{D.4}$$

As before, if an arbitrary function, $f(x)$ can be represented as an infinite series of the functions,

$$f(x) = C_1 g_1(x) + C_2 g_2(x) + \cdots + C_n g_n(x) + \cdots + C_m g_m(x) + \cdots$$

$$= \sum_{n=1}^{\infty} C_n g_n(x),$$

the constants are given by

$$C_n = \frac{\int_a^b p(x)f(x)g_n(x) \, dx}{\int_a^b p(x)g_n^2(x) \, dx}. \tag{D.5}$$

Some of the orthogonal sets of functions used in Chapter 4 will now be discussed as examples.

SINE AND COSINE FUNCTIONS

Consider the following set of functions in the interval $0 \leq x \leq L$:

$$\sin \frac{\pi x}{L}, \ \sin \frac{2\pi x}{L}, \ \sin \frac{3\pi x}{L}, \ \ldots, \ \sin \frac{n\pi x}{L}, \ \ldots$$

This may also be expressed as

$$\sin \lambda_1 x, \ \sin \lambda_2 x, \ \sin \lambda_3 x, \ \ldots, \ \sin \lambda_n x, \ \ldots,$$

$$\lambda_n = \frac{n\pi}{L}; \ n = 1, 2, 3, \ldots \tag{D.6}$$

Now, in the interval $0 \le x \le L$:

$$\int_a^b \sin \lambda_n x \sin \lambda_m x \, dx = \left[-\frac{\sin (\lambda_n + \lambda_m)x}{2(\lambda_m + \lambda_n)} + \frac{\sin (\lambda_n - \lambda_m)x}{2(\lambda_n - \lambda_m)} \right]_0^L \quad \text{(D.7)}$$

$$= 0 \quad \text{for } \lambda_m \ne \lambda_n,$$

since

$$\lambda_n = \frac{n\pi}{L}, \qquad \lambda_m = \frac{m\pi}{L}.$$

Thus, the set of functions in Eq. (D.6) is an orthogonal set. Also, for $m = n$:

$$\int_0^L \sin^2 \lambda_n x \, dx = \frac{1}{2\lambda_n}(\lambda_n x - \sin \lambda_n x \cos \lambda_n x)\bigg|_0^L \quad \text{(D.8)}$$

$$= \frac{L}{2}.$$

Thus, an arbitrary function, $f(x)$, may, if the series converges, be represented as a series of the functions of Eq. (D.6):

$$f(x) = C_1 \sin \lambda_1 x + C_2 \sin \lambda_2 x + \cdots,$$

or

$$f(x) = \sum_{n=1}^\infty C_n \sin \lambda_n x. \quad \text{(D.9)}$$

The C_n's will be, from Eqs. (D.3) and (D.8),

$$C_n = \frac{2}{L} \int_0^L f(x) \sin \lambda_n x \, dx, \quad \text{(D.10)}$$

$$\lambda_n = \frac{n\pi}{L}, \qquad n = 1, 2, 3, \ldots$$

Thus, the function $f(x)$ is representable by the series

$$f(x) = \frac{2}{L} \sum_{n=1}^\infty \sin \lambda_n x \int_0^L f(x) \sin \lambda_n x \, dx. \quad \text{(D.11)}$$

In a similar fashion, one can show that the set of functions

$$\{\cos \lambda_n x\}, \; \lambda_n = \frac{n\pi}{L}, \qquad n = 0, 1, 2, 3, \ldots, \quad \text{(D.12)}$$

is an orthogonal set in $0 \le x \le L$. Also, then, an arbitrary function $f(x)$, may be represented as a convergent series of these functions:

$$f(x) = \frac{A_0}{2} + \sum_{n=1}^\infty A_n \cos \lambda_n x, \quad \text{(D.13)}$$

if

$$A_n = \frac{2}{L} \int_0^L f(x) \cos \lambda_n x \, dx \qquad (D.14)$$

with

$$\lambda_n = \frac{n\pi}{L}, \qquad n = 0, 1, 2, 3, \ldots.$$

Thus,

$$f(x) = \frac{1}{L} \int_0^L f(x) \, dx + \frac{2}{L} \sum_{n=1}^{\infty} \cos \lambda_n x \int_0^L f(x) \cos \lambda_n x \, dx. \qquad (D.15)$$

In many instances in heat conduction problems it may be necessary to express a function as an infinite series of the sines or cosines, such as in Eqs. (D.9) or (D.13), but in which the λ_n's are defined by relations other than that specified by Eqs. (D.6) and (D.12). These characteristic equations defining the λ_n's arise out of the application of the boundary conditions of the particular problem under consideration. One such case is that wherein one wishes to represent a function as a sine series in the interval $0 \leq x \leq L$ as in Eq. (D.9), when the λ_n's are defined as the roots of the equation:

$$(\lambda_n L) \cot (\lambda_n L) + B = 0, \qquad n = 1, 2, 3, \ldots, B = \text{constant.} \qquad (D.16)$$

Now, since Eq. (D.16) may be written

$$(\lambda_n L) \cos \lambda_n L = -B \sin \lambda_n L,$$

some algebra will show that the integral expressed in Eq. (D.7) will again vanish. Also, Eq. (D.8) gives, instead of $L/2$, that

$$\int_0^L \sin^2 \lambda_n x \, dx = \frac{L}{2} - \frac{\sin \lambda_n L \cos \lambda_n L}{2\lambda_n}.$$

Thus, if λ_n is a root of

$$(\lambda_n L) \cot \lambda_n L + B = 0,$$

then an arbitrary function, $f(x)$, may be expressed as a sine series

$$f(x) = \sum_{n=1}^{\infty} C_n \sin \lambda_n x, \qquad (D.17)$$

where

$$C_n = \frac{\int_0^L f(x) \sin \lambda_n x \, dx}{\dfrac{L}{2} - \dfrac{\sin \lambda_n L \cos \lambda_n L}{2\lambda_n}}. \qquad (D.18)$$

Thus, in terms of these constants, $f(x)$ may be represented by

$$f(x) = 2 \sum_{n=1}^{\infty} \sin \lambda_n x \frac{\lambda_n \int_0^L f(x) \sin \lambda_n x \, dx}{\lambda_n L - \sin \lambda_n L \cos \lambda_n L}.$$

Similarly, it may be shown that if the λ_n's are the roots of

$$(\lambda_n L) \tan \lambda_n L - B = 0, \tag{D.19}$$

then an expansion in terms of cosines may be made:

$$f(x) = \sum_{n=1}^{\infty} A_n \cos \lambda_n x,$$

where

$$A_n = \frac{\displaystyle\int_0^L f(x) \cos \lambda_n x \, dx}{\dfrac{L}{2} + \dfrac{\sin \lambda_n L \cos \lambda_n L}{2\lambda_n}}. \tag{D.20}$$

The corresponding representation of $f(x)$ is, then,

$$f(x) = 2 \sum_{n=1}^{\infty} \cos \lambda_n L \frac{\lambda_n \displaystyle\int_0^L f(x) \cos \lambda_n x \, dx}{\lambda_n L + \sin \lambda_n L \cos \lambda_n L}.$$

BESSEL FUNCTIONS

In heat conduction problems in cylindrical coordinate systems, the solutions are often expressed in terms of the Bessel functions (Appendix C). To express an arbitrary function as an infinite series of such functions it will be necessary to show their orthogonality. The Bessel functions are orthogonal with respect to the weighting factor: $p(x) = x$. For example, considering J_0, it will be shown that an arbitrary function may be expressed, in an interval, as a linear combination of the set $J_0(\lambda_1 x)$, $J_0(\lambda_2 x)$, $J_0(\lambda_3 x)$, . . . , $J_0(\lambda_n x)$, . . . , where the parameters denoted by λ_n are defined, in some way, by the boundary conditions of the problem.

In other words, it will be shown that a function $f(x)$ may be represented in the following way, provided that the λ_n's are properly defined:

$$f(x) = C_1 J_0(\lambda_1 x) + C_2 J_0(\lambda_2 x) + \cdots \tag{D.21}$$

$$= \sum_{n=1}^{\infty} C_n J_0(\lambda_n x).$$

In order to be able to do this, Eq. (D.4) shows that [for $p(x) = x$] the following condition must be satisfied:

$$\int_a^b x J_0(\lambda_n x) J_0(\lambda_m x) \, dx = 0, \qquad m \neq n. \tag{D.22}$$

Then Eq. (D.5) shows that the constants C_n are

$$C_n = \frac{\int_a^b xf(x)J_0(\lambda_n x)\,dx}{\int_a^b xJ_0^2(\lambda_n x)\,dx}. \tag{D.23}$$

In order to prove the orthogonality condition of Eq. (D.22) and to evaluate the constants given by Eq. (D.23), one needs expressions for

$$\int_a^b xJ_0(\lambda_n x)J_0(\lambda_m x)\,dx \qquad \text{and} \qquad \int_a^b xJ_0^2(\lambda_n x)\,dx.$$

These two integrals may readily be evaluated by repeated "integration by parts," utilizing the following formulas resulting from Eq. (C.11):

$$\frac{d}{dx}[J_0(\lambda_n x)] = -\lambda_n J_1(\lambda_n x), \qquad \frac{d}{dx}[xJ_1(\lambda_n x)] = \lambda_n xJ_0(\lambda_n x),$$

$$\int J_1(\lambda_n x)\,dx = -\frac{1}{\lambda_n}J_0(\lambda_n x), \qquad \int xJ_0(\lambda_n x)\,dx = \frac{x}{\lambda_n}J_1(\lambda_n x).$$

The results are

$$\int xJ_0(\lambda_n x)J_0(\lambda_m x)\,dx = \frac{x}{\lambda_n^2 - \lambda_m^2}[\lambda_n J_0(\lambda_m x)J_1(\lambda_n x)$$

$$- \lambda_m J_0(\lambda_n x)J_1(\lambda_m x)], \tag{D.24}$$

$$\int xJ_0^2(\lambda_n x)\,dx = \frac{x^2}{2}[J_0^2(\lambda_n x) + J_1^2(\lambda_n x)]. \tag{D.25}$$

As a particular example, consider the set of functions $J_0(\lambda_1 x)$, $J_0(\lambda_2 x)$, . . . , $J_0(\lambda_n x)$, . . . , in which the λ_n's are defined in the following way for the interval $0 \le x \le R$. Let the λ_n's be the roots of the equation

$$J_0(\lambda_n R) = 0. \tag{D.26}$$

Examination of the tables of $J_0(x)$ given in Appendix C shows that J_0 has a succession of zeros that differ by an interval approaching π as $x \to \infty$. Hence, there are a countably infinite set of the λ_n's defined in Eq. (D.26). For the interval $0 \le x \le R$, Eq. (D.24) reduces to

$$\int_0^R xJ_0(\lambda_n x)J_0(\lambda_m x)\,dx = \frac{R}{\lambda_n^2 - \lambda_m^2}[\lambda_n J_0(\lambda_m R)J_1(\lambda_n R)$$

$$- \lambda_m J_0(\lambda_n R)J_1(\lambda_m R)].$$

By virtue of Eq. (D.26), $J_0(\lambda_n R) = J_0(\lambda_m R) = 0$, so

$$\int_0^R xJ_0(\lambda_n x)J_0(\lambda_m x)\,dx = 0.$$

Thus, the functions $J_0(\lambda_1 x)$, $J_0(\lambda_2 x)$, . . . , are orthogonal in the interval $0 \le x \le R$ if λ_n is a root of Eq. (D.26). To use Eq. (D.23) to obtain the constants of the linear series expansion, Eq. (D.25) must be evaluated for the particular definition of λ_n. Thus,

$$\int_0^R xJ_0^2(\lambda_n x)\,dx = \frac{R^2}{2}[J_0^2(\lambda_n R) + J_1^2(\lambda_n R)]$$

$$= \frac{R^2}{2}[0 + J_1^2(\lambda_n R)]$$

$$= \frac{R^2}{2} J_1^2(\lambda_n R).$$

Summarizing, an arbitrary function may be expressed, in the interval $0 \le x \le R$, as a series of J_0's:

$$f(x) = \sum_{n=1}^{\infty} C_n J_0(\lambda_n x).$$

The constants, C_n, will be given by Eq. (D.23):

$$C_n = \frac{\displaystyle\int_0^R xf(x)J_0(\lambda_n x)\,dx}{\dfrac{R^2}{2} J_1^2(\lambda_n R)} \tag{D.27}$$

if the λ_n's are the roots of

$$J_0(\lambda_n R) = 0.$$

In a similar fashion it may be shown that the same expression,

$$f(x) = \sum_{n=1}^{\infty} C_n J_0(\lambda_n x), \tag{D.28}$$

may be written in the interval $0 \le x \le R$ if the λ_n's are the roots of

$$J_1(\lambda_n R) = 0. \tag{D.29}$$

In this case, the C_n's are given by

$$C_n = \frac{\displaystyle\int_0^R xf(x)J_0(\lambda_n x)\,dx}{\dfrac{R^2}{2} J_0^2(\lambda_n R)}. \tag{D.30}$$

As a final example, Sec. 4.7 considers the possibility of expressing an arbitrary function, $f(x)$, as a series expansion in $J_0(\lambda_n x)$, when λ_n is defined as the nth root of the transcendental equation

$$\lambda_n R \frac{J_1(\lambda_n R)}{J_0(\lambda_n R)} - B = 0. \tag{D.31}$$

In the latter equation B is a constant. That the functions with λ_n thus defined are orthogonal in the interval $0 \le x \le R$ can be seen by substitution of Eq.

(D.31) into Eq. (D.24):

$$\int_0^R x J_0(\lambda_n x) J_0(\lambda_m x)\, dx$$

$$= \frac{R}{\lambda_m^2 - \lambda_n^2}\left[\lambda_n\left(\lambda_m R\, \frac{J_1(\lambda_m R)}{B}\right) J_1(\lambda_n R) - \lambda_m \lambda_n R\left(\frac{J_1(\lambda_n R)}{B}\right) J_1(\lambda_m R)\right]$$

$$= 0.$$

The fact that this equation equals zero results from the definition of the λ_n's (and λ_m's) given in Eq. (D.31). Equation (D.25) yields, then,

$$\int_0^R x J_0^2(\lambda_n x)\, dx = \frac{R^2}{2}[J_0^2(\lambda_n R) + J_1^2(\lambda_n R)],$$

so Eq. (D.23) gives the C_n's to be

$$C_n = \frac{\dfrac{2}{R^2}\displaystyle\int_0^R x f(x) J_0(\lambda_n x)\, dx}{J_0^2(\lambda_n R) + J_1^2(\lambda_n R)}. \qquad (D.32)$$

Determination of the Shape Factor

The shape factor, F_{1-2}, was introduced in Sec. 11.6 as the fraction of diffuse radiation leaving one surface, A_1, which directly strikes a second surface, A_2. This definition led to the following expression for the shape factor,

$$F_{1-2} = \frac{1}{A_1} \int_{A_2} \int_{A_1} \frac{\cos \theta_1 \cos \theta_2 \, dA_1 \, dA_2}{\pi r^2}. \tag{E.1}$$

The geometrical variables θ_1, θ_2 r, A_1, and A_2 are illustrated in Fig. 11.14. The shape factor is seen to be a purely geometrical quantity—dependent only upon the size, shape, and relative orientation of the two surfaces.

The discussions of Sec. 11.6, and succeeding sections, emphasized the importance of this shape factor in the calculation of radiant exchange. Whether dealing with individual surfaces, enclosures, etc., the shape factor entered into almost every consideration. Thus, it is important to have at hand numerical values of the shape factor for geometric configurations of practical significance. Some of the fundamental configurations will be considered here. References 1, 2, 3, and 4 may be consulted for more extensive tabulations of the shape factor and for a great variety of cases. In addition to consideration of basic configurations, application may be made of the reciprocal and additive properties of the shape factor, discussed in Sec. 11.6, to calculate F for other, more complex, configurations.

THE SHAPE FACTOR FOR FINITE PARALLEL, OPPOSED RECTANGLES

Of very great application in engineering is the configuration of two equal, parallel, directly opposed rectangles. Figure E.1 shows two rectangles $W \times L$ in size and spaced a distance D apart. By arbitrarily denoting the upper area by A_1 and the lower by A_2, the two coordinate systems (x_1, y_1) and (x_2, y_2) may be chosen as shown. Selecting area elements dA_1 and dA_2 on each surface, one finds that their centers have coordinates (x_1, y_1) and (x_2, y_2), respectively. The distance r between dA_1 and dA_2 is then given by

$$r^2 = D^2 + (x_1 - x_2)^2 + (y_1 - y_2)^2.$$

The cosines of the two angles θ_1 and θ_2 are identical and equal to $\cos \theta_1 = \cos \theta_2 = D/r$. Equation (E.1) for the shape factor F_{1-2} (or F_{2-1} in this case of equal areas) gives

$$F_{1-2} = \frac{1}{WL} \int_0^W \int_0^L \int_0^W \int_0^L \frac{D^2\, dx_1\, dy_1\, dx_2\, dy_2}{\pi[D^2 + (x_1 - x_2)^2 + (y_1 - y_2)^2]^2}.$$

The result of the integration of the above equation may be written in terms of the dimensionless ratios W/D and L/D. The result is given below with the following symbols introduced for simplicity:

$$R_1 = \frac{L}{D},$$

$$R_2 = \frac{W}{D},$$

$$
\begin{aligned}
F_{1-2} = \frac{1}{\pi}\Bigg[& \frac{1}{R_1 R_2} \ln \frac{(1 + R_1^2)(1 + R_2^2)}{(1 + R_1^2 + R_2^2)} - \frac{2}{R_1}\tan^{-1} R_2 - \frac{2}{R_2}\tan^{-1} R_1 \\
& + 2\sqrt{1 + \frac{1}{R_1^2}}\,\tan^{-1}\frac{R_2}{\sqrt{1 + R_1^2}} + 2\sqrt{1 + \frac{1}{R_2^2}}\,\tan^{-1}\frac{R_1}{\sqrt{1 + R_2^2}}\Bigg].
\end{aligned}
$$
(E.2)

For rapid determination of the shape factor expressed by this equation, Fig. 11.15 presents F_{1-2} as a function of $R_1 = L/D$ and $R_2 = W/D$.

FIGURE E.1

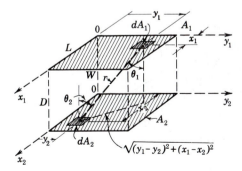

THE SHAPE FACTOR FOR PERPENDICULAR RECTANGLES HAVING A COMMON EDGE

Another configuration of particular engineering importance is that of two perpendicular rectangles with a common edge. Such a configuration exists between the walls of a room (or furnace) and the floor or ceiling.

The approach to the evaluation of the shape factor is identical to that used in the preceding section. Figure E.2 depicts a rectangle, call it A_1, of dimensions $D \times L$ located normal to A_2 with dimensions $D \times W$. The dimension D is, then, the length of the common edge. Figure E.2 shows the coordinate system selected, and the various quantities needed to evaluate F_{1-2} are indicated. Without discussion, the shape factor is expressed by

$$r^2 = y_1^2 + y_2^2 + (x_1 - x_2)^2,$$

$$\cos \theta_1 = \frac{y_2}{r}, \qquad \cos \theta_2 = \frac{y_1}{r}.$$

In terms of the two dimensionless parameters,

$$R_1 = \frac{L}{D},$$

$$R_2 = \frac{W}{D},$$

integration gives

$$F_{1-2} = \frac{1}{\pi R_1} \left\{ R_1 \tan^{-1} \frac{1}{R_1} + R_2 \tan^{-1} \frac{1}{R_2} - \sqrt{R_1^2 + R_2^2} \tan^{-1} \frac{1}{\sqrt{R_1^2 + R_2^2}} \right.$$

$$+ \frac{1}{4} \ln \left[\frac{(1 + R_1^2)(1 + R_2^2)}{(1 + R_1^2 + R_2^2)} \left(\frac{R_2^2(1 + R_1^2 + R_2^2)}{(1 + R_2^2)(R_1^2 + R_2^2)} \right)^{R_2^2} \right.$$

$$\left. \left. \times \left(\frac{R_1^2(1 + R_1^2 + R_2^2)}{(1 + R_1^2)(R_1^2 + R_2^2)} \right)^{R_1^2} \right] \right\}$$

$$\tag{E.3}$$

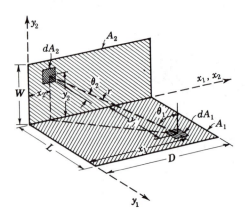

The results of this expression are presented graphically in Fig. 11.16. As should be intuitively apparent, Fig. 11.16 shows that F_{1-2} must always be less than 0.5 for the configuration under consideration.

EXAMPLE

A square black plane, 10 ft × 10 ft, is parallel and directly opposed to another square plane of the same size located 5 ft away. Compare the amount of radiation exchanged in this case to that obtained for a square plane, 10 ft × 10 ft, located normal to another rectangle, 10 ft × 5 ft, with its 10 ft side a common edge with the square.

Solution.   For $R_1 = R_2 = 2$, Fig. 11.15 or Eq. (E.2) gives F_{1-2} in the parallel case to be $F_{1-2} = 0.415$. In the perpendicular case, for $R_1 = 1$, $R_2 = 0.5$, Fig. 11.16 or Eq. (E.3) gives $F_{1-2} = 0.146$. These calculations show that a plane radiates more of its energy directly overhead than to the sides.   ∎

OTHER CONFIGURATIONS DERIVABLE FROM PERPENDICULAR RECTANGLES WITH A COMMON EDGE

As examples of the application of the additive property of shape factors, this section will consider a few cases of practical value for which the shape factor can be deduced from the results obtained in Eq. (E.3) and Fig. 11.16. The same principles of analysis may be applied to other basic configurations.

Case 1.   Let it be desired to find the shape factor F_{1-2} for the configuration shown in Fig. E.3(a) with one rectangle displaced from the common intersection

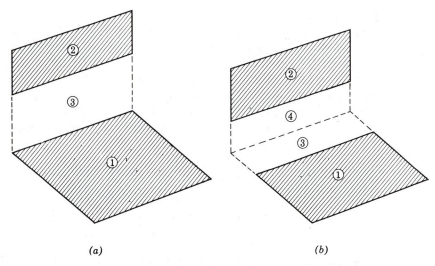

(a)   (b)

FIGURE E.3

line. Considering the area A_2 plus the fictitious area A_3 to compose a single area, call it $A_{(2,3)}$, then one obtains

$$A_1 F_{1-(2,3)} = A_1 F_{1-2} + A_1 F_{1-3}. \qquad (E.4)$$

both $F_{1-(2,3)}$ and F_{1-3} may be found from the results of Eq. (E.3) or Fig. 11.16. To obtain F_{2-1} for this configuration one could use the reciprocal property

$$F_{2-1} = F_{1-2} \frac{A_1}{A_2},$$

or one could do the following:

$$A_{(2,3)} F_{(2,3)-1} = A_2 F_{2-1} + A_3 F_{3-1}, \qquad (E.5)$$

$$F_{2-1} = \frac{A_{(2,3)} F_{(2,3)-1} - A_3 F_{3-1}}{A_2}.$$

Case 2. By similar reasoning the shape factor F_{1-2} for the situation depicted in Fig. E.3(b) is

$$F_{1-2} = \frac{A_{(1,3)} F_{(1,3)-(2,4)} + A_3 F_{3-4} - A_3 F_{3-(2,4)} - A_{(1,3)} F_{(1,3)-4}}{A_1}. \qquad (E.6)$$

All the F's in this equation are of the type given in Fig. 11.16. The areas A_3 and A_4 are, of course, fictitious surfaces.

Case 3. The determination of the shape factor F_{1-2} for the configuration shown in Fig. E.4(a) is a little more involved than the above two cases. If one defines the fictitious areas A_3 and A_4 as shown, the additive principle gives

$$F_{1-2} = \frac{1}{A_1} [A_{(1,3)} F_{(1,3)-(2,4)} - A_1 F_{1-4} - A_3 F_{3-2} - A_3 F_{3-4}]. \qquad (E.7)$$

One notes now that the shape factor F_{3-4} required in this expression is for a configuration of the same form as that for F_{1-2}—which is what is sought. This dilemma may be solved by developing another reciprocal relation of considerable importance.

The integral representation of the two shape factors F_{1-2} and F_{3-4} must be formed. Figure E.4(b) and (c) shows the dimensions and coordinates used for this purpose. The integrands of these expressions will be the same as for the basic configuration discussed in the preceding section, only the limits of integration will be different. The resulting expressions for F_{1-2} and F_{3-4} are

$$A_1 F_{1-2} = \frac{1}{\pi} \int_0^d \int_0^a \int_0^c \int_a^{a+b} \frac{y_1 y_2 \, dx_1 \, dy_1 \, dx_2 \, dy_2}{[y_1^2 + y_2^2 + (x_1 - x_2)^2]^2},$$

$$A_3 F_{3-4} = \frac{1}{\pi} \int_0^d \int_a^{a+b} \int_0^c \int_0^a \frac{y_3 y_4 \, dx_3 \, dy_3 \, dx_4 \, dy_4}{[y_3^2 + y_4^2 + (x_3 - x_4)^2]^2}.$$

These two integrals are of identical form except for the order of integration. Since the nature of the integrand permits the interchange of the order of the

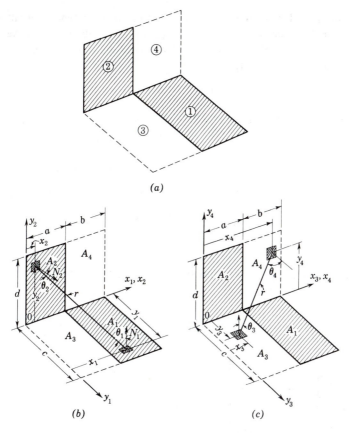

(a)

(b)   (c)

FIGURE E.4

integration, the following reciprocal formula is obtained for the configuration of Fig. E.4(a):

$$A_1 F_{1-2} = A_3 F_{3-4}.$$

This reciprocal relation then enables one to obtain F_{1-2} since Eq. (E.7) now reduces to

$$F_{1-2} = \frac{1}{2A_1} [A_{(1,3)} F_{(1,3)-(2,4)} - A_1 F_{1-4} - A_3 F_{3-2}]. \qquad \text{(E.8)}$$

This involves only shape factors obtainable from Fig. 11.16.

GENERAL RELATIONS FOR PERPENDICULAR AND PARALLEL RECTANGLES

The methods of the foregoing sections may be applied to obtain expressions for the shape factor for perpendicular and parallel rectangles oriented in rather general ways. The results obtained by Hamilton and Morgan (Ref. 1) are given below. These relations will enable one to calculate shape factors between any portions (windows, doors, walls, etc.) of a parallelepiped structure.

Perpendicular Rectangles. Figure E.5(a) illustrates a general orientation of two rectangles located in perpendicular planes. The following reciprocal relations exist [see Fig. E.5(a) for the explanation of the subscripts used to denote the various areas involved]:

$$A_1 F_{1-3'} = A_3 F_{3-1'} = A_{3'} F_{3'-1} = A_{1'} F_{1'-3}. \tag{E.9}$$

The shape factor, $F_{1-3'}$, is given by

$$A_1 F_{1-3'} = K_{1-3'} = \tfrac{1}{2}[K_{(1,2,3,4,5,6)^2} - K_{(2,3,4,5)^2} - K_{(1,2,5,6)^2} + K_{(4,5,6)^2}$$

$$- K_{(4,5,6)-(1',2',3',4',5',6')} - K_{(1,2,3,4,5,6)-(4',5',6')}$$

$$+ K_{(1,2,5,6)-(5',6')} + K_{(2,3,4,5)-(4',5')} + K_{(5,6)-(1',2',5',6')}$$

$$+ K_{(4,5)-(2',3',4',5')} + K_{(2,5)^2} - K_{(2,5)-5'} - K_{(5,6)^2}$$

$$- K_{(4,5)^2} - K_{5-(2'.5')} + K_{5^2}]. \tag{E.10}$$

In Eq. (E.10) K_{m-n} is used to symbolize $K_{m-n} = A_m F_{m-n}$, and $K_{(m)^2} = A_m F_{m-m'}$.

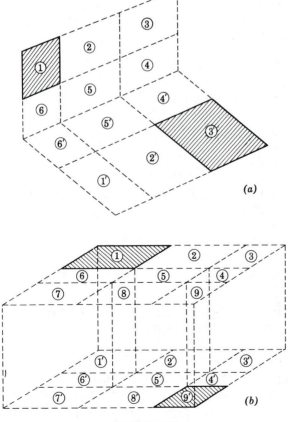

FIGURE E.5

Parallel Rectangles. Similar relations for opposed parallel rectangles are given below, using the notation depicted in Fig. E.5(b). The following reciprocal relations exist:

$$A_1F_{1-9'} = A_3F_{3-7'} = A_9F_{9-1'} = A_7F_{7-3'}. \tag{E.11}$$

The shape factor $F_{1-9'}$ is given by

$$A_1F_{1-9'} = K_{1-9'} = \tfrac{1}{4}\{K_{(1,2,3,4,5,6,7,8,9)^2} - K_{(1,2,5,6,7,8)^2} - K_{(2,3,4,5,8,9)^2}$$

$$- K_{(1,2,3,4,5,6)^2} + K_{(1,2,5,6)^2} + K_{(2,3,4,5)^2} + K_{(4,5,8,9)^2}$$

$$- K_{(4,5)^2} - K_{(5,8)^2} - K_{(5,6)^2} - K_{(4,5,6,7,8,9)^2}$$

$$+ K_{(5,6,7,8)^2} + K_{(4,5,6)^2} + K_{(2,5,8)^2} - K_{(2,5)^2}\}. \tag{E.12}$$

OTHER CONFIGURATIONS OF INTEREST

The number of configurations which may be considered is nearly limitless. However, three other situations of practical value will be presented here. No derivations are given; only the analytical results are presented, with corresponding graphical displays given in the main body of the text.

Directly Opposed Disks. For two disks, radii r_1 and r_2, placed L apart in parallel planes and having collinear centers

$$F_{1-2} = \tfrac{1}{2}[x - \sqrt{x^2 - 4(R_2/R_1)^2}],$$

$$x = 1 + (1 + R_2^2)/R_1^2, \tag{E.13}$$

$$R_1 = \frac{r_1}{L}, \qquad R_2 = \frac{r_2}{L}.$$

Figure 11.17 presents this shape factor in graphical form.

Concentric Cylinders. Two concentric cylinders of equal length L form a configuration of considerable engineering interest. If A_1 represents the outer surface of the inner cylinder of radius r_1, and A_2 represents the inner surface of the outer cylinder of radius r_2, the various shape factors involved may be expressed in terms of the ratios $R_1 = r_1/r_2$ and $R_2 = L/r_2$. Equations for F_{2-1} and F_{2-2} (the self-shape factor of the outer cylinder) in terms of R_1 and R_2 are reported in Ref. 2 but are not quoted here for the sake of brevity. These shape factors are displayed graphically in Fig. 11.18.

Small Spheres or Planes Irradiated by a Large Sphere. Present-day applications in space technology require knowledge of the amounts of heat acquired by satellites and spacecraft from planetary bodies. In most instances the size of the receiving body may be taken to be quite small compared to that of the planetary body. Figure 11.19 illustrates the geometrical parameters involved in consideration of the radiant exchange between a large planetary body and a small sphere or plane element. The small receiving surface is denoted by dA_1 and the large emitting one by A_2. In the case of the plane element (which might be part of a larger space structure), the direction of the surface normal with respect to the line drawn to the planetary center, λ, must be known. Because the expressions for the shape factor $F_{dA_1 - 2}$ cannot be written in simple closed forms, only the results of numerical integrations are given in Fig. 11.19 (Ref. 5) for the two cases.

One must note that the shape factors given are valid only if the large sphere has a uniform radiosity. In the event that only part of the sphere is radiating (as might be the case for reflected solar radiation) one must take this fact into account. Results of such calculations are given in Refs. 3 and 5.

REFERENCES

1. HAMILTON, D. C., and W. R. MORGAN, "Radiant Interchange Configuration Factors," *NACA Tech. Note 2836,* Dec. 1952.
2. SIEGEL, R., and J. R. HOWELL, *Thermal Radiation Heat Transfer,* 2nd ed., New York, McGraw-Hill, 1981.
3. STEVENSON, J. A., and J. C. GRAFTON, "Radiation Heat Transfer Analysis for Space Vehicles," *Rep. ASD 61–119, Part I,* Flight Accessories Laboratory, Aeronautical Systems Division, Wright-Patterson Air Force Base, Ohio, Dec. 1961.
4. PLAMONDON, J. A., "Numerical Determination of Radiation Configuration Factors for Some Common Geometrical Situations," *Tech. Rep. 32–127,* Jet Propulsion Laboratory, Calif. Inst. Technology, 1961.
5. CLARK, L. G., and E. C. ANDERSON, "Geometric Shape Factors for Planetary-Thermal and Planetary-Reflected Radiation Incident upon Spinning and Non-Spinning Spacecraft," *NASA Tech. Note D-2835,* May 1965.

SI-Unit Conversion Table for Heat Transfer Calculations

The discussion presented in Section 1.7 describes the SI and English engineering systems of units. This appendix presents conversion factors between these two systems to facilitate computations and the use of the various tables of physical properties. These conversions are limited to those quantities occurring commonly in heat transfer calculations (i.e., no conversions are given for electrical quantities, etc.). These conversions may be obtained from various sources (e.g., Ref. 1):

Dimension	English Units	SI Units
Acceleration	1 ft/s^2 1 ft/h^2	$= 3.0480 \times 10^{-1}$ m/s^2 $= 2.3519 \times 10^{-8}$ m/s^2
Area	1 ft^2 1 in.^{2}	$= 9.2903 \times 10^{-2}$ m^2 $= 6.4516 \times 10^{-4}$ m^2
Conductance, thermal	1 Btu/h-ft^2-°F	$= 5.6784$ W/m^2-°C
Conductivity, thermal	1 Btu/h-ft-°F	$= 1.7308$ W/m-°C
Density	1 lb$_m$/ft^3	$= 1.6018 \times 10$ kg/m^3
Diffusivity, thermal	1 ft^2/s 1 ft^2/h	$= 9.2903 \times 10^{-2}$ m^2/s $= 2.5806 \times 10^{-5}$ m^2/s
Energy	1 Btu 1 kW-h 1 ft-lb$_f$ 1 hp-h	$= 1.0551$ kJ $= 3.6000 \times 10^3$ kJ $= 1.3558 \times 10^{-3}$ kJ $= 2.6845 \times 10^3$ kJ
Force	1 lb$_f$	$= 4.4482$ N
Heat	1 Btu	$= 1.0551$ kJ
Heat flow rate	1 Btu/s 1 Btu/h	$= 1.0551 \times 10^3$ W $= 2.9308 \times 10^{-1}$ W
Heat flux (unit area) (unit length)	1 Btu/h-ft^2 1 Btu/h-ft	$= 3.1546$ W/m^2 $= 9.6152 \times 10^{-1}$ W/m
Heat generation rate (unit mass) (unit volume)	1 Btu/h-lb$_m$ 1 Btu/h-ft^3	$= 6.4612 \times 10^{-1}$ W/kg $= 1.0350 \times 10$ W/m^3
Heat transfer coefficient	1 Btu/h-ft^2-°F	$= 5.6784$ W/m^2-°C
Latent heat	1 Btu/lb$_m$	$= 2.3260$ kJ/kg
Length	1 ft 1 μm 1 in. 1 mile	$= 3.0480 \times 10^{-1}$ m $= 1.0000 \times 10^{-6}$ m $= 2.5400 \times 10^{-2}$ m $= 1.6093 \times 10^3$ m
Mass	1 lb$_m$	$= 4.5359 \times 10^{-1}$ kg
Mass flow rate	1 lb$_m$/s 1 lb$_m$/h	$= 4.5359 \times 10^{-1}$ kg/s $= 1.2600 \times 10^{-4}$ kg/s
Mass flux	1 lb$_m$/s-ft^2 1 lb$_m$/h-ft^2 1 lb$_m$/s-in.^{2} 1 lb$_m$/h-in.^{2}	$= 4.8824$ kg/s-m^2 $= 1.3562 \times 10^{-3}$ kg/s-m^2 $= 7.0362 \times 10^2$ kg/s-m^2 $= 1.9545 \times 10^{-1}$ kg/s-m^2
Momentum, linear	1 lb$_m$-ft/s 1 lb$_m$-ft/h	$= 1.3825 \times 10^{-1}$ kg-m/s $= 3.8404 \times 10^{-5}$ kg-m/s
Power	1 Btu/s 1 ft-lb$_f$/s 1 Btu/h 1 hp	$= 1.0551 \times 10^3$ W $= 1.3558$ W $= 2.9308 \times 10^{-1}$ W $= 7.4570 \times 10^2$ W

Dimension	English Units	SI Units
Pressure	1 lb_f/ft^2	= 4.7880×10^{-2} kN/m^2
	1 $lb_f/in.^2$	= 6.8948 kN/m^2
	1 standard atmosphere	= 1.0133×10^2 kN/m^2
	1 in. water	= 2.4909×10^{-1} kN/m^2
	1 ft water	= 2.9891 kN/m^2
	1 in. mercury	= 3.3866 kN/m^2
Resistance, thermal		
(total)		= 1.8956°C/W
(unit)	1 h-°F/Btu	= 1.7611×10^{-1}
	1 h-ft^2-°F/Btu	m^2-°C/W
Specific energy	1 Btu/lb_m	= 2.3260 kJ/kg
	1 $ft-lb_f/lb_m$	= 2.9891×10^{-3} kJ/kg
Specific heat	1 Btu/lb_m-°F	= 4.1868 kJ/kg-°C
Specific volume	1 ft^3/lb_m	= 6.2428×10^{-2} m^3/kg
Surface tension	1 lb_f/in.	= 1.7513×10^2 N/m
Temperature	°R	°K = $\frac{5}{9} \times$ °R
	°F	°C = $\frac{5}{9} \times$ (°F − 32)
Temperature difference	1°F(°R)	= $\frac{5}{9}$°C(°K)
Time	1 h	= 3.6000×10^3 s
	1 min	= 6.0000×10 s
Velocity	1 ft/s	= 3.0480×10^{-1} m/s
	1 ft/h	= 8.4667×10^{-5} m/s
	1 mph	= 4.4704×10^{-1} m/s
Viscosity, dynamic	1 poise (g/cm-s)	= 1.0000×10^{-1}
		kg/m-s ($N-s/m^2$)
	1 lb_m/ft-s	= 1.4882 kg/m-s
	1 lb_m/ft-h	= 4.1338×10^{-4} kg/m-s
	1 lb_f-s/in.^{2}	= 6.8947×10^3 kg/m-s
	1 lb_f-h/ft^2	= 1.7237×10^5 kg/m-s
Viscosity, kinematic	1 stoke (cm^2/s)	= 1.0000×10^{-4} m^2/s
	1 ft^2/s	= 9.2903×10^{-2} m^2/s
	1 ft^2/h	= 2.5806×10^{-5} m^2/s
Volume	1 ft^3	= 2.8317×10^{-2} m^3
	1 in.^{3}	= 1.6387×10^{-5} m^3
Volume flow rate	1 ft^3/s	= 2.8317×10^{-2} m^3/s
	1 ft^3/min	= 4.7195×10^{-4} m^3/s
	1 ft^3/h	= 7.8658×10^{-6} m^3/s

REFERENCE

1. MECHTLY, E. A., "The International System of Units, Physical Constants and Conversion Factors," *NASA SP-7012*, National Aeronautics and Space Administration, Washington, D.C., 1964.

INDEX

Absorption coefficient, 426
Absorptivity, 360
  of gases, 426
  hemispherical, 378
  monochromatic, 360, 378
  solar, 381, 384–385
  total, 361, 379
Air space, combined losses in, 512
Akers, W. W., 342
Analogy
  Colburn, 244, 261
  Prandtl, 244, 261
  Reynolds, 241, 242, 260
  von Kármán, 246, 261
Arnold, J. N., 321
Aspect ratio, 321
Azer, N. Z., 286

Beckmann, W., 308
Bernstein, M., 294
Bessel equation, 66, 68, 95, 112, 576
Bessel functions, 66, 68, 95, 112, 576–581
  tabulated values of, 582, 583
Biot number, 98, 107, 115
Blackbody. *See* Radiation
Blasius, H., 213, 234, 253
Boelter, L. M. K., 281
Boiling, 330, 344–353
  film, 348, 350
  forced convection, 345, 352
  free convection, 347
  nucleate, 347, 348–350
  peak heat flux in, 348, 349
  pool, 345, 346, 348, 350
  radiation effects in, 351
Boltzmann, L., 14, 367
Boundary layer, 10–13, 189–209
  energy equation, 194
  equation of motion, 191–193
  in forced convection. *See* Forced convection
  in free convection. *See* Free convection
  integral equations
    of energy, 198–200

  of momentum, 196–198
  laminar, 189–195
    on flat plates, 211–216, 220–223
    with viscous dissipation, 227–232,
      272–279
  separation, 291
  thermal, 12, 194
  turbulent, 200–209
    on flat plates, 232–240
    mixing length, 207
    shear velocity, 237, 254
    velocity distribution, 234–238, 253
    with viscous dissipation, 274, 276
Bowman, R. A., 471
Bromley, L. A., 350, 351
Buffer zone, 207, 242, 243, 246
Bulk temperature, 255, 267

Capacitance, thermal, 157
Capacity rate, 452
Capacity ratio, 469, 474
Chao, B. T., 286
Chapman, A. J., 339
Chu, H. H. S., 316, 318, 319
Churchill, S. W., 269, 294, 316, 318, 319
Clifton, J. V., 339
Coefficient
  absorption, 426
  combined radiation and convection,
    509–514
  drag, 216, 247
  heat transfer. *See* Heat transfer coefficient
  overall heat transfer. *See* Heat transfer coef-
    ficient
  radiation, 351, 509–514
  skin friction. *See* Skin friction coefficient
  of thermal expansion, 19, 27, 305
    of gases, 28
    of liquids, 28
Colburn, A. P., 244, 281
Colburn analogy
  on flat plate, 244
  in pipe flow, 261